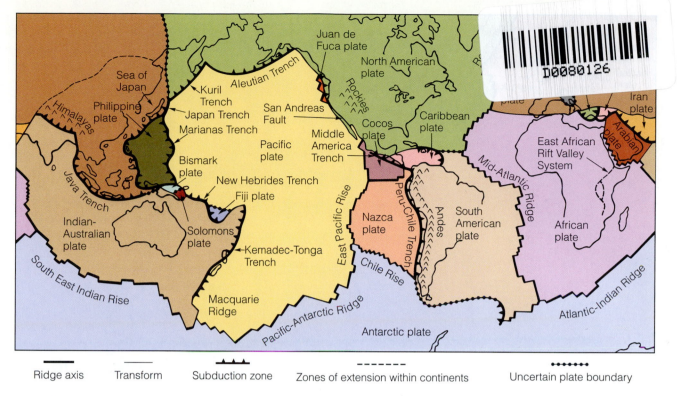

Ridge axis   Transform   Subduction zone   Zones of extension within continents   Uncertain plate boundary

Earth's Plates

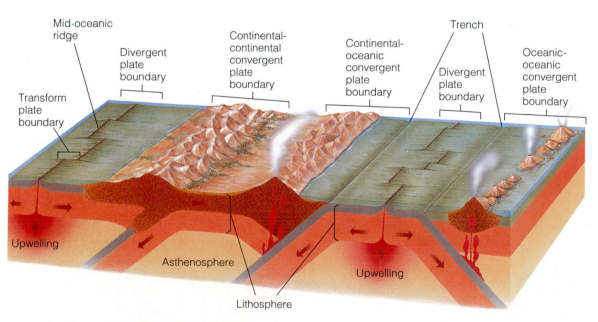

Three Principal Types of Plate Boundaries

www.wadsworth.com

wadsworth.com is the World Wide Web site for Wadsworth Publishing Company and is your direct source to dozens of online resources.

At *wadsworth.com* you can find out about supplements, demonstration software, and student resources. You can also send e-mail to many of our authors and preview new publications and exciting new technologies.

wadsworth.com
Changing the way the world learns®

Essentials *of* Geology

SECOND EDITION

Reed Wicander & James S. Monroe

CENTRAL MICHIGAN UNIVERSITY

Wadsworth Publishing Company
I(T)P® An International Thomson Publishing Company

Belmont, CA • Albany, NY • Boston • Cincinnati • Johannesburg • London • Madrid • Melbourne
Mexico City • New York • Pacific Grove, CA • Scottsdale, AZ • Singapore • Tokyo • Toronto

Earth Science Editor: Stacey Purviance
Editorial Assistant: Erin Conlon
Marketing Manager: Christine Henry
Project Editor: Tanya Nigh
Print Buyer: Stacey Weinberger
Permissions Editor: Bob Kauser
Production: Ruth Cottrell
Designer: Design Office, San Francisco
Copy Editor: Betty Duncan
Illustrator: Precision Graphics and Carlyn Iverson
Cover Design: Stephen Rapley
Cover Image: Thomas Moran (1837–1926). The Lower Gorge of the Grand Canyon. Art Resource, New York.
Compositor: GTS Graphics
Printer: R. R. Donnelley & Sons/Roanoke

Printed in the United States of America
1 2 3 4 5 6 7 8 9 10

For more information, contact Wadsworth Publishing Company,
10 Davis Drive, Belmont, CA 94002, or electronically at
http://www.wadsworth.com

International Thomson Publishing Europe
Berkshire House
168-173 High Holborn
London, WC1V 7AA, United Kingdom

International Thomson Editores
Seneca, 53
Colonia Polanco
11560 México D.F. México

Nelson ITP, Australia
102 Dodds Street
South Melbourne
Victoria 3205 Australia

International Thomson Publishing Asia
60 Albert Street
#15-01 Albert Complex
Singapore 189969

Nelson Canada
1120 Birchmount Road
Scarborough, Ontario
Canada M1K 5G4

International Thomson Publishing Japan
Hirakawa-cho Kyowa Building, 3F
2-2-1 Hirakawa-cho, Chiyoda-ku
Tokyo 102 Japan

International Thomson Publishing Southern Africa
Building 18, Constantia Square
138 Sixteenth Road, P.O. Box 2459
Halfway House, 1685 South Africa

Library of Congress Cataloging-in-Publication Data

Wicander, Reed
        Essentials of geology / Reed Wicander, James S. Monroe. — 2nd ed.
            p.   cm.
        Includes bibliographical references and index.
        ISBN 0-534-54774-5
            1. Geology. I. Monroe, James S. (James Stewart).
    II. Title.
QE26.2.W53 1998   98-28421
  ⁵0—dc21

REED WICANDER

REED WICANDER is a geology professor at Central Michigan University where he teaches physical geology, historical geology, prehistoric life, and invertebrate paleontology. He has co-authored several geology textbooks with James S. Monroe. His main research interests involve various aspects of Paleozoic palynology, specifically the study of acritarchs, on which he has published many papers. He is a past president of the American Association of Stratigraphic Palynologists and currently a councillor of the International Federation of Palynological Societies.

JAMES S. MONROE

JAMES S. MONROE is professor emeritus of geology at Central Michigan University where he taught physical geology, historical geology, prehistoric life, and stratigraphy and sedimentology since 1975. He has co-authored several textbooks with Reed Wicander and has interests in Cenozoic geology and geologic education.

Contents

CHAPTER 1

Understanding Earth: An Introduction to Physical Geology 2

CHAPTER 2

Plate Tectonics: A Unifying Theory 22

CHAPTER 3

Minerals

46

CHAPTER 4

Igneous Rocks and Intrusive Igneous Activity

68

CHAPTER 5
Volcanism

88

CHAPTER 6
Weathering, Erosion, and Soil

112

CHAPTER 7

Sediment and Sedimentary Rocks   134

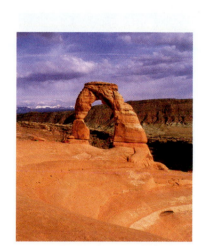

CHAPTER 8

Metamorphism and Metamorphic Rocks   156

CHAPTER 9

Earthquakes and Earth's Interior 174

CHAPTER 10

Deformation and Mountain Building 200

CHAPTER 15

The Work of Wind and Deserts 314

CHAPTER 16

The Seafloor, Shorelines, and Shoreline Processes 334

CHAPTER 17

Geologic Time: Concepts and Principles

360

CHAPTER 18

Earth History

382

CHAPTER 19
Life History

EPILOGUE

Preface

Earth is a dynamic planet

that has changed continuously during its 4.6 billion years of existence. The size, shape, and geographic distribution of the continents and ocean basins have changed through time, as have the atmosphere and biota. Over the past 20 years, bold new theories and discoveries concerning Earth's origin and how it works have sparked a renewed interest in geology. We have become increasingly aware of how fragile our planet is and, more importantly, how interdependent all of its various systems are. We have learned that we cannot continually pollute our environment and that our natural resources are limited and, in most cases, nonrenewable. Furthermore, we are coming to realize how central geology is to our everyday lives. For these and other reasons, geology is one of the most important college or university courses a student can take.

Essentials of Geology is designed for a one-semester introductory course and is written with the student in mind. One of the problems with any introductory science course is that students are overwhelmed by the amount of material that must be learned. Furthermore, most of the material does not seem to be linked by any unifying theme and does not always appear to be relevant to their lives.

One of the goals of this book is to provide students with a basic understanding of geology and its processes and, more importantly,

with an understanding of how geology relates to the human experience: that is, how geology affects not only individuals but society in general. With these goals in mind, we introduce the major themes of the book in the first chapter to provide students with an overview of the subject and to enable them to see how the various systems of Earth are interrelated. We also discuss the economic and environmental aspects of geology throughout the book rather than treating these topics in separate chapters. In this way students can see, through relevant and interesting examples, how geology impacts our lives.

New Features in the Second Edition

THE second edition has undergone considerable rewriting and updating to produce a book that is easier to read with a high level of current information and many new photographs, prologues, and perspectives. Drawing on the comments and suggestions of reviewers, we have incorporated many new features into this edition. One of the more noticeable changes is that we have added an Epilogue that summarizes the grand aspects of the topics covered in this book and places them in perspective in terms of Earth systems.

New material in this edition of

Essentials of Geology includes a section on Earth systems and sustainable development in Chapter 1 as well as updated information on recent earthquakes and the seismic-moment magnitude scale in Chapter 9. Several parts of Chapter 10 (Deformation and Mountain Building) have been rewritten and clarified, especially the section on mountains. The chapters on surface processes (Chapters 11–16) are still largely descriptive, but many of the sections of these chapters have been rewritten to emphasize the dynamic nature of these processes. Chapters 18 (Earth History) and 19 (Life History) have been updated and revised, including an expanded treatment of the Precambrian.

Current information on mineral and energy resources and environmental issues has been added to many chapters, such as a discussion of natural resources for each Era in Chapter 18 and new information on methane hydrate in the Exclusive Economic Zone in Chapter 16. Updated Perspectives relating to mineral and energy resources and environmental issues include Gold (Chapter 3), Asbestos (Chapter 8), and Radioactive Waste Disposal (Chapter 16).

Other important changes include a number of new Prologues, such as gemstones (Chapter 3), the A.D. 79 eruption of Mt. Vesuvius (Chapter 5), differential weathering and erosion, highlighting Bryce and Badlands National Parks (Chapter 6), and mountains and how their heights are measured (Chapter 10). New Perspectives also appear, such as the abundance of fossils (Chapter 7), surging glaciers (Chapter 14), Cape Cod, Massachusetts (Chapter 16), petroglyphs (Chapter 18), and Mary Anning's contribution to paleontology (Chapter 19).

Many photographs in the first edition have been replaced, including most of the chapter-opening photographs. In addition, a number of photographs within the chapters have been enlarged to enhance their visual impact. We feel the rewriting and updating done in the text as well as the addition of new photographs greatly improves the second edition by making it easier to read and comprehend, as well as a more effective teaching tool. Additionally, improvements have been made in the ancillary package that accompanies the book.

World Wide Web Activities

ONE of the new features of this book is the section at the end of each chapter titled World Wide Web Activities. Each section lists several web sites relevant to the chapter. These sites were selected because they expand the topic coverage of the chapter or supply updates on recent research or events, such as volcanic eruptions, earthquakes, or dinosaurs. Each site has a short summary as well as some questions for students to answer. Furthermore, most sites contain links to other sites students may find of interest.

Rather than publishing the URL (universal resource locator) for each site, we have instructed students to log on to *http://www.wadsworth.com/geo* for direct links to the sites listed as well as additional related web sites. The advantage of logging on to the Wadsworth web site is that the URL for each site is kept current and new sites can be added between editions.

We think students will find the World Wide Web Activities section a valuable and useful resource in their geological education.

CD-ROM Explorations

THE second edition of *Essentials of Geology* is accompanied *by In-Terra-Active Version 2.0: The Wadsworth Geology CD-ROM for Students*. At the end of each chapter, students will find sections titled CD-ROM Exploration. These engaging, end-of-chapter sections will encourage students to explore the corresponding interactive CD-ROM module(s) to gain greater understanding of the key concepts and processes presented in each chapter.

Text Organization

PLATE tectonic theory is the unifying theme of geology and this book. This theory has revolutionized geology because it provides a global perspective of Earth and allows geologists to treat many seemingly unrelated geologic phenomena as part of a total planetary system. Because plate tectonic theory is so important, it is covered in Chapter 2 and is discussed in most subsequent chapters in terms of the subject matter of that chapter.

We have organized *Essentials of Geology* into several informal categories. Chapter 1 is an introduction to geology and Earth systems, its relevance to the human experience, plate tectonic theory, the rock cycle, geologic time and uniformitarianism, and the origin of the solar system and Earth. Chapter 2 deals with plate tectonics, while Chapters 3–8 examine Earth's materials (minerals and igneous, sedimentary, and metamorphic rocks) and the geo-

logic processes associated with them, including the role of plate tectonics in their origin and distribution. Chapters 9 and 10 deal with the related topics of Earth's interior, earthquakes, deformation, and mountain building. Chapters 11–16 cover Earth's surface processes, and Chapter 17 discusses geologic time, introduces several dating methods, and explains how geologists correlate rocks. Chapters 18 and 19 provide an overview of the geologic history of Earth and its biota. A short Epilogue summarizes the broad aspects of the topics covered in this book and places them in perspective in terms of Earth systems.

Chapter Organization

ALL chapters have the same organizational format. Each chapter opens with a small photograph set within a large background photograph that relates to the chapter material, a detailed outline, and a Prologue, which is intended to stimulate interest in the chapter by discussing some aspect of the material.

The text is written in a clear informal style, making it easy for students to comprehend. Numerous color diagrams and photographs complement the text, providing a visual representation of the concepts and information presented. Each chapter contains a Perspective that presents a brief discussion of an interesting aspect of geology or geological research. Mineral and energy resources are discussed in the final sections of a number of chapters to provide interesting, relevant information in the context of the chapter topics.

The end-of-chapter materials begin with a concise review of important concepts and ideas in the Chapter Summary. The Important Terms, which are printed in bold-face type in the chapter text, are listed at the end of each chapter for easy review, and a full glossary of important terms appears at the end of the text. The Review Questions are another important feature of this book; they include multiple-choice questions with answers as well as short answer and essay questions, and thought-provoking and quantitative questions under the Points to Ponder heading. Many new multiple-choice questions as well as short answer and essay questions have been added in each chapter for this edition. Each chapter concludes with World Wide Web Activities and a CD-ROM Exploration. An up-to-date list of references for each chapter is provided in the Additional Readings section located at the text-specific web site. Go to http://www.wadsworth.com/geo. Most of these references are written at a level appropriate for beginning students interested in pursuing a particular topic.

Special Features

THIS book contains a number of special features that set it apart from other geology textbooks. Among them are a critical thinking and study skills section, the chapter Prologues, the integration of economic and environmental geologic issues throughout the book, a set of multiple-choice questions with answers for each chapter, a Points to Ponder section that contains thought-provoking and quantitative questions, and two new sections, World Wide Web Activities and CD-ROM Exploration.

STUDY SKILLS

Immediately following the Preface is a section devoted to developing critical thinking and study skills. This section contains hints to help students improve their study habits, prepare for exams, and generally get the most out of every course they take. While these tips can be helpful in any course, many of them are particularly relevant to geology. Whether students are just beginning college or about to graduate, it will pay them to take a few minutes to read this section.

PROLOGUES

Many of the introductory Prologues focus on the human aspects of geology, such as the eruption of Krakatau (Chapter 1), the A.D. 79 eruption of Mount Vesuvius (Chapter 5), the 1994 Northridge earthquake (Chapter 9), and the Flood of '93 (Chapter 12).

ECONOMIC AND ENVIRONMENTAL GEOLOGY

The topics of environmental and economic geology are discussed throughout the text. Integrating economic and environmental geology with the chapter material helps students see the importance and relevance of geology to their lives. In addition, many chapters close with a section on resources, further emphasizing the importance of geology in today's world.

ILLUSTRATIONS

Many of the illustrations depicting geologic processes or events are block diagrams rather than cross sections so that students can more easily visualize the salient features of these processes and events. Some illustrations have also been updated and revised to make them easier to understand. Photos and art have been used together to clarify the geologic features being discussed. Our color Paleogeographic maps in Chapter 18 are designed to illustrate clearly and accurately the geography during the various geologic periods. Full-color scenes showing associations of plants and animals in Chapter 19 are based on the most current interpretations. Great care has been taken to ensure that the art and captions provide an attractive, informative, and accurate illustration program.

PERSPECTIVES

The chapter Perspectives focus on aspects of environmental, economic, or planetary geology, as well as National Parks, Monuments, and Seashores. They include The Aral Sea (Chapter 1), asbestos (Chapter 8), radioactive waste disposal (Chapter 13), and Cape Cod, Massachusetts (Chapter 16). The topics for the Perspectives were chosen to provide students with an overview of the many fascinating aspects of geology. The Perspectives can be assigned as part of the chapter reading, used as the basis for lecture or discussion topics, or even used as the starting point for student papers.

Ancillary Materials

TO accompany *Essentials of Geology, 2e,* we are pleased to offer a full suite of text and multimedia products.

► THE WADSWORTH EARTH SCIENCE RESOURCE CENTER ON THE WORLD WIDE WEB

http://www.wadsworth.com/geo
An award-winning site that makes an encyclopedia's worth of online resources easy to find. It includes map resources and online field trips.

► THE *ESSENTIALS OF GEOLOGY,* 2E BOOK SPECIFIC WEB SITE

http://www.wadsworth.com/geo
Through the Wadsworth main Earth Science Resource Center site, you can also access resources specific to the book, including hyper-contents (where links for each chapter expand the book's coverage), critical thinking questions, and self-quizzes for students.

► THE WADSWORTH GEOLINK: GEOLOGY AND OCEANOGRAPHY PRESENTATION TOOL

The Wadsworth Geolink is the one instructor's tool that will allow you to build years' worth of multimedia presentations. Through a friendly interface, users are able to assemble, edit, publish, and present custom lectures, consisting of your choice of images, animation, QuickTime movies, and text you create. Once done, you can save a lecture presentation of a web site or export to any word processor. The content and organization of the media files correspond to major Wadsworth titles in earth science, including Ahrens' *Meteorology Today,* 5e, Pipkin/Trent's *Geology and the Environment, 2e* and Wicander/Monroe's *Essentials of Geology, 2e.*
GeoLink 2.0 ISBN 0-534-54225-5
GeoLink 3.0 ISBN 0-534-52190-8

► CNN'S PHYSICAL GEOLOGY TODAY

Wadsworth has partnered with CNN to bring you videotapes with 90 minutes worth of recent news coverage of major topics in physical geology: volcanoes, earthquakes, natural resources, and more. The news clips, organized to follow the topic sequence of *Physical Geology 3e* and *Essentials of Geology 2e,* both deepen and broaden topic coverage of the text, often adding an environmental focus. Printed worksheets accompanying the videos ask students to analyze information contained in the video and relate it to what they already know. Produced by Turner Learning, Inc.
Volume One, ISBN 0-534-53783-9
Volume Two, ISBN 0-534-54780-X

► INSTRUCTOR'S MANUAL WITH TEST BANK

A full array of teaching tips and resources: chapter outlines/overviews, learning objectives, lecture suggestions, important terms, and a list of key resources. This manual also contains approximately fifty test questions per chapter.
ISBN 0-534-54776-1

► THOMSON WORLD CLASS TESTING TOOLS

A fully integrated collection of test creation, delivery, and classroom management tools that features all the test items found in the Instructor's Manual. Allows you to deliver tests on paper, LAN (local area network) and on ITP's Internet server.
Macintosh ISBN 0-534-54778-8
Windows ISBN 0-534-54777-X

► THOMSON WORLD CLASS COURSE

Post your own course information, office hours, lesson information, assignments, sample tests and link to rich web content, including student review and enrichment material from Wadsworth. Updates are quick and easy to make and customer support is available twenty-four hours a day, seven days a week. More information is available at *http://www.worldclasslearning. com*

► INFOTRAC COLLEGE EDITION

Updated daily, this massive online library of authoritative sources gives you access to full-text articles from over 600 periodicals such as *Discover, Science, Astronomy, Conservationist, Human Ecology, Oceanus,* and *Sierra,* going back as far as four years. Accompanied by an online student guide that correlates each chapter in this text to InfoTrac College Edition articles. For access information, please contact your ITP representative or the Wadsworth Marketing Dept. at 1-800-876-2350, ext. 332.

► TRANSPARENCIES

One hundred transparencies selected by the authors, providing clear and effective illustrations of important artwork and maps from the text.
ISBN 0-534-54779-6

► EARTH SCIENCE VIDEO LIBRARY

Videotape series on topics in geology, oceanography, and meteorology. Minimum adoption required. Call Wadsworth Marketing Department for details or orders: 1-800-876-2350, ext. 332.

► GREAT IDEAS FOR TEACHING GEOLOGY

Dozens of teaching ideas, demonstrations, and analogies that have been used successfully in classrooms throughout the world.
ISBN 0-314-00394-0

FOR STUDENTS

► STUDY GUIDE

Contains self-tests, learning objectives, key terms, and critical thinking questions to guide study.
ISBN 0-534-54775-3

► CURRENT PERSPECTIVES IN GEOLOGY, 1998 EDITION

Michael McKinney, Parri Shariff, and Robert Tolliver (University of Tennessee, Knoxville)
This book of 50 current readings is designed to supplement any geology textbook and is ideal for instructors who include a writing component in their course. Available for sale to students or bundled at a discount with any Wadsworth geology text.
ISBN 0-314-20617-5

► STUDY SKILLS FOR SCIENCE STUDENTS

Daniel Chiras (University of Colorado, Denver)
Designed to accompany any introductory science text. It offers tips on improving your memory, learning more quickly, getting the most out of lectures, preparing for tests, producing first-rate term papers, and improving critical thinking skills.
ISBN 0-314-03983-X

► IN-TERRA-ACTIVE VERSION 2.0: WADSWORTH GEOLOGY CD-ROM FOR STUDENTS.

Packaged with every copy of *Essentials of Geology, 2e,* this interactive CD-ROM will allow students to create a landslide, predict a volcanic eruption, locate the epicenter of an earthquake, and explore other key topics in geology through some 40 interactive exercises. In each case, students can manipulate variables and data and view the results of their selections. Over 35 minutes of full motion video clips and animations help illustrate difficult concepts.

► GEOLOGY WORKBOOK FOR THE WEB

Bruce Blackerby (California State University, Fresno)
Over twenty chapters designed to encourage students to explore geology-related sites on the Internet. Chapter exercises, on topics from minerals to Cenozoic Earth, focus most heavily on physical geology. The workbook comes three-hole punched with perforated pages for easy tearing, so worksheets can be handed in as assignments.
ISBN 0-314-21072-5

► EARTH ONLINE

Michael Ritter (University of Wisconsin, Steven's Point)
An inexpensive, hands-on Internet guide written for the novice. It provides a tool for students to get "up and running" on the Internet with homework exercises, lab exercises, web searches, and more. To keep the book as useful as possible, the author maintains an Earth Online homepage with exercises, tips, new links, and constant updates of the exercises and reference sites. Access it through the Wadsworth Earth Science Resource Center at http://www.wadsworth.com/geo.
ISBN 0-534-51707-2

Acknowledgements

As the authors, we are, of course, responsible for the organization, style, and accuracy of the text, and any mistakes, omissions, or errors are our responsibility. During the preparation of our previous geology textbooks, we received numerous comments and advice from many geologists who reviewed various parts of the text. We wish to express our sincere appreciation to all of the many reviewers whose contributions were invaluable during the writing of those books. Because *Essentials of Geology 2e* combines material from our previous books, the reviewers of those books thus contributed to this book. More specifically we would like to thank those individuals who reviewed the first edition of *Essentials of Geology* and made many useful and insightful comments that were incorporated into the second edition of this book.

Norman Fox
Hocking Technical College

Joan Fryxell
California State University, San Bernardino

David Gibson
University of Maine

Patrick Hicks
La Grange College

Paul Morgan
Northern Arizona University

Susan Morgan
Utah State University

John Ritter
Wittenburg University

Robert J. Smith
Seattle University

James L. Talbot
Western Washington University

In addition, we would also like to thank the following reviewers who advised us on developing the first edition of *Essentials*. Their thoughts and suggestions were instrumental in organizing and developing the text.

Lawrence Balthaser
California Polytechnic State University, San Luis Obispo

Fredric R. Goldstein
Trenton State College

Brian Grant
Brock University

William F. Kean
University of Wisconsin, Milwaukee

Susan Morgan
Utah State University

Louis Pinto
Monroe Community College

We also wish to thank Richard V. Dietrich (Professor Emeritus), Eric L. Johnson, David J. Matty, Jane M. Matty, Wayne E. Moore (Professor Emeritus), and Stephen D. Stahl of the Geology Department of Central Michigan University and Bruce M. C. Pape of the Geography Department for providing us with photographs. We also thank Pam Iacco of the Geology Department, whose general efficiency was invaluable during the preparation of this book. We are also grateful for the generosity of the various agencies and individuals from many countries who provided photographs. Special thanks must go to our editor, Stacey Purviance, and to our publisher for the sciences at Wadsworth, Gary Carlson, for guiding us through this project. We are equally indebted to our production editor, Ruth Cottrell, whose attention to detail and consistency is greatly appreciated. We would also like to thank Betty Duncan for her copyediting skills. We appreciate her help in improving our manuscript. We thank Roberta Broyer and Robert Kauser who coordinated the permissions portion of the text production, and Kathleen Garcia who did the index. Because geology is such a visual science, we extend special thanks to Carlyn Iverson who rendered the reflective art and to the artists at Precision Graphics who were responsible for much of the rest of the art program. They did an excellent job, and we enjoyed working with them.

Our families were very patient and encouraging when most of our spare time and energy were devoted to this book. We thank them for their support and understanding.

Developing Critical Thinking and Study Skills

Introduction

COLLEGE is a demanding and important time, a time when your values will be challenged, and you will try out new ideas and philosophies. You will make personal and career decisions that will affect your entire life. One of the most important lessons you can learn in college is how to balance your time among work, study, and recreation. If you develop good time management and study skills early in your college career, you will find that your college years will be successful and rewarding.

This section offers some suggestions to help you maximize your study time and develop critical thinking and study skills that will benefit you, not only in college, but throughout your life. While mastering the content of a course is obviously important, learning how to study and to think critically is, in many ways, far more important. Like most things in life, learning to think critically and study efficiently will initially require additional time and effort, but once mastered, these skills will save you time in the long run.

You may already be familiar with many of the suggestions and may find that others do not directly apply to you. Nevertheless, if you take the time to read this section and apply the appropriate suggestions to your own situation, we are confident that you will become a better and more efficient student, find your classes more rewarding, have more time for yourself, and get better grades. We have found that the better students are usually also the busiest. Because these students are busy with work or extracurricular activities, they have had to learn to study efficiently and manage their time effectively.

One of the keys to success in college is avoiding procrastination. While procrastination provides temporary satisfaction because you have avoided doing something you did not want to do, in the long run it leads to stress. While a small amount of stress can be beneficial, waiting until the last minute usually leads to mistakes and a subpar performance. By setting clear, specific goals and working toward them on a regular basis, you can greatly reduce the temptation to procrastinate. It is better to work efficiently for short periods of time than to put in long, unproductive hours on a task, which is usually what happens when you procrastinate.

Another key to success in college is staying physically fit. It is easy to fall into the habit of eating junk food and never exercising. To be mentally alert, you must be physically fit. Try to develop a program of regular exercise. You will find that you have more energy, feel better, and study more efficiently.

General Study Skills

MOST courses, and geology in particular, build upon previous material, so it is extremely important to keep up with the coursework and set aside regular time for study in each of your courses. Try to follow these hints, and you will find you do better in school and have more time for yourself:

- Develop the habit of studying on a daily basis.
- Set aside a specific time each day to study. Some people are day people, and others are night people. Determine when you are most alert and use that time for study.
- Have an area dedicated for study. It should include a well-lighted space with a desk and the study materials you need, such as a dictionary, thesaurus, paper, pens, and pencils, and a computer if you have one.
- Study for short periods and take frequent breaks, usually after an hour of study. Get up and move around and do something completely different. This will help you stay alert, and you'll return to your studies with renewed vigor.
- Try to review each subject every day or at least the day of the class. Develop the habit of reviewing lecture material from a class the same day.
- Become familiar with the vocabulary of the course. Look up any unfamiliar words in the glossary of your textbook or in a dictionary. Learning the language of the discipline will help you learn the material.

Getting the Most from Your Notes

IF you are to get the most out of a course and do well on exams, you must learn to take good notes. Taking good notes does not mean you should try to write down every word your professor says. Part of being a good note taker is knowing what is important and what you can safely leave out.

Early in the semester, try to determine whether the lecture will follow the textbook or be predominantly new material. If much of the material is covered in the textbook, your notes do not have to be as extensive or detailed as when the material is new. In any case, the following suggestions should make you a better note taker and enable you to derive the maximum amount of information from a lecture:

▶ Regardless of whether the lecture discusses the same material as the textbook or supplements the reading assignment, read or scan the chapter the lecture will cover *before* class. This way you will be somewhat familiar with the concepts and can listen critically to what is being said rather than trying to write down everything. Later a few key words or phrases will jog your memory about what was said.

▶ Before each lecture, briefly review your notes from the previous lecture. Doing this will refresh your memory and provide a context for the new material.

▶ Develop your own style of note taking. Do not try to write down every word. These are notes you're taking, not a transcript. Learn to abbreviate and develop your own set of abbreviations and symbols for common words and phrases: for example, w/o (without), w (with), = (equals), ∧ (above or increases), ∨ (below or decreases), < (less than), > (greater than), & (and), u (you).

▶ Geology lends itself to many abbreviations that can increase your note-taking capability: for example, pt (plate tectonics), ig (igneous), meta (metamorphic), sed (sedimentary), rx (rock or rocks), ss (sandstone), my (million years), and gts (geologic time scale).

▶ Rewrite your notes soon after the lecture. Rewriting your notes helps reinforce what you heard and gives you an opportunity to determine whether you understand the material.

▶ By learning the vocabulary of the discipline before the lecture, you can cut down on the amount you have to write—you won't have to write down a definition if you already know the word.

▶ Learn the mannerisms of the professor. If he or she says something is important or repeats a point, be sure to write it down and highlight it in some way. Students have told me (RW) that when I stated something twice during a lecture, they knew it was important and probably would appear on a test. (They were usually right!)

▶ Check any unclear points in your notes with a classmate or look them up in your textbook. Pay particular attention to the professor's examples, which usually elucidate and clarify an important point and are easier to remember than an abstract concept.

▶ Go to class regularly and sit near the front of the class if possible. It is easier to hear and see what is written on the board or projected onto the screen, and there are fewer distractions.

▶ If the professor allows it, tape record the lecture, but don't use the recording as a substitute for notes. Listen carefully to the lecture and write down the important points; then fill in any gaps when you replay the tape.

▶ If your school allows it, and if they are available, buy class lecture notes. These are usually taken by a graduate student who is familiar with the material; typically they are quite comprehensive. Again use these notes to supplement your own.

▶ Ask questions. If you don't understand something, ask the professor. Many students are reluctant to do this, especially in a large lecture hall, but if you don't understand a point, other people are probably confused as well. If you can't ask questions during a lecture, talk to the professor after the lecture or during office hours.

Getting the Most Out of What You Read

THE old adage that "you get out of something what you put into it" is true when it comes to reading textbooks. By carefully reading your text and following these suggestions, you can greatly increase your understanding of the subject:

▶ Look over the chapter outline to see what the material is about and how it flows from topic to topic. If you have time, skim through the chapter before you start to read in depth.

- Pay particular attention to the tables, charts, and figures. They contain a wealth of information in abbreviated form and illustrate important concepts and ideas. Geology, in particular, is a visual science, and the figures and photographs will help you visualize what is being discussed in the text and provide actual examples of features such as faults or unconformities.
- As you read your textbook, highlight or underline key concepts or sentences, but make sure you don't highlight everything. Make notes in the margins. If you don't understand a term or concept, look it up in the glossary.
- Read the chapter summary carefully. Be sure you understand all the key terms, especially those in boldface or italic type. Because geology builds on previous material, it is imperative that you understand the terminology.
- Go over the end-of-chapter questions. Write your answers as if you were taking a test. Only when you see your answer in writing will you know if you really understood the material.
- Access the latest geologic information on the Internet. The end-of-chapter World Wide Web Activities will enhance your understanding of the chapter concepts and the way geologic information is disseminated today. Knowing how to search the Internet is essential now, and it will be more so in coming years.

Developing Critical Thinking Skills

FEW things in life are black and white, and it is important to be able to examine an issue from all sides and come to a logical conclusion. One of the most important things you will learn in college is to think critically and not accept everything you read and hear at face value. Thinking critically is particularly important in learning new material and relating it to what you already know. Although you can't know everything, you can learn to question effectively and arrive at conclusions consistent with the facts. Thus, these suggestions for critical thinking can help you in all your courses:

- Whenever you encounter new facts, ideas, or concepts, be sure you understand and can define all of the terms used in the discussion.

- Determine how the facts or information was derived. If the facts were derived from experiments, were the experiments well executed and free of bias? Can they be repeated? The controversy over cold fusion is an excellent example. Two scientists claimed to have produced cold fusion reactions using simple experimental laboratory apparatus, yet other scientists have never been able to achieve the same reaction by repeating the experiments.
- Do not accept any statement at face value. What is the source of the information? How reliable is the source?
- Consider whether the conclusions follow from the facts. If the facts do not appear to support the conclusions, ask questions and try to determine why they don't. Is the argument logical or is it somehow flawed?
- Be open to new ideas. After all, the underlying principles of plate tectonic theory were known early in this century yet were not accepted until the 1970s despite overwhelming evidence.
- Look at the big picture to determine how various elements are related. For example, how will constructing a dam across a river that flows to the sea affect the stream's profile? What will be the consequences to the beaches that will be deprived of sediment from the river? One of the most important lessons you can learn from your geology course is how interrelated the various systems of Earth are. When you alter one feature, you affect numerous other features as well.

Improving Your Memory

WHY do you remember some things and not others? The reason is that the brain stores information in different ways and forms, making it easy to remember some things and difficult to remember others. Because college requires that you learn a vast amount of information, any suggestions that can help you retain more material will help you in your studies:

- Pay attention to what you read or hear. Focus on the task at hand and avoid daydreaming. Repetition of any sort will help you remember material. Review the previous lecture before going to class, or look over the last chapter before beginning the next. Ask yourself questions as you read.

- Use mnemonic devices to help you learn unfamiliar material. For example, the order of the Paleozoic periods (Cambrian, Ordovician, Silurian, Devonian, Mississippian, Pennsylvanian, and Permian) of the geologic time scale can be remembered by the phrase, **C**ampbell's **O**nion **S**oup **D**oes **M**ake **P**eter **P**ale, or the order of the Cenozoic Epochs (Paleocene, Eocene, Oligocene, Miocene, Pliocene, and Pleistocene) can be remembered by the phrase, **P**ut **E**ggs **O**n **M**y **P**late **P**lease. Using rhymes can also be helpful.

- Look up the roots of important terms. If you understand where a word comes from, its meaning will be easier to remember. For example, *pyroclastic* comes from *pyro,* meaning "fire," and *clastic,* meaning "broken pieces." Hence a pyroclastic rock is one formed by volcanism and composed of pieces of other rocks. We have provided the roots of many important terms throughout this text to help you remember their definitions.

- Outline the material you are studying. This practice will help you see how the various components are interrelated. Learning a body of related material is much easier than learning unconnected and discrete facts. Looking for relationships is particularly helpful in geology because so many things are interrelated. For example, plate tectonics explains how mountain building, volcanism, and earthquakes are all related. The rock cycle relates the three major groups of rocks to each other and to subsurface and surface processes (Chapter 1).

- Use deductive reasoning to tie concepts together. Remember that geology builds on what you learned previously. Use that material as your foundation and see how the new material relates to it.

- Draw a picture. If you can draw a picture and label its parts, you probably understand the material. Geology lends itself very well to this type of memory device because so much is visual. For example, instead of memorizing a long list of glacial terms, draw a picture of a glacier and label its parts and the type of topography it forms.

- Focus on what is important. You can't remember everything, so focus on the important points of the lecture or the chapter. Try to visualize the big picture and use the facts to fill in the details.

Preparing for Exams

FOR most students, tests are the critical part of a course. To do well on an exam, you must be prepared. These suggestions will help you focus on preparing for examinations:

- The most important advice is to study regularly rather than try to cram everything into one massive study session. Get plenty of rest the night before an exam, and stay physically fit to avoid becoming susceptible to minor illnesses that sap your strength and lessen your ability to concentrate on the subject at hand.

- Set up a schedule so that you cover small parts of the material on a regular basis. Learning some concrete examples will help you understand and remember the material.

- Review the chapter summaries. Construct an outline to make sure you understand how everything fits together. Drawing diagrams will help you remember key points. Make flash cards to help you remember terms and concepts.

- Form a study group, but make sure your group focuses on the task at hand, not on socializing. Quiz each other and compare notes to be sure you have covered all the material. We have found that students dramatically improved their grades after forming or joining a study group.

- Write the answers to all the Review Questions. Before doing so, however, become thoroughly familiar with the subject matter by reviewing your lecture notes and reading the chapter. Otherwise, you will spend an inordinate amount of time looking up answers.

- If you have any questions, visit the professor or teaching assistant. If review sessions are offered, be sure to attend. If you are having problems with the material, ask for help as soon as you have difficulty. Don't wait until the end of the semester.

- If old exams are available, look at them to see what is emphasized and what types of questions are asked. Find out whether the exam will be all objective or all essay or a combination. If you have trouble with a particular type of question (such as multiple choice or essay), practice answering questions of that type—your study group or a classmate may be able to help.

Taking Exams

THE most important thing to remember when taking an exam is not to panic. This, of course, is easier said than done. Almost everyone suffers from test anxiety to some degree. Usually, it passes as soon as the exam begins, but in some cases, it is so debilitating that an individual does not perform as well as he or she could. If you are one of those people, get help as soon as possible. Most colleges and universities have a program to help students overcome test anxiety or at least keep it in check. Don't be afraid to seek help if you suffer test anxiety. Your success in college depends to a large extent on how well you perform on exams, so by not seeking help, you are only hurting yourself. In addition, the following suggestions may be helpful:

▶ First of all, relax. Then look over the exam briefly to see its format and determine which questions are worth the most points. If it helps, quickly jot down any information you are afraid you might forget or particularly want to remember for a question.

▶ Answer the questions that you know the best first. Make sure, however, that you don't spend too much time on any one question or on one that is worth only a few points.

▶ If the exam is a combination of multiple choice and essay, answer the multiple-choice questions first. If you are not sure of an answer, go on to the next one. Sometimes the answer to one question can be found in another question. Furthermore, the multiple-choice questions may contain many of the facts needed to answer some of the essay questions.

▶ Read the question carefully and answer only what it asks. Save time by not repeating the question as your opening sentence to the answer. Get right to the point. Jot down a quick outline for longer essay questions to make sure you cover everything.

▶ If you don't understand a question, ask the examiner. Don't assume anything. After all, it is your grade that will suffer if you misinterpret the question.

▶ If you have time, review your exam to make sure you covered all the important points and answered all the questions.

▶ If you have followed our suggestions, by the time you finish the exam, you should feel confident that you did well and will have cause for celebration.

Concluding Comments

WE hope that the suggestions we have offered will be of benefit to you, not only in this course but throughout your college career. Though it is difficult to break old habits and change a familiar routine, we are confident that following these suggestions will make you a better student. Furthermore, many of the suggestions will help you work more efficiently, not only in college, but also throughout your career. Learning is a lifelong process that does not end when you graduate. The critical thinking skills that you learn now will be invaluable throughout your life, both in your career and as an informed citizen.

Essentials *of* Geology

1

Understanding Earth:

An Introduction

Outline

As a result of numerous eruptions like the one shown here, Anak Krakatau emerged above sea level in 1928 from the 275-meter-deep caldera formed by the 1883 eruption of Krakatau.

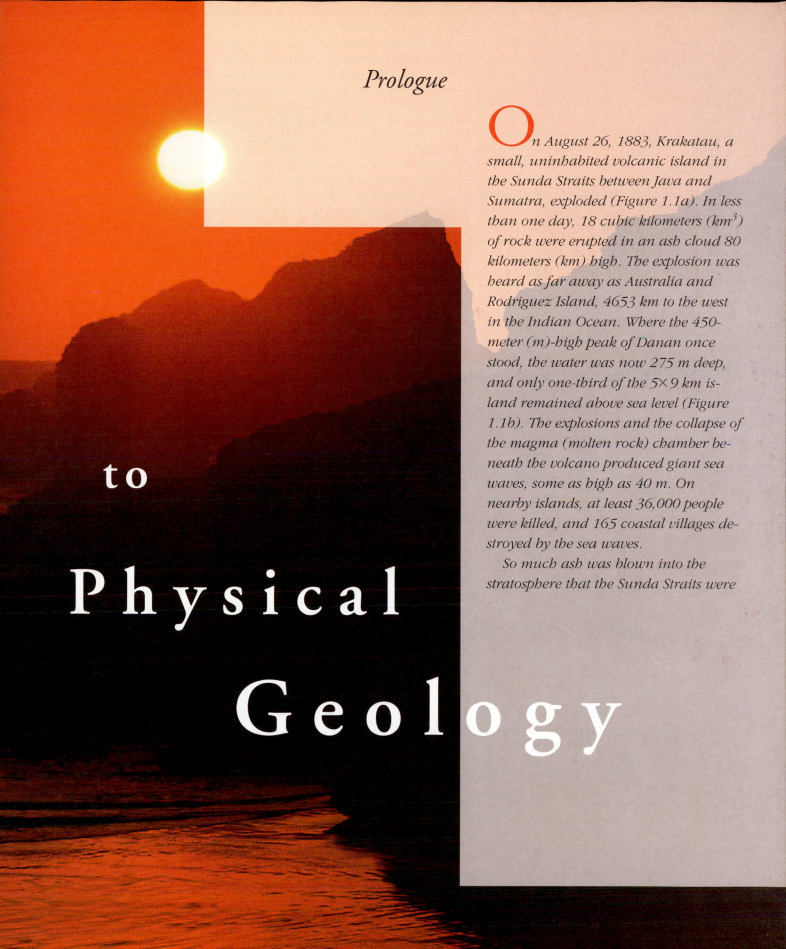

Prologue

to Physical Geology

On August 26, 1883, Krakatau, a small, uninhabited volcanic island in the Sunda Straits between Java and Sumatra, exploded (Figure 1.1a). In less than one day, 18 cubic kilometers (km³) of rock were erupted in an ash cloud 80 kilometers (km) high. The explosion was heard as far away as Australia and Rodriguez Island, 4653 km to the west in the Indian Ocean. Where the 450-meter (m)-high peak of Danan once stood, the water was now 275 m deep, and only one-third of the 5×9 km island remained above sea level (Figure 1.1b). The explosions and the collapse of the magma (molten rock) chamber beneath the volcano produced giant sea waves, some as high as 40 m. On nearby islands, at least 36,000 people were killed, and 165 coastal villages destroyed by the sea waves.

So much ash was blown into the stratosphere that the Sunda Straits were

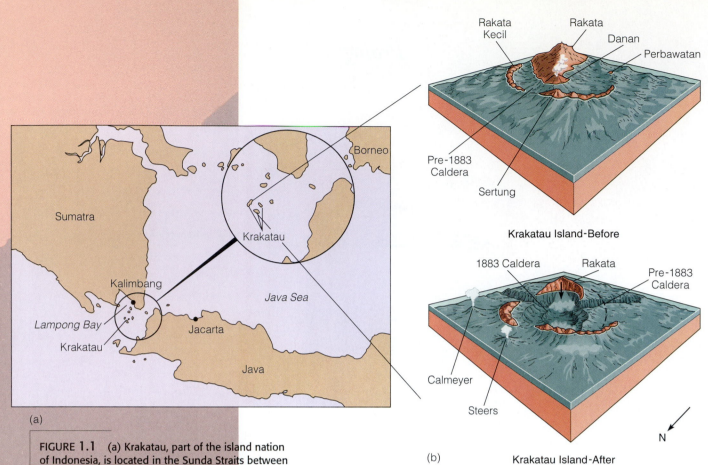

FIGURE 1.1 (a) Krakatau, part of the island nation of Indonesia, is located in the Sunda Straits between Java and Sumatra. (b) Krakatau before and after the 1883 eruption.

completely dark from 10 A.M., August 27, until dawn the next day. Ash was reported falling on ships as far away as 6076 km. For the next three years, vivid red sunsets were common around the world due to these airborne products. The volcanic dust also reflected incoming solar radiation back into space; the average global temperature dropped as much as 1/2°C during the following year and did not return to normal until 1888.

Of course, all animal and plant life was destroyed on Krakatau. A year after the eruption, however, a few shoots of grass appeared, and three years later 26 species of plants had colonized the island, thus providing a suitable habitat for the animals that soon followed.

Why have we chosen the eruption of Krakatau as an introduction to geology? The reason is that it illustrates several of the aspects of geology that we will examine, including the way Earth's interior, surface, and atmosphere are interrelated.

Sumatra, Java, Krakatau, and the Lesser Sunda Islands are part of a 3000-km-long chain of volcanic islands comprising the nation of Indonesia. Their location is a result of a collision between two pieces of Earth's outer layer, generally called the lithosphere. The theory that Earth's lithosphere is divided into rigid plates that move over a plastic zone is known as plate tectonics *(see Chapter 2). This unifying theory explains and ties together such apparently unrelated geologic phenomena as volcanic eruptions, earthquakes, and the origin of mountain ranges.*

In tropical areas such as Indonesia, physical and chemical processes rapidly break down ash falls and lava flows, converting them into rich soils that are agriculturally productive and can support large populations (see Chapter 6). Despite the dangers of living in a region of active volcanism, a strong correlation exists between volcanic activity and population density. Indonesia has experienced 972 eruptions during historic time, 83 of which have caused fatalities. Yet these same eruptions are also ultimately responsible for the high food production that supports large numbers of people.

Volcanic eruptions also affect weather patterns; recall that the eruption of Krakatau caused a global cooling of 1/2°C. More recently, the 1991 eruption of Mount Pinatubo in the Philippines resulted in lower global temperatures and abnormal weather patterns the following summer (see Chapter 5).

As you read this book, keep in mind that the different topics you are studying are parts of dynamic interrelated systems, not isolated pieces of information. Volcanic eruptions such as Krakatau are the result of complex interactions involving Earth's interior and surface. These eruptions not only have an immediate effect on the surrounding area but also contribute to climatic changes that affect the entire planet.

Introduction to Earth Systems

IN this book, we focus on Earth as a complex, dynamic planet that has changed continually since its origin some 4.6 billion years ago. These changes are the result of internal and external processes that interact and affect each other, leading to the present-day features we observe. In fact, Earth is unique among the planets of our solar system in that it supports life and has oceans of water, a hospitable atmosphere, and a variety of climates. It is ideally suited for life as we know it because of a combination of factors, including its distance from the Sun and the evolution of its interior, crust, oceans, and atmosphere. Over time, changes in Earth's atmosphere, oceans, and to some extent, its crust have been influenced by life processes. In turn, these physical changes have affected the evolution of life.

If we view Earth as a whole, we can see innumerable interactions occurring between its various components. Furthermore, these components do not act in isolation but are interconnected—such that when one part of the system changes, it affects the other parts of the system.

One way to help understand the complexity of Earth is to think of it as a system. A **system** is defined as a combination of related parts that interact in an organized fashion (Figure 1.2). Earth, when considered as a system, consists of a collection of various subsystems, or related parts, interacting with each other in complex ways. Information, materials, and energy entering the system from the outside are *inputs,* whereas information, materials, and energy that leave the system are *outputs.*

An automobile is a good example of a system. Its various subsystems include the engine, transmission, steering, and brakes. These subsystems are interconnected in such a way that a change in any one subsystem affects other subsystems. The main input into the automobile system is gasoline, and its outputs are movement, heat, and pollutants.

Thus, we can view Earth in the same way we view an automobile—that is, as a system of interconnected components that interact and affect each other in many ways. Its complex interactions result in a dynamically changing body that exchanges matter and energy and recycles them into different forms (Table 1.1). These systems can interact with themselves, and many of the interactions are two-way. Also, chains of interactions frequently occur. For example, solar heating warms the land; unequal heating of land and water drive the wind, which in turn drives the ocean currents. When examined in this manner, the continuous evolution of Earth and its life makes geology an exciting and ever-changing science in which new discoveries are continually being made.

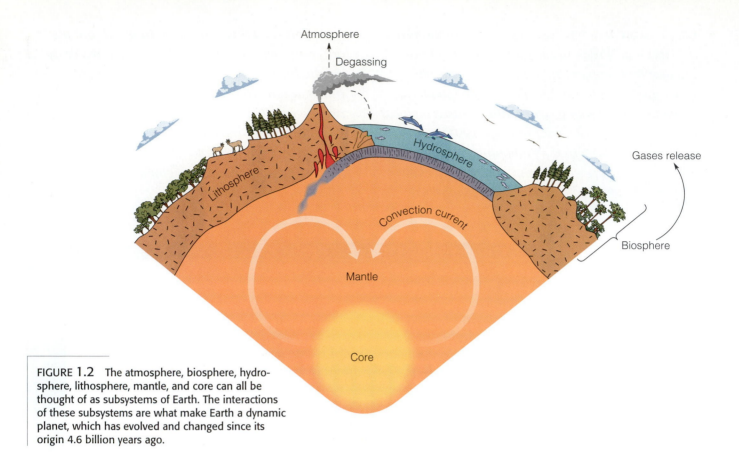

FIGURE 1.2 The atmosphere, biosphere, hydrosphere, lithosphere, mantle, and core can all be thought of as subsystems of Earth. The interactions of these subsystems are what make Earth a dynamic planet, which has evolved and changed since its origin 4.6 billion years ago.

TABLE 1.1

Interactions Between Major Earth Systems

	EARTH–UNIVERSE SYSTEM	ATMOSPHERE	HYDROSPHERE	BIOSPHERE	SOLID EARTH
Earth–Universe System	Gravitational interaction Recycling of stellar materials	Solid energy input Creation of ozone layer	Heating by solar energy Tides	Solar energy for photosynthesis Daily/seasonal rhythms	Heating by solar energy Meteor impacts
Atmosphere	Escape of heat radiation to space	Interaction of air masses	Winds drive surface currents Evaporation	Gases for respiration Transport of seeds, spores	Wind erosion Transport of water vapor for precipitation
Hydrosphere	Tidal friction affects Moon's orbit	Input of water vapor and stored solar heat	Mixing of oceans Deep ocean circulation Hydrologic cycle	Water for cell fluids Medium for aquatic organisms Transport of organisms	Precipitation Water and glacial erosion Solution of minerals
Biosphere	Biogenic gases affect escape of heat radiation to space	Gases from respiration	Removal of dissolved materials by organisms	Predator–prey interactions Food cycles	Modification of weathering and erosion processes Formation of soil
Solid Earth	Gravity of Earth affects other bodies	Input of stored solar heat Mountains divert air movements	Source of sediment and dissolved materials	Source of mineral nutrients Modification of habitats by crustal movements	Plate tectonics Crustal movements

SOURCE: Adapted by permission from Stephen Dutch, James S. Monroe, and Joseph Moran, *Earth Science* (Minneapolis/St. Paul: West Publishing Co.).

What Is Geology?

JUST what is geology and what is it that geologists do? **Geology,** from the Greek *geo* and *logos,* is defined as the study of Earth. It is generally divided into two broad areas—physical geology and historical geology. *Physical geology* is the study of Earth materials, such as minerals and rocks, as well as the processes operating within Earth and upon its surface. *Historical geology* examines the origin and evolution of Earth, its continents, oceans, atmosphere, and life.

The discipline of geology is so broad that it is subdivided into many different fields or specialties. Table 1.2 shows many of the diverse fields of geology and their relationship to the sciences of astronomy, physics, chemistry, and biology.

Nearly every aspect of geology has some economic or environmental relevance. Many geologists are involved in exploration for mineral and energy resources, using their specialized knowledge to locate the natural resources on which our industrialized society is based. As the demand for these nonrenewable resources increases, geologists are applying the basic principles of geology in increasingly sophisticated ways to help focus their attention on areas with a high potential for economic success (Figure 1.3).

Although locating mineral and energy resources is extremely important, geologists are also being asked to use their expertise to help solve many environmental problems. Some geologists are involved in finding groundwater for the ever-burgeoning needs of communities and industries or in monitoring surface and underground water pol-

FIGURE 1.3 Geologists increasingly use computers in their search for petroleum and other natural resources.

lution and suggesting ways to clean it up. Geological engineers help find safe locations for dams, waste-disposal sites, and power plants and design earthquake-resistant buildings.

Geologists are also involved in making short- and long-range predictions about earthquakes and volcanic eruptions and the potential destruction that may result. In addition, they are working with civil defense planners to help draw up contingency plans should such natural disasters occur.

TABLE 1.2		
Specialties of Geology and Their Broad Relationship to the Other Sciences		
SPECIALTY	AREA OF STUDY	RELATED SCIENCE
Geochronology	Time and history of Earth	Astronomy
Planetary geology	Geology of the planets	
Paleontology	Fossils	Biology
Economic geology	Mineral and energy resources	
Environmental geology	Environment	
Geochemistry	Chemistry of Earth	Chemistry
Hydrogeology	Water resources	
Mineralogy	Minerals	
Petrology	Rocks	
Geophysics	Earth's interior	
Structural geology	Rock deformation	Physics
Seismology	Earthquakes	
Geomorphology	Landforms	
Oceanography	Oceans	
Paleogeography	Ancient geographic features and locations	
Stratigraphy/sedimentology	Layered rocks and sediments	

As this brief survey illustrates, geologists are employed in a wide variety of pursuits. As the world's population increases and greater demands are made on Earth's limited resources, the need for geologists and their expertise will become even greater.

Geology and the Human Experience

MANY people are surprised at the extent to which we depend on geology in our everyday lives and also at the numerous references to geology in the arts, music, and literature. Rocks and landscapes are realistically represented in many sketches and paintings. Examples by famous artists include Leonardo da Vinci's *Virgin of the Rocks* and *Virgin and Child with Saint Anne,* Giovanni Bellini's *Saint Francis in Ecstasy* and *Saint Jerome,* and Asher Brown Durand's *Kindred Spirits* (Figure 1.4).

In the field of music, Ferde Grofé's *Grand Canyon Suite* was, no doubt, inspired by the grandeur and timelessness of Arizona's Grand Canyon and its vast rock exposures. The rocks on the Island of Staffa in the Inner Hebrides provided the inspiration for Felix Mendelssohn's famous *Hebrides* Overture.

FIGURE 1.4 *Kindred Spirits* by Asher Brown Durand (1849) realistically depicts the layered rocks occurring along gorges in the Catskill Mountains of New York State. Durand was one of numerous artists of the nineteenth-century Hudson River School, who were known for their realistic landscapes.

References to geology abound in *The German Legends of the Brothers Grimm,* and Jules Verne's *Journey to the Center of the Earth* describes an expedition into Earth's interior. On one level, the poem "Ozymandias" by Percy B. Shelley deals with the fact that nothing lasts forever and even solid rock eventually disintegrates under the ravages of time and weathering. References to geology can even be found in comics, two of the best known being "B.C." by Johnny Hart and "The Far Side" by Gary Larson.

Geology has also played an important role in history. Wars have been fought for the control of such natural resources as oil, gas, gold, silver, diamonds, and other valuable minerals. Empires throughout history have risen and fallen on the distribution and exploitation of natural resources. The configuration of Earth's surface, or its topography, which is shaped by geologic agents, plays a critical role in military tactics. Natural barriers such as mountain ranges and rivers have frequently served as political boundaries.

How Geology Affects Our Everyday Lives

DESTRUCTIVE volcanic eruptions, devastating earthquakes, disastrous landslides, large sea waves, floods, and droughts are headline-making events that affect many people (Figure 1.5). Although we cannot prevent most of these natural disasters, the more we know about them, the better we are able to predict, and possibly control, the severity of their impact. The environmental movement has forced everyone to take a closer look at our planet and the delicate balance between its various systems (see Perspective 1.1).

The increasing complexity and the technological orientation of society will force us, as citizens, to better understand science so that we can make informed choices about those things that affect our lives. We are already aware of some of the negative aspects of an industrialized society, such as problems relating to solid-waste disposal, contaminated groundwater, and acid rain. We are also learning the impact that humans, in increasing numbers, have on the environment and that we can no longer ignore the role that we play in the dynamics of the global ecosystem.

Most people are unaware of the extent to which geology affects their lives. For many people, the connection between geology and such well-publicized problems as nonrenewable energy and mineral resources, let alone waste disposal and pollution, is simply too far removed or too complex to be fully appreciated. But consider for a moment just how dependent we are on geology in our daily routines.

Much of the electricity for our appliances comes from the burning of coal, oil, or natural gas or from uranium consumed in nuclear-generating plants. It is geologists who locate the coal, petroleum, and uranium. The copper or other metal wires through which electricity travels are manufactured from materials found as the result of mineral

FIGURE 1.5 As these headlines from various newspapers indicate, geology affects our everyday lives.

act in a responsible manner, based on sound scientific knowledge, so future generations will inherit a habitable environment.

The concept of *sustainable development* has received increasing attention, particularly since the United Nations Conference on Environment and Development met in Rio de Janeiro, Brazil, during the summer of 1992. This important concept links satisfying basic human needs with safeguarding our environment to ensure continued economic development.

If we are to have a world in which poverty is not widespread, then we must develop policies that encourage management of our natural resources along with continuing economic development. A growing global population will mean increased demand for food, water, and natural resources, particularly nonrenewable mineral and energy resources. In meeting these demands, geologists will play an important role in locating the needed resources, as well as in ensuring protection of the environment for the benefit of future generations.

exploration. The buildings we live and work in owe their very existence to geologic resources. A few examples are the concrete foundation (concrete is a mixture of clay, sand, or gravel, and limestone), the drywall (made largely from the mineral gypsum), the windows (the mineral quartz is the principal ingredient in the manufacture of glass), and the metal or plastic plumbing fixtures inside the building (the metals are from ore deposits, and the plastics are most likely manufactured from petroleum distillates of crude oil).

Furthermore, when we go to work, the car or public transportation we use is powered and lubricated by some type of petroleum by-product and is constructed of metal alloys and plastics. And the roads or rails we ride over come from geologic materials, such as gravel, asphalt, concrete, or steel. All these items are the result of processing geologic resources.

As individuals and societies, the standard of living we enjoy is obviously directly dependent on the consumption of geologic materials. Therefore, we need to be aware of geology and of how our use and misuse of geologic resources may affect the delicate balance of nature and irrevocably alter our culture as well as our environment.

When such environmental issues as acid rain, the greenhouse effect, and the depletion of the ozone layer are discussed and debated, it is important to remember that they are not isolated topics but part of a larger system that involves Earth. Accordingly, we must understand that changes we make in the global ecosystem can have wide-ranging effects that we might not be aware of. For this reason, an understanding of geology, and science in general, can help minimize disruption to the ecosystem caused by these changes. We must also remember that humans are part of the ecosystem, and like all other life-forms, our presence alone affects the ecosystem. We must therefore

The Origin of the Solar System and the Differentiation of Early Earth

ACCORDING to the currently accepted theory for the origin of the solar system (Figure 1.6), interstellar material in a spiral arm of the Milky Way Galaxy condensed and began collapsing. As this cloud gradually collapsed under the influence of gravity, it flattened and began rotating counterclockwise, with about 90% of its mass concentrated in the central part of the cloud. The rotation and concentration of material continued, and an embryonic Sun formed, surrounded by a turbulent, rotating cloud of material called a *solar nebula*.

The turbulence in this solar nebula formed localized eddies where gas and solid particles condensed. During the condensation process gaseous, liquid, and solid particles began accreting into ever-larger masses called *planetesimals* that eventually became true planetary bodies. While the planets were accreting, material that had been pulled into the center of the nebula also condensed, collapsed, and was heated to several million degrees by gravitational compression. The result was the birth of a star, our Sun.

Some 4.6 billion years ago, enough material eventually gathered together in one of the turbulent eddies that swirled around the early Sun to form the planet Earth. Scientists think that this early Earth was rather cool, so the accreting elements and nebular rock fragments were solids rather than gases or liquids. This early Earth is also thought to have been of generally uniform composition and density throughout (Figure 1.7a). It was composed mostly of compounds of silicon, iron, magnesium, oxygen, aluminum, and smaller amounts of all the other chemical elements. Subsequently, when Earth underwent heating, this

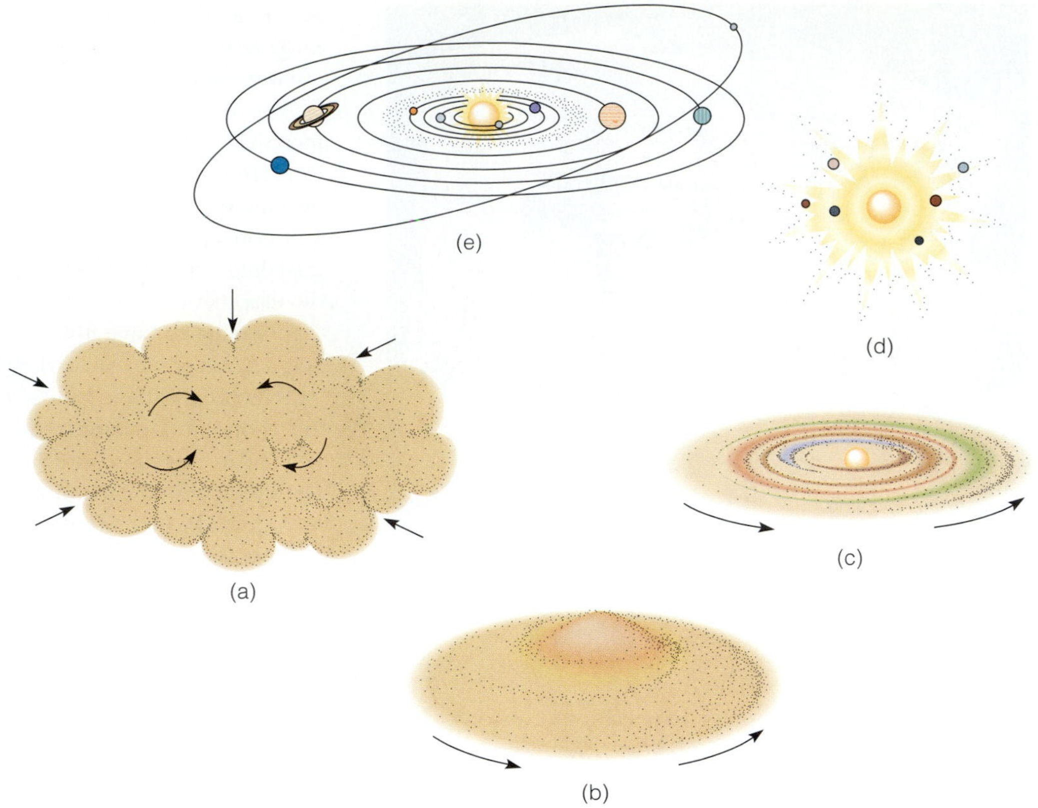

FIGURE 1.6 The currently accepted theory for the origin of our solar system involves (a) a huge nebula condensing under its own gravitational attraction, then (b) contracting, rotating, and (c) flattening into a disk, with the Sun forming in the center and eddies gathering up material to form planets. As the Sun contracted and began to visibly shine, (d) intense solar radiation blew away unaccreted gas and dust until finally, (e) the Sun began burning hydrogen and the planets completed their formation.

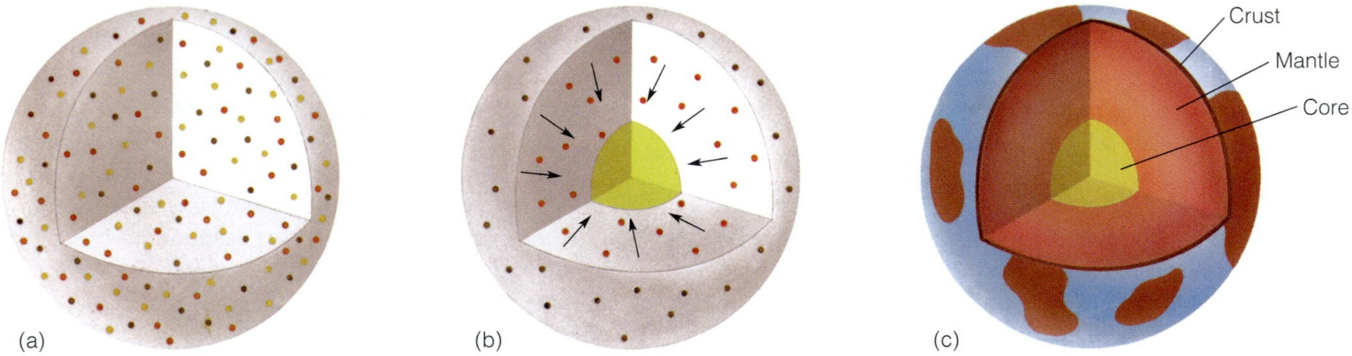

FIGURE 1.7 (a) Early Earth was probably of uniform composition and density throughout. (b) Heating of early Earth reached the melting point of iron and nickel, which, being denser than silicate minerals, settled to Earth's center. At the same time, the lighter silicates flowed upward to form the mantle and the crust. (c) In this way, a differentiated Earth formed, consisting of a dense iron–nickel core, an iron-rich silicate mantle, and a silicate crust with continents and ocean basins.

homogeneous composition disappeared (Figure 1.7b), and the result was a differentiated planet, consisting of a series of concentric layers of differing composition and density (Figure 1.7c). This differentiation into a layered planet is probably the most significant event in Earth history. Not only did it lead to the formation of a crust and eventually to continents, but it was also probably responsible for the emission of gases from the interior, which eventually led to the formation of the oceans and the atmosphere.

Earth as a Dynamic Planet

EARTH is a dynamic planet that has continuously changed during its 4.6-billion-year existence. The size, shape, and geographic distribution of continents and ocean basins have changed through time, the composition of the atmosphere has evolved, and life-forms existing today differ from those that lived during the past. We can easily visualize how mountains and hills are worn down by erosion and how landscapes are changed by the forces of wind, water, and ice. Volcanic eruptions and earthquakes reveal an active interior, and folded and fractured rocks indicate the tremendous power of Earth's internal forces.

Earth consists of three concentric layers: the core, the mantle, and the crust (Figure 1.8). This orderly division results from density differences between the layers as a function of variations in composition, temperature, and pressure.

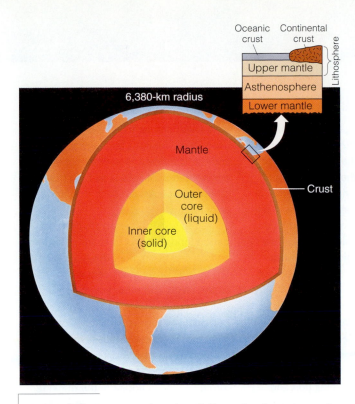

FIGURE 1.8 A cross section of Earth illustrating the core, mantle, and crust. The enlarged portion shows the relationship between the lithosphere, composed of the continental crust, oceanic crust, and upper mantle, and the underlying asthenosphere and lower mantle.

The **core** has a calculated density of 10 to 13 grams per cubic centimeter (g/cm^3) and occupies about 16% of Earth's total volume. Seismic (earthquake) data indicate that the core consists of a small, solid inner core and a larger, apparently liquid, outer core. Both are thought to consist largely of iron and a small amount of nickel.

The **mantle** surrounds the core and comprises about 83% of Earth's volume. It is less dense than the core (3.3–5.7 g/cm^3) and is thought to be composed largely of *peridotite*, a dark, dense igneous rock containing abundant iron and magnesium. The mantle can be divided into three distinct zones based on physical characteristics. The lower mantle is solid and forms most of the volume of Earth's interior. The **asthenosphere** surrounds the mantle. It has the same composition as the lower mantle but behaves plastically and slowly flows. Partial melting within the asthenosphere generates *magma* (molten material), some of which rises to the surface because it is less dense than the rock from which it was derived. The upper mantle surrounds the asthenosphere. The solid upper mantle and the overlying crust constitute the **lithosphere,** which is broken into numerous individual pieces called **plates** that move over the asthenosphere as a result of underlying *convection cells* (Figure 1.9). Interactions of these plates are responsible for such phenomena as earthquakes, volcanic eruptions, and the formation of mountain ranges and ocean basins.

The **crust,** the outermost layer of Earth, consists of two types. *Continental crust* is thick (20–90 km), has an average density of 2.7 g/cm^3, and contains considerable silicon and aluminum. *Oceanic crust* is thin (5–10 km), denser than continental crust (3.0 g/cm^3), and is composed of the dark igneous rock *basalt*.

Since the widespread acceptance of plate tectonic theory about 25 years ago, geologists have viewed Earth from a global perspective in which all of its systems are interconnected. Thus, the distribution of mountain chains, major fault systems, volcanoes and earthquakes, the origin of new ocean basins, the movement of continents, and several other geologic processes and features are perceived to be interrelated.

Geology and the Formulation of Theories

THE term **theory** has various meanings. In colloquial usage, it means a speculative or conjectural view of something—hence the widespread belief that scientific theories

FIGURE 1.9 Earth's plates are thought to move as a result of underlying mantle convection cells in which warm material from deep within Earth rises toward the surface, cools, and then, upon losing heat, descends back into the interior. The movement of these convection cells is thought to be the mechanism responsible for the movement of Earth's plates, as shown in this diagrammatic cross section.

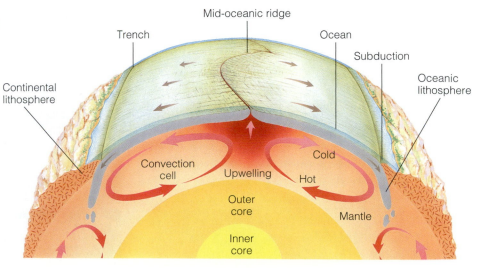

The Aral Sea

The Aral Sea in the Central Asian desert region of the former Soviet Union (Figure 1) is a continuing environmental and human disaster. In 1960 it was the world's fourth largest lake. Since then its size has steadily decreased from 67,000 km² to about 34,000 km², and its volume has fallen from 1090 km³ to 300 km³, so that today it is only the sixth largest lake. Presently, it is divided into two separate basins, a small northern basin and a larger southern basin. At its present rate of reduction, the Aral Sea could disappear within the next 30 years.

What could have caused such a disaster? For thousands of years, the Aral Sea was fed by two large rivers, the Amu Dar'ya and Syr Dar'ya. The headwaters of both rivers begin in the mountains more than 2000 km to the southeast and flow northward through the Kyzyl Kum and Kara Kum deserts into the Aral Sea. Until the early part of this century, a balance existed between the water supplied by the Amu Dar'ya and the Syr Dar'ya rivers

and the rate of evaporation of the Aral Sea, which has no outlet. In 1918, however, it was decreed that waters from the two rivers supplying the Aral Sea would be diverted to irrigate millions of acres of cotton so that the Soviet Union could become self-sufficient in cotton production. As a result of this decision, the Aral Sea has become an ecological and environmental nightmare affecting some 35 million people.

To gain an appreciation of the magnitude of this disaster, consider that as late as the 1960s the Aral Sea was home to 24 native species of fish and supported a major fishing industry (Figure 2). The town of Muynak was the major fishing port and produced 3% of the Soviet Union's annual catch. The fishing industry provided about 60,000 jobs, with approximately 10,000 fishermen working out of Muynak. Today, Muynak is more than 20 km from the shoreline of the Aral Sea, and there are no native fish species left in the sea.

As the Aral Sea shrinks, vast areas of the former sea bottom are exposed. This

new sediment, far from being rich and productive, contains large amounts of sodium chloride and sodium sulfate. The concentration of salts is so high that only one plant species grows in it. It is too salty for anything else.

As the winds blow across the near-barren land, salt and dust are picked up and carried throughout the Aral region. These salts cause great damage to the cotton crops and other vegetation of the Aral basin and also exact a heavy toll on humans. The dry, salty dust has caused respiratory and eye diseases to increase dramatically during the past 30 years, as well as the reported number of cases of throat cancer. In addition, the drinking water supply has become so polluted that many people suffer from intestinal disorders.

As a result of the diversion of water from the Amu Dar'ya and the Syr Dar'ya rivers and the resulting reduction in the Aral Sea, desertification has become a major problem in the Aral basin. This has caused a change in the weather pat-

Figure 1 Location (left) and 1985 space shuttle view of the Aral Sea (right).

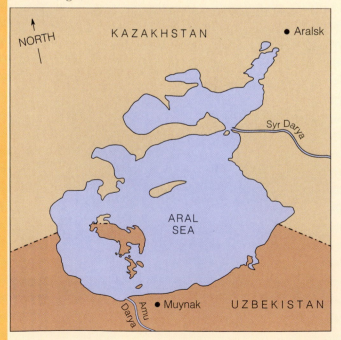

Figure 2 Fishing boats lie abandoned in the dry seabed that was once part of the Aral Sea. As recently as the 1960s, the Aral Sea supported a large fishing industry, but today the water is too saline for any fish to survive. Consequently, frozen fish are shipped in from the Pacific for processing in plants that are now more than 20 km from the Aral Sea.

lower stream and groundwater levels, requiring more irrigation to maintain the same production levels. This means that even more water is diverted from the Aral Sea, causing it to shrink still further.

Can the Aral Sea be saved? Only by a concerted and cooperative effort by the countries that comprise the Aral basin. Realizing this, the now independent states of Kazakhstan, Uzbekistan, Kyrgyzstan, Tadzhikistan, and Turkmenistan took the first steps in 1992 by signing a formal agreement on sharing the waters of the Amu Dar'ya and Syr Dar'ya. In addition, these states entered into discussions to create a council to oversee and coordinate the management of the basin's resources. While achieving full restoration of the Aral Sea is probably not possible, maintaining its current surface level and even raising the surface level of the small northern Aral Sea is feasible. If this could be accomplished, salinity levels would be reduced, making it possible to reintroduce native fish species and revive commercial fishing.

terns of the region, so it is now colder in the winter and hotter and drier in the summer.

Despite the damage done to the region by the dying of the Aral Sea, irrigation for cotton production continues. However, the soil in many of the fields has become increasingly salty due to

are little more than unsubstantiated wild guesses. In scientific usage, however, a theory is a coherent explanation for one or several related natural phenomena supported by a large body of objective evidence. From a theory are derived predictive statements that can be tested by observation and/or experiment so that their validity can be assessed. The law of universal gravitation is an example of a theory describing the attraction between masses (an apple and Earth in the popularized account of Newton and his discovery).

Theories are formulated through the process known as the **scientific method.** This method is an orderly, logical approach that involves gathering and analyzing the facts or data about the problem under consideration. Tentative explanations, or **hypotheses,** are then formulated to explain the observed phenomena. Next, the hypotheses are tested to see if what they predicted actually occurs in a given situation. Finally, if one of the hypotheses is found, after repeated tests, to explain the phenomena, then that hypothesis is proposed as a theory. One should remember, however, that in science even a theory is still subject to further testing and refinement as new data become available.

The fact that a scientific theory can be tested and is subject to such testing separates science from other forms of human inquiry. Because scientific theories can be tested, they have the potential of being supported or even proved wrong. Accordingly, science must proceed without any appeal to beliefs or supernatural explanations, not because such beliefs or explanations are necessarily untrue, but because we have no way to investigate them. For this reason, science makes no claim about the existence or nonexistence of a supernatural or spiritual realm.

Each scientific discipline has certain theories that are of particular importance for that discipline. In geology, the formulation of plate tectonic theory has changed the way geologists view Earth. Geologists now view Earth history in terms of interrelated events that are part of a global pattern of change.

Plate Tectonic Theory

THE acceptance of **plate tectonic theory** is recognized as a major milestone in the geologic sciences. It is comparable to the revolution caused by Darwin's theory of evolution in biology. Plate tectonics has provided a framework for interpreting the composition, structure, and internal processes of Earth on a global scale. It has led to the realization that the continents and ocean basins are part of a lithosphere–atmosphere–hydrosphere (water portion of the planet) system that evolved together with Earth's interior (Table 1.3).

According to plate tectonic theory, the lithosphere is divided into plates that move over the asthenosphere (Figure 1.10). Zones of volcanic activity, earthquake activity, or both mark most plate boundaries. Along these boundaries, plates diverge, converge, or slide sideways past each other.

At **divergent plate boundaries,** plates move apart as magma rises to the surface from the asthenosphere (Figure 1.11). The magma solidifies to form rock, which attaches to the moving plate. The margins of divergent plate boundaries are marked by mid-oceanic ridges in oceanic crust, such as the Mid-Atlantic Ridge, and are recognized by linear rift valleys, where newly forming divergent boundaries occur beneath continental crust.

Plates move toward one another along **convergent plate boundaries** where one plate sinks beneath another plate along what is known as a **subduction zone** (Figure 1.11). As the plate descends into Earth, it becomes hotter until it melts, or partially melts, thus generating a magma. As this magma rises, it may erupt at Earth's surface, forming a chain of volcanoes. The Andes Mountains on the west coast of South America are a good example of a volcanic mountain range formed as a result of subduction along a convergent plate boundary (Figure 1.10).

TABLE 1.3	
Plate Tectonics and Earth Systems	
Solid Earth	Plate tectonics is driven by convection in the mantle and in turn drives mountain-building and associated igneous and metamorphic activity.
Atmosphere	Arrangement of continents affects solar heating and cooling, and thus winds and weather systems. Rapid plate spreading and hot-spot activity may release volcanic carbon dioxide and affect global climate.
Hydrosphere	Continental arrangement affects ocean currents. Rate of spreading affects volume of mid-oceanic ridges and hence sea level. Placement of continents may contribute to onset of ice ages.
Biosphere	Movement of continents creates corridors or barriers to migration, the creation of ecological niches, and transport of habitats into more-or-less favorable climates.
Extraterrestrial	Arrangement of continents affects free circulation of ocean tides and influences tidal slowing of Earth's rotation.

SOURCE: Adapted by permission from Stephen Dutch, James S. Monroe, and Joseph Moran, *Earth Science* (Minneapolis/St. Paul: West Publishing Co.).

Transform plate boundaries are sites where plates slide sideways past each other (Figure 1.11). The San Andreas fault in California is a transform plate boundary separating the Pacific plate from the North American plate (Figure 1.10). The earthquake activity along the San Andreas fault results from the Pacific plate moving northward relative to the North American plate.

FIGURE **1.10** Earth's lithosphere is divided into rigid plates of various sizes that move over the asthenosphere.

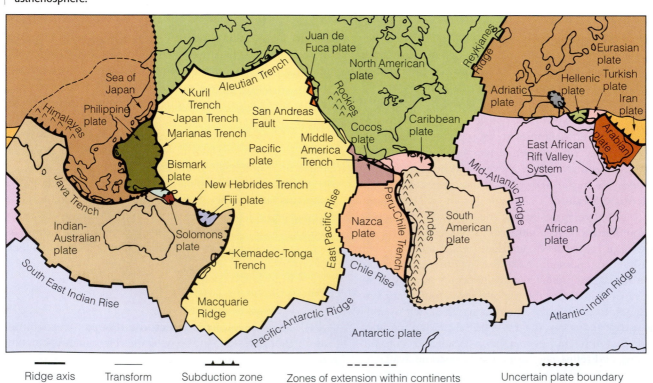

Ridge axis   Transform   Subduction zone   Zones of extension within continents   Uncertain plate boundary

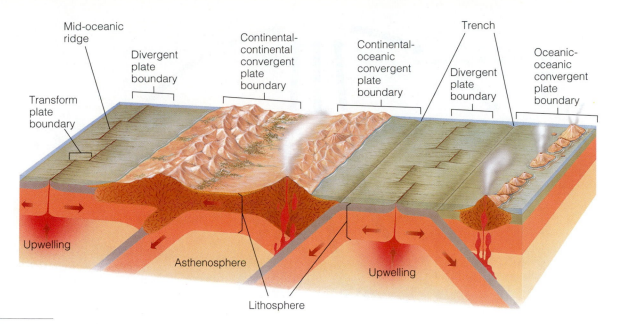

FIGURE 1.11 An idealized cross section illustrating the relationship between the lithosphere and the underlying asthenosphere and the three principal types of plate boundaries: divergent, convergent, and transform.

A revolutionary concept when it was proposed in the 1960s, plate tectonic theory has had significant and far-reaching consequences in all fields of geology because it provides the basis for relating many seemingly unrelated geologic phenomena. Besides being responsible for the major features of Earth's crust, plate movements also affect the formation and distribution of Earth's natural resources, as well as influence the distribution and evolution of the world's biota.

The impact of plate tectonic theory has been particularly notable in the interpretation of Earth history. For example, the Appalachian Mountains in eastern North America and the mountain ranges of Greenland, Scotland, Norway, and Sweden are not the result of unrelated mountain-building episodes but, rather, are part of a larger mountain-building event that involved the closing of an ancient "Atlantic Ocean" and the formation of the supercontinent Pangaea about 245 million years ago (see Chapter 18).

The Rock Cycle

A **rock** is an aggregate of **minerals,** which are naturally occurring, inorganic, crystalline solids with definite physical and chemical properties. Minerals are composed of elements such as oxygen, silicon, and aluminum, and elements are made up of atoms, the smallest particles of matter that still retain the characteristics of an element. Over 3500 minerals have been identified and described, but only about a dozen make up the bulk of the rocks in the crust.

Geologists recognize three major groups of rocks—*igneous, sedimentary,* and *metamorphic*—each of which is characterized by its mode of formation. Each group contains a variety of individual rock types that differ from one another on the basis of composition or texture (the size, shape, and arrangement of mineral grains).

The **rock cycle** is a way of viewing the interrelationships between Earth's internal and external processes (Figure 1.12). It relates the three rock groups to each other; to surficial processes such as weathering, transportation, and deposition; and to internal processes such as magma generation and metamorphism. Plate movement is the mechanism responsible for recycling rock materials and therefore drives the rock cycle.

Igneous rocks result from the crystallization of magma or the accumulation and consolidation of volcanic ejecta such as ash. As a magma cools, minerals crystallize, and the resulting rock is characterized by interlocking mineral grains. Magma that cools slowly beneath the surface produces *intrusive igneous rocks* (Figure 1.13a); magma that cools at the surface produces *extrusive igneous rocks* (Figure 1.13b).

Rocks exposed at Earth's surface are broken into particles and dissolved by various weathering processes. The particles and dissolved material may be transported by wind, water, or ice and eventually deposited as *sediment*. This sediment may then be compacted or cemented into sedimentary rock.

Sedimentary rocks originate by consolidation of rock fragments, precipitation of mineral matter from solution, or compaction of plant or animal remains (Figure 1.13c and d). Because sedimentary rocks form at or near Earth's surface, geologists can make inferences about the environment in which they were deposited, the type of transporting agent, and perhaps even something about the source

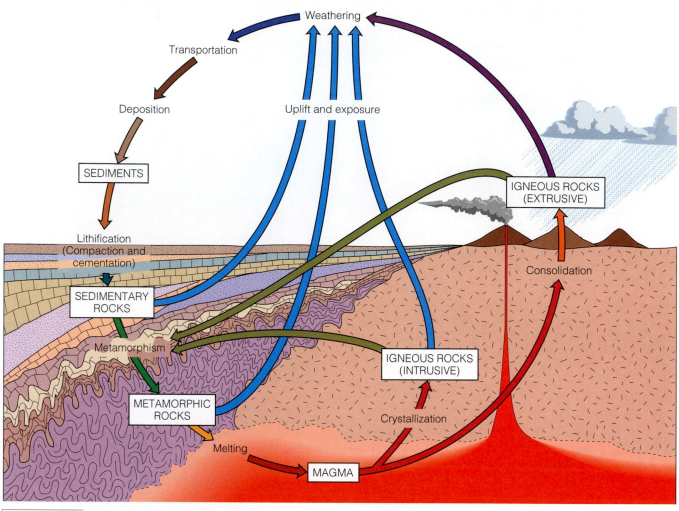

FIGURE 1.12 The rock cycle showing the interrelationships between Earth's internal and external processes and how each of the three major rock groups is related to the others. (Modified from Fig. 12, Dietrich, R. V. 1979. *Geology and Michigan: Forty-nine Questions and Answers*)

FIGURE 1.13 Hand specimens of common igneous (a, b), sedimentary (c, d), and metamorphic (e, f) rocks. (a) Granite, an intrusive igneous rock. (b) Basalt, an extrusive igneous rock. (c) Conglomerate, a sedimentary rock formed by the consolidation of rock fragments. (d) Limestone, a sedimentary rock formed by the extraction of mineral matter from seawater by organisms or by the inorganic precipitation of the mineral calcite from seawater. (e) Gneiss, a foliated metamorphic rock. (f) Quartzite, a nonfoliated metamorphic rock. *(Photos courtesy of Sue Monroe.)*

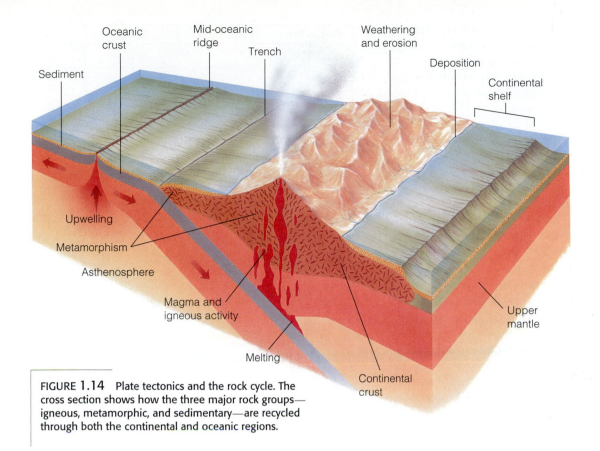

FIGURE 1.14 Plate tectonics and the rock cycle. The cross section shows how the three major rock groups—igneous, metamorphic, and sedimentary—are recycled through both the continental and oceanic regions.

Labels in figure: Sediment, Oceanic crust, Mid-oceanic ridge, Trench, Weathering and erosion, Deposition, Continental shelf, Upwelling, Metamorphism, Asthenosphere, Magma and igneous activity, Melting, Continental crust, Upper mantle

from which the sediments were derived (see Chapter 7). Accordingly, sedimentary rocks are very useful for interpreting Earth history.

Metamorphic rocks result from the alteration of other rocks, usually beneath the surface, by heat, pressure, and the chemical activity of fluids. For example, marble, a rock preferred by many sculptors and builders, is a metamorphic rock produced when the agents of metamorphism are applied to the sedimentary rocks limestone or dolostone. Metamorphic rocks are either *foliated* (Figure 1.13e) or *nonfoliated* (Figure 1.13f). Foliation, the parallel alignment of minerals due to pressure, gives the rock a layered or banded appearance.

THE ROCK CYCLE AND PLATE TECTONICS

Interactions among plates determine, to a certain extent, which of the three rock groups will form (Figure 1.14). For example, weathering and erosion produce sediments that are transported by agents such as running water from the continents to the oceans, where they are deposited and accumulate. These sediments, some of which may be lithified and become sedimentary rock, become part of a moving plate along with the underlying oceanic crust. When plates converge, heat and pressure generated along the plate boundary may lead to igneous activity and metamorphism within the descending oceanic plate, thus producing various igneous and metamorphic rocks.

Some of the sediment and sedimentary rock is subducted and melts, while other sediments and sedimentary rocks along the boundary of the nonsubducted plate are metamorphosed by the heat and pressure generated along the converging plate boundary. Later, the mountain range or chain of volcanic islands formed along the convergent plate boundary will once again be weathered and eroded, and the new sediments will be transported to the ocean to begin yet another cycle.

Geologic Time and Uniformitarianism

AN appreciation of the immensity of geologic time is central to understanding the evolution of Earth and its biota. Indeed, time is one of the main aspects that sets geology apart from the other sciences. Most people have difficulty comprehending geologic time because they tend to think in terms of the human perspective—seconds, hours, days, and years. Ancient history is what occurred hundreds or even thousands of years ago. When geologists talk of ancient geologic history, however, they are referring to events that happened hundreds of millions or even billions of years ago. To a geologist, recent geologic events are those that occurred within the last million years or so.

It is also important to remember that Earth goes through cycles of a much longer duration than the human perspective of time. Although they may have disastrous effects on the human species, global warming and cooling are part of a larger cycle that has resulted in numerous glacial advances and retreats during the past 1.6 million years. In fact, geologists can make important contributions to the debate on global warming because of their geologic perspective.

The **geologic time scale** resulted from the work of many nineteenth-century geologists who pieced together information from numerous rock exposures and constructed a sequential chronology based on changes in Earth's biota through time. Subsequently, with the discovery of radioactivity in 1895 and the development of various radiometric dating techniques, geologists have been able to assign absolute age dates in years to the subdivisions of the geologic time scale (Figure 1.15).

One of the cornerstones of geology is the **principle of uniformitarianism.** It is based on the premise that present-day processes have operated throughout geologic time. Therefore, to understand and interpret the rock record, we must first understand present-day processes and their results.

Uniformitarianism is a powerful principle that allows us to use present-day processes as the basis for interpreting the past and for predicting potential future events. We should keep in mind that uniformitarianism does not exclude such sudden or catastrophic events as volcanic eruptions, earthquakes, landslides, or flooding. These are processes that shape our modern world, and, in fact, some geologists view the history of Earth as a series of such short-term or punctuated events. Such a view is certainly in keeping with the modern principle of uniformitarianism.

Furthermore, uniformitarianism does not require that the rates and intensities of geologic processes be constant through time. We know that volcanic activity was more intense in North America 5 to 10 million years ago than it is today, and that glaciation has been more prevalent during the last several million years than in the previous 300 million years.

What uniformitarianism means is that even though the rates and intensities of geologic processes have varied during the past, the physical and chemical laws of nature have remained the same. Although Earth is in a dynamic state of change and has been ever since it was formed, the processes that shaped it during the past are the same ones that are in operation today.

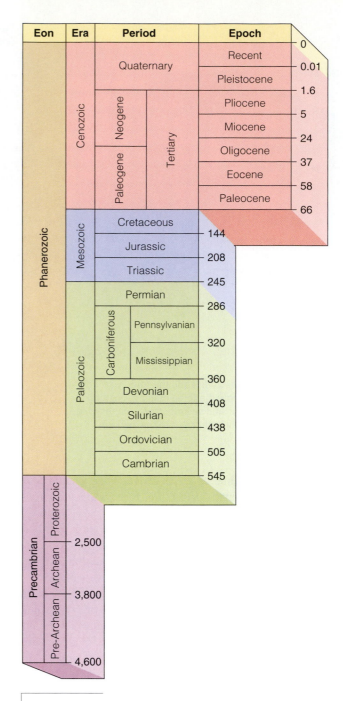

FIGURE **1.15** The geologic time scale. Numbers to the right of the columns are ages in millions of years before the present.

1. Geology, the study of Earth, is divided into two broad areas: physical geology is the study of Earth materials as well as the processes that operate within and upon Earth's surface; historical geology examines the origin and evolution of Earth, its continents, oceans, atmosphere, and life.

2. Geology is part of the human experience. We can find examples of it in the arts, music, and literature. A basic understanding of geology is also important for dealing with the many environmental problems and issues facing society.

3. Geologists engage in a variety of occupations, the main one being exploration for mineral and energy resources. They are also becoming increasingly involved in environmental issues and making short- and long-range predictions of the potential dangers from such natural disasters as volcanic eruptions and earthquakes.

4. About 4.6 billion years ago, the solar system formed from a rotating cloud of interstellar matter. Eventually, as this cloud condensed, it collapsed under the influence of gravity and flattened into a rotating disk. Within this rotating disk, the Sun, planets, and moons formed from the turbulent eddies of nebular gases and solids.

5. Earth is differentiated into layers. The outermost layer is the crust, which is divided into continental and oceanic portions. Below the crust is the upper mantle. The crust and upper mantle, or lithosphere, overlie the asthenosphere, a zone that slowly flows. The asthenosphere is underlain by the solid lower mantle. Earth's core consists of an outer liquid portion and an inner solid portion.

6. The lithosphere is broken into a series of plates that diverge, converge, and slide sideways past one another.

7. The scientific method is an orderly, logical approach that involves gathering and analyzing facts about a particular phenomenon, formulating hypotheses to explain the phenomenon, testing the hypotheses, and finally proposing a theory. A theory is a testable explanation for some natural phenomenon that has a large body of supporting evidence.

8. Plate tectonic theory provides a unifying explanation for many geologic features and events. The interaction between plates is responsible for volcanic eruptions, earthquakes, the formation of mountain ranges and ocean basins, and the recyling of rock material.

9. Igneous, sedimentary, and metamorphic rocks comprise the three major groups of rocks. Igneous rocks result from the crystallization of magma or the consolidation of volcanic ejecta. Sedimentary rocks are formed mostly by the consolidation of rock fragments, precipitation of mineral matter from solution, or compaction of plant or animal remains. Metamorphic rocks are produced from other rocks, generally beneath Earth's surface, by heat, pressure, and chemically active fluids.

10. The rock cycle illustrates the interactions between internal and external Earth processes and shows how the three rock groups are interrelated.

11. Time sets geology apart from the other sciences, except astronomy. The geologic time scale is the calendar geologists use to date past events.

12. The principle of uniformitarianism is basic to the interpretation of Earth history. This principle holds that the laws of nature have been constant through time and that the same processes operating today have operated in the past, although at different rates.

Important **Terms**

asthenosphere	igneous rock	rock
convergent plate boundary	lithosphere	rock cycle
core	mantle	scientific method
crust	metamorphic rock	sedimentary rock
divergent plate boundary	mineral	subduction zone
geologic time scale	plate	system
geology	plate tectonic theory	theory
hypothesis	principle of uniformitarianism	transform plate boundary

1. A combination of related parts interacting in an organized fashion is:
   a. _____ sustainable development;
   b. _____ a system;
   c. _____ uniformitarianism;
   d. _____ a theory;
   e. _____ none of these.

2. Which of the following statements about a mineral is *not* true?
   a. _____ it is organic;
   b. _____ it has definite physical and chemical properties;
   c. _____ it is naturally occurring;
   d. _____ it is a crystalline solid;
   e. _____ none of these.

3. A plate is composed of the:
   a. _____ core and lower mantle;
   b. _____ lower mantle and asthenosphere;
   c. _____ asthenosphere and upper mantle;
   d. _____ upper mantle and crust;
   e. _____ continental and oceanic crust.

4. The layer between the core and the crust is the:
   a. _____ mantle;
   b. _____ lithosphere;
   c. _____ hydrosphere;
   d. _____ biosphere;
   e. _____ asthenosphere.

5. What fundamental process is thought to be responsible for plate motion?
   a. _____ hot-spot activity;
   b. _____ subduction;
   c. _____ spreading ridges;
   d. _____ convection cells;
   e. _____ density differences.

6. Which of the following statements about a scientific theory is *not* true?
   a. _____ it is an explanation for some natural phenomenon;
   b. _____ it has a large body of supporting evidence;
   c. _____ it is a conjecture or guess;
   d. _____ it is testable;
   e. _____ none of these.

7. Mid-oceanic ridges are examples of what type of boundary?
   a. _____ divergent;
   b. _____ convergent;
   c. _____ transform;
   d. _____ subduction;
   e. _____ answers (b) and (d).

8. The Andes Mountains of South America are a good example of what type of plate boundary?
   a. _____ divergent;
   b. _____ transform;
   c. _____ convergent;
   d. _____ hot spot;
   e. _____ answers (a) and (d).

9. Which rocks result from the alteration of other rocks, usually beneath Earth's surface, by heat, pressure, and the chemical activity of fluids?
   a. _____ igneous;
   b. _____ sedimentary;
   c. _____ metamorphic;
   d. _____ all of these;
   e. _____ none of these.

10. The premise that present-day processes have operated throughout geologic time is known as the principle of:
    a. _____ plate tectonics;
    b. _____ seafloor spreading;
    c. _____ continental drift;
    d. _____ volcanism;
    e. _____ uniformitarianism.

11. The rock cycle implies that:
    a. _____ metamorphic rocks are derived from magma;
    b. _____ any rock type can be derived from any other rock type;
    c. _____ igneous rocks only form beneath Earth's surface;
    d. _____ sedimentary rocks only form from the weathering of igneous rocks;
    e. _____ all of these.

12. Why is it important for people to have a basic understanding of geology?

13. Describe some of the ways in which geology affects our everyday lives.

14. Name the major layers of Earth, and describe their general composition.

15. Describe the scientific method, and explain how it may lead to a scientific theory.

16. Briefly describe plate tectonic theory, and explain why it is a unifying theory in geology.

17. Describe the rock cycle, and explain how it may be related to plate tectonics.

18. Does the principle of uniformitarianism allow for catastrophic events? Explain.

Points to Ponder

1. Propose a pre–plate tectonic hypothesis explaining the formation and distribution of mountain ranges.

2. Provide several examples of how a knowledge of geology would be useful in planning a military campaign against another country.

For these web site addresses, along with current updates and exercises, log on to

http://www.wadsworth.com/geo

► **EARTH SCIENCE RESOURCES ON THE INTERNET**

This site, maintained by the Geology Department at the University of North Carolina at Chapel Hill, is an excellent starting place to learn how to find geoscience items on the World Wide Web. As you will see, a tremendous amount of information, graphics, and maps is available, and this site will get you started finding many of them and practicing your Net skills.

1. Read the *Starting up* and *How to find things* sections.

2. Visit several of the geology sites mentioned to see the variety of information, videos, and images available.

3. Visit the home page of several of the different geologic societies and organizations to see what the societies do and what information is available from them.

4. Check the home page of several of the different schools and universities to find information on course offerings, faculty, and requirements to earn a degree. Compare the home pages and information available from different universities around the world.

► **ONLINE RESOURCES FOR EARTH SCIENTISTS (ORES)**

This large and comprehensive Earth science resource list is operated by Bill Thoen and Ted Smith and provides links to many other resource lists. It is continually updated and is certainly a site worth visiting.

1. Check out the various headings under *Contents* to see what is available in the different geoscience and related areas and explore the various links to other resource lists.

2. When you find a particular link interesting and would like to return to it in the future, make a "Bookmark" of the page (that is, save its URL). This way when you want to go back to the site, all you will have to do is find its name in the "Bookmark Menu" and click it, and your browser will automatically connect you to that site.

► **WEST'S GEOLOGY DIRECTORY**

This comprehensive directory of geologic Internet links is an excellent place to find links on virtually any geologic topic. The Contents–Index lists numerous topics. Just click on a topic and a list of sites, which you can then click on, will appear. The links are regularly updated, and sites are checked, reviewed, and rated where possible. In addition, there is also a link to field-trip guides and bibliographies.

CD-ROM Exploration

Explore the following *In-Terra-Active 2.0* CD-ROM module(s) and increase your understanding of key concepts and processes presented in this chapter.

► **SECTION: MATERIALS**
MODULE: ROCK CYCLE

During a visit to Thailand, you find a specimen of limestone and another of marble in the same area. **Question: Which rock is older, and how can you tell?**

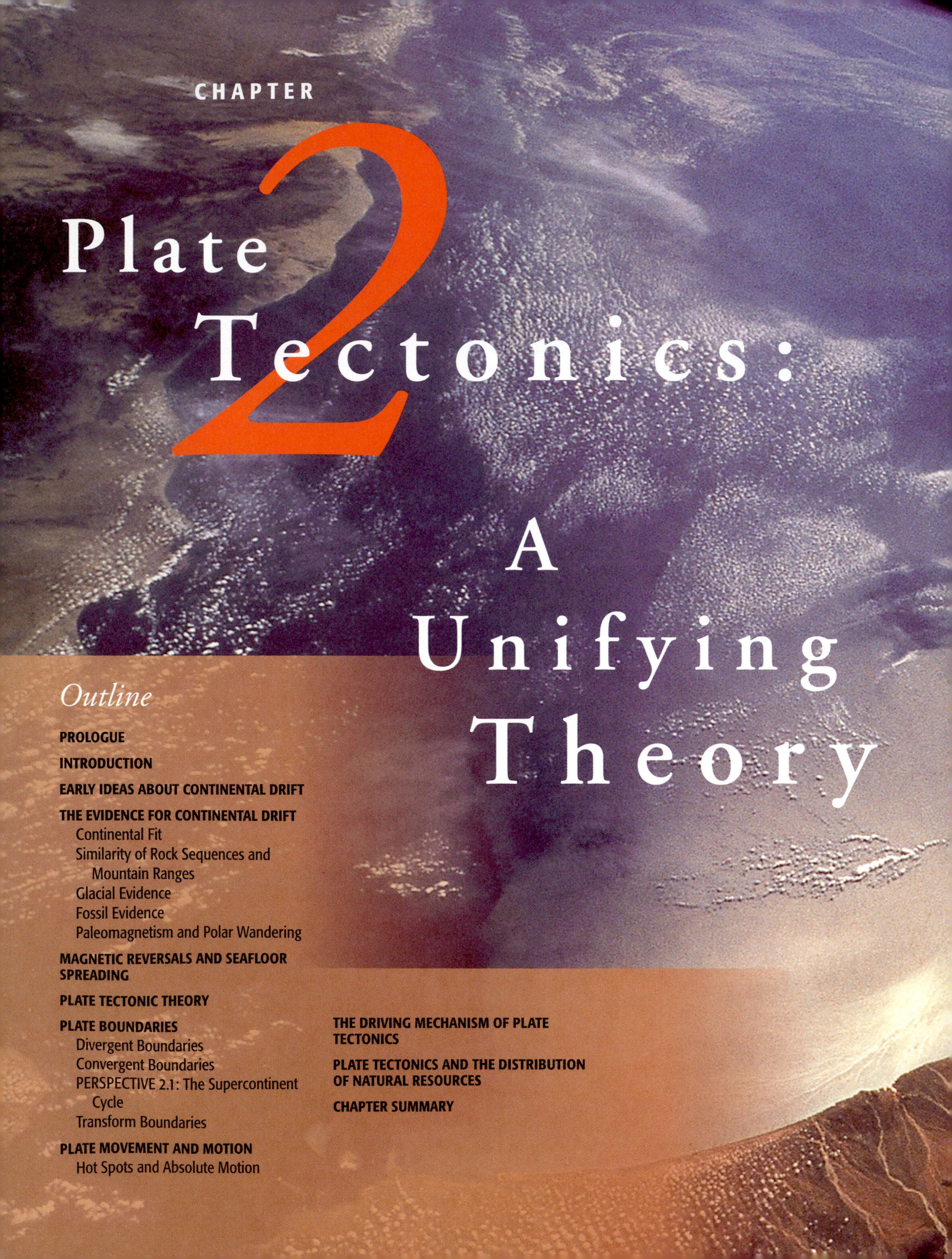

Plate

2

Tectonics:

A Unifying Theory

Outline

Prologue

The four inner, or terrestrial, planets—Mercury, Venus, Earth, and Mars—all had a similar early history involving accretion, differentiation into a metallic core and silicate mantle and crust, and formation of an early atmosphere by outgassing. Their early history was also marked by widespread volcanism and meteorite impacts, both of which helped modify their surfaces. The volcanic and tectonic activity and resultant surface features (other than meteorite craters) of these planets are clearly related to the way they transport heat from their interiors to their surfaces.

Earth appears to be unique in that its surface is broken up into a series of plates. The creation and destruction of these plates at spreading ridges and subduction zones transfer the majority of Earth's internally produced heat. In addition, movement of the plates—together with life-forms, the formation of sedimentary rocks, and water—is responsible for the cycling of carbon dioxide between the atmosphere and lithosphere and thus the maintenance of a habitable climate on Earth.

Heat is transferred between the interior and surface of both Mercury and Mars mainly by lithospheric conduction. This method is sufficient for these planets because both are significantly smaller than Earth or Venus. Because

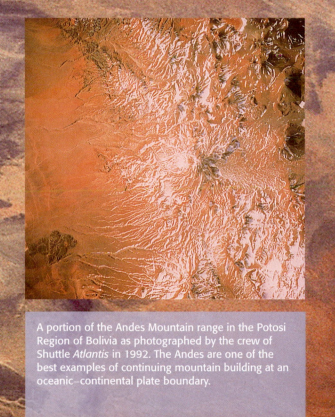

A portion of the Andes Mountain range in the Potosi Region of Bolivia as photographed by the crew of Shuttle *Atlantis* in 1992. The Andes are one of the best examples of continuing mountain building at an oceanic–continental plate boundary.

Mercury and Mars have a single, globally continuous plate, they have exhibited fewer types of volcanic and tectonic activity than has Earth. The initial interior warming of Mercury and Mars produced tensional features such as normal faults and widespread volcanism, while their subsequent cooling produced folds and faults resulting from compressional forces, as well as volcanic activity.

Mercury's surface is heavily cratered and shows little in the way of primary volcanic structures. It does, however, have a global system of lobate scarps. These have been interpreted as evidence that Mercury shrank soon after its crust hardened, resulting in crustal cracking.

Mars has numerous features that indicate an extensive early period of volcanism. These include Olympus Mons, the solar system's largest volcano, lava flows, and uplifted regions thought to have resulted from mantle convection. In addition to volcanic features, Mars also displays abundant evidence of tensional tectonics, including numerous faults and large fault-produced valley structures. Although Mars was tectonically active during the past, no evidence indicates that plate tectonics comparable to that on Earth has ever occurred there.

Venus underwent essentially the same early history as the other terrestrial planets, including a period of volcanism, but it is more Earth-like in its tectonics than either Mercury or Mars. Initial radar mapping in 1990 by the Magellan spacecraft revealed a surface of extensive lava flows, volcanic domes, folded mountain ranges, and an extensive and intricate network of faults, all of which attest to an internally active planet (Figure 2.1).

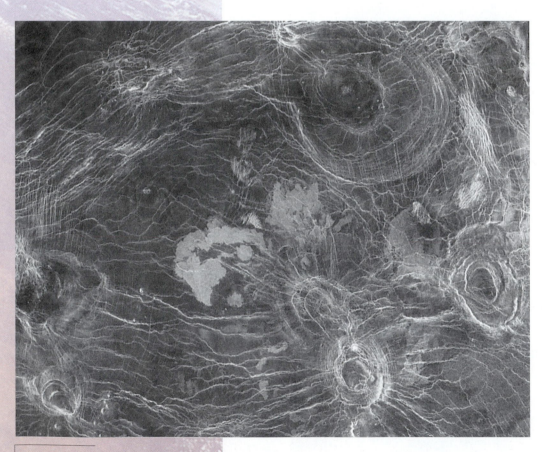

FIGURE 2.1   This radar image of Venus made by the *Magellan* spacecraft reveals circular and oval-shaped volcanic features. A complex network of cracks and fractures extends outward from the volcanic features. Geologists think these features were created by blobs of magma rising from the interior of Venus with magma filling some of the cracks.

Introduction

THE recognition that Earth's geography has changed continuously through time has led to a revolution in the geologic sciences, forcing geologists to greatly modify the way they view Earth. Although many people have only a vague notion of what plate tectonic theory is, plate tectonics has a profound effect on all of our lives. We now realize that most earthquakes and volcanic eruptions occur near plate margins and are not merely random occurrences. Furthermore, the formation and distribution of many important natural resources, such as metallic ores, are related to plate boundaries, and geologists are now incorporating plate tectonic theory into their prospecting efforts.

The interaction of plates determines the location of continents, ocean basins, and mountain systems, which in turn affects the atmospheric and oceanic circulation patterns that ultimately determine global climates. Furthermore, plate movements have profoundly influenced the geographic distribution, evolution, and extinction of plants and animals.

Plate tectonic theory is now almost universally accepted among geologists, and its application has led to a greater understanding of how Earth has evolved. This powerful, unifying theory accounts for many apparently unrelated geologic events, allowing geologists to view such phenomena as part of a continuing story rather than as a series of isolated incidents.

Early Ideas About Continental Drift

THE idea that Earth's geography was different during the past is not new. During the late nineteenth century, the Austrian geologist Edward Suess noted the similarities between the Late Paleozoic plant fossils of India, Australia, Africa, Antarctica, and South America, as well as evidence of glaciation in the rock sequences of these southern continents. In 1885 he proposed the name **Gondwanaland** (or **Gondwana** as we will use here) for a supercontinent composed of these southern landmasses. Gondwana is a province in east-central India where evidence exists for extensive glaciation as well as abundant fossils of the *Glossopteris* flora (Figure 2.2), an association of Late Paleozoic plants found only in India and the Southern Hemisphere continents. Suess thought the distribution of plant fossils and glacial deposits was a consequence of extensive land bridges that once connected the continents and later sank beneath the ocean. The distribution of glacial deposits was also consistent with this interpretation.

Alfred Wegener, a German meteorologist (Figure 2.3), is generally credited with developing the hypothesis of **continental drift**. In his monumental book, *The Origin of Continents and Oceans* (first published in 1915), Wegener proposed that all landmasses were originally united into a single supercontinent that he named **Pangaea**, from the Greek meaning "all land." Wegener portrayed his grand concept of continental movement in a series of maps showing the breakup of Pangaea and the movement of the various continents to their present-day

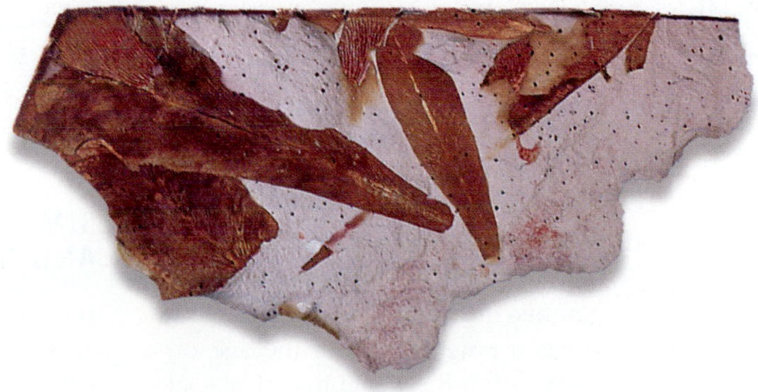

FIGURE 2.2 *Glossopteris* leaves from the Upper Permian Dunedoo Formation, Australia. Fossils of the *Glossopteris* flora are found on all five of the Gondwana continents. *(Photo courtesy of Patricia G. Gensel, University of North Carolina.)*

FIGURE 2.3 Alfred Wegener, a German meteorologist, proposed the continental drift hypothesis in 1912 based on a tremendous amount of geologic, paleontologic, and climatologic evidence. He is shown here waiting out the Arctic winter in an expedition hut.

locations. Wegener had amassed a tremendous amount of geologic, paleontologic, and climatologic evidence in support of continental drift, but the initial reaction of scientists to his then-heretical ideas can best be described as mixed.

Nevertheless, the eminent South African geologist Alexander du Toit further developed Wegener's arguments and gathered more geologic and paleontologic evidence in support of continental drift. In 1937 du Toit published *Our Wandering Continents,* in which he contrasted the glacial deposits of Gondwana with coal deposits of the same age found in the continents of the Northern Hemisphere. To resolve this apparent climatologic paradox, du Toit moved the Gondwana continents to the South Pole and brought the northern continents together such that the coal deposits were located at the equator. He named this northern landmass **Laurasia**. It consisted of present-day North America, Greenland, Europe, and Asia (except for India).

Despite what seemed to be overwhelming evidence, most geologists still refused to accept the idea that continents moved. Not until the 1960s, when oceanographic research provided convincing evidence that the continents had once been joined together and subsequently separated, did the hypothesis of continental drift finally become widely accepted.

The Evidence for Continental Drift

THE evidence used by Wegener, du Toit, and others to support the hypothesis of continental drift includes the fit of the shorelines of continents, the appearance of the same rock sequences and mountain ranges of the same age on continents now widely separated, the matching of glacial deposits and paleoclimatic zones, and the similarities of many extinct plant and animal groups whose fossil remains are found today on widely separated continents.

CONTINENTAL FIT

Wegener, like some before him, was impressed by the close resemblance between the coastlines of continents on opposite sides of the Atlantic Ocean, particularly between South America and Africa. He cited these similarities as partial evidence that the continents were at one time joined together as a supercontinent that subsequently split apart. As his critics pointed out, though, the configuration of coastlines results from erosional and depositional processes and therefore is continually being modified. So even if the continents had separated during the Mesozoic Era, as Wegener proposed, it is not likely that the coastlines would fit exactly.

A more realistic approach is to fit the continents together along the continental slope where erosion would be minimal. In 1965 Sir Edward Bullard, an English geophysicist, and two associates showed that the best fit between the continents occurs at a depth of about 2000 m (Figure 2.4). Since then, other reconstructions using the latest ocean basin data have confirmed the close fit between continents when they are reassembled to form Pangaea.

SIMILARITY OF ROCK SEQUENCES AND MOUNTAIN RANGES

If the continents were at one time joined together, then the rocks and mountain ranges of the same age in adjoining locations on the opposite continents should closely match.

FIGURE 2.4 The best fit between continents occurs along the continental slope, where erosion would be minimal.

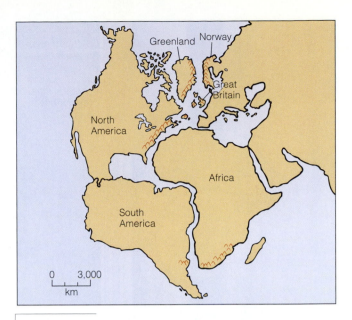

FIGURE 2.5 When continents are brought together, their mountain ranges form a single continuous range of the same age and style of deformation throughout. Such evidence indicates the continents were at one time joined together and were subsequently separated.

Such is the case for the Gondwana continents. Marine, nonmarine, and glacial rock sequences of Pennsylvanian to Jurassic age are almost identical for all five Gondwana continents, strongly indicating that they were joined together at one time.

The trends of several major mountain ranges also support the hypothesis of continental drift. These mountain ranges seemingly end at the coastline of one continent only to apparently continue on another continent across the ocean. The folded Appalachian Mountains of North America, for example, trend northeastward through the eastern United States and Canada and terminate abruptly at the Newfoundland coastline. Mountain ranges of the same age and deformational style occur in eastern Greenland, Ireland, Great Britain, and Norway. Even though these mountain ranges are currently separated by the Atlantic Ocean, they form an essentially continuous mountain range when the continents are positioned next to each other (Figure 2.5).

GLACIAL EVIDENCE

During the Late Paleozoic Era, massive glaciers covered large continental areas of the Southern Hemisphere. Evidence for this glaciation includes layers of till (sediments deposited by glaciers) and striations (scratch marks) in the bedrock beneath the till. Fossils and sedimentary rocks of the same age from the Northern Hemisphere, however, give no indication of glaciation. Fossil plants found in coals indicate that the Northern Hemisphere had a tropical cli-mate during the time that the Southern Hemisphere was glaciated.

All the Gondwana continents except Antarctica are currently located near the equator in subtropical to tropical climates. Mapping of glacial striations in bedrock in Australia, India, and South America indicates that the glaciers moved from the areas of the present-day oceans onto land. This would be highly unlikely because large continental glaciers flow outward from their central area of accumulation toward the sea.

If the continents did not move during the past, one would have to explain how glaciers moved from the oceans onto land and how large-scale continental glaciers formed near the equator. But if the continents are reassembled as a single landmass with South Africa located at the South Pole, the direction of movement of Late Paleozoic glaciers makes sense. Furthermore, this geographic arrangement places the northern continents nearer the tropics, which is consistent with the fossil and climatologic evidence from Laurasia (Figure 2.6).

FOSSIL EVIDENCE

Some of the most compelling evidence for continental drift comes from the fossil record (Figure 2.7). Fossils of the *Glossopteris* flora are found in equivalent Pennsylvanian- and Permian-aged coal deposits on all five Gondwana continents. The *Glossopteris* flora is characterized by the seed fern *Glossopteris* (Figure 2.2) as well as by many other distinctive and easily identifiable plants. Pollen and spores of

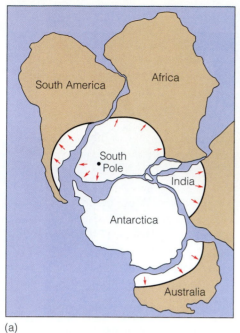

(a)

Glaciated area
Arrows indicate
the direction of
glacial movement
based on striations
preserved in
bedrock.

(b)

FIGURE 2.6 (a) If the Gondwana continents are brought together so that South Africa is located at the South Pole, then the glacial movements indicated by the striations make sense. In this situation, the glacier, located in a polar climate, moved radially outward from a thick central area toward its periphery. (b) Permian-aged glacial striations in bedrock exposed at Hallet's Cove, Australia, indicate the direction of glacial movement more than 200 million years ago.

FIGURE 2.7 Some of the animals and plants whose fossils are found today on the widely separated continents of South America, Africa, India, Australia, and Antarctica. These continents were joined together during the Late Paleozoic to form Gondwana, the southern landmass of Pangaea. *Glossopteris* and similar plants are found in Pennsylvanian- and Permian-aged deposits on all five continents. *Mesosaurus* is a freshwater reptile whose fossils are found in Permian-aged rocks in Brazil and South Africa. *Cynognathus* and *Lystrosaurus* are land reptiles who lived during the Early Triassic Period. Fossils of *Cynognathus* are found in South America and Africa, while fossils of *Lystrosaurus* have been recovered from Africa, India, and Antarctica.

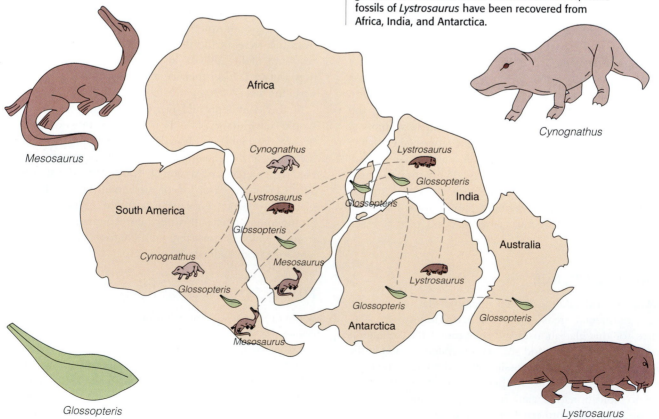

plants can be dispersed over great distances by wind, but *Glossopteris*-type plants produced seeds that are too large to have been carried by winds. Even if the seeds had floated across the ocean, they probably would not have remained viable for any length of time in salt water.

The present-day climates of South America, Africa, India, Australia, and Antarctica range from tropical to polar and are much too diverse to support the type of plants that compose the *Glossopteris* flora. Wegener therefore reasoned that these continents must once have been joined such that these widely separated localities were all in the same latitudinal climatic belt.

The fossil remains of animals also provide strong evidence for continental drift. One of the best examples is *Mesosaurus,* a freshwater reptile whose fossils are found in Permian-aged rocks in certain regions of Brazil and South Africa and nowhere else in the world. Because the physiology of freshwater and marine animals is completely different, it is hard to imagine how a freshwater reptile could have swum across the Atlantic Ocean and found a freshwater environment nearly identical to its former habitat. Moreover, if *Mesosaurus* could have swum across the ocean, its fossil remains should be widely dispersed. It is more logical to assume that *Mesosaurus* lived in lakes in what are now adjacent areas of South America and Africa, but were then united into a single continent.

Lystrosaurus and *Cynognathus* are both land-dwelling reptiles that lived during the Triassic Period; their fossils are found only on the present-day continental fragments of Gondwana. Because they are both land animals, they certainly could not have swum across the oceans currently separating the Gondwana continents. Therefore, the continents must once have been connected.

PALEOMAGNETISM AND POLAR WANDERING

Interest in continental drift revived in the 1950s as a result of new evidence from studies of Earth's ancient magnetic field. **Paleomagnetism** is the remanent magnetism in ancient rocks recording the direction of the magnetic poles at the time of the rock's formation. Earth can be thought of as a giant dipole magnet in which the magnetic poles correspond closely to the location of the geographic poles (Figure 2.8). This arrangement means that the strength of the magnetic field is not constant but varies, being weakest at the equator and strongest at the poles. Earth's magnetic

FIGURE 2.8 (a) The magnetic field of Earth has lines of force just like those of a bar magnet. (b) The strength of the magnetic field changes uniformly from the magnetic equator to the magnetic poles. This change in strength causes a dip needle to parallel Earth's surface only at the magnetic equator, whereas its inclination with respect to the surface increases to 90° at the magnetic poles.

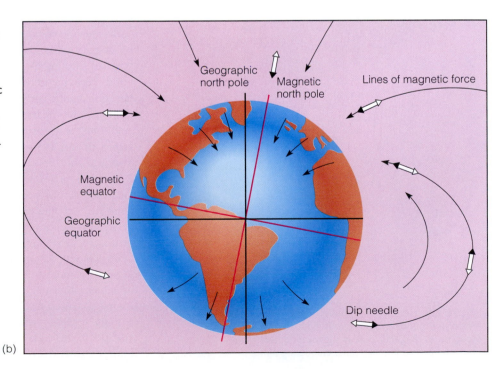

(a)

(b)

field is thought to result from convection within the liquid outer core.

When a magma cools, the iron-bearing minerals align themselves with Earth's magnetic field, recording both its direction and strength. The temperature at which iron-bearing minerals gain their magnetization is called the **Curie point**. As long as a rock is not subsequently heated above the Curie point, it will preserve that remanent magnetism. Thus, an ancient lava flow provides a record of the orientation and strength of Earth's magnetic field at the time the lava flow cooled.

As paleomagnetic research progressed in the 1950s, some unexpected results emerged. When geologists measured the magnetism of recent rocks, they found it was generally consistent with Earth's current magnetic field. The paleomagnetism of ancient rocks, however, showed different orientations. For example, paleomagnetic studies of Silurian lava flows in North America indicated that the north magnetic pole was located in the western Pacific Ocean at that time, while the paleomagnetic evidence from Permian lava flows pointed to yet another location in northern Asia. When plotted on a map, the paleomagnetic readings of numerous lava flows from all ages in North America trace the apparent movement of the magnetic pole through time (Figure 2.9). This paleomagnetic evidence from a single continent could be interpreted in three ways: the continent remained fixed and the north magnetic pole moved; the north magnetic pole stood still and the continent moved; or both the continent and the north magnetic pole moved.

Upon analysis, magnetic minerals from European Silurian and Permian lava flows pointed to a different magnetic pole location than those of the same age from North America (Figure 2.9). Furthermore, analysis of lava flows from all continents indicated each continent had its own series of magnetic poles. Does this mean there were different north magnetic poles for each continent? That would be highly unlikely and difficult to reconcile with the theory accounting for Earth's magnetic field.

The best explanation for such data is that the magnetic poles have remained at their present locations near the geographic north and south poles and the continents have moved. When the continental margins are fitted together so that the paleomagnetic data point to only one magnetic pole, we find, just as Wegener did, that the rock sequences and glacial deposits match, and that the fossil evidence is consistent with the reconstructed paleogeography.

Magnetic Reversals and Seafloor Spreading

GEOLOGISTS refer to Earth's present magnetic field as being normal, that is, with the north and south magnetic poles located approximately at the north and south geographic poles. At numerous times in the geologic past, Earth's magnetic field has completely reversed. The existence of such **magnetic reversals** was discovered by dating and determining the orientation of the remanent magnetism in lava flows on land (Figure 2.10). Once their existence was well established, magnetic reversals were also discovered in igneous rocks of the oceanic crust as part of the extensive mapping of the ocean basins during the 1960s. Although the cause of magnetic reversals is still uncertain, their occurrence in the geologic record is well documented.

Besides the discovery of magnetic reversals, mapping of the ocean basins also revealed a ridge system 65,000 km long, constituting the most extensive mountain range in the world. Perhaps the best-known part of the ridge system is the Mid-Atlantic Ridge, which divides the Atlantic Ocean basin into two nearly equal parts (Figure 2.11).

In 1962, as a result of the oceanographic research conducted in the 1950s, Harry Hess of Princeton University proposed the theory of **seafloor spreading** to account for continental movement. Hess suggested that continents do not move across oceanic crust, but rather that the continents and oceanic crust move together. He suggested that the seafloor separates at oceanic ridges where new crust is formed by upwelling magma. As the magma cools, the newly formed oceanic crust moves laterally away from the ridge. As a mechanism to drive this system, Hess revived the idea of **thermal convection cells** in the mantle; that is, hot magma rises from the mantle, intrudes along fractures defining oceanic ridges and thus forms new crust. Cold crust is subducted back into the mantle at oceanic trenches (long, narrow, deep features, along which subduction oc-

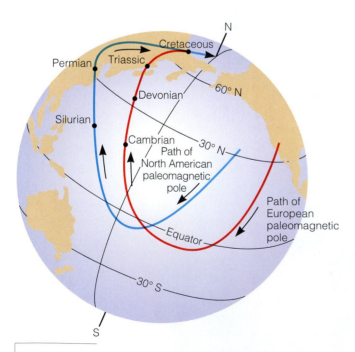

FIGURE 2.9 The apparent paths of polar wandering for North America and Europe. The apparent location of the north magnetic pole is shown for different periods on each continent's polar wandering path.

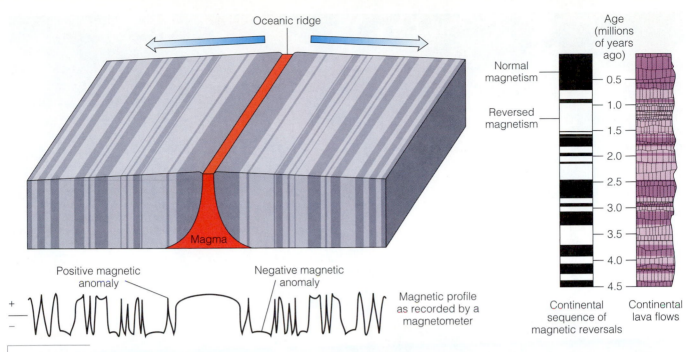

Age (millions of years ago)

Normal magnetism

Reversed magnetism

Positive magnetic anomaly

Negative magnetic anomaly

Magnetic profile as recorded by a magnetometer

Continental sequence of magnetic reversals

Continental lava flows

FIGURE 2.10 The sequence of magnetic anomalies preserved within the oceanic crust on both sides of an oceanic ridge is identical to the sequence of magnetic reversals already known from continental lava flows. Magnetic anomalies are formed when magma intrudes into oceanic ridges; when the magma cools below the Curie point, it records Earth's magnetic polarity at the time. Seafloor spreading splits the previously formed crust in half, so that it moves laterally away from the oceanic ridge. Repeated intrusions record a symmetrical series of magnetic anomalies that reflect periods of normal and reversed polarity. The magnetic anomalies are recorded by a magnetometer, which measures the strength of the magnetic field.

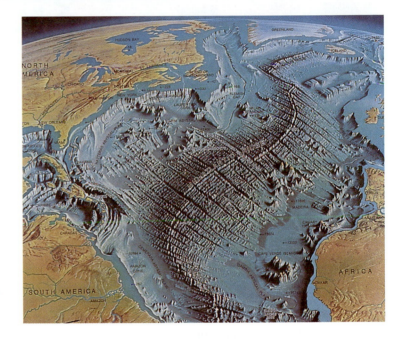

FIGURE 2.11 Artistic view of what the Atlantic Ocean basin would look like without water. The major feature is the Mid-Atlantic Ridge. (Photo courtesy of ALCOA.)

curs), where it is heated and recycled, thus completing a thermal convection cell (see Figure 1.9).

How could Hess's hypothesis be confirmed? Magnetic surveys of the oceanic crust revealed striped **magnetic anomalies** (deviations from the average strength of Earth's magnetic field) in the rocks that were both parallel to and symmetrical with the oceanic ridges (Figure 2.10). Furthermore, the pattern of oceanic magnetic anomalies matched the pattern of magnetic reversals already known from studies of continental lava flows. When magma wells up and cools along a ridge summit, it records Earth's magnetic field at that time as either normal or reversed. As new crust forms at the summit, the previously

formed crust moves laterally away from the ridge. These magnetic stripes, representing times of normal or reversed polarity, are parallel to and symmetric around oceanic ridges (where upwelling magma forms new oceanic crust), conclusively confirming Hess's theory of seafloor spreading.

One of the consequences of the seafloor spreading theory is its confirmation that ocean basins are geologically young features whose openings and closings are partially responsible for continental movement (Figure 2.11). Radiometric dating reveals that the oldest oceanic crust is less than 180 million years old, whereas the oldest continental crust is 3.96 billion years old (Figure 2.12).

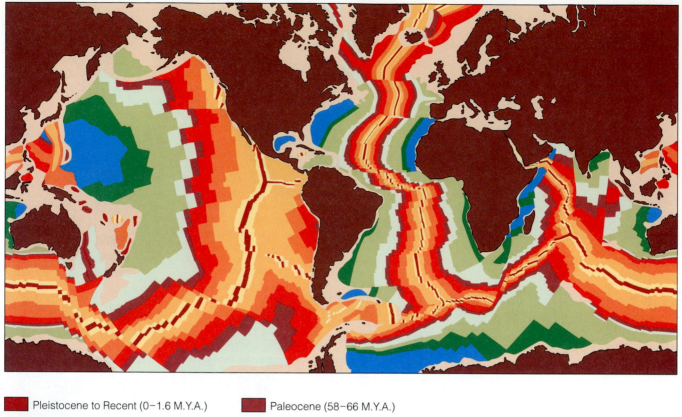

Color	Period
Pleistocene to Recent (0–1.6 M.Y.A.)	Paleocene (58–66 M.Y.A.)
Pliocene (1.6–5 M.Y.A.)	Late Cretaceous (66–88 M.Y.A.)
Miocene (5–24 M.Y.A.)	Middle Cretaceous (88–118 M.Y.A.)
Oligocene (24–37 M.Y.A.)	Early Cretaceous (118–144 M.Y.A.)
Eocene (37–58 M.Y.A.)	Late Jurassic (144–161 M.Y.A.)

FIGURE 2.12 The age of the world's ocean basins established from magnetic anomalies demonstrates that the youngest oceanic crust is adjacent to the oceanic ridges and that its age increases away from the ridge axis.

Plate Tectonic Theory

PLATE tectonic theory is based on a simple model of Earth. The rigid lithosphere, consisting of both oceanic and continental crust, as well as the underlying upper mantle, consists of numerous variable-sized pieces called **plates** (Figure 2.13). The plates vary in thickness; those composed of upper mantle and continental crust are as much as 250 km thick, whereas those composed of upper mantle and oceanic crust are up to 100 km thick.

The lithosphere overlies the hotter and weaker semi-plastic asthenosphere. It is thought that movement resulting from some type of heat transfer system within the asthenosphere causes the overlying plates to move. As plates move over the asthenosphere, they separate, mostly at oceanic ridges; in other areas such as at oceanic trenches, they collide and are subducted back into the mantle.

Most geologists accept plate tectonic theory, in part because the evidence for it is overwhelming and because it is a unifying theory that accounts for a variety of apparently unrelated geologic features and events. Consequently, geologists now view many geologic processes, such as mountain building, earthquake activity, and volcanism, from the perspective of plate tectonics (see Perspective 2.1). Furthermore, because all inner planets have had a similar origin and early history (see the Prologue to this chapter), geologists are interested in determining whether plate tectonics is unique to Earth or whether it operates in the same way on other planets.

Plate Boundaries

PLATES move relative to one another such that their boundaries can be characterized as *divergent, convergent,* and *transform.* Interaction of plates at their boundaries accounts for most of Earth's volcanic and earthquake activity and, as will be apparent in Chapter 10, the origin of mountain systems.

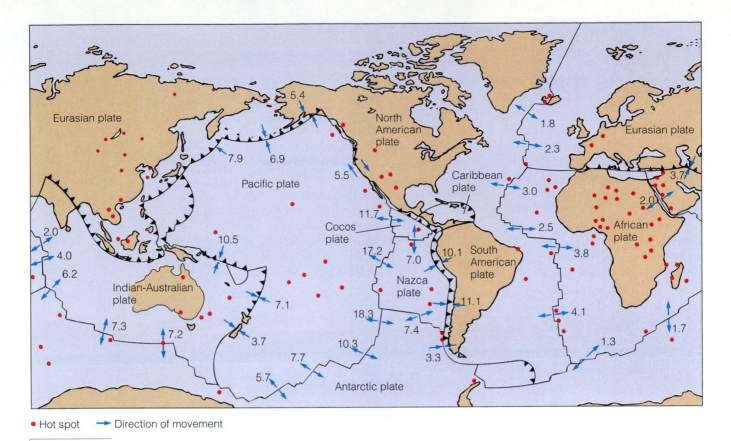

• Hot spot   → Direction of movement

FIGURE 2.13   A map of the world showing the plates, their boundaries, relative motion and rates of movement in centimeters per year, and hot spots.

DIVERGENT BOUNDARIES

Divergent plate boundaries or *spreading ridges* occur where plates are separating and new oceanic lithosphere is forming. Divergent boundaries are places where the crust is being extended, thinned, and fractured as magma, derived from the partial melting of the mantle, rises to the surface, intrudes into vertical fractures, and flows out onto the seafloor—forming pillow lavas (see Figure 5.7). As successive injections of magma cool and solidify, they form new oceanic crust and record the intensity and orientation of Earth's magnetic field (Figure 2.10). Divergent boundaries most commonly occur along the crests of oceanic ridges, for example, the Mid-Atlantic Ridge. Oceanic ridges are thus characterized by rugged topography with high relief resulting from displacement of rocks along large fractures, shallow earthquakes, high heat flow, and pillow lavas.

Divergent boundaries are also present under continents during the early stages of continental breakup (Figure 2.14). When magma wells up beneath a continent, the crust is initially elevated, stretched, and thinned, producing fractures and rift valleys. During this stage, magma typically intrudes into the fractures and flows onto the valley floor. The East African rift valleys are an excellent

example of this stage of continental breakup (Figure 2.15). As rifting proceeds, the continental crust eventually breaks. If magma continues welling up, the two parts of the continent will move away from each other, as is happening today beneath the Red Sea. As this newly formed narrow sea continues enlarging, it may eventually become an expansive ocean basin such as the Atlantic Ocean basin is today.

CONVERGENT BOUNDARIES

Whereas new crust is formed at divergent plate boundaries, old crust must be destroyed and recycled in order for the entire surface area of Earth to remain constant. Otherwise, we would have an expanding Earth. Such plate destruction occurs at **convergent plate boundaries** where two plates collide and one plate is subducted under another and eventually resorbed in the asthenosphere.

Convergent boundaries are characterized by deformation, volcanism, mountain building, metamorphism, earthquake activity, and important mineral deposits. Three types of convergent plate boundaries are recognized: *oceanic–oceanic, oceanic–continental,* and *continental–continental.*

The Supercontinent Cycle

At the end of the Paleozoic Era, all continents were amalgamated into the supercontinent Pangaea. Pangaea began fragmenting during the Triassic Period and continues to do so, thus accounting for the present distribution of continents and oceans. It now appears that another supercontinent existed at the end of the Proterozoic Eon, and there is some evidence for even earlier supercontinents. It has been proposed that supercontinents consisting of all or most of Earth's landmasses form, break up, and reform in a cycle spanning about 500 million years.

The supercontinent cycle hypothesis is an expansion on the ideas of the Canadian geologist J. Tuzo Wilson. During the early 1970s, Wilson proposed a cycle (now known as the Wilson cycle) that includes continental fragmentation, the opening and closing of an ocean basin, and reassembly of the continent. According to the *supercontinent cycle hypothesis,* heat accumulates beneath a supercontinent because rocks of continents are poor conductors of heat. As a result of the heat accumulation, the supercontinent domes upward and fractures. Magma rising from below fills the fractures. As a fracture widens, it begins subsiding and forms a long narrow ocean such as the present-day Red Sea. Continued rifting eventually forms an expansive ocean basin such as the Atlantic.

According to proponents of the supercontinent cycle, one of the most convincing arguments for their hypothesis is the "surprising regularity" of mountain building caused by compression during continental collisions. Such mountain-building

episodes occur about every 400 to 500 million years and are followed by an episode of rifting about 100 million years later. In other words, a supercontinent fragments and its individual plates disperse following a rifting episode, an interior ocean forms, and then the dispersed fragments reassemble to form another supercontinent (Figure 1).

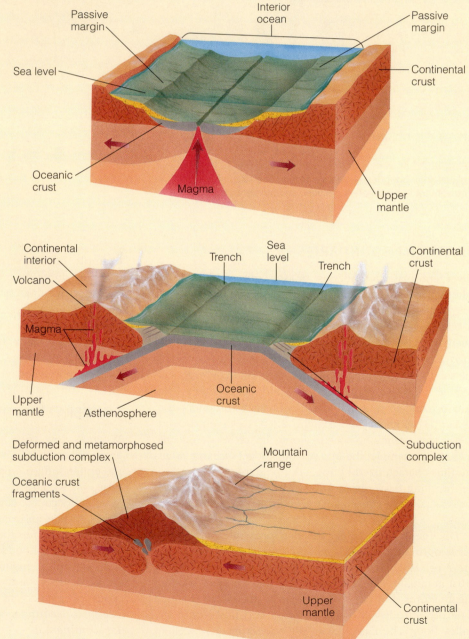

Figure 1 *The supercontinent cycle.*
(a) Breakup of a supercontinent and the formation of an ocean basin.
(b) Subduction along the margins of the ocean basin begins approximately 200 million years later, resulting in volcanic activity and deformation along an active oceanic–continental plate boundary.
(c) Continental collisions and the formation of a new supercontinent result when all of the oceanic crust of the ocean basin is subducted.

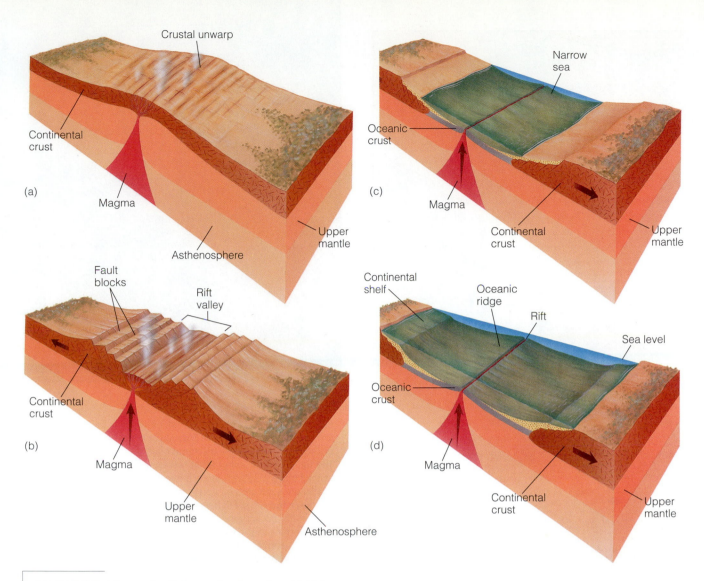

FIGURE 2.14 History of a divergent plate boundary. (a) Rising magma beneath a continent pushes the crust up, producing numerous cracks and fractures. (b) As the crust is stretched and thinned, rift valleys develop, and lava flows onto the valley floors. (c) Continued spreading further separates the continent until a narrow seaway develops. (d) As spreading continues, an oceanic ridge system forms, and an ocean basin develops and grows.

OCEANIC–OCEANIC BOUNDARIES When two oceanic plates converge, one is subducted beneath the other along an **oceanic–oceanic plate boundary** (Figure 2.16). The subducting plate bends downward as it descends into the mantle and is heated and partially melted, thus generating magma. This magma is less dense than the surrounding mantle rocks and rises to the surface of the nonsubducted plate, forming a curved chain of volcanic islands called a **volcanic island arc** (any plane intersecting a sphere makes an arc). This arc is nearly parallel to the oceanic trench and is separated from it by a distance of up to several hundred kilometers—the distance depending on the angle of dip of the subducting plate.

In those areas where the rate of subduction is faster than the forward movement of the overriding plate, the lithosphere on the landward side of the volcanic island arc

may be subjected to tensional stress and stretched and thinned, resulting in the formation of a *back-arc basin*. This back-arc basin may grow by spreading if magma breaks through the thin crust and forms new oceanic crust (Figure 2.16). A good example of a back-arc basin associated with an oceanic–oceanic plate boundary is the Sea of Japan between the Asian continent and the islands of Japan.

Most present-day active volcanic island arcs are in the Pacific Ocean basin and include the Aleutian Islands, the Kermadec–Tonga arc, and the Japanese and Philippine Islands. The Scotia and Antillean (Caribbean) island arcs are present in the Atlantic Ocean basin.

OCEANIC–CONTINENTAL BOUNDARIES An **oceanic–continental plate boundary** occurs when denser oceanic

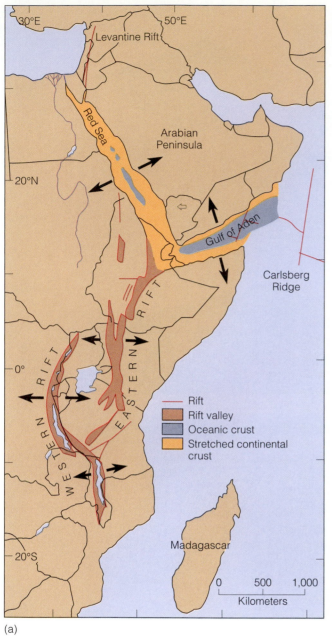

(a)

(b)

FIGURE 2.15 (a) The East African rift valley is being formed by the separation of East Africa from the rest of the continent along a divergent plate boundary. The Red Sea and Gulf of Aden represent a more advanced stage of rifting in which two continental blocks are separated by a narrow sea. (b) View looking down the Great Rift Valley of Africa. Little Magadi, seen in the background, is one of numerous soda lakes forming in the valley. Because of high evaporation rates and lack of any drainage outlets, these lakes are very saline. The Great Rift Valley is part of the system of rift valleys resulting from stretching of the crust as plates move away from each other in East Africa.

FIGURE 2.16 Oceanic–oceanic plate boundary. An oceanic trench forms where one oceanic plate is subducted beneath another. On the nonsubducted plate, a volcanic island arc forms from the rising magma generated from the subducting plate.

crust is subducted under continental crust (Figure 2.17). The magma generated by subduction rises beneath the continent and either crystallizes as a large igneous body before reaching the surface or erupts at the surface, producing a chain of volcanoes (also called a volcanic arc). An excellent example of an oceanic–continental plate boundary is the Pacific coast of South America where the oceanic Nazca plate is currently being subducted under South America (Figure 2.13). The Peru–Chile Trench marks the site of subduction, and the Andes Mountains are the resulting volcanic mountain chain on the nonsubducting plate.

CONTINENTAL–CONTINENTAL BOUNDARIES Two continents approaching each other will initially be separated by an ocean floor that is being subducted under one continent. The edge of that continent will display the features

characteristic of oceanic–continental convergence. As the ocean floor continues to be subducted, the two continents will come closer together until they eventually collide. Because continental lithosphere, which consists of continental crust and the upper mantle, is less dense than oceanic lithosphere (oceanic crust and the upper mantle), it cannot sink into the asthenosphere. Although one continent may partly slide under the other, it cannot be pulled or pushed down into a subduction zone (Figure 2.18).

When two continents collide, they are welded together along a zone marking the former site of subduction. At this **continental–continental plate boundary**, an interior mountain belt is formed consisting of deformed sedimentary rocks, igneous intrusions, metamorphic rocks, and fragments of oceanic crust. In addition, the entire region is subjected to numerous earthquakes. The Himalayas in central

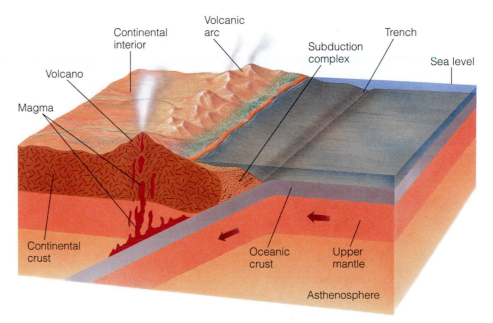

FIGURE 2.17 Oceanic–continental plate boundary. When an oceanic plate is subducted beneath a continental plate, a volcanic mountain range is formed on the continental plate as a result of rising magma.

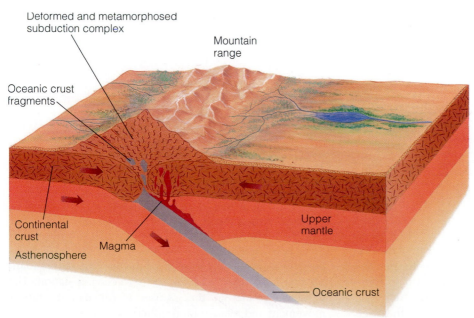

FIGURE 2.18 Continental–continental plate boundary. When two continental plates converge, neither is subducted because of their great thickness and low and equal densities. As the two continental plates collide, a mountain range is formed in the interior of a new and larger continent.

Plate Boundaries   37

Asia are the result of a continental–continental collision between India and Asia that began about 40 to 50 million years ago and is still continuing.

TRANSFORM BOUNDARIES

The third type of plate boundary is a **transform plate boundary**. These occur along fractures in the seafloor, known as *transform faults,* where plates slide laterally past one another roughly parallel to the direction of plate movement. Although lithosphere is neither created nor destroyed along a transform boundary, the movement between plates results in a zone of intensely shattered rock and numerous shallow earthquakes.

Transform faults are particular types of faults that "transform" or change one type of motion between plates into another type of motion. The majority of transform faults connect two oceanic ridge segments, but they can also connect ridges to trenches and trenches to trenches (Figure 2.19). While the majority of transform faults are in oceanic crust and are marked by distinct fracture zones, they may also extend into continents.

One of the best-known transform faults is the San Andreas fault in California. It separates the Pacific plate from the North American plate and connects spreading ridges in the Gulf of California and the Juan de Fuca and Pacific plates off the coast of northern California (Figure 2.20). Many of the earthquakes affecting California are the result of movement along this fault.

Plate Movement and Motion

How fast and in what direction are Earth's various plates moving, and do they all move at the same rate? Rates of plate movement can be calculated in several ways. The least accurate method is to determine the age of the sediments immediately above any portion of the oceanic crust and divide that age by the distance from the spreading ridge. Such calculations give an average rate of movement.

A more accurate method of determining both the average rate of movement and relative motion is by dating the magnetic reversals in the crust of the seafloor. The distance from an oceanic ridge axis to any magnetic reversal indicates the width of new seafloor that formed during that time interval. Thus, for a given interval of time, the wider the strip of seafloor, the faster the plate has moved. In this way not only can the present average rate of movement and relative motion be determined (Figure 2.13), but the average rate of movement during the past can also be calculated by dividing the distance between reversals by the amount of time elapsed between reversals.

The average rate of movement as well as the relative motion between any two plates can also be determined by satellite–laser ranging techniques. Laser beams from a station on one plate are bounced off a satellite (in geosynchronous orbit) and returned to a station on a different plate. As the plates move away from each other, the laser

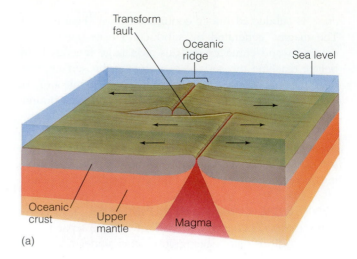

(a)

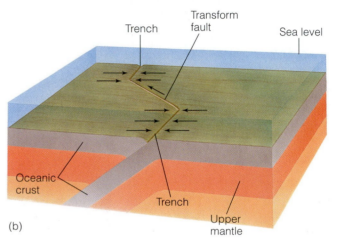

(b)

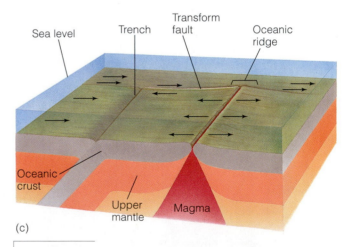

(c)

FIGURE 2.19 Horizontal movement between plates occurs along a transform fault. (a) The majority of transform faults connect two oceanic ridge segments. Note that relative motion between the plates only occurs between the two ridges. (b) A transform fault connecting two trenches. (c) A transform fault connecting a ridge and a trench.

beam takes more time to go from the sending station to the stationary satellite and back to the receiving station. This difference in elapsed time is used to calculate the rate of movement and relative motion between plates.

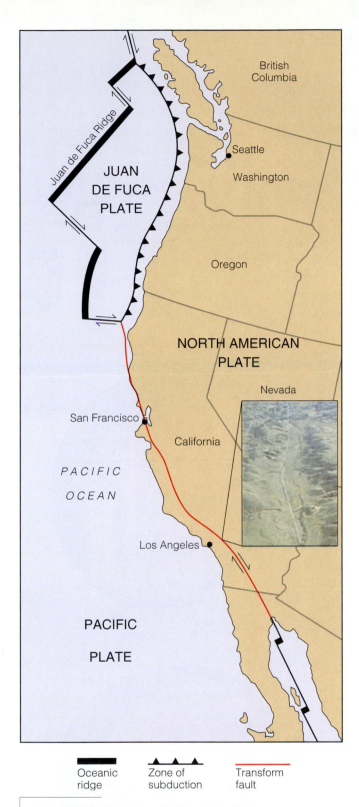

FIGURE 2.20 Transform plate boundary. The San Andreas fault is a transform fault separating the Pacific plate from the North American plate. Movement along this fault has caused numerous earthquakes. The photograph shows a segment of the San Andreas fault as it cuts through the Leona Valley, California.
(Photo courtesy of Eleanora I. Robbins, U.S. Geological Survey.)

Map legend:
- Oceanic ridge
- Zone of subduction
- Transform fault

HOT SPOTS AND ABSOLUTE MOTION

Plate motions determined from magnetic reversals and satellite lasers give only the relative motion of one plate with respect to another. To determine absolute motion, we must have a fixed reference from which the rate and direction of plate movement can be determined. **Hot spots**, which may provide reference points, are locations where stationary columns of magma, originating deep within the mantle (mantle plumes), slowly rise to the surface and form volcanoes or flood basalts (Figure 2.13).

One of the best examples of hot-spot activity is that over which the Emperor Seamount–Hawaiian Island chain formed (Figure 2.21). Currently, the only active volcanoes in this island chain are on the island of Hawaii and the Loihi Seamount. The rest of the islands and seamounts (structures of volcanic origin rising more than 1 km above the seafloor) of the chain are also of volcanic origin and are progressively older west-northwestward along the Hawaiian chain and north-northwestward along the Emperor Seamount chain.

These islands and seamounts are progressively older as you move toward the north and northwest because the Pacific plate has moved over an apparently stationary mantle plume. Thus, a line of volcanoes was formed near the middle of the Pacific plate, marking the direction of the plate's movement. In the case of the Emperor Seamount–Hawaiian Island chain, the Pacific plate moved first north-northwesterly and then west-northwesterly over a single mantle plume.

Mantle plumes and hot spots are useful to geologists in helping explain some of the geologic activity occurring within plates as opposed to that occurring at or near plate boundaries. In addition, if mantle plumes are essentially fixed with respect to Earth's rotational axis—and some recent research suggests they might not be—they may prove useful as reference points for determining paleolatitude.

The Driving Mechanism of Plate Tectonics

A major obstacle to the acceptance of continental drift was the lack of a driving mechanism to explain continental movement. When it was shown that continents and ocean floors moved together and not separately and that new crust formed at spreading ridges by rising magma, most geologists accepted some type of convective heat system as the basic process responsible for plate motion. The question still remains however: What exactly drives the plates?

Two models involving thermal convection cells have been proposed to explain plate movement (Figure 2.22). In one model, thermal convection cells are restricted to the asthenosphere; in the second model, the entire mantle is involved. In both models, spreading ridges mark the ascending limbs of adjacent convection cells, while trenches are present where convection cells descend back into Earth's interior. The locations of spreading ridges and

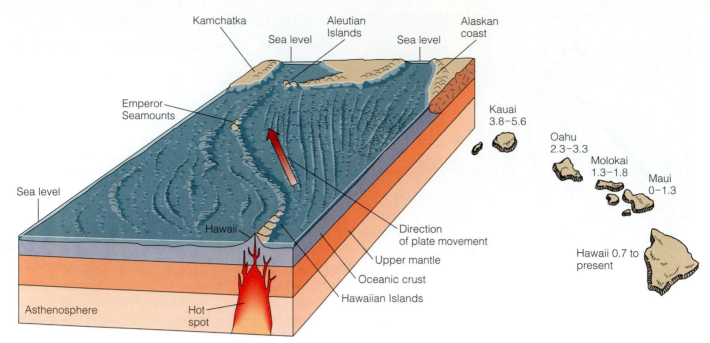

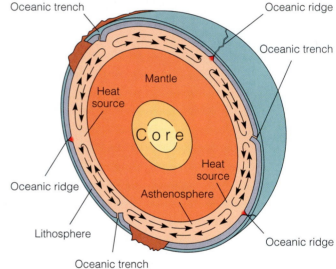

FIGURE 2.21   The Emperor Seamount–Hawaiian Island chain formed as a result of movement of the Pacific plate over a hot spot. The line of the volcanic islands traces the direction of plate movement. The numbers indicate the age of the islands in millions of years.

trenches are therefore determined by the convection cells themselves, and the lithosphere is considered to be the top of the thermal convection cell. Each plate thus corresponds to a single convection cell.

Although most geologists agree that Earth's internal heat plays an important role in plate movement, problems are inherent in both models. The major problem associated with the first model is the difficulty in explaining the source of heat for the convection cells and why they are restricted to the asthenosphere. In the second model, the source of heat comes from the outer core, but it is still not known how heat is transferred from the outer core to the mantle. Nor is it clear how convection can involve both the lower mantle and the asthenosphere.

Some geologists think that, besides thermal convection, plate movement also occurs, in part, because of a mechanism involving "slab-pull" or "ridge-push" (Figure 2.23). Both mechanisms are gravity driven but still depend on thermal differences within Earth. In slab-pull, the subducting cold slab of lithosphere is denser than the surrounding warmer asthenosphere and thus pulls the rest of the plate

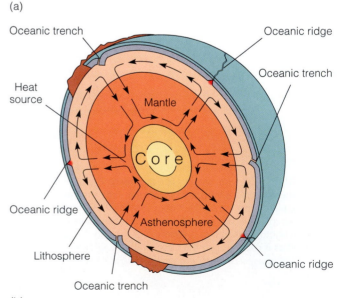

FIGURE 2.22   Two models involving thermal convection cells have been proposed to explain plate movement. (a) In one model, thermal convection cells are restricted to the asthenosphere. (b) In the other model, thermal convection cells involve the entire mantle.

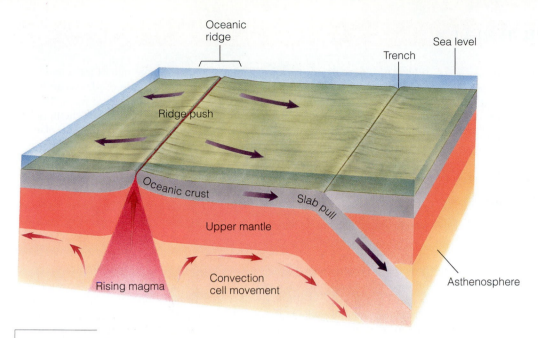

Oceanic ridge

Sea level

Trench

Ridge push

Oceanic crust

Slab pull

Upper mantle

Rising magma

Convection cell movement

Asthenosphere

FIGURE 2.23 Plate movement is also thought to occur because of gravity-driven "slab-pull," or "ridge-push" mechanisms. In slab-pull, the edge of the subducting plate descends into the interior, and the rest of the plate is pulled downward. In ridge-push, rising magma pushes the oceanic ridges higher than the rest of the oceanic crust. Gravity thus pushes the oceanic lithosphere away from the ridges and toward the trenches.

along with it as it descends into the asthenosphere. As the lithosphere moves downward, there is a corresponding upward flow back into the spreading ridge.

Operating in conjunction with slab-pull is the ridge-push mechanism. As a result of rising magma, the oceanic ridges are higher than the surrounding oceanic crust. It is thought that gravity pushes the oceanic lithosphere away from the higher spreading ridges and toward the trenches.

Currently, geologists are fairly certain that some type of convective system is involved in plate movement, and the extent to which other mechanisms such as slab-pull and ridge-push are involved is still unresolved. Consequently, a comprehensive theory of plate movement has not as yet been developed, and much still remains to be learned about Earth's interior.

Plate Tectonics and the Distribution of Natural Resources

BESIDES being responsible for the major features of Earth's crust, plate movements also affect the formation and distribution of some natural resources. Consequently, geologists are using plate tectonic theory in their search for new mineral deposits and in explaining the occurrence of known deposits.

Many metallic mineral deposits such as copper, gold, lead, silver, tin, and zinc are related to igneous and associated hydrothermal activity, so it is not surprising that a close relationship exists between plate boundaries and the occurrence of these valuable deposits.

The magma generated by partial melting of a subducting plate rises toward the surface, and as it cools, it precipitates and concentrates various metallic ores. Many of the world's major metallic ore deposits, such as the copper deposits of western North and South America, are excellent examples of the relationship between convergent plate boundaries and the distribution, concentration, and exploitation of metallic ores.

Divergent plate boundaries also yield valuable resources. The island of Cyprus in the Mediterranean is rich in copper and has been supplying all or part of the world's needs for the last 3000 years. The concentration of copper on Cyprus formed as a result of precipitation adjacent to hydrothermal vents along a divergent plate boundary. This deposit was brought to the surface when the copper-rich seafloor collided with the European plate, warping the seafloor and forming Cyprus.

Studies indicate that minerals containing such metals as copper, gold, iron, lead, silver, and zinc are currently forming in the Red Sea. The Red Sea is opening as a result of plate divergence and represents the earliest stage in the growth of an ocean basin (see Figures 2.14c and 2.15a).

It is becoming increasingly clear that if we are to keep up with the continuing demands of a global industrialized society, the application of plate tectonic theory to the origin and distribution of mineral resources is essential.

Chapter Summary

1. The concept of continental movement is not new. Alfred Wegener is generally credited with developing the hypothesis of continental drift. He provided abundant geologic and paleontologic evidence to show that the continents were once united into one supercontinent he named Pangaea. Unfortunately, Wegener could not explain how the continents moved, and most geologists ignored his ideas.

2. The hypothesis of continental drift was revived during the 1950s when paleomagnetic studies indicated the presence of multiple magnetic north poles instead of just one as there is today. This paradox was resolved by moving the continents into different positions, making the paleomagnetic data consistent with a single magnetic north pole.

3. Magnetic surveys of the oceanic crust reveal magnetic anomalies in the rocks indicating that Earth's magnetic field has reversed itself in the past. Because the anomalies are parallel and form symmetric belts adjacent to the oceanic ridges, new oceanic crust must have formed as the seafloor was spreading.

4. Seafloor spreading has been confirmed by radiometric dating of rocks on oceanic islands. Such dating reveals that the oceanic crust becomes older with distance from spreading ridges.

5. Plate tectonic theory became widely accepted by the 1970s because of the overwhelming evidence supporting it and because it provides geologists with a powerful theory for explaining such phenomena as volcanism, earthquake activity, mountain building, global climatic changes, past and present animal and plant distribution, and the distribution of some mineral resources.

6. Three types of plate boundaries are recognized: divergent boundaries, where plates move away from each other; convergent boundaries, where two plates collide; and transform boundaries, where two plates slide past each other.

7. The average rate of movement and relative motion of plates can be calculated in several ways. The results of these different methods all agree and indicate that the plates move at different average velocities.

8. Absolute motion of plates can be determined by the movement of plates over mantle plumes. A mantle plume is an apparently stationary column of magma that rises to the surface where it becomes a hot spot and forms a volcano.

9. Although a comprehensive theory of plate movement has yet to be developed, geologists think that some type of convective heat system is involved.

10. A close relationship exists between the formation of some mineral deposits and plate boundaries. Furthermore, the formation and distribution of some natural resources are related to plate movements.

Important Terms

continental–continental plate boundary
continental drift
convergent plate boundary
Curie point
divergent plate boundary
Glossopteris flora
Gondwana

hot spot
Laurasia
magnetic anomaly
magnetic reversal
oceanic–continental plate boundary
oceanic–oceanic plate boundary
paleomagnetism
Pangaea

plate
plate tectonic theory
seafloor spreading
thermal convection cell
transform fault
transform plate boundary
volcanic island arc

Review Questions

1. The man credited with developing the continental drift hypothesis is:
   a. _____ Wilson;
   b. _____ Hess;
   c. _____ Vine;
   d. _____ Wegener;
   e. _____ du Toit.

2. The southern part of Pangaea, consisting of South America, Africa, India, Australia, and Antarctica, is called:
   a. _____ Gondwana;
   b. _____ Laurasia;
   c. _____ Atlantis;
   d. _____ Laurentia;
   e. _____ Pacifica.

3. Which of the following has been used as evidence for continental drift?
   a. _____ continental fit;
   b. _____ fossil plants and animals;
   c. _____ similarity of rock sequences;
   d. _____ paleomagnetism;
   e. _____ all of these.

4. The formatin of and distribution of copper deposits are associated with what type of boundaries?
   a. _____ divergent;
   b. _____ convergent;
   c. _____ transform;
   d. _____ answers (a) and (b);
   e. _____ answers (b) and (c).

5. Divergent boundaries are areas where:
   a. _____ new continental lithosphere is forming;
   b. _____ new oceanic lithosphere is forming;
   c. _____ two plates come together;
   d. _____ two plates slide past each other;
   e. _____ answers (b) and (d).

6. Along what type of plate boundary does subduction occur?
   a. _____ divergent;
   b. _____ transform;
   c. _____ convergent;
   d. _____ answers (a) and (b);
   e. _____ answers (b) and (c).

7. Iron-bearing minerals in a magma gain their magnetism and align themselves with the magnetic field when they cool through the:
   a. _____ negative magnetic anomaly;
   b. _____ Curie point;
   c. _____ positive magnetic reversal;
   d. _____ thermal convection point;
   e. _____ hot-spot point.

8. Back-arc basins are associated with _____ plate boundaries.
   a. _____ divergent;
   b. _____ convergent;
   c. _____ transform;
   d. _____ answers (a) and (b);
   e. _____ answers (b) and (c).

9. The San Andreas fault is an example of a(n) _____ boundary.
   a. _____ divergent;
   b. _____ convergent;
   c. _____ transform;
   d. _____ oceanic–continental;
   e. _____ continental–continental.

10. Which of the following will allow you to determine the absolute motion of plates?
    a. _____ hot spots;
    b. _____ the age of the sediment directly above any portion of the ocean crust;
    c. _____ magnetic reversals in the seafloor crust;
    d. _____ satellite–laser ranging techniques;
    e. _____ all of these.

11. The formation of the island of Hawaii and the Loihi Seamount are the result of:
    a. _____ oceanic-oceanic plate boundaries;
    b. _____ hot spots;
    c. _____ divergent plate boundaries;
    d. _____ transform boundaries;
    e. _____ oceanic–continental plate boundaries.

12. The driving mechanism of plate movement is thought to be:
    a. _____ composition;
    b. _____ magnetism;
    c. _____ thermal convection cells;
    d. _____ rotation of Earth;
    e. _____ none of these.

13. The Mid-Atlantic Ridge is an example of what type of plate boundary?
    a. _____ oceanic–oceanic convergent;
    b. _____ oceanic–continental convergent;
    c. _____ continental–continental convergent;
    d. _____ divergent;
    e. _____ hot spot.

14. What evidence convinced Wegener that the continents were once joined together and subsequently broke apart?

15. What is the significance of polar wandering in relation to continental drift?

16. How can magnetic anomalies be used to show that the seafloor has been spreading?

17. Summarize the geologic features characterizing the three different types of plate boundaries.

18. What are some of the positive and negative features of the various models proposed to explain plate movement?

1. If movement along the San Andreas fault, which separates the Pacific plate from the North American plate, averages 5.5 cm/year, how long will it take before Los Angeles is opposite San Francisco?

2. Using the age for each of Hawaiian Islands in Figure 2.21, calculate the average rate of movement per year for the Pacific plate since each island formed. Is the average rate of movement the same for each island? Would you expect it to be? Explain why it may not be.

3. What features would an astronaut look for on the Moon or another planet to find out if plate tectonics is currently active or if it was active during the past?

World Wide Web Activities

For these web site addresses, along with current updates and exercises, log on to

http://www.wadsworth.com/geo

▶ACTIVE TECTONICS

This site functions mainly to disseminate information and links to other sites related to plate tectonics. Check out some of the other links listed.

▶PLATE TECTONICS

This site contains pages of tutorial material from the University of Nevada's Seismological Laboratory at Reno. It contains the same information as found in this chapter but also contains many images not seen in this chapter and is worth the visit.

▶TECTONIC PLATE MOTION

Maintained by NASA, this site contains a wealth of information on how the various space geodetic technologies are used for calculating and tracking current plate movements.

1. What are the different ways plate movements can be determined and calculated?

2. Click on the various geographic sites listed to see what the average velocity and direction of movement of various plates are.

Explore the following *In-Terra-Active 2.0* CD-ROM module(s) and increase your understanding of key concepts and processes presented in this chapter.

▶ **SECTION: INTERIOR**

MODULE: PLATE TECTONICS: EVIDENCE

Question: If you were an astronaut landing on Venus, what evidence would you look for to confirm or deny the existence of plate tectonics on Venus?

▶ **SECTION: INTERIOR**

MODULE: PLATE TECTONICS: PLATE BOUNDARIES

The photograph shows a view of the Butte, Montana, copper deposit. Much of the copper minerals at Butte are found in porphyritic (big mineral grains in a dominantly fine-grained rock), silica-rich, igneous rock. These kinds of igneous rocks and associated copper deposits are typically formed in Arizona, the Andes Mountains, volcanic islands of the South Pacific, and the coast ranges of western Canada and the United States. **Question: What does the copper deposit tell us about the plate tectonic setting of Butte, Montana, at the time this deposit formed 50–60 million years ago?**

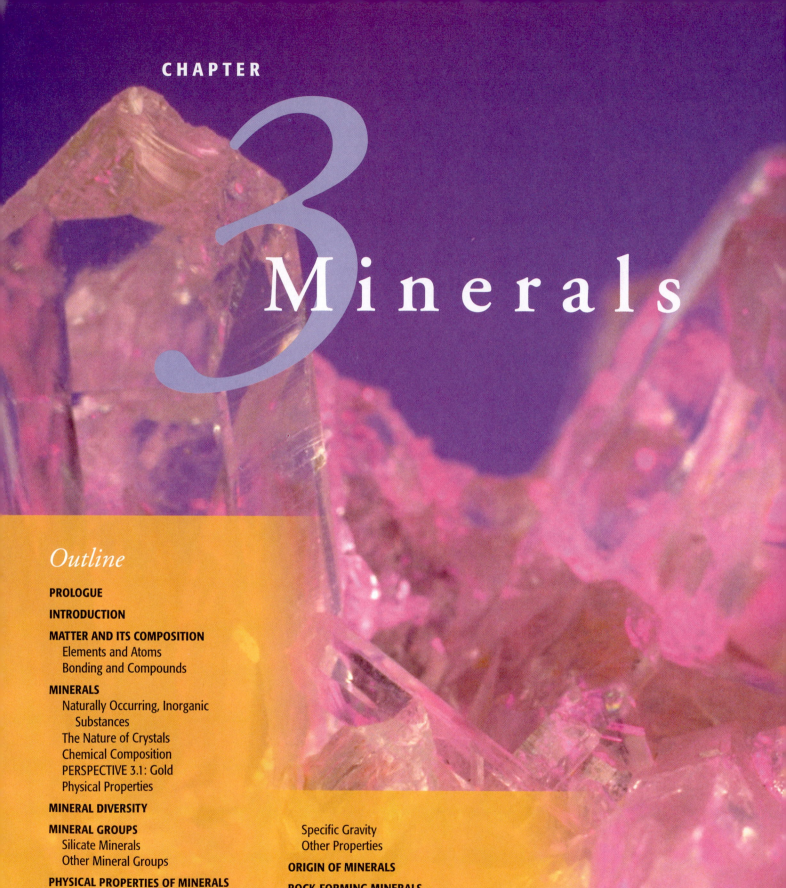

CHAPTER

3

Minerals

Outline

Prologue

This mineral specimen, called the Steamboat, consists of red and green tourmaline and colorless quartz crystals. From the Tourmaline Mine, near Pala, San Diego County, California, the specimen is about 28 cm high. National Museum of Natural History specimen #R51. *(Photo by D. Penland, courtesy of Smithsonian Institution.)*

Humans have been fascinated by minerals and stones for thousands of years. Indeed, archaeological evidence indicates that people in Spain and France were carving objects from bone, ivory, horn, and various stones at least 75,000 years ago. The ancient Egyptians mined turquoise more than 5000 years ago, and by 3400 B.C. they were making ornaments from rock crystal (colorless quartz), amethyst (purple quartz), lapis lazuli (a rock with a variety of minerals), and several other stones. Turquoise remains a popular gemstone in many cultures, including those of the Native Americans of the southwestern United States (Figure 3.1).

Most gemstones are minerals, more rarely rocks, that are cut and polished for jewelry. To qualify as a gemstone, a mineral or rock must be appealing for some reason. Desirable qualities of gemstones include brilliance, beauty, durability, and scarcity. They are even more desirable when some kind of lore is associated with them. According to one legend, diamond wards off evil spirits, sickness, and floods, whereas topaz was once thought to avert mental disorders, and ruby was believed to preserve its owner's health and warn of imminent bad luck. Relating gemstones to birth month gives them an added appeal to many people (Table 3.1).

FIGURE 3.1 Turquoise, a sky blue, blue-green, or light green hydrated copper aluminum phosphate, is a semiprecious gemstone used for jewelry and as a decorative stone.

Many minerals and rocks are attractive but fail to qualify as gemstones. Cut and polished fluorite is beautiful, but it is fairly common and too soft to be durable. Diamond, on the other hand, meets most of the criteria for a gemstone. In fact, only about two dozen gemstones are widely used, some of which are considered precious, *and others are recognized as* semiprecious. *The precious gemstones are diamond (Figure 3.2); ruby and sapphire, which are red and blue varieties of the mineral corundum, respectively; emerald, a bright green variety of the mineral beryl; and precious opal. Among the semiprecious gemstones are garnet, jade, tourmaline, topaz, peridot (olivine), aquamarine (light bluish green beryl), turquoise, and several varieties of quartz such as amethyst, agate, and tiger's eye.*

Also included among the semiprecious gemstones is amber, a fossil resin from coniferous trees that often contains insects. Recall that insects preserved in amber played a pivotal role in the novel and movie Jurassic Park. *Because amber is an organic substance, it is not truly a mineral or rock,*

TABLE 3.1

Birthstones

MONTH	BIRTHSTONE	MINERAL	SYMBOLIZES
January	Garnet	Garnet	Constancy
February	Amethyst	Purple quartz	Sincerity
March	Aquamarine	Light green, blue-green beryl	Courage
	Bloodstone	Greenish chalcedony	
April	Diamond	Diamond	Innocence
May	Emerald	Beryl and several other green minerals	Love, success
June	Pearl	Dense, spherical white or light-colored calcareous concretion produced by organisms, especially mollusks	Health, longevity
	Alexandrite	Greenish chrysoberyl	
	Moonstone	Type of potassium feldspar	
July	Ruby	Red corundum	Contentment
August	Peridot	Pale, clear, yellowish green olivine	Married happiness
	Sardonyx	Gem variety of chalcedony	
September	Sapphire	Blue transparent corundum	Clear thinking
October	Opal	Opal	Hope
	Tourmaline	Tourmaline	
November	Topaz	Topaz	Fidelity
December	Turquoise	Turquoise	Prosperity
	Zircon	Zircon	

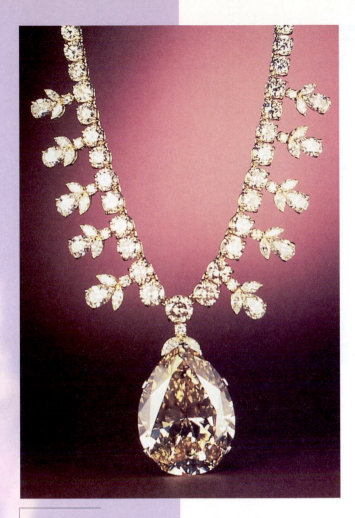

FIGURE 3.2 Gem-quality diamonds. Their hardness, brilliance, beauty, durability, and scarcity make them the most sought-after gemstones. The pendant in this necklace, housed in the Smithsonian Institution, is the Victoria Transvaal diamond from South Africa. National Museum of Natural History, specimen #G7101. *(Photo by D. Penland, courtesy of Smithsonian Institution.)*

but it is nevertheless considered to be a gemstone. Pearl is an unusual gemstone in that it is produced by organisms and is essentially ready to use when found. It forms as successive layers of tiny crystals are deposited around some irritant in a mollusk such as an oyster or clam. Most pearls are lustrous white, but some are silver gray, green, or black.

Transparent gemstones are most often cut to yield small, polished plane surfaces known as facets, which enhance the quality of reflected or refracted light. For instance, the brilliance of diamonds is maximized by faceting the back of the stone so that as much light as possible is reflected (Figure 3.2). Opaque and translucent gemstones are rarely faceted. Rather, they are cut and polished into dome-shaped cabochons to emphasize their most interesting features, or they are simply polished by tumbling (Figure 3.1).

Off all gemstones, diamond probably has the widest appeal. Many people are surprised to learn that diamond and the soft, gray mineral graphite are composed of the same chemical element, carbon. But other than composition, they share little in common: diamond is the hardest mineral, whereas graphite can be scratched by a fingernail. Their dissimilarities are accounted for by differences in the way their atoms are arranged. In diamond, carbon atoms are tightly bonded to one another in a three-dimensional framework, but in graphite the atoms form sheets held together by weak bonds. Incidentally, many people believe that to determine whether the stone in a ring is a diamond is to see if it will scratch glass. Diamond will indeed scratch glass, but so will several other minerals—including common quartz.

"Cutting" a diamond, the hardest substance known, is actually done by several processes, one of which is cleaving it along planes of weakness. Diamond possesses four internal planes of weakness, or cleavage planes, so that if a diamond is cleaved perfectly along these planes the resulting "stone" will be shaped like two pyramids placed base to base. Large diamonds are commonly preshaped by cleaving them into smaller pieces that are then further shaped by sawing and grinding with diamond dust.

Gemstones have traditionally been used for many purposes, especially for personal adornment and as symbols of wealth and power. In addition, gemstones along with other minerals, rocks, and fossils have also served as religious symbols and talismans, or they have been worn or carried for their presumed mystical or curative powers. Many people own small gemstones, but most of the truly magnificent ones are in museums or collections of crown jewels.

Introduction

THE term *mineral* commonly brings to mind dietary components essential for good nutrition, such as calcium, iron, potassium, and magnesium, which are actually chemical elements, not minerals in the geologic sense. Mineral is also sometimes used to refer to any material that is neither animal nor vegetable. Such usage implies that minerals are inorganic, which is correct, but not all inorganic substances are minerals. Water is not a mineral even though it is inorganic and is composed of the same chemical elements as ice, which is a mineral. Ice is of course a solid, whereas water is a liquid; minerals are solids rather than liquids or gases. In fact, geologists use a specific definition of the term **mineral**: a naturally occurring, inorganic, crystalline solid. Crystalline means a mineral has a regular internal structure. Furthermore, minerals have a narrowly defined chemical composition and characteristic physical properties such as density, color, and hardness. Most rocks are aggregates of one or more minerals, so minerals are the building blocks of rocks.

Obviously, minerals as constituents of rocks are important to geologists, but they are also important for other reasons. Gemstones such as diamond and topaz are minerals, and rubies are simply red-colored varieties of the mineral corundum. The sand used in the manufacture of glass is composed of the mineral quartz, and ore deposits are natural concentrations of economically valuable minerals. Indeed, industrialized societies depend directly upon finding and using mineral resources such as iron, copper, gold, and many others. More than $34 billion worth of minerals, excluding fuels, were produced in the United States during 1994.

Matter and Its Composition

ANYTHING that has mass and occupies space is *matter*. The atmosphere, water, plants and animals, and minerals and rocks are all composed of matter. Matter occurs in one of three states or phases, all of which are important in geology: *solids, liquids,* and *gases*. Atmospheric gases and liquids such as surface water and groundwater will be discussed later in this book, but here we are concerned chiefly with solids because all minerals are solids.

ELEMENTS AND ATOMS

Matter is made up of chemical **elements**, each of which is composed of incredibly small particles known as **atoms**. Atoms are the smallest units of matter that retain the characteristics of an element. Ninety-two naturally occurring elements have been discovered, some of which are listed in Table 3.2, and more than a dozen additional elements have been made in laboratories. Each natural element and most artificially produced ones have a name and a chemical symbol.

Atoms consist of a compact **nucleus** composed of one or more **protons**, which are particles with a positive electrical charge, and **neutrons**, which are electrically neutral (Figure 3.3). The nucleus of an atom makes up most of its mass. Encircling the nucleus are negatively charged **electrons**, which orbit rapidly around the nucleus at specific distances in one or more **electron shells**. The number of protons in the nucleus of an atom determines what the element is and determines the **atomic number** for that element. For example, each atom of the element hydrogen (H) has 1 proton in its nucleus and thus has an atomic number of 1. Helium (He) possesses 2 protons, carbon (C) has 6, and uranium (U) has 92, so their atomic numbers are 2, 6, and 92, respectively (Table 3.2).

TABLE 3.2

Symbols, Atomic Numbers, and Electron Configurations for Some Naturally Occurring Elements

ELEMENT	SYMBOL	ATOMIC NUMBER	NUMBER OF ELECTRONS IN EACH SHELL			
			1	2	3	4
Hydrogen	H	1	1			
Helium	He	2	2			
Lithium	Li	3	2	1		
Beryllium	Be	4	2	2		
Boron	B	5	2	3		
Carbon	C	6	2	4		
Nitrogen	N	7	2	5		
Oxygen	O	8	2	6		
Fluorine	F	9	2	7		
Neon	Ne	10	2	8		
Sodium	Na	11	2	8	1	
Magnesium	Mg	12	2	8	2	
Aluminum	Al	13	2	8	3	
Silicon	Si	14	2	8	4	
Phosphorus	P	15	2	8	5	
Sulfur	S	16	2	8	6	
Chlorine	Cl	17	2	8	7	
Argon	Ar	18	2	8	8	
Potassium	K	19	2	8	8	1
Calcium	Ca	20	2	8	8	2

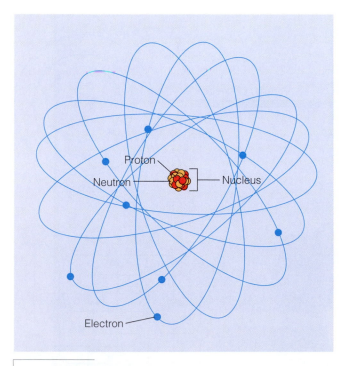

FIGURE 3.3 The structure of an atom. The dense nucleus consisting of protons and neutrons is surrounded by a cloud of orbiting electrons.

Atoms are also characterized by their **atomic mass number**, which is determined by adding together the number of protons and neutrons in the nucleus (electrons contribute negligible mass to an atom). Not all atoms of the same element have the same number of neutrons in their nuclei. In other words, atoms of the same element may have different atomic mass numbers. Different carbon (C) atoms, for instance, have atomic mass numbers of 12, 13, and 14. All of these atoms possess 6 protons—otherwise, they would not be carbon—but the number of neutrons varies. Forms of the same element with different atomic mass numbers are *isotopes* (Figure 3.4).

A number of elements have a single isotope, but many—such as uranium and carbon—have several (Figure 3.4). All isotopes of an element behave the same chemically. For example, both carbon 12 and carbon 14 are present in carbon dioxide (CO_2).

BONDING AND COMPOUNDS

Atoms are joined to other atoms through the process of **bonding**. When atoms of two or more elements are bonded, the resulting substance is a **compound**. A chemical substance such as gaseous oxygen, which consists entirely of oxygen atoms, is an element, whereas ice, consisting of hydrogen and oxygen, is a compound. Most

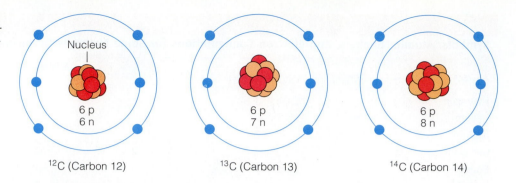

FIGURE 3.4 Schematic representation of isotopes of carbon. A carbon atom has an atomic number of 6 and an atomic mass number of 12, 13, or 14—depending on the number of neutrons in its nucleus.

Nucleus

6 p
6 n

6 p
7 n

6 p
8 n

^{12}C (Carbon 12)

^{13}C (Carbon 13)

^{14}C (Carbon 14)

minerals are compounds although gold and silver and several others are important exceptions.

To understand bonding, it is necessary to delve deeper into the structure of atoms. Recall that negatively charged electrons orbit the nuclei of atoms in electron shells. With the exception of hydrogen, which has only one proton and one electron, the innermost electron shell of an atom contains only two electrons. The other shells contain various numbers of electrons, but the outermost shell never contains more than eight (Table 3.2). The electrons in the outermost shell are the ones that are usually involved in chemical bonding.

Two types of chemical bonds, *ionic* and *covalent,* are particularly important in minerals, and many minerals contain both types of bonds. Two other types of chemical bonds, *metallic* and *van der Waals,* are much less common but are extremely important in determining the properties of some useful minerals.

IONIC BONDING   Notice in Table 3.2 that most atoms have fewer than eight electrons in their outermost electron shell. A few elements, such as neon and argon, have complete outer shells containing eight electrons; they are known as the *noble gases.* The noble gases do not react readily with other elements to form compounds because of this electron configuration. Interactions among atoms tend to produce electron configurations similar to those of the noble gases. That is, atoms interact so that their outermost electron shell is filled with eight electrons, unless the first shell (with two electrons) is also the outermost electron shell, as in helium.

One way that the noble gas configuration can be attained is by the transfer of one or more electrons from one atom to another. Common salt is composed of the elements sodium (Na) and chlorine (Cl), each of which is poisonous, but when combined chemically forms the compound sodium chloride (NaCl), the mineral halite. Notice in Figure 3.5a that

FIGURE 3.5   (a) Ionic bonding. The electron in the outermost shell of sodium is transferred to the outermost electron shell of chlorine. Once the transfer has occurred, sodium and chlorine are positively and negatively charged ions, respectively. (b) The crystal structure of sodium chloride, the mineral halite. The diagram on the left shows the relative sizes of the sodium and chlorine ions, and the diagram on the right shows the locations of the ions in the crystal structure.

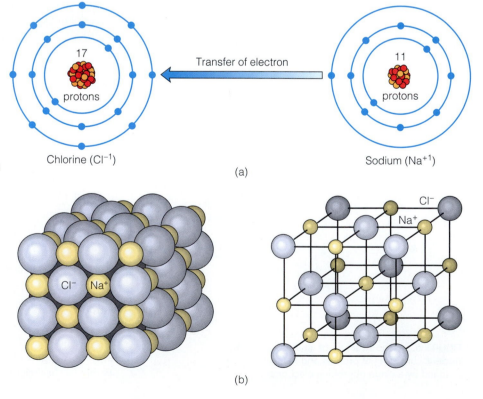

17
protons

Transfer of electron

11
protons

Chlorine (Cl^{-1})

Sodium (Na^{+1})

(a)

Cl$^-$

Na$^+$

Cl$^-$ Na$^+$

(b)

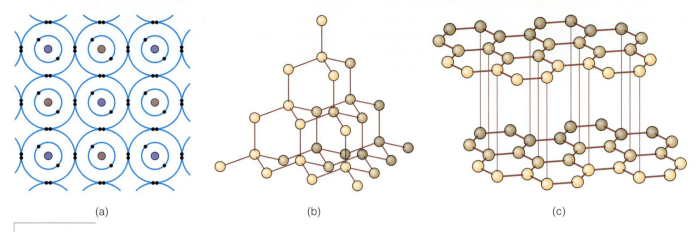

(a)   (b)   (c)

FIGURE 3.6 (a) Covalent bonds formed by adjacent atoms sharing electrons in diamond. (b) The three-dimensional framework of carbon atoms in diamond. (c) Covalent bonding also occurs in graphite, but here the carbon atoms are bonded together to form sheets that are held to one another by van der Waals bonds. The sheets themselves are strong, but the bonds between sheets are weak.

sodium has 11 protons and 11 electrons; thus, the positive electrical charges of the protons are exactly balanced by the negative charges of the electrons, and the atom is electrically neutral. Likewise, chlorine with 17 protons and 17 electrons is electrically neutral (Figure 3.5a). But neither sodium nor chlorine has 8 electrons in its outermost electron shell; sodium has only 1, whereas chlorine has 7. To attain a stable configuration, sodium loses the electron in its outermost electron shell, leaving its next shell with 8 electrons as the outermost one (Figure 3.5a). Sodium now has one fewer electron (negative charge) than it has protons (positive charge), so it is an electrically charged **ion** and is symbolized Na^{+1}.

The electron lost by sodium is transferred to the outermost electron shell of chlorine, which originally had 7 electrons. This addition of one more electron gives chlorine an outermost electron shell of 8 electrons, the configuration of a noble gas. But now its total number of electrons is 18, which exceeds by one the number of protons. Accordingly, chlorine also becomes an ion, but it is negatively charged (Cl^{-1}). An **ionic bond** forms between sodium and chlorine because of the attractive force between the positively charged sodium ion and the negatively charged chlorine ion (Figure 3.5a).

In ionic compounds, such as sodium chloride (the mineral halite), the ions are arranged in a three-dimensional framework that results in overall electrical neutrality. In halite, sodium ions are bonded to chlorine ions on all sides, and chlorine ions are surrounded by sodium ions (Figure 3.5b).

COVALENT BONDING **Covalent bonds** form between atoms when their electron shells overlap and electrons are shared. Atoms of the same element cannot bond by transferring electrons from one atom to another. Carbon (C), which forms the minerals graphite and diamond, has four electrons in its outermost electron shell (Figure 3.6a). If

these four electrons were transferred to another carbon atom, the atom receiving the electrons would have the noble gas configuration of eight electrons in its outermost electron shell, but the atom contributing the electrons would not.

In such situations, adjacent atoms share electrons by overlapping their electron shells. A carbon atom in diamond, for instance, shares all four of its outermost electrons with a neighbor to produce a stable noble gas configuration (Figure 3.6a).

Covalent bonds are not restricted to substances composed of atoms of a single kind. Among the most common minerals, the silicates (discussed later in this chapter), the element silicon forms partly covalent and partly ionic bonds with oxygen.

METALLIC AND VAN DER WAALS BONDS *Metallic bonding* results from an extreme type of electron sharing. The electrons of the outermost electron shell of such metals as gold, silver, and copper are readily lost and move about from one atom to another. This electron mobility accounts for the fact that metals have a metallic luster (their appearance in reflected light), provide good electrical and thermal conductivity, and can be easily reshaped. Only a few minerals possess metallic bonds, but those that do are very useful; copper, for example, is used for electrical wiring because of its high electrical conductivity.

Some electrically neutral atoms and molecules* have no electrons available for ionic, covalent, or metallic bonding. Nevertheless, a weak attractive force exists between them when they are in proximity. This weak attractive force is a *van der Waals* or *residual bond*. The carbon atoms in the mineral graphite are covalently bonded to form sheets, but

---

*A molecule is the smallest unit of a substance having the properties of that substance. A water molecule (H_2O), for example, possesses two hydrogen atoms and one oxygen atom.

the sheets are weakly held together by van der Waals bonds (Figure 3.6b). This type of bonding makes graphite useful for pencil leads; when pencil lead is moved across a piece of paper, small pieces of graphite flake off along the planes held together by van der Waals bonds and adhere to the paper.

Minerals

BEFORE we discuss minerals in more detail, let us recall our formal definition: a mineral is a naturally occurring, inorganic, crystalline solid, with a narrowly defined chemical composition and characteristic physical properties. The next sections will examine each part of this definition.

NATURALLY OCCURRING, INORGANIC SUBSTANCES

Naturally occurring excludes from minerals all substances manufactured by humans. Accordingly, synthetic diamonds and rubies and a number of other artificially synthesized substances are not regarded as minerals by most geologists.

Some geologists think the term *inorganic* in the mineral definition is superfluous. It does, however, remind us that animal matter and vegetable matter are not minerals. Nevertheless, some organisms such as corals and clams construct their shells of the compound calcium carbonate ($CaCO_3$), which is either aragonite or calcite, both of which are minerals.

THE NATURE OF CRYSTALS

By definition a mineral is a **crystalline solid**—that is, a solid in which the constituent atoms are arranged in a regular, three-dimensional framework (Figure 3.5b). Under ideal conditions, such as in a cavity, mineral crystals can grow and form perfect crystals that possess planar surfaces (crystal faces), sharp corners, and straight edges (Figure 3.7). In other words, the regular geometric shape of a well-formed mineral crystal is the exterior manifestation of an ordered internal atomic arrangement. Not all rigid substances are crystalline solids; natural and manufactured glass lack the ordered arrangement of atoms and are said to be *amorphous,* meaning "without form."

As early as 1669, the well-known Danish scientist Nicholas Steno determined that the angles of intersection of equivalent crystal faces on different specimens of quartz are identical. Since then the *constancy of interfacial angles* has been demonstrated for many other minerals, regardless of their size, shape, or geographic occurrence (Figure 3.8). Steno postulated that mineral crystals are composed of very small, identical building blocks and that the arrangement of these blocks determines the external form of the crystals. Such regularity of the external form of minerals must surely mean that external crystal form is controlled by internal structure.

Crystalline structure can be demonstrated even in minerals lacking obvious crystals. For example, many minerals possess a property known as *cleavage,* meaning that they break or split along closely spaced, smooth planes. The fact that these minerals can be split along such smooth planar surfaces indicates that the mineral's internal structure controls such breakage. The behavior of light and X-ray beams transmitted through minerals also provides compelling evidence for an orderly arrangement of atoms within minerals.

CHEMICAL COMPOSITION

Mineral composition is generally shown by a chemical formula, which is a shorthand way of indicating the number of atoms of different elements composing the mineral. The mineral quartz consists of one silicon (Si) atom for every two oxygen (O) atoms, and thus has the formula SiO_2; the subscript number indicates the number of atoms. Orthoclase is composed of one potassium, one aluminum, three silicon, and eight oxygen atoms, so its formula is $KAlSi_3O_8$. A few minerals are composed of a single element. Known as **native elements**, they include such minerals as platinum

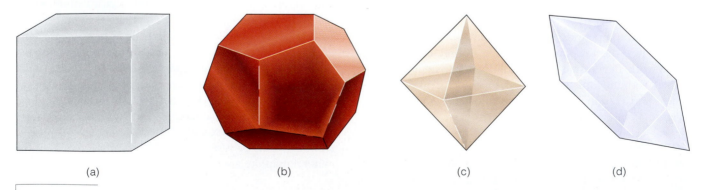

(a) (b) (c) (d)

FIGURE 3.7 Mineral crystals occur in a variety of shapes, several of which are shown here. (a) Cubic crystals typically develop in the minerals halite, galena, and pyrite. (b) Dodecahedron crystals such as those of garnet have 12 sides. (c) Diamond has octahedral, or 8-sided, crystals. (d) A prism terminated by a pyramid is found in quartz.

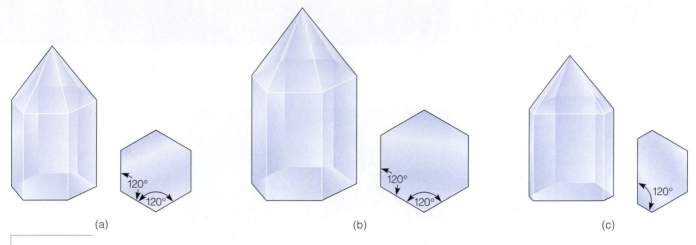

FIGURE **3.8** Side views and cross sections of three quartz crystals showing the constancy of interfacial angles: (a) a well-shaped crystal; (b) a larger crystal; and (c) a poorly shaped crystal. The angles formed between equivalent crystal faces on different specimens of the same mineral are the same regardless of the size or shape of the specimens.

(Pt); graphite and diamond, both composed of carbon (C); silver (Ag); and gold (Au) (Perspective 3.1).

The definition of a mineral contains the phrase *a narrowly defined chemical composition,* because some minerals actually have a range of compositions. For many minerals the chemical composition is constant, as in quartz (SiO_2) and halite (NaCl). Other minerals have a range of compositions because one element can substitute for another if the atoms of two or more elements are nearly the same size and the same charge. Notice in Figure 3.9 that iron and magnesium atoms are about the same size; therefore, they can substitute for one another. The chemical formula for the mineral olivine is $(Mg,Fe)_2SiO_4$, meaning that, in addition to silicon and oxygen, it may contain only magnesium, only iron, or a combination of both. A number of other minerals also have ranges of compositions, so these are actually mineral groups with several members.

FIGURE **3.9** Electrical charges and relative sizes of ions common in minerals. The numbers within the ions are the radii shown in Ångstrom units ($1 Å = 10^{-10}$ m).

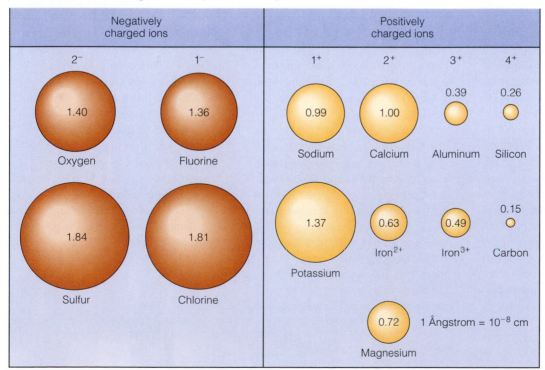

Gold

Among the hundreds of minerals used by humans, none is so highly prized and eagerly sought as gold (Figure 1). This deep-yellow mineral has been the cause of feuds and wars and was one of the incentives for the exploration of the Americas. Gold has been mined for at least 6000 years, and archaeological evidence indicates that people in Spain possessed small quantities of gold 40,000 years ago. Probably no other substance has caused so much misery while providing so many benefits for those who possessed it.

Why is gold so highly prized? Certainly not for use in tools or weapons, for it is too soft and pliable to hold a cutting edge. Furthermore, it is too heavy to be practical for most utilitarian purposes (it weighs more than twice as much as lead). During most of historic time, gold has been used for jewelry, ornaments, and ritual objects and has served as a symbol of wealth and as a monetary standard. Gold is desired for several reasons: (1) its pleasing appearance, (2) the ease with which it can be worked, (3) its durability, and (4) its scarcity (it is much rarer than silver).

Central and South American natives used gold extensively long before the arrival of Europeans. In fact, the Europeans' lust for gold was responsible for the ruthless conquest of the natives in those areas. In the United States, gold was first profitably mined in North Carolina in 1801 and in Georgia in 1829, but the truly spectacular finds were in California in 1848. This latter discovery culminated in the great gold rush of 1849 when tens of thousands of people flocked to California to find riches. Unfortunately, only a few found what they sought. Nevertheless, during the five years from 1848 to 1853, which constituted the gold rush proper, more than $200 million in gold was recovered.

Figure 1
Specimen of gold from Grass Valley, California—National Museum of Natural History (NMNH) specimen #R121297. (Photo by D. Penland, courtesy of Smithsonian Institution.)

Another gold rush took place in 1876 following a report by Lieutenant Colonel George Armstrong Custer that "gold in satisfactory quantities can be obtained in the Black Hills [South Dakota]." The flood of miners into the Black Hills, the Holy Wilderness of the Sioux Indians, resulted in the Indian War during which Custer and some 260 of his men were annihilated at the Battle of the Little Bighorn in Montana in June 1876. Despite this stunning victory, the Sioux could not sustain a war against the U.S. Army, and in September 1876, they were forced to relinquish the Black Hills.

Canada, too, has had its gold rushes. The first discovery came in 1850 in the Queen Charlotte islands on the Pacific coast, and by 1858 about 10,000 people were panning for gold there. The greatest Canadian gold rush was between 1897 and 1899 when as many as 35,000 men and women traveled to the remote, hostile Klondike region in the Yukon Territory. In fact, Dawson City grew so rapidly that hundreds of people had to be evacuated during the winter of 1897 because of food shortages. As in other gold rushes, local merchants made out better than the miners, most of whom barely eked out a living.

For 50 years following the California gold rush, the United States led the world in gold production, and it still produces a considerable amount, mostly from mines in Nevada, California, and South Dakota. Currently, the leading producer is South Africa with the United States a distant second, followed by Russia, Australia, and Canada.

Although much gold still goes into jewelry, it is now used in the chemical industry and for gold plating, electrical circuitry, and glass making. Consequently, the quest for gold has not ceased or even abated. In many industrialized nations, including the United States, domestic production cannot meet the demand, and much of the gold used must be imported.

PHYSICAL PROPERTIES

The last criterion in our definition of a mineral, *characteristic physical properties,* refers to such properties as hardness, color, and crystal form. These properties are controlled by composition and structure. We shall have more to say about physical properties of minerals later in this chapter.

Mineral Diversity

MORE than 3500 minerals have been identified and described, but only a few—perhaps two dozen—are particularly common. One might think that an extremely large number of minerals could be formed from 92 elements, but several factors limit the number possible. For one thing, many combinations of elements are chemically impossible; no compounds consist of only potassium and sodium or of silicon and iron, for example. Another important factor restricting the number of common minerals is that only eight chemical elements make up the bulk of Earth's crust

(Table 3.3). Oxygen and silicon constitute more than 74% (by weight) of the crust and nearly 84% of the atoms available to form compounds. By far the most common minerals consist of silicon and oxygen, combined with one or more of the other elements listed in Table 3.3.

Mineral Groups

GEOLOGISTS recognize mineral classes or groups, each with members sharing the same negatively charged ion or ion group (Table 3.4). We mentioned in a previous section that ions are atoms having either a positive or negative electrical charge resulting from the loss or gain of electrons in their outermost shell. In addition to ions, some minerals contain tightly bonded, complex groups of different atoms known as *radicals* that act as single units within minerals. A good example is the carbonate ion consisting of a carbon atom bonded to three oxygen atoms, thus having the formula CO_3 and a -2 electrical charge. Table 3.4 shows other common radicals and their charges.

TABLE 3.3

Common Elements in Earth's Crust

ELEMENT	SYMBOL	PERCENTAGE OF CRUST (BY WEIGHT)	PERCENTAGE OF CRUST (BY ATOMS)
Oxygen	O	46.6%	62.6%
Silicon	Si	27.7	21.2
Aluminum	Al	8.1	6.5
Iron	Fe	5.0	1.9
Calcium	Ca	3.6	1.9
Sodium	Na	2.8	2.6
Potassium	K	2.6	1.4
Magnesium	Mg	2.1	1.8
All others		1.5	0.1

TABLE 3.4

Some of the Mineral Groups Recognized by Geologists

MINERAL GROUP	NEGATIVELY CHARGED ION OR ION GROUP	EXAMPLES	COMPOSITION
Carbonate	$(CO_3)^{-2}$	Calcite	$CaCO_3$
		Dolomite	$CaMg(CO_3)_2$
Halide	Cl^{-1}, F^{-1}	Halite	$NaCl$
Hydroxide	$(OH)^{-1}$	Limonite	$FeO(OH) \cdot H_2O$
Native element	—	Gold	Au
		Diamond	C
Oxide	O^{-2}	Hematite	Fe_2O_3
Silicate	$(SiO_4)^{-4}$	Quartz	SiO_2
		Olivine	$(Mg,Fe)_2SiO_4$
Sulfate	$(SO_4)^{-2}$	Gypsum	$CaSO_4 \cdot 2H_2O$
Sulfide	S^{-2}	Galena	PbS

SILICATE MINERALS

Because silicon and oxygen are the two most abundant elements in the crust, it is not surprising that many minerals contain these elements. A combination of silicon and oxygen is known as **silica**, and the minerals containing silica are **silicates**. Quartz (SiO_2) is composed entirely of silicon and oxygen, so it is pure silica. Most silicates, however, have one or more additional elements, as in orthoclase ($KAlSi_3O_8$) and olivine [$(Mg,Fe)_2SiO_4$]. Silicate minerals include about one-third of all known minerals, but their abundance is even more impressive when one considers that they make up perhaps as much as 95% of Earth's crust.

The basic building block of all silicate minerals is the **silica tetrahedron**, which consists of one silicon atom and four oxygen atoms (Figure 3.10). These atoms are arranged so that the four oxygen atoms surround a silicon atom, which occupies the space between the oxygen atoms; thus, a four-faced pyramidal structure is formed. The silicon atom has a positive charge of 4, while each of the four oxygen atoms has a negative charge of 2, resulting in an ion group with a total negative charge of 4 $(SiO_4)^{-4}$.

Because the silica tetrahedron has a negative charge, it does not exist in nature as an isolated ion group; rather, it combines with positively charged ions or shares its oxygen atoms with other silica tetrahedra. In the simplest silicate minerals, the silica tetrahedra exist as single units bonded to positively charged ions. In minerals containing isolated tetrahedra, the silicon to oxygen ratio is 1:4, and the negative charge of the silica ion is balanced by positive ions (Figure 3.11a). Olivine [$(Mg,Fe)_2SiO_4$], for example, has either two magnesium (Mg^{+2}) ions, two iron (Fe^{+2}) ions, or one of each to offset the -4 charge of the silica ion.

Silica tetrahedra may also be arranged so that they join together to form chains of indefinite length (Figure 3.11b). Single chains, as in the pyroxene minerals, form when each tetrahedron shares two of its oxygens with adjacent tetrahedra; the result is a silicon to oxygen ratio of 1:3. Enstatite, a pyroxene group mineral, reflects this ratio in its

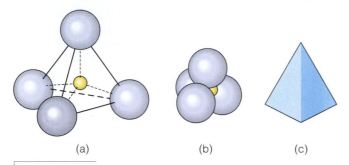

(a)   (b)   (c)

FIGURE 3.10   The silica tetrahedron. (a) Expanded view showing oxygen atoms at the corners of a tetrahedron and a small silicon atom at the center. (b) View of the silica tetrahedron as it really exists with the oxygen atoms touching. (c) The silica tetrahedron represented diagramatically; the oxygen atoms are at the four points of the tetrahedron.

FIGURE 3.11   Structures of some of the common silicate minerals shown by various arrangements of silica tetrahedra: (a) isolated tetrahedra, (b) continuous chains, (c) continuous sheets, and (d) networks. The arrows adjacent to single-chain, double-chain, and sheet silicates indicate that these structures continue indefinitely in the directions shown.

			Formula of negatively charged ion group	Silicon to oxygen ratio	Example
(a)	Isolated tetrahedra		$(SiO_4)^{-4}$	1:4	Olivine
(b)	Continuous chains of tetrahedra	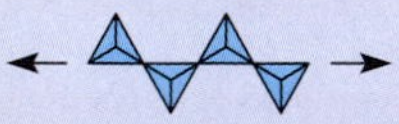 Single chain	$(SiO_3)^{-2}$	1:3	Pyroxene group
		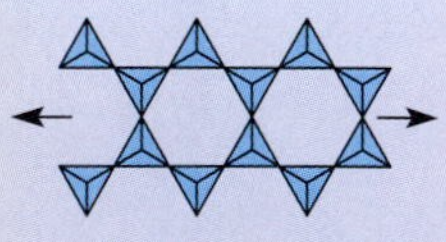 Double chain	$(Si_4O_{11})^{-6}$	4:11	Amphibole group
(c)	Continuous sheets	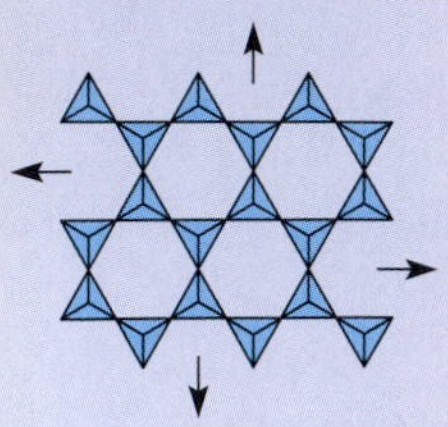	$(Si_4O_{10})^{-4}$	2:5	Micas
(d)	Three-dimensional networks	Too complex to be shown by a simple two-dimensional drawing	$(SiO_2)^0$	1:2	Quartz

chemical formula, MgSiO$_3$. Individual chains, however, possess a net −2 electrical charge, so they are balanced by positive ions, such as Mg^{+2}, that link parallel chains together (Figure 3.11b).

The amphibole group of minerals is characterized by a double-chain structure in which alternate tetrahedra in two parallel rows are cross-linked (Figure 3.11b). The formation of double chains results in a silicon to oxygen ratio of 4:11 so that each double chain possesses a −6 electrical charge. Mg^{+2}, Fe^{+2}, and Al^{+2} are usually involved in linking the double chains together.

In sheet structure silicates, three oxygens of each tetrahedron are shared by adjacent tetrahedra (Figure 3.11c). Such structures result in continuous sheets of silica tetrahedra with silicon to oxygen ratios of 2:5. Continuous sheets also possess a negative electrical charge satisfied by positive ions located between the sheets. This particular structure accounts for the characteristic sheet structure of the *micas,* such as biotite and muscovite, and the *clay minerals.*

Three-dimensional frameworks of silica tetrahedra form when all four oxygens of the silica tetrahedron are shared by adjacent tetrahedra (Figure 3.11d). Such sharing of oxygen atoms results in a silicon to oxygen ratio of 1:2, which is electrically neutral. Quartz is a common framework silicate.

Two main subgroups of silicates are recognized, ferromagnesian and nonferromagnesian silicates. The **ferromagnesian silicates** are those containing iron (Fe), magnesium (Mg), or both. These minerals are commonly dark colored and more dense than nonferromagnesian silicates. Some of the common ferromagnesian silicate minerals are olivine, the pyroxenes, the amphiboles, and biotite (Figure 3.12).

The **nonferromagnesian silicates** lack iron and magne-

FIGURE **3.13** Common nonferromagnesian silicates: (a) quartz, (b) the potassium feldspar orthoclase, (c) plagioclase feldspar, and (d) muscovite mica. *(Photo courtesy of Sue Monroe.)*

sium, are generally light colored, and are less dense than ferromagnesian silicates (Figure 3.13). The most common minerals in the crust are nonferromagnesian silicates known as *feldspars.* Feldspar is a general name for two distinct groups, each of which includes several species (Figure 3.13b and c). The *potassium feldspars* are represented by microcline and orthoclase (KAlSi$_3$O$_8$). The second group of feldspars, the *plagioclase feldspars,* range from calcium-rich (CaAl$_2$Si$_2$O$_8$) to sodium-rich (NaAlSi$_3$O$_8$) varieties.

Quartz (SiO$_2$) is another common nonferromagnesian silicate. It is a framework silicate that can usually be recognized by its glassy appearance and hardness (Figure 3.13a). Another fairly common nonferromagnesian silicate is muscovite, which is a mica (Figure 3.13d).

Various clay minerals also possess the sheet structure typical of the micas, but their crystals are so small that they can be seen only with extremely high magnification. These clay minerals are important constituents of several types of rocks and are essential components of soils.

OTHER MINERAL GROUPS

Besides silicates, several other mineral groups are recognized (Table 3.4). One of these, the *native elements* such as gold and silver, has already been discussed. Another group consists of the **carbonate minerals,** all of which contain the negatively charged carbonate ion (CO$_3$)$^{-2}$. An example is calcium carbonate (CaCO$_3$), the mineral calcite, which is the main constituent of the sedimentary rock limestone. A number of other carbonate minerals are known, but only dolomite [CaMg (CO$_3$)$_2$], the mineral in the rock dolostone, need concern us.

Physical Properties of Minerals

THE characteristic physical properties of minerals are determined by their internal structure and chemical composition. Many physical properties are remarkably constant for

FIGURE **3.12** Common ferromagnesian silicates; (a) olivine; (b) augite, a pyroxene group mineral; (c) hornblende, an amphibole group mineral; and (d) biotite mica. *(Photo courtesy of Sue Monroe.)*

FIGURE 3.14 (a) Streak is the color of a powdered mineral. Note that these two varieties of hematite yield similar streaks although the samples are different colors. (b) Hematite (left) has the appearance of a metal and is said to have a metallic luster, whereas orthoclase (right) has a nonmetallic luster.

a given mineral species, but some, especially color, may vary. Though a professional geologist may use sophisticated techniques in studying and identifying minerals, most common minerals can be identified by using the following physical properties (see Appendix C).

COLOR AND LUSTER

Although the color of some minerals varies because of minute amounts of impurities, some generalizations can be made. Ferromagnesian silicates are typically black, brown, or dark green, although olivine is olive green (Figure 3.12a). Nonferromagnesian silicates, on the other hand, can vary considerably in color, but are only rarely dark (Figure 3.13). Minerals that have the appearance of metals are rather consistent in color.

Some of the color variations of minerals can be evened out by crushing the mineral to a fine powder. A convenient way to make this test is to rub a mineral on a streak plate, an unglazed ceramic tile, thereby leaving a trail of fine powder. The iron mineral hematite has a variable appearance—it may be red and earthy or silver gray and metallic—but it always leaves a reddish brown streak (Figure 3.14a).

Luster (not to be confused with color) is the appearance of a mineral in reflected light. Two basic types of luster are recognized: *metallic* and *nonmetallic* (Figure 3.14b). They are distinguished by observing the quality of light reflected from a mineral and determining if it has the appearance of a metal or a nonmetal. Several types of nonmetallic luster are recognized, including glassy or vitreous, greasy, waxy, brilliant (as in diamond), and dull or earthy.

CRYSTAL FORM

As previously noted, minerals are crystalline solids, but many form under conditions that do not allow perfect crystals to develop. For some minerals, though, well-formed

FIGURE 3.15 Mineral crystals. (a) Cubic crystals of fluorite. (b) A calcite crystal. *(Photos courtesy of Sue Monroe.)*

crystals are common (Figure 3.15), such as 12-sided garnet crystals and 6- and 12-sided pyrite crystals. Minerals that form and grow in cavities or are precipitated from hot water (hydrothermal solutions) in cracks and crevices in rocks also commonly occur as well-formed crystals.

Crystal form can be useful for identifying minerals, but a number of different minerals have the same crystal form. For instance, cubic crystals of pyrite (FeS_2), galena (PbS_2), and halite ($NaCl$) are quite common. However, these minerals can be easily differentiated from one another by other properties such as color, luster, hardness, and density. For minerals lacking obvious crystals, these other physical properties are used for identification.

CLEAVAGE AND FRACTURE

Cleavage is a property of individual mineral crystals. Not all minerals possess cleavage, but those that do tend to break, or split, along a smooth plane or planes of weakness determined by the strength of the bonds within the mineral crystal. Cleavage can be characterized in terms of quality (perfect, good, poor), direction, and angles of intersection of cleavage planes. Biotite, a common ferromagnesian silicate, has perfect cleavage in one direction (Figure 3.16a). The fact that biotite preferentially cleaves along a number of closely spaced, parallel planes is related to its structure; it is a sheet silicate with the sheets of silica tetrahedra weakly bonded to one another by iron and magnesium ions (Figure 3.16c).

Feldspars possess two directions of cleavage that intersect at right angles (Figure 3.16b), and the mineral halite has three directions of cleavage, all of which intersect at right angles (Figure 3.16c). Calcite also possesses three directions of cleavage, but none of the intersection angles is a right angle, so cleavage fragments of calcite are rhombohedrons (Figure 3.16d). Minerals with four directions of cleavage include fluorite and diamond (Figure 3.16e). Ironically, diamond, the hardest mineral, can be easily cleaved. A few minerals such as sphalerite, an ore of zinc, have six directions of cleavage (Figure 3.16f).

Cleavage is an important diagnostic property of minerals, and its recognition is essential in distinguishing between some minerals. The pyroxene mineral augite and the amphibole mineral hornblende look much alike: Both are generally dark green to black, have the same hardness, and possess two directions of cleavage. But the cleavage planes of augite intersect at about 90°, whereas the cleavage planes of hornblende intersect at angles of 56° and 124° (Figure 3.17).

In contrast to cleavage, *fracture* is mineral breakage along irregular surfaces, indicating that planes of weakness are absent. Any mineral can be fractured if enough force is applied, but the fracture surfaces will not be smooth; they are commonly uneven or conchoidal (smoothly curved) (Figure 3.18).

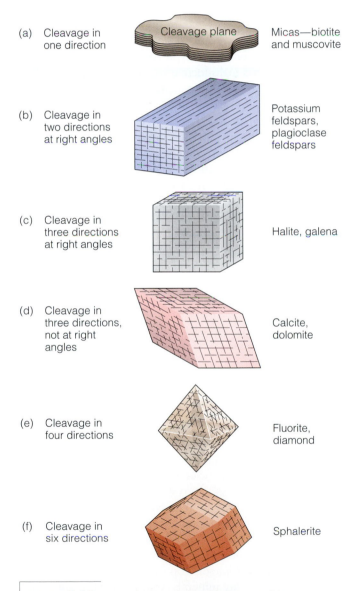

(a) Cleavage in one direction — Micas—biotite and muscovite

(b) Cleavage in two directions at right angles — Potassium feldspars, plagioclase feldspars

(c) Cleavage in three directions at right angles — Halite, galena

(d) Cleavage in three directions, not at right angles — Calcite, dolomite

(e) Cleavage in four directions — Fluorite, diamond

(f) Cleavage in six directions — Sphalerite

FIGURE **3.16** Several types of mineral cleavage: (a) one direction; (b) two directions at right angles; (c) three directions at right angles; (d) three directions, not at right angles; (e) four directions; and (f) six directions.

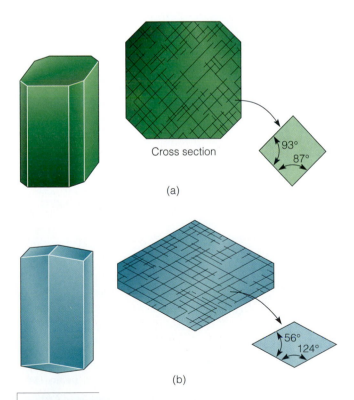

Cross section

93°
87°

(a)

56°
124°

(b)

FIGURE **3.17** Cleavage in augite and hornblende. (a) Augite crystal and cross section of crystal showing cleavage. (b) Hornblende crystal and cross section of crystal showing cleavage.

FIGURE 3.18 This specimen shows a conchoidal fracture, which consists of a smoothly curved surface resembling a clam shell. (*Photo courtesy of Sue Monroe.*)

HARDNESS

Hardness is the resistance of a mineral to abrasion. The Austrian geologist Friedrich Mohs devised a relative hardness scale for ten minerals. He arbitrarily assigned a hardness value of 10 to diamond, the hardest mineral known, and lesser values to the other minerals. Relative hardness can be determined easily by using Mohs hardness scale (Table 3.5). Quartz will scratch fluorite but cannot be scratched by fluorite, gypsum can be scratched by a fingernail, and so on. Hardness is controlled mostly by internal structure. Both graphite and diamond are composed of carbon, but the former has a hardness of 1 to 2, whereas the latter has a hardness of 10.

SPECIFIC GRAVITY

The *specific gravity* of a mineral is the ratio of its weight to the weight of an equal volume of water. A mineral with a specific gravity of 3.0 is three times as heavy as water. Like all ratios, specific gravity is not expressed in units such as grams per cubic centimeter—it is a dimensionless number.

Specific gravity varies in minerals depending upon their composition and structure. Among the common silicates, the ferromagnesian silicates have specific gravities ranging from 2.7 to 4.3, whereas the nonferromagnesian silicates vary from 2.6 to 2.9. Obviously, the ranges of values overlap somewhat, but for the most part ferromagnesian silicates have greater specific gravities than nonferromagnesian silicates. In general, the metallic minerals, such as galena (7.58) and hematite (5.26), are heavier than nonmetals. Pure gold has a specific gravity of 19.3, making it about 2½ times more dense than lead. Structure as a control of specific gravity is illustrated by the native element carbon (C): the specific gravity of graphite varies from 2.09 to 2.33; that of diamond is 3.5.

OTHER PROPERTIES

A number of other physical properties characterize some minerals. Talc has a distinctive soapy feel, graphite writes on paper, halite tastes salty, and magnetite is magnetic. Some minerals are plastic and, when bent into a new shape, will retain that shape, whereas others are flexible and, if bent, will return to their original position when the forces that bent them are removed.

The minerals calcite and dolomite can be identified by a simple chemical test in which a drop of dilute hydrochloric acid is applied to the mineral specimen. If the mineral is calcite, it will react vigorously with the acid and release carbon dioxide, which causes the acid to bubble, or effervesce. Dolomite, on the other hand, will not react with hydrochloric acid unless it is powdered.

Origin of Minerals

THUS far we have discussed the composition, structure, and physical properties of minerals but have not fully addressed how they form. One common phenomenon accounting for the origin of some minerals is cooling of molten rock material known as *magma;* magma that reaches Earth's surface is lava. As magma or lava cools, minerals begin to crystallize and grow, thereby determining the mineral composition of igneous rocks (see Chapter 4). Minerals also crystallize from hot-water (hydrothermal) solutions that invade cracks and crevices in rocks; many well-formed quartz (SiO_2) crystals are found in these so-called hydrothermal veins. Dissolved substances in seawater, more rarely lake water, might combine to form various minerals such as halite (NaCl), gypsum ($CaSO_4 \cdot 2H_2O$), and others when the water evaporates, or calcite ($CaCO_3$) may be extracted from solution by various organisms (see Chapter 7). Some clay minerals form when other minerals, especially feldspars, are changed compositionally and structurally by chemical weathering processes (see Chapter 6), and other minerals originate when rocks are altered by metamorphism (see Chapter 8).

TABLE 3.5		
Mohs Hardness Scale		
HARDNESS	MINERAL	HARDNESS OF SOME COMMON OBJECTS
10	Diamond	
9	Corundum	
8	Topaz	
7	Quartz	
		Steel file (6½)
6	Orthoclase	
		Glass (5½–6)
5	Apatite	
4	Fluorite	
3	Calcite	Copper penny (3)
		Fingernail (2½)
2	Gypsum	
1	Talc	

Rock-Forming Minerals

ROCKS are generally defined as aggregates of one or more minerals. Two important exceptions to this definition are natural glass such as obsidian (see Chapter 4) and the sedimentary rock coal (see Chapter 7). Although it is true that many minerals are present in various kinds of rocks, only a few varieties are common enough to be designated as **rock-forming minerals**. Most of the others constitute such small proportions of rocks that they can be disregarded in identification and classification; these are generally called *accessory minerals.*

Most common rocks are composed of silicate minerals. The igneous rock basalt is made up largely of ferromagnesian silicates, such as pyroxene group minerals and olivine (Figure 3.12), and plagioclase feldspar, a nonferromagnesian silicate (Figure 3.13c). Granite, another igneous rock, consists mostly of the nonferromagnesian silicates potassium feldspar and quartz (Figure 3.19). Both of these rocks contain a variety of accessory minerals, most of which are silicates as well. The minerals noted in the preceding examples are also common in metamorphic rocks, and many sedimentary rocks are composed of quartz, various feldspars, and clay minerals. Among the nonsilicate minerals, only the carbonates calcite ($CaCO_3$) and dolomite [$CaMg(CO_3)_2$]—the primary constituents of the sedimentary rocks limestone and dolostone, respectively—are particularly common as rock-forming minerals.

Mineral Resources and Reserves

GEOLOGISTS of the U.S. Geological Survey and the U.S. Bureau of Mines define a **resource** as follows:

> A concentration of naturally occurring solid, liquid, or gaseous material in or on the Earth's crust in such form

and amount that economic extraction of a commodity from the concentration is currently or potentially feasible.

Accordingly, resources include such substances as metals *(metallic resources);* sand, gravel, crushed stone, and sulfur *(nonmetallic resources);* and uranium, coal, oil, and natural gas *(energy resources).* An important distinction, however, must be made between a resource, which is the total amount of a commodity whether discovered or undiscovered, and a **reserve**, which is that part of the resource base that can be extracted economically.

When resources are mentioned, many people think of precious metals and gemstones. These are indeed resources, but so are many common minerals such as quartz, gypsum, clays, and feldspars. Most sand in sand dunes, on beaches, and in streams is quartz, yet this common mineral has a variety of uses including the manufacture of glass, optical instruments, sandpaper, and steel alloys. Gypsum is used to make wallboard, clay minerals are needed to make ceramics and paper, and potassium feldspars such as orthoclase are used for porcelain, ceramics, enamel, and glass.

The amount of mineral resources used has steadily increased since Europeans settled North America. In the United States alone, nearly 90 million metric tons of iron and steel were consumed in 1990, and about 70 million metric tons of cement and more than 5 million metric tons of aluminum were used. According to one estimate, the yearly per capita used of mineral resources by North Americans is about 14 metric tons; much of this is bulk items such as crushed stone and sand and gravel. It is no exaggeration to say that industrialized societies are totally dependent on mineral resources. Unfortunately, they are being used at rates far faster than they form. Thus, mineral resources are nonrenewable, meaning that once the resources from a deposit have been exhausted, new supplies or suitable substitutes must be found.

For some mineral resources, adequate supplies are available for indefinite periods (sand and gravel, for

(a)

FIGURE 3.19 The igneous rock granite is composed largely of potassium feldspar and quartz, lesser amounts of plagioclase feldspar, and accessory minerals such as biotite mica. (a) Hand specimen of granite. (b) Photomicrograph of a thin slice of granite, showing the various minerals.

(b)

example), whereas for others, supplies are limited or must be imported from other parts of the world (Figure 3.20). The United States is almost totally dependent on imports of manganese, an essential element in the manufacture of steel. Even though the United States is a leading producer of gold, it still depends on imports for more than half of its gold needs. About half of the crude oil used in the United States is imported, much of it from the Middle East where more than 50% of the proven reserves exist. A poignant reminder of our dependence on the availability of resources was the United States' response to the takeover of Kuwait by Iraq in August 1990.

In terms of mineral and energy resources, Canada is more self-reliant than the United States. It meets most of its domestic needs for mineral resources, although it must import phosphate, chromium, manganese, and bauxite, the ore of aluminum. Canada also produces more crude oil and natural gas than it uses, and it is the world leader in the production and export of uranium.

What constitutes a resource as opposed to a reserve depends on several factors. For example, iron-bearing minerals are found in many rocks, but in quantities or ways that make their recovery uneconomical. As a matter of fact, most minerals concentrated in economic quantities are mined in only a few areas; 75% of all the metals mined in the world come from about 150 locations. Geographic location is also an important consideration. A mineral resource in a remote region may not be mined because transportation costs are too high, and what may be considered a resource in the United States or Canada may be a reserve in a less-developed country where labor costs are low and it can be economically mined. The market price of a commodity is of course important in evaluating a potential resource. From 1935 to 1968, the U.S. government maintained the price of gold at $35 per troy ounce (= 31.1 g). When this restriction was removed and the price of gold became subject to supply and demand, the price rose (it reached an all-time high of $843 per troy ounce during January 1980). As a result, many marginal deposits became reserves, and many abandoned mines were reopened.

Technological developments can also change the status of a resource. The rich iron ores of the Great Lakes region of the United States and Canada had been depleted by World War II (1939–1945). However, the development of a method of separating the iron from previously unusable rocks and shaping it into pellets that are ideal for use in blast furnaces made it feasible to mine poorer grade ores.

Most of the largest and richest mineral deposits have probably already been discovered and, in some cases, depleted. To ensure continued supplies of essential minerals, geologists are using increasingly sophisticated geophysical and geochemical mineral-exploration techniques. The U.S. Geological Survey and the U.S. Bureau of Mines continually assess the status of resources in view of changing economic and political conditions and developments in science and technology. In the following chapters, we will discuss the origin and distribution of various mineral resources and reserves.

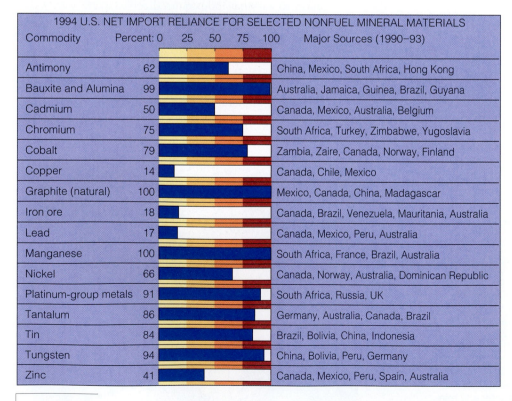

1994 U.S. NET IMPORT RELIANCE FOR SELECTED NONFUEL MINERAL MATERIALS		
Commodity	Percent	Major Sources (1990–93)
Antimony	62	China, Mexico, South Africa, Hong Kong
Bauxite and Alumina	99	Australia, Jamaica, Guinea, Brazil, Guyana
Cadmium	50	Canada, Mexico, Australia, Belgium
Chromium	75	South Africa, Turkey, Zimbabwe, Yugoslavia
Cobalt	79	Zambia, Zaire, Canada, Norway, Finland
Copper	14	Canada, Chile, Mexico
Graphite (natural)	100	Mexico, Canada, China, Madagascar
Iron ore	18	Canada, Brazil, Venezuela, Mauritania, Australia
Lead	17	Canada, Mexico, Peru, Australia
Manganese	100	South Africa, France, Brazil, Australia
Nickel	66	Canada, Norway, Australia, Dominican Republic
Platinum-group metals	91	South Africa, Russia, UK
Tantalum	86	Germany, Australia, Canada, Brazil
Tin	84	Brazil, Bolivia, China, Indonesia
Tungsten	94	China, Bolivia, Peru, Germany
Zinc	41	Canada, Mexico, Peru, Spain, Australia

FIGURE 3.20 The percentages of some mineral resources imported by the United States. The lengths of the blue bars correspond to the percentage of resources imported.

Chapter Summary

1. All matter is composed of chemical elements, each of which consists of atoms. Individual atoms have a nucleus made up of protons and neutrons, and electrons that orbit the nucleus in electron shells.

2. Atoms are characterized by their atomic number (the number of protons in the nucleus) and their atomic mass number (the number of protons plus the number of neutrons in the nucleus).

3. Bonding is the process whereby atoms are joined to other atoms. If atoms of different elements are bonded, they form a compound. Ionic and covalent bonds are most common in minerals, but metallic and van der Waals bonds occur in a few.

4. Most minerals are compounds, but a few, including gold and silver, are composed of a single element and are called native elements.

5. All minerals are crystalline solids, meaning that they possess an orderly internal arrangement of atoms.

6. Some minerals vary in chemical composition because atoms of different elements can substitute for one another provided that the electrical charge is balanced and the atoms are about the same size.

7. Most of the more than 3500 known minerals are silicates. Ferromagnesian silicates contain iron (Fe) and magnesium (Mg), and nonferromagnesian silicates lack these elements.

8. In addition to silicates, several other mineral groups are recognized, including native elements and carbonates.

9. The physical properties of minerals such as color, hardness, cleavage, and crystal form are controlled by composition and structure.

10. A few minerals are common enough constituents of rocks to be designated rock-forming minerals.

11. Many resources are concentrations of minerals of economic importance.

12. Reserves are that part of the resource base that can be extracted economically.

Important Terms

atom
atomic mass number
atomic number
bonding
carbonate mineral
cleavage
compound
covalent bond
crystalline solid
electron

electron shell
element
ferromagnesian silicate
ion
ionic bond
mineral
native element
neutron
nonferromagnesian silicate
nucleus

proton
reserve
resource
rock
rock-forming mineral
silica
silica tetrahedron
silicate

Review Questions

1. The number of protons and neutrons in an atom's nucleus determines its:
   a. _____ silica content;
   b. _____ ionic bonding potential;
   c. _____ atomic mass number;
   d. _____ type of cleavage;
   e. _____ color and luster.

2. The most common carbonate minerals are:
   a. _____ olivine and plagioclase feldspar;
   b. _____ amphibole and quartz;
   c. _____ galena and gypsum;
   d. _____ calcite and dolomite;
   e. _____ biotite and muscovite.

3. To be considered crystalline, a solid must have:
   a. _____ at least three elements bonded together;
   b. _____ an orderly internal arrangement of atoms;
   c. _____ good cleavage in two directions;
   d. _____ covalent and ionic bonds;
   e. _____ silicon and oxygen in equal amounts.

4. The type of bonding in which electrons are transferred from one atom to another is known as:
   a. _____ ionic;
   b. _____ van der Waals;
   c. _____ tetrahedral;
   d. _____ covalent;
   e. _____ silicate.

5. An atom of magnesium has 12 protons and 12 neutrons in its nucleus and 12 electrons orbiting its nucleus. Therefore, its atomic number is _____ and its atomic mass number is _____.
   a. _____ 12 and 12;
   b. _____ 24 and 24;
   c. _____ 12 and 36;
   d. _____ 36 and 24;
   e. _____ 12 and 24.

6. Minerals possessing the property known as cleavage:
   a. _____ have a higher specific gravity than other minerals;
   b. _____ are invariably dark and shiny;
   c. _____ break along closely spaced smooth planes;
   d. _____ are made up of at least three chemical elements;
   e. _____ are characterized as noble gases.

7. The specific gravity of a mineral is a measure of its:
   a. _____ melting temperature at sea level;
   b. _____ hardness and luster;
   c. _____ atomic mass number divided by its electrical charge;
   d. _____ ability to combine with other minerals to form rocks;
   e. _____ weight relative to the weight of an equal volume of water.

8. The two most abundant elements in Earth's crust are:
   a. _____ iron and magnesium;
   b. _____ carbon and potassium;
   c. _____ sodium and nitrogen;
   d. _____ silicon and oxygen;
   e. _____ manganese and beryllium.

9. All silicate minerals are made up of basic building blocks known as:
   a. _____ silicon sheets;
   b. _____ oxygen silicon cubes;
   c. _____ silica tetrahedra;
   d. _____ silicate pyramids;
   e. _____ silicon frameworks.

10. A good example of a nonferromagnesian silicate mineral is:
    a. _____ calcite;
    b. _____ quartz;
    c. _____ hematite;
    d. _____ native copper;
    e. _____ pyroxene.

11. How does a crystalline solid differ from a noncrystalline substance such as glass?

12. The minerals pyrite, galena, and halite might all have cubic crystals. How could you differentiate one from another? (See Appendix C.)

13. What is the chemical composition of the mineral olivine, and how is it possible for its composition to vary?

14. Briefly describe three ways that minerals form.

15. Why are the angles between the same crystal faces on all specimens of quartz always the same? Explain how this constancy of angles is consistent with the principle of uniformitarianism. (See Chapter 1).

16. Define a compound. Are all compounds minerals? Explain.

17. How does a native element differ from a compound, and why are native elements considered to be minerals?

18. How does a rock differ from a mineral? Are there any exceptions to the general definition of a rock? If so, what are they?

19. What is an isotope? Give an example and explain how a chemical element can have more than one isotope.

20. Compare covalent and ionic bonding. Give an example of a mineral possessing each of these bonds.

Points to Ponder

1. Why must the United States, a natural resource–rich nation, import a large part of the resources it needs? What are some of the problems created by dependence on imports?

2. Explain how the composition and structure of minerals control such mineral properties as hardness, cleavage, color, and specific gravity.

World Wide Web Activities

▶ **U.S. GEOLOGICAL SURVEY MINERAL RESOURCE SURVEYS PROGRAM**

Maintained by the U.S. Geological Survey (USGS), this site contains information on the USGS Mineral Resource Survey Program, Fact Sheets, Contacts, and numerous links to a variety of sources on mineral resources.

1. Go to the *Contacts* section. How would you use this section if you were interested in learning more about asbestos?

2. Go to the *Publications and Information Products* section in the *Mineral Information* link under the *Links* section. How would you use this section if you were interested in finding out about the supply and demand of gold in the world?

World Wide Web Activities

For these web site addresses, along with current updates and exercises, log on to

http://www.wadsworth.com/geo

▶ SMITHSONIAN GEM & MINERAL COLLECTION

This site contains a variety of photos and information about various minerals in the Smithsonian National Museum of Natural History.

▶ MINERALOGY AND PETROLOGY RESEARCH ON THE WEB

This site contains links to mineralogy and petrology journals, professional societies, databases, research groups, and mineral collecting and commercial sites.

1. Go to the *Mineralogical Databases* section and click on the *Alphabetical Mineral Reference* heading. It will take you to a listing of minerals and information about them. Look up information about the common rock-forming mineral feldspar. How many different types of feldspar are there? What are the different uses of feldspar? What are the major differences between orthoclase and plagioclase?

2. Click on the *Gem Database* link found at the beginning of *Alphabetical Mineral Reference*. Look up your birthstone or any gem that interests you. What did you learn about it that you did not already know?

▶ AMETHYST GALLERIES

Amethyst Galleries maintains this site, which has a wealth of information on and images of hundreds of minerals. It lists minerals alphabetically by name and class and has minerals in interesting groups such as gemstones and birthstones. A full text search for mineral identification by keyword is also available.

▶ WEST'S GEOLOGY DIRECTORY

This site was introduced in Chapter 1 as a comprehensive directory of geologic Internet links and an excellent place to find links on virtually any geologic topic. Scroll down and click *Minerals* and then click on *Minerals and Mineralogy*. Here you can find information on all common minerals and many others, including gemstones and jewelry. Click on any of the mineral group names listed. For example, go to the *mica group* of minerals:

1. What are the uses of biotite, muscovite, and lepidolite? Which one of these minerals is least common?

2. What types of rocks are these minerals found in?

Also go to *quartz*. What types of quartz are listed, and what are their properties?

▶ THE MINERAL GALLERY

This site has images of dozens of minerals, along with information on their chemistry, physical properties, uses, associations with other minerals, and locales where they are found. Click on *sphalerite*. What is its chemical formula, its hardness, and uses? Click on *malachite* and see where it is found, the type of rock it is found in, and its uses. Click on *topaz* and give its chemical formula, specific gravity, and uses. Also, what is the origin of the name *topaz*?

 # CD-ROM Exploration

Explore the following *In-Terra-Active 2.0* CD-ROM module(s) and increase your understanding of key concepts and processes presented in this chapter.

▷ SECTION: MATERIALS
MODULE: MINERALS

Study the five varieties of quartz shown. **Question: If you want to determine scientifically whether they are indeed the same mineral, how would you go about it? How would you defend your answer?**

▷ VIRTUAL REALITY FIELD TRIP: CHALICE GOLD MINE, WESTERN AUSTRALIA

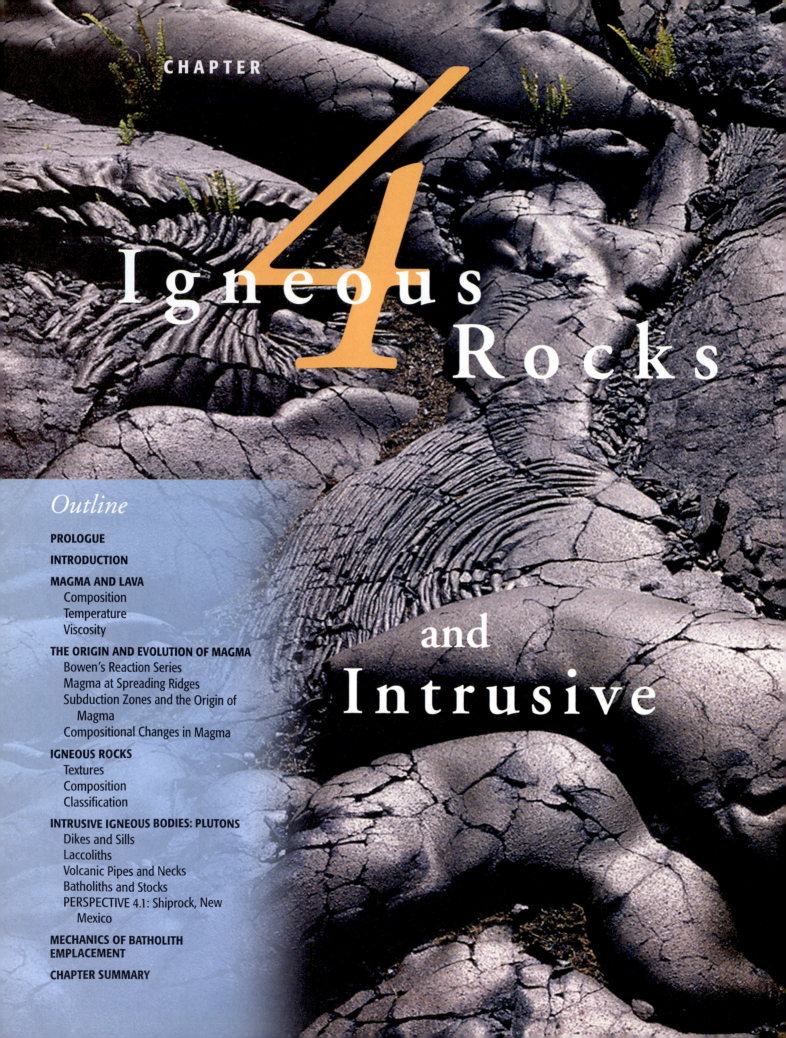

CHAPTER

4

Igneous Rocks

and Intrusive

Igneous Activity

Devil's Tower in northeastern Wyoming originated as a small igneous intrusion. It rises about 260 m above its base and can be seen from 48 km away. Some geologists think it represents an eroded remnant of magma that solidified in the conduit leading from a magma chamber to a volcano's summit. The vertical striations result from the intersection of fractures known as columnar joints. According to Cheyenne legend, however, the striations are deep scratches made by a gigantic grizzly bear.

About 45 to 50 million years ago in what is now northeastern Wyoming, several small masses of molten rock cooled and solidified in the crust, thus forming igneous rock bodies. Erosion has now exposed these rock bodies at the surface, the best known of which, Devil's Tower, was established as our first national monument by President Theodore Roosevelt in 1906. Devil's Tower is a remarkable and prominent feature of the landscape; it rises nearly 260 m above its base and is visible from 48 km away (see the photo at the left).

Devil's Tower and other similar nearby bodies are important in the legends of the Cheyenne and Lakota Sioux. These Native Americans call Devil's Tower Mateo Tepee, *which means "Grizzly Bear Lodge." It has also been called the "Bad God's Tower," and reportedly, "Devil's Tower" is a translation of this phrase. According to one legend, the tower formed when the Great Spirit caused it to rise up from the ground, carrying with it several children who were trying to escape from a gigantic grizzly bear. Another legend tells of six brothers and a woman who were also being pursued by a grizzly bear. The youngest brother carried a small rock,* and when he sang a song, the rock grew to the present size of Devil's Tower. In both legends, the bear's attempts to reach his human prey left deep scratch marks in the tower's rocks (Figure 4.1).

Geologists have a less dramatic explanation for the tower's origin. The near vertical striations (the bear's scratch marks) are simply the lines formed by the intersections of columnar joints. Columnar joints form in response to cooling and contraction in some igneous bodies and in some lava flows. Many of the columns are six sided, but columns with four, five, and seven sides are present as well. The larger columns measure about 2.5 m across. A pile of rubble at the tower's base is an accumulation of columns that have fallen from the tower.

Geologists agree that Devil's Tower originated as a small body that cooled and solidified from molten rock, and that subsequent erosion exposed it in its present form. But the type of igneous body and the extent of its modification by erosion are debatable. Some geologists think that Devil's Tower is the eroded remnant of a more extensive body of intrusive rock, whereas others think it is simply the remnant of the rock that solidified in a pipelike conduit of a volcano and that it has been little modified by erosion.

FIGURE 4.1 An artist's rendition of a Cheyenne legend about the origin of Devil's Tower.

Introduction

Rocks resulting from volcanic eruptions are widespread, but they represent only a small portion of the total rocks formed by the cooling and crystallization of molten rock material called magma. Most magma cools below the surface and forms bodies of rock known as *plutons*. The same types of magmas are involved in both volcanism and the origin of plutons, although some magmas are more mobile and more commonly reach the surface. Plutons typically underlie areas of extensive volcanism and were the sources of the overlying lavas and fragmental materials ejected from volcanoes during explosive eruptions. Furthermore, like volcanism, most plutonism takes place at or near plate margins. In this chapter we are concerned primarily with the textures, composition, and classification of igneous rocks and with plutonic or intrusive igneous activity. Volcanism will be discussed in Chapter 5.

Magma and Lava

Geologists use the term **magma** for molten rock material below the surface and **lava** for magma that reaches the surface. Because magma is less dense than the rock that melted, it tends to rise toward the surface where it may be erupted as **lava flows** or be forcefully ejected into the atmosphere as particles known as **pyroclastic materials** (from the Greek *pyro* = fire and *klastos* = broken).

Igneous rocks (from the Latin *ignis* = fire) form when magma cools and crystallizes or when pyroclastic materials such as volcanic ash (particles measuring less than 2 mm) become consolidated. Magma extruded onto the surface as lava and pyroclastic materials forms **volcanic** or **extrusive igneous** rocks, whereas magma that crystallizes within the crust forms **plutonic** or **intrusive igneous** rocks.

COMPOSITION

Recall from Chapter 3 that the most abundant minerals in Earth's crust are silicates, composed of silicon, oxygen, and the other elements listed in Table 3.3. Accordingly, when crustal rocks melt and form magma, the magma is typically silica rich and also contains considerable aluminum, calcium, sodium, iron, magnesium, and potassium, as well as many other elements in lesser quantities. Not all magmas originate by melting of crustal rocks, however; some are derived from upper mantle rocks that are composed largely of ferromagnesian silicates. A magma from this source contains comparatively less silica and more iron and magnesium.

Although silica is the primary constituent of nearly all magmas, silica content varies and serves to distinguish **felsic**, **intermediate**, and **mafic magmas** (Table 4.1). A felsic magma, for example, contains more than 65% silica and considerable sodium, potassium, and aluminum but little calcium, iron, and magnesium. In contrast to felsic magmas, mafic magmas are silica poor and contain proportionately more calcium, iron, and magnesium. As one would expect, intermediate magmas have mineral compositions intermediate between those of mafic and felsic magmas (Table 4.1).

TEMPERATURE

No direct measurements of magma temperatures below the surface have been made. Erupting lavas generally range from 1000° to 1200°C, although temperatures of 1350°C have been recorded above Hawaiian lava lakes where volcanic gases reacted with the atmosphere.

Most direct temperature measurements have been taken at volcanoes characterized by little or no explosive activity where geologists can safely approach the lava. Therefore,

little is known of the temperatures of felsic lavas because eruptions of such lavas are rare, and when they do occur, they tend to be explosive. The temperatures of some lava domes, most of which are bulbous masses of felsic magma, have been measured at a distance by using an instrument called an *optical pyrometer*. The surfaces of these domes have temperatures up to 900°C, but the exterior of a dome is probably much cooler than its interior.

When Mount St. Helens erupted in 1980, it ejected felsic magma as particulate matter in pyroclastic flows. Two weeks later, these flows still had temperatures between 300° and 420°C.

VISCOSITY

Magma is also characterized by its **viscosity**, or resistance to flow. The viscosity of some liquids, such as water, is very low; they are highly fluid and flow readily. The viscosity of some other liquids is so high that they flow much more slowly. Motor oil and syrup flow readily when hot but become stiff and flow slowly when cold. Thus, one might expect that temperature controls the viscosity of magma, and such an inference is partly correct. We can generalize and say that hot lava flows more readily than cooler lava.

Magma viscosity is also strongly controlled by silica content. In a felsic lava, numerous networks of silica tetrahedra retard flow because the strong bonds of the networks must be ruptured for flow to occur. Mafic lavas, on the other hand, contain fewer silica tetrahedra networks and consequently flow more readily. Felsic lavas form thick, slow-moving flows, whereas mafic lavas tend to form thinner flows that move rather rapidly over great distances. One mafic flow in Iceland in 1783 flowed about 80 km, and some ancient flows in the state of Washington can be traced for more than 500 km.

The Origin and Evolution of Magma

MOST people are familiar with magma that reaches Earth's surface either as lava flows or pyroclastic materials ejected during explosive volcanic eruptions. The lava flows issuing from Kilauea in Hawaii fascinate many observers, and the activity at Mount St. Helens, Washington, in 1980 and Mount Pinatubo in the Philippines in 1991 remind us of the violence of some eruptions. Yet most people have little understanding of how magma originates in the first place, how it can change or evolve, or how it reaches the surface. Indeed, many believe the misconceptions that magma comes from Earth's molten core or a continuous layer of molten material beneath the crust.

Some magma rises from depths as great as 100 to 300 km, but most of it forms at much shallower depths in the upper mantle or lower crust and accumulates in reservoirs known as **magma chambers**. The volume of a magma chamber may be several cubic kilometers of molten rock

TABLE 4.1

The Most Common Types of Magmas

TYPE OF MAGMA	SILICA CONTENT (%)
Mafic	45–52
Intermediate	53–65
Felsic	>65

within the otherwise solid lithosphere. This magma might simply cool and crystallize in place, thus forming various intrusive igneous rock bodies, but some migrates to the surface giving rise to several types of volcanism.

BOWEN'S REACTION SERIES

During the early part of this century, N. L. Bowen hypothesized that mafic, intermediate, and felsic magmas could all derive from a parent mafic magma. He knew that minerals do not all crystallize simultaneously from a cooling magma, but rather crystallize in a predictable sequence. Based on his observations and laboratory experiments, Bowen proposed a mechanism, now called **Bowen's reaction series**, to account for the derivation of intermediate and felsic magmas from a basaltic (mafic) magma (Figure 4.2). Bowen's reaction series consists of two branches: a *discontinuous branch* and a *continuous branch*. Crystallization of minerals occurs along both branches simultaneously, but for convenience we will discuss them separately.

In the discontinuous branch, which contains only ferromagnesian silicates, one mineral changes to another over specific temperature ranges (Figure 4.2). As the temperature decreases, it reaches a range where a given mineral begins to crystallize. Once a mineral forms, it reacts with

the remaining liquid magma (the melt) such that it forms the next mineral in the sequence. For example, olivine $[(Mg,Fe)_2SiO_3]$ is the first ferromagnesian silicate to crystallize. As the magma continues to cool, it reaches the temperature range at which pyroxene is stable; a reaction occurs between the olivine and the remaining melt, and pyroxene forms.

With continued cooling, a similar reaction takes place between pyroxene and the melt, and the pyroxene structure is rearranged to form amphibole. Further cooling causes a reaction between the amphibole and the melt, and its structure is rearranged so that the sheet structure typical of biotite mica forms. Although the reactions just described tend to convert one mineral to the next in the series, the reactions are not always complete. Olivine might have a rim of pyroxene, indicating an incomplete reaction. If a magma cools rapidly enough, the early-formed minerals do not have time to react with the melt, and all the ferromagnesian silicates in the discontinuous branch can be in one rock. In any case, by the time biotite has crystallized, essentially all magnesium and iron present in the original magma have been used up.

Plagioclase feldspars, which are nonferromagnesian silicates, are the only minerals in the continuous branch of Bowen's reaction series (Figure 4.2). Calcium-rich plagio-

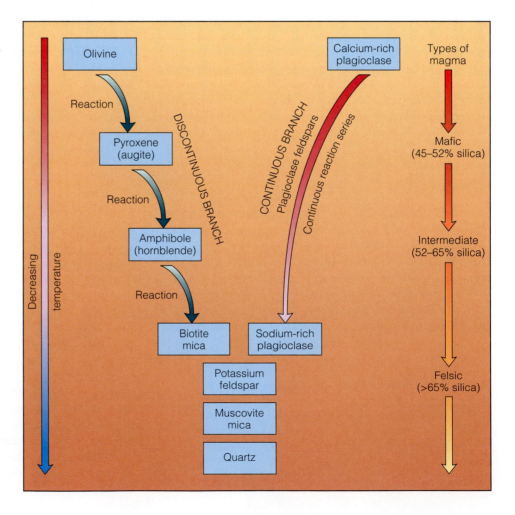

FIGURE 4.2 Bowen's reaction series. Note that it consists of a discontinuous branch and a continuous branch.

clase crystallizes first. As cooling of the magma proceeds, calcium-rich plagioclase reacts with the melt, and plagioclase containing proportionately more sodium crystallizes until all the calcium and sodium are used up. In many cases, cooling is too rapid for a complete transformation from calcium-rich to sodium-rich plagioclase to occur. Plagioclase forming under these conditions is *zoned,* meaning that it has a calcium-rich core surrounded by zones progressively richer in sodium.

Magnesium and iron, on the one hand, and calcium and sodium, on the other, are used up as crystallization occurs along the two branches in Bowen's reaction series. Accordingly, any magma left over is enriched in potassium, aluminum, and silicon. These elements combine to form potassium feldspar ($KAlSi_3O_8$), and if the water pressure is high, the sheet silicate muscovite mica will form. Any remaining magma is predominantly silicon and oxygen (silica) and forms the mineral quartz (SiO_2). The crystallization of potassium feldspar and quartz is not a true reaction series because they form independently rather than from a reaction of the orthoclase with the remaining melt.

MAGMA AT SPREADING RIDGES

One fundamental observation we can make regarding the origin of magma is that Earth's temperature increases with depth. This temperature increase, known as the *geothermal gradient,* averages about 25°C/km. Accordingly, rocks at depth are hot but remain solid because their melting temperature rises with increasing pressure. However, beneath spreading ridges, the temperature locally exceeds the melting temperature, at least in part, because pressure decreases. That is, plate separation at ridges probably causes a decrease in pressure on the already hot rocks at depth, thus initiating melting (Figure 4.3a). In addition, the presence of water can also decrease the melting temperature beneath spreading ridges because water aids thermal energy in breaking the chemical bonds in minerals (Figure 4.3b).

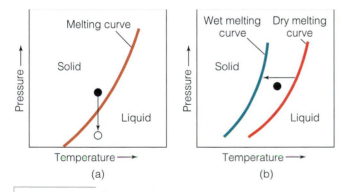

FIGURE 4.3 (a) Melting temperature rises with increasing pressure, so a decrease in pressure on already hot rocks can initiate melting. (b) The melting curve shifts to the left when water is present because it provides an additional agent to break chemical bonds.

Another explanation for the origin of spreading-ridge magmas is that localized, cylindrical plumes of hot mantle material, called *mantle plumes,* rise beneath the ridges and spread outward in all directions. Perhaps localized concentrations of radioactive minerals within the crust and upper mantle decay and provide the heat necessary to melt rocks and thus generate magma.

The magmas formed beneath spreading ridges are invariably mafic (45–52% silica). But the upper mantle rocks from which these magmas are derived are characterized as ultramafic (<45% silica), consisting largely of ferromagnesian silicates and lesser amounts of nonferromagnesian silicates. To explain how mafic magma originates from ultramafic rock, geologists propose that the magma is formed from source rock that only partially melts. This phenomenon of partial melting takes places because various minerals have different melting temperatures.

Recall the sequence of minerals in Bowen's reaction series (Figure 4.2). The order in which these minerals melt is the opposite of their order of crystallization. Accordingly, quartz, potassium feldspar, and sodium-rich plagioclase melt before most of the ferromagnesian silicates and the calcic varieties of plagioclase. So when ultramafic rock of the upper mantle begins to melt, the minerals richest in silica melt first, followed by those containing less silica. Therefore, if melting is not complete, a mafic magma containing proportionately more silica than the source rock results. Once this mafic magma forms, some of it rises to the surface where it forms lava flows, and some simply cools beneath the surface to form various intrusive igneous bodies.

SUBDUCTION ZONES AND THE ORIGIN OF MAGMA

We can make another basic observation regarding igneous activity: Where an oceanic plate is subducted beneath either a continental plate or another oceanic plate, a belt of volcanoes and plutons is found near the leading edge of the overriding plate (Figure 4.4). It would seem then that subduction and the origin of magma must be related in some way, and indeed they are. Furthermore, magma at these convergent plate boundaries is intermediate (53–65% silica) or felsic (>65% silica).

Once again, geologists invoke the phenomenon of partial melting to explain the origin and composition of magma in these areas. As a subducted plate descends toward the asthenosphere, it eventually reaches the depth where the temperature is high enough to initiate partial melting. Additionally, the wet oceanic crust descends to a depth at which dewatering takes place; as the water rises into the overlying mantle, it enhances melting, and magma forms (Figure 4.3b).

Recall that partial melting of ultramafic rock at spreading ridges yields mafic magma. Likewise, partial melting of mafic rocks of the oceanic crust yields intermediate (53–65% silica) and felsic (>65% silica) magmas, both of which are richer in silica than the source rock. Addition-

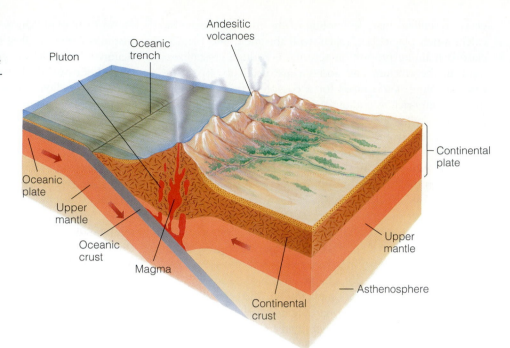

FIGURE 4.4 The subduction of an oceanic plate beneath a continental plate produces magma. Some of the magma forms intrusive igneous bodies of rock, and some is erupted to form volcanoes.

ally, some of the silica-rich sediments and sedimentary rocks of continental margins are probably carried downward with the subducted plate and contribute their silica to the magma. Also, mafic magma rising through the lower continental crust may be contaminated with silica-rich materials, which changes its composition.

COMPOSITIONAL CHANGES IN MAGMA

A magma's composition may change by **crystal settling**, which involves the physical separation of minerals by crystallization and gravitational settling (Figure 4.5). Olivine, the first ferromagnesian silicate to form in the discontinuous branch of Bowen's reaction series, has a specific gravity greater than that of the remaining magma and tends to sink downward in the melt. Accordingly, the remaining melt becomes relatively rich in silica, sodium, and potas-

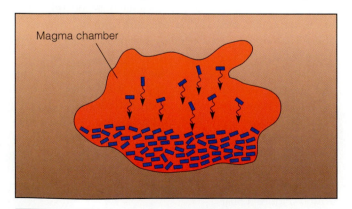

FIGURE 4.5 Chemical change in a magma by crystal settling. Early-formed ferromagnesian silicates have a specific gravity greater than that of the magma, so they settle and accumulate in the lower part of the magma chamber.

sium because much of the iron and magnesium were removed when minerals containing these elements crystallized.

Although crystal settling does take place, it does not do so on the scale envisioned by Bowen. In some thick, tabular, intrusive igneous bodies called *sills,* the first formed minerals in the reaction series are indeed concentrated. The lower parts of these bodies contain more olivine and pyroxene than the upper parts, which are less mafic. But, even in these bodies, crystal settling has yielded little felsic magma from an original mafic magma.

If felsic magma could be derived on a large scale from mafic magma as Bowen thought, there should be far more mafic magma than felsic magma. For crystal settling to yield a particular volume of granite (a felsic igneous rock), about ten times as much mafic magma would have to be present initially. If this were so, then mafic intrusive igneous rocks should be much more common than felsic ones. Just the opposite is the case, however, so it appears that mechanisms other than crystal settling must account for the large volume of felsic magma. Partial melting of mafic oceanic crust and silica-rich sediments of continental margins during subduction yields magma richer in silica than the source rock. Furthermore, magma rising through the continental crust can absorb some felsic materials by assimilation and become more enriched in silica.

The composition of a magma can also be changed by **assimilation**, a process whereby a magma reacts with preexisting rock, called *country rock,* with which it comes in contact (Figure 4.6). The walls of a volcanic conduit or magma chamber are of course heated by the adjacent magma, which may reach temperatures of 1300°C. Some of these rocks can be partly or completely melted, provided their melting temperature is less than that of the magma. Because the assimilated rocks seldom have the same com-

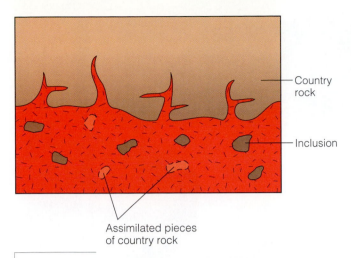

FIGURE 4.6 As magma moves upward, fragments of country rock are dislodged and settle into the magma. If they have a lower melting temperature than the magma, they may be incorporated into the magma by assimilation. Incompletely assimilated pieces of country rock are inclusions.

position as the magma, the composition of the magma is changed.

The fact that assimilation takes place can be demonstrated by *inclusions,* incompletely melted pieces of rock that are fairly common within igneous rocks. Many inclusions were simply wedged loose from the country rock as the magma forced its way into preexisting fractures (Figures 4.6 and 4.7).

No one doubts that assimilation occurs, but its effect on the bulk composition of most magmas must be slight. The reason is that the heat for melting must come from the magma itself, and this would have the effect of cooling the magma. Consequently, only a limited amount of rock can

be assimilated by a magma, and that amount is usually insufficient to bring about a major compositional change.

Neither crystal settling nor assimilation can produce a significant amount of felsic magma from a mafic one. But the two processes operating concurrently can change the composition of a mafic magma much more than either process acting alone. Some geologists think that this is one way that many intermediate magmas form where oceanic lithosphere is subducted beneath continental lithosphere.

The fact that a single volcano can erupt lavas of different composition indicates that magmas of differing composition must be present. It seems likely that some of these magmas would come into contact and mix with one another. If this is the case, we would expect that the composition of the magma resulting from **magma mixing** would be a modified version of the parent magmas. Suppose a rising mafic magma mixes with a felsic magma of about the same volume (Figure 4.8). The resulting "new" magma would have a more intermediate composition.

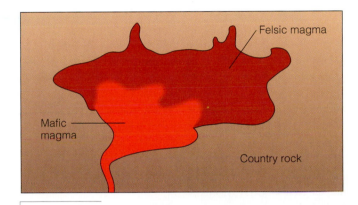

FIGURE 4.8 Magma mixing. Two rising magmas mix and produce a magma with a composition different from either of the parent magmas.

FIGURE 4.7 Dark-colored inclusions in granitic rock in California. *(Photo courtesy of David J. Matty.)*

Igneous Rocks

ALL intrusive and many extrusive igneous rocks form when minerals crystallize from magma. The process of crystallization involves the formation and subsequent growth of crystal nuclei. The atoms in a magma are in constant motion, but when cooling begins, some atoms bond to form small groups, or nuclei, whose arrangement of atoms corresponds to the arrangement in mineral crystals. As other atoms in the liquid chemically bond to these nuclei, they do so in an ordered geometric arrangement, and the nuclei grow into crystalline *mineral grains,* the individual particles that comprise a rock. During rapid cooling, the rate of nuclei formation exceeds the rate of growth, and an aggregate of many small grains results (Figure 4.9a). With slow cooling, the rate of growth exceeds the rate of nucleation, so relatively large grains form (Figure 4.9b).

TEXTURES

Several textures of igneous rocks are related to the cooling history of a magma or lava. Rapid cooling, as occurs in lava flows or some near-surface intrusions, results in a fine-grained texture termed **aphanitic**. In an aphanitic texture, individual mineral grains are too small to be observed without magnification (Figure 4.10a). In contrast, igneous rocks with a coarse-grained, or **phaneritic**, texture have mineral grains that are easily visible without magnification (Figure 4.10b). Such large mineral grains indicate slow cool-

FIGURE 4.10 Textures of igneous rocks. (a) Aphanitic (fine-grained) texture in which individual minerals are too small to be seen without magnification. (b) Phaneritic (coarse-grained) texture in which minerals are easily discerned without magnification. (c) Porphyritic texture consisting of minerals of markedly different sizes. *(Photos courtesy of Sue Monroe.)*

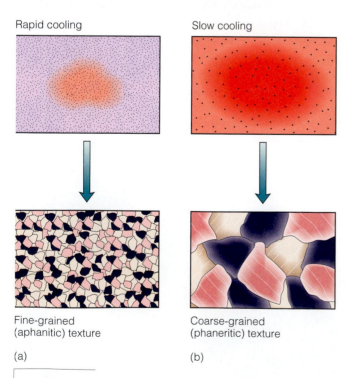

Fine-grained (aphanitic) texture

Coarse-grained (phaneritic) texture

(a) (b)

FIGURE 4.9 The effect of the cooling rate of a magma on nucleation and growth of crystals. (a) Rapid cooling results in many small grains and a fine-grained, or aphanitic, texture. (b) Slow cooling results in a coarse-grained, or phaneritic, texture.

ing and generally an intrusive origin; a phaneritic texture can develop in the interiors of some thick lava flows as well.

Rocks with a **porphyritic** texture, a combination of mineral grains of markedly different sizes, have a somewhat more complex cooling history. The larger grains are *phenocrysts,* and the smaller ones are referred to as *groundmass* (Figure 4.10c). Suppose a magma begins cooling slowly as an intrusive body and some mineral crystal nuclei form and begin to grow. Suppose further, before the magma has completely crystallized, the remaining liquid phase and solid mineral grains within it are extruded onto the surface where it cools rapidly, forming an aphanitic texture. The resulting igneous rock would have large mineral grains (phenocrysts) suspended in a finely crystalline groundmass, and the rock would be characterized as a *porphyry.*

A lava may cool so rapidly that its constituent atoms do not have time to become arranged in the ordered, three-dimensional frameworks typical of minerals. As a result of such rapid cooling, a *natural glass* such as *obsidian* forms (Figure 4.11a). Even though obsidian with its glassy texture is not composed of minerals, it is still considered to be an igneous rock.

Some magmas contain large amounts of water vapor and other gases. These gases may be trapped in cooling

A **pyroclastic**, or **fragmental**, **texture** characterizes igneous rocks formed by explosive volcanic activity. Ash may be discharged high into the atmosphere and eventually settle to the surface where it accumulates; if it is turned into rock, it is considered to be a pyroclastic igneous rock.

COMPOSITION

Igneous rocks, just as magmas, are characterized as mafic (45–52% silica), intermediate (53–65% silica), or felsic (>65% silica). The parent magma plays an important role in determining the mineral composition of igneous rocks, yet the same magma can yield a variety of igneous rocks because its composition can change as a result of the sequence in which minerals crystallize, crystal settling, assimilation, and magma mixing.

CLASSIFICATION

Most igneous rocks are classified on the basis of their textures and composition. Notice in Figure 4.12 that all of the rocks, except peridotite, constitute pairs; the members of a pair have the same composition but different textures. Basalt and gabbro, andesite and diorite, and rhyolite and granite are compositional (mineralogical) equivalents, but basalt, andesite, and rhyolite are aphanitic and most commonly extrusive, whereas gabbro, diorite, and granite have phaneritic textures that generally indicate an intrusive origin.

The igneous rocks shown in Figure 4.12 are also differentiated by composition. Reading across the chart from rhyolite to andesite to basalt, for example, the relative proportions of nonferromagnesian and ferromagnesian silicates differ. The differences in composition are gradual, however, so that a compositional continuum exists. In

FIGURE **4.11** (a) The natural glass obsidian forms when lava cools too quickly for mineral crystals to form. (b) Vesicular texture. *(Photos courtesy of Sue Monroe.)*

lava where they form numerous small holes or cavities known as **vesicles**; rocks possessing numerous vesicles are termed *vesicular,* as in vesicular basalt (Figure 4.11b).

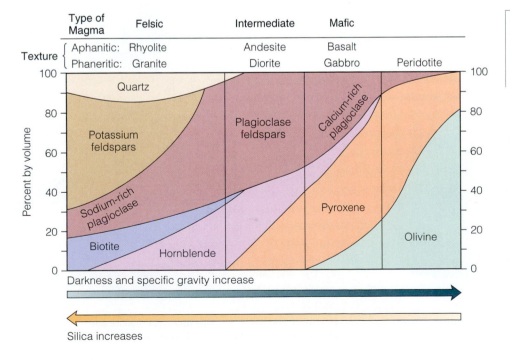

FIGURE **4.12** Classification of igneous rocks. The diagram illustrates the relative proportions of the chief mineral components of common igneous rocks.

other words, rocks exist with compositions intermediate between rhyolite and andesite, and so on.

ULTRAMAFIC ROCKS Ultramafic rocks (<45% silica) are composed largely of ferromagnesian silicate minerals. The ultramafic rock *peridotite* contains mostly olivine, lesser amounts of pyroxene, and generally a little plagioclase feldspar (Figure 4.12). Another ultramafic rock *(pyroxenite)* is composed predominantly of pyroxene. Because these minerals are dark colored, the rocks are generally black or dark green. Peridotite is thought to be the rock type composing the upper mantle, but ultramafic rocks are rare at the surface. Ultramafic rocks are generally thought to have originated by concentration of the early-formed ferromagnesian silicates that separated from mafic magmas.

BASALT-GABBRO *Basalt* and *gabbro* are the fine-grained and coarse-grained rocks, respectively, that crystallize from mafic magmas (45–52% silica) (Figure 4.13). Both have the same composition—mostly calcium-rich plagioclase and pyroxene, with smaller amounts of olivine and amphibole (Figure 4.12). Because they contain a large proportion of ferromagnesian silicates, basalt and gabbro are dark colored; those that are porphyritic typically contain calcium plagioclase or olivine phenocrysts.

Basalt is generally considered to be the most common extrusive igneous rock. Extensive basalt lava flows were erupted in vast areas in Washington, Oregon, Idaho, and northern California (see Chapter 5). Oceanic islands such as Iceland, the Galapagos, the Azores, and the Hawaiian Islands are composed mostly of basalt, as is the upper part of the oceanic crust.

Gabbro is much less common than basalt, at least in the continental crust or where it can be easily observed. Small intrusive bodies of gabbro are present in the continental crust, but intermediate to felsic intrusive rocks such as diorite and granite are

(a)

(b)

FIGURE 4.13 Mafic igneous rocks: (a) basalt and (b) gabbro. *(Photos courtesy of Sue Monroe.)*

much more common. The lower part of the oceanic crust is composed of gabbro, however.

ANDESITE-DIORITE Magmas intermediate in composition (53–65% silica) crystallize to form *andesite* and *diorite,* which are compositionally equivalent fine- and coarse-grained igneous rocks. Andesite and diorite are composed predominantly of plagioclase feldspar, with the typical ferromagnesian component being amphibole or biotite (Figure 4.12).

Andesite is a common extrusive igneous rock formed from lavas erupted in volcanic island arcs at convergent plate margins. The volcanoes of the Andes Mountains of South America and the Cascade Range in western North America are composed in part of andesite. Intrusive bodies composed of diorite are fairly common in the continental crust.

RHYOLITE-GRANITE *Rhyolite* and *granite* (>65% silica) crystallize from felsic magmas and are therefore silica-rich rocks (Figure 4.14). They consist largely of potassium feldspar, sodium-rich plagioclase, and quartz, with perhaps some biotite and rarely amphibole (Figure 4.12). Because nonferromagnesian silicates predominate, these rocks are generally light colored. Rhyolite is fine grained, although most often it contains phenocrysts of potassium feldspar or quartz, and granite is coarse grained. Granite porphyry is also fairly common.

Rhyolite lava flows are much less common than andesite and basalt flows. Recall that the greatest control of viscosity in a magma is the silica content. Thus, if a felsic magma rises to the surface, it begins to cool, the pressure on it decreases, and gases are released explosively, usually yielding rhyolitic pyroclastic materials. The rhyolitic lava flows that do occur are thick and highly viscous and move only short distances.

Granitic rocks are by far the

(a)

(b)

FIGURE 4.14 Felsic igneous rocks: (a) rhyolite and (b) granite. *(Photos courtesy of Sue Monroe.)*

most common intrusive igneous rocks, although they are restricted to the continents. Most granitic rocks were intruded at or near convergent plate margins during episodes of mountain building. When these mountainous regions are uplifted and eroded, the vast bodies of granitic rocks forming their cores are exposed. The granitic rocks of the Sierra Nevada of California form a composite body measuring about 640 km long and 110 km wide, and the granitic rocks of the Coast Ranges of British Columbia, Canada, are much more voluminous.

PEGMATITE *Pegmatite* is a very coarsely crystalline igneous rock. It contains minerals measuring at least 1 cm across, and many crystals are much larger (Figure 4.15). The name pegmatite refers to texture rather than a specific composition, but most pegmatites are composed largely of quartz, potassium feldspar, and sodium-rich plagioclase—a composition similar to granite. Many pegmatites are associated with granite plutons and appear to represent the minerals that formed from the fluid and vapor phases that remained after most of the granite crystallized.

The water-rich vapor phase that exists after most of a magma has crystallized as granite has properties that differ from the magma from which it separated. It has a lower density and viscosity and commonly invades the country rock where it crys-

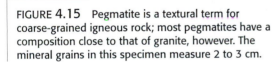

FIGURE 4.15 Pegmatite is a textural term for coarse-grained igneous rock; most pegmatites have a composition close to that of granite, however. The mineral grains in this specimen measure 2 to 3 cm.

tallizes. The water-rich vapor phase ordinarily contains a number of elements that rarely enter into the common minerals that form granite.

The formation and growth of mineral crystal nuclei in pegmatites are similar to those processes in magma, but with one critical difference: The vapor phase from which pegmatites crystallize inhibits the formation of nuclei. Some nuclei do form, however, and because the appropriate ions in the liquid can move easily and attach themselves to a growing crystal, individual mineral grains have the opportunity to grow to very large sizes, several meters long in some cases.

OTHER IGNEOUS ROCKS Some igneous rocks, including tuff, volcanic breccia, obsidian, pumice, and scoria, are identified solely by their textures (Figure 4.16). Much of the fragmental material erupted by volcanoes is *ash*, a desig-

Composition		Felsic ⟷ Mafic	
Texture	Vesicular	Pumice	Scoria
	Glassy	Obsidian	
	Pyroclastic or Fragmental	⟵ Volcanic Breccia ⟶	
		Tuff/welded tuff	

FIGURE 4.16 Classification of igneous rocks for which texture is the main consideration.

nation for pyroclastic materials less than 2.0 mm in diameter, most of which is broken pieces or shards of volcanic glass. The consolidation of ash forms the pyroclastic rock *tuff*. Some ash flows are so hot that as they come to rest, the ash particles fuse together and form a *welded tuff*. Consolidated deposits of larger pyroclastic materials such as cinders, blocks, and bombs, are *volcanic breccia*.

Both *obsidian* and *pumice* are varieties of volcanic glass. Obsidian may be black, dark gray, red, or brown, the color depending on the presence of tiny particles of iron minerals (Figure 4.17a). Analyses of numerous samples

FIGURE 4.17 (a) Obsidian, (b) pumice, and (c) scoria. *(Photos courtesy of Sue Monroe.)*

indicate that most obsidian has a high silica content and is compositionally similar to rhyolite.

Pumice is a variety of volcanic glass containing numerous bubble-shaped vesicles that develop when gas escapes through lava and forms a froth (Figure 4.17b). Some pumice forms as crusts on lava flows, and some forms as particles erupted from explosive volcanoes. If pumice falls into water, it can be carried great distances because it is so porous and light that it floats. Another vesicular rock is scoria, which has more vesicles than solid rock (Figure 4.17c).

Intrusive Igneous Bodies: Plutons

Intrusive igneous bodies, or **plutons**, form when magma cools and crystallizes within the crust (Figure 4.18). Geologists face a special challenge in studying the origins of plutons for, unlike extrusive or volcanic activity that can be observed, intrusive igneous activity can be studied only indirectly. Although plutons can be observed after erosion has exposed them at the surface, we cannot duplicate the conditions that existed deep in the crust when they formed, except in small-scale laboratory experiments.

Several types of plutons are recognized, all of which are defined by their geometry (three-dimensional shape) and their relationship to the intruded rock (Figure 4.18). The shapes of plutons are characterized as massive or irregular, tabular, cylindrical, or mushroom shaped. Plutons are also described as concordant or discordant. A **concordant** pluton, such as a sill, has boundaries that parallel the layering in the intruded rock, or what is commonly called the *country rock*. A **discordant** pluton, such as a dike, has boundaries that cut across the layering of the country rock (Figure 4.18).

DIKES AND SILLS

Both **dikes** and **sills** are tabular or sheetlike plutons, but dikes are discordant whereas sills are concordant (Figure 4.18). Dikes are common intrusive features, most of which are small—measuring 1 or 2 m across—but they range from a few centimeters to more than 100 m thick. Dikes are emplaced within zones of weakness where fractures exist or where the fluid pressure is great enough for them to form their own fractures during emplacement.

Erosion of the Hawaiian volcanoes exposes dikes in rift zones, the large fractures that cut across these volcanoes.

FIGURE **4.18** Block diagram showing the various types of plutons. Notice that some of these plutons cut across the layering in the country rock and are thus discordant, whereas others parallel the layering and are concordant.

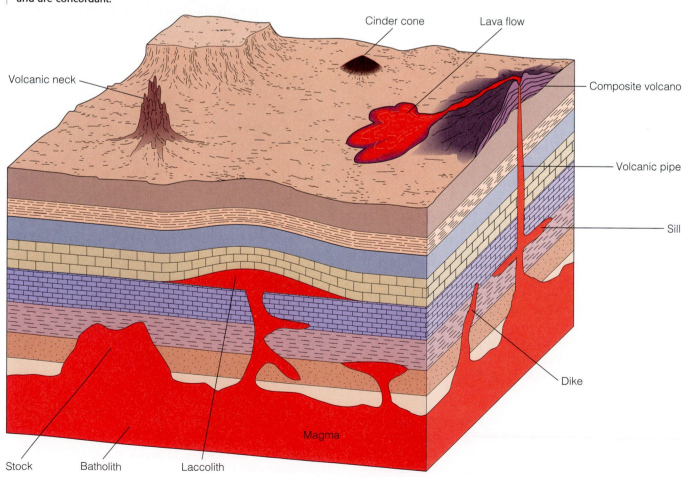

Volcanic neck

Cinder cone

Lava flow

Composite volcano

Volcanic pipe

Sill

Dike

Magma

Stock   Batholith   Laccolith

The Columbia River basalts in Washington issued from long fissures, and the magma that cooled in the fissures formed dikes. Some of the large historic fissure eruptions are underlain by dikes; for example, dikes underlie both the Laki fissure which erupted during 1783 in Iceland and the Eldgja fissure, also in Iceland, where eruptions occurred in A.D. 950 from a fissure 300 km long.

Sills are concordant plutons, many of which are a meter or less thick, although some are much thicker (Figure 4.18). A well-known sill in the United States is the Palisades sill that forms the Palisades along the west side of the Hudson River in New York and New Jersey. It is exposed for 60 km along the river and is up to 300 m thick.

Most sills have been intruded into sedimentary rocks, but eroded volcanoes also reveal that sills are commonly injected into piles of volcanic rocks. In fact, some of the inflation of volcanoes preceding eruptions may be caused by the injection of sills.

In contrast to dikes, which follow zones of weakness, sills are emplaced when the fluid pressure is so great that the intruding magma actually lifts the overlying rocks. Because emplacement requires fluid pressure exceeding the force exerted by the weight of the overlying rocks, sills are typically shallow intrusive bodies.

LACCOLITHS

Laccoliths are similar to sills in that they are concordant, but instead of being tabular, they have a mushroomlike geometry (Figure 4.18). They tend to have a flat floor and are domed up in their central part. Like sills, laccoliths are rather shallow intrusive bodies that actually lift up the overlying rocks. In this case, however, the overlying rock layers are arched upward over the pluton (Figure 4.18). Most laccoliths are rather small bodies. The best-known laccoliths in the United States are in the Henry Mountains of southeastern Utah.

VOLCANIC PIPES AND NECKS

The conduit connecting the crater of a volcano with an underlying magma chamber is a **volcanic pipe** (Figure 4.18). In other words, it is the structure through which magma rises to the surface. When a volcano ceases to erupt, it is eroded as it is attacked by water, gases, and acids. The volcanic mountain eventually erodes away, but the magma that solidified in the pipe is commonly more resistant to weathering and erosion and is often left as an erosional remnant, a **volcanic neck** (Figure 4.18). A number of volcanic necks are present in the southwestern United States, especially in Arizona and New Mexico (see Perspective 4.1), and others are recognized elsewhere.

BATHOLITHS AND STOCKS

Batholiths are the largest intrusive bodies. By definition they must have at least 100 km² of surface area, and most

FIGURE **4.19** View of granitic rocks in part of the Sierra Nevada batholith in Yosemite National Park, California. The near vertical cliff is El Capitan, meaning "The Chief." It rises more than 900 m above the valley floor, making it the tallest unbroken cliff in the world! *(Photo courtesy of Richard L. Chambers.)*

are much larger than this (Figure 4.18). **Stocks** have the same general features as batholiths but are smaller, although some stocks are simply the exposed parts of much larger intrusions. Batholiths are generally discordant, and most consist of multiple intrusions. In other words, a batholith is a large composite body produced by repeated, voluminous intrusions of magma in the same area. The coastal batholith of Peru, for instance, was emplaced over a period of 60 to 70 million years and consists of perhaps as many as 800 individual plutons.

The igneous rocks composing batholiths are mostly granitic, although diorite may also be present. Most batholiths are emplaced along convergent plate margins. One example is the Coast Range batholith of British Colombia, Canada. It was emplaced over a period of millions of years. Later uplift and erosion exposed this huge composite pluton at the surface. Other large batholiths in North America include the Idaho batholith and the Sierra Nevada batholith in California (Figure 4.19).

Shiprock, New Mexico

According to Navajo legend, a young man named Nayenezgani asked his grandmother where the mythical birdlike creatures known as Tse'na'hale lived. She replied, "They dwell at Tsaebidahi," which means Winged Rock or Rock with Wings. We know Winged Rock as Shiprock, a volcanic neck rising nearly 550 m above the surrounding plain (Figure 1). Radiating outward from this conical volcanic neck are three dikes. Navajo legend holds that Winged Rock represents a giant bird that brought the Navajo people from the north and that the dikes are snakes that have turned to stone.

Shiprock is the most impressive of many volcanic necks exposed in the Four Corners region of the southwestern United States. (Four Corners is a designation for the point where the boundaries of Colorado, Utah, Arizona, and New Mexico converge.) Shiprock is visible from as far as 160 km and was a favorite with rock climbers for many years until the Navajo ended all climbing on the reservation.

The country rock penetrated by this volcanic neck includes ancient metamorphic and igneous rocks and about 1000 m of overlying sedimentary rocks. The rock unit exposed at the surface is the Mancos Shale, a sedimentary rock unit composed mostly of mud that was deposited in an arm of the sea that existed in North America during the Cretaceous Period. Absolute dating of one of the dikes indicates that the magma that solidified to form the dike was emplaced about 27 million years ago.

Shiprock is one of several volcanic necks in the Navajo volcanic field that formed as a result of explosive eruptions. During these eruptions, volcanic materials and large pieces of country rock torn from the vent walls were hurled high into the air and fell randomly around the area. The material composing Shiprock itself is characterized as a tuff-breccia

Figure 1 Shiprock, a volcanic neck in northwestern New Mexico, rises nearly 550 m above the surrounding plain. This view shows one of the dikes radiating from Shiprock. (Photo courtesy of Frank Hanna.)

consisting of fragmental volcanic debris, along with inclusions of various sedimentary rocks and some granite and metamorphic rocks. Because Shiprock now stands about 550 m above the surrounding plain, at least that much erosion must have taken place to expose it in its present form. We can only speculate on how much higher and larger it was when it was part of an active volcano.

The dikes radiating from Shiprock (Figure 1) formed when magma ascended rather quietly and was emplaced

in the country rock. The fractures along which this magma rose, however, may have formed as a result of the explosive emplacement of the tuff-breccia that filled the volcanic vent. The dike on the northeast side of Shiprock extends for more than 2900 m outward from the vent and averages 2.3 m thick. Because the dike rock, like the material composing the volcanic neck, is more resistant to erosion than the adjacent Mancos Shale, the dikes stand as near-vertical walls above the surrounding plain (Figure 1).

A number of mineral resources are found in rocks of batholiths and stocks and in the country rocks adjacent to them. Granitic rocks are the primary source of gold, which forms from mineral-rich solutions moving through cracks and fractures of the igneous body. The copper deposits at Butte, Montana, are in rocks near the margins of the granitic rocks of the Boulder batholith. Near Salt Lake City, Utah, copper is mined from the mineralized rocks adjacent to the Bingham stock, a composite pluton composed of granite and granite porphyry.

Mechanics of Batholith Emplacement

GEOLOGISTS realized long ago that the emplacement of batholiths posed a space problem; that is, what happened to the rock that formerly occupied the space now occupied by a batholith? One proposed answer was that no displacement had occurred but that batholiths had been formed in place by alteration of the country rock through a process called *granitization*. According to this view, granite did not originate as a magma but rather from hot, ion-rich solutions that simply altered the country rock and transformed it into granite. Granitization is a solid-state phenomenon so it is essentially an extreme type of metamorphism (see Chapter 8).

Most geologists think that only small quantities of granite are formed by granitization and that it cannot account for the huge granite batholiths of the world. These geologists think an igneous origin for granite is clear, but then they must deal with the space problem. One solution is that these large igneous bodies melted their way into the crust. In other words, they simply assimilated the country rock as they moved up (Figure 4.6). The presence of inclusions, especially near the tops of such intrusive bodies, indicates that assimilation does occur. Nevertheless, as we noted previously, assimilation is a limited process because magma is cooled as country rock is assimilated; calcula-

tions indicate that far too little heat is available in a magma to assimilate the huge quantities of country rock necessary to make room for a batholith.

Geologists now generally agree that batholiths were emplaced by forceful injection as magma moved up toward the surface. Recall that granite is derived from viscous felsic magma and therefore rises slowly. It appears that the magma deforms and shoulders aside the country rock, and as it rises further, some of the country rock fills the space beneath the magma. A somewhat analogous situation occurs when large masses of sedimentary rock known as rock salt rise through the overlying rocks to form salt domes.

Salt domes are recognized in several areas of the world, including the Gulf Coast of the United States. Layers of rock salt exist at some depth, but salt is less dense than most other types of rock materials. When under pressure, it rises toward the surface even though it remains solid, and as it moves up, it pushes aside and deforms the country rock (Figure 4.20). Natural examples of rock salt flowage are known, and it can easily be demonstrated experimentally. In the arid Middle East, for example, salt moving upward in the manner described actually flows out at the surface.

Some batholiths do indeed show evidence of having been emplaced forcefully by shouldering aside and deforming the country rock. This mechanism probably occurs in the deeper parts of the crust where temperature and pressure are high and the country rocks are easily deformed in the manner described. At shallower depths, however, the crust is more rigid and tends to deform by fracturing. In this environment, batholiths may be emplaced by **stoping**, a process in which magma detaches and engulfs pieces of country rock (Figure 4.21). According to this concept, magma moves up along fractures and the planes separating layers of country rock. Eventually, pieces of country rock are detached and settle into the magma. No new room is created during stoping; the magma simply fills the space formerly occupied by country rock (Figure 4.21).

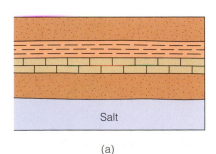
(a)

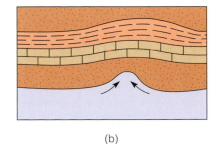
(b)

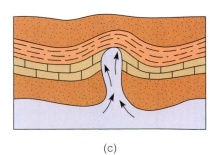
(c)

FIGURE 4.20 Three stages in the origin of a salt dome. Rock salt is a low-density sedimentary rock that (a) when deeply buried (b) tends to rise toward the surface, (c) pushing aside and deforming the country rock and forming a dome. Salt domes are thought to originate in much the same manner as batholiths are intruded into Earth's crust.

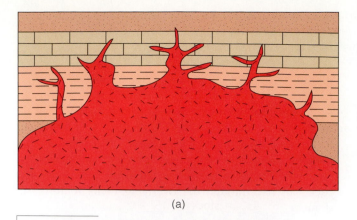
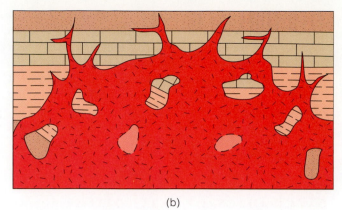

(a)

(b)

FIGURE 4.21 Emplacement of a batholith by stoping. (a) The magma is injected into the country rock along fractures and planes between layers. (b) Blocks of country rock are detached and engulfed in the magma by stoping. Some of these blocks may be assimilated.

Chapter Summary

1. Molten rock material below the surface is magma, whereas lava is magma that reaches the surface. The silica content of magma varies and serves to differentiate felsic, intermediate, and mafic magmas.

2. The viscosity of magma depends on its temperature and especially its composition. Silica-rich (felsic) magma is much more viscous than silica-poor (mafic) magma.

3. Minerals crystallize from magma and lava when small crystal nuclei form and grow.

4. Under ideal cooling conditions, a mafic **magma yields a sequence of different miner**als that are stable within specific temperature ranges. This sequence, called Bowen's re-action series, consists of a discon-tinuous branch and a continuous branch.
   a. The discontinuous branch con-tains only ferromagnesian sili-cates, each of which reacts with the melt to form the next min-eral in the sequence.
   b. The continuous branch involves changes only in plagioclase feldspar as sodium replaces cal-cium in the crystal structure.

5. The ferromagnesian silicates that form first in Bowen's reaction se-ries can settle and become concen-trated near the base of a magma chamber or intrusive body. Such settling of iron- and magnesium-rich minerals causes a chemical change in the remaining melt.

6. A magma can be changed compo-sitionally when it assimilates coun-try rock, but this process usually has only a limited effect. Magma mixing may also bring about com-positional changes in magmas.

7. Volcanic rocks generally have aphanitic textures because of rapid cooling, whereas slow cooling and phaneritic textures characterize plutonic rocks. Igneous rocks with a porphyritic texture have mineral crystals of markedly different sizes. Other igneous rock textures in-clude glassy, vesicular, and pyro-clastic.

8. The composition of igneous rocks is determined largely by the com-position of the parent magma. It is possible, however, for an individ-ual magma to yield igneous rocks of differing compositions.

9. Most igneous rocks are classified on the basis of their textures and composition. Two fundamental groups of igneous rocks are recog-nized: volcanic or extrusive rocks, and plutonic or intrusive rocks.
   a. Common volcanic rocks include rhyolite, andesite, and basalt.

   b. Common plutonic rocks include granite, diorite, and gabbro.

10. A few volcanic rocks, such as tuff, obsidian, and pumice, are classified by their textures only.

11. Pegmatites are coarse-grained ig-neous rocks, most of which have an overall composition similar to that of granite. Crystallization from a vapor-rich phase left over after the crystallization of granite ac-counts for the very large mineral crystals in pegmatites.

12. Plutons are igneous bodies that formed in place or were intruded into the crust. Various types of plutons are classified by their geometry and whether they are concordant or discordant.

13. Common plutons include dikes (tabular geometry, discordant); sills (tabular geometry, concordant); volcanic necks (cylindrical geome-try, discordant); laccoliths (mush-room shaped, concordant); and batholiths and stocks (irregular geometry, discordant).

14. By definition batholiths must have at least 100 km^2 of surface area; stocks are similar to batholiths but smaller. Many batholiths are large composite bodies consisting of many plutons emplaced over a long period of time.

15. Most batholiths appear to have formed in the cores of mountain ranges during episodes of mountain building.

16. Some geologists think that granite batholiths are emplaced when felsic magma moves up and shoulders aside and deforms the country

rock. The upward movement of rock salt and the formation of salt domes provide a somewhat analogous situation.

Important Terms

aphanitic
assimilation
batholith
Bowen's reaction series
concordant
crystal settling
dike
discordant
felsic magma
igneous rock
intermediate magma

laccolith
lava
lava flow
mafic magma
magma
magma chamber
magma mixing
phaneritic
pluton
plutonic (intrusive igneous) rock
porphyritic

pyroclastic materials
pyroclastic (fragmental) texture
sill
stock
stoping
vesicle
viscosity
volcanic neck
volcanic pipe
volcanic (extrusive igneous) rock

Review Questions

1. A sill is a _____ pluton having a(an) _____ geometry.
   a. _____ discordant/irregular;
   b. _____ concordant/tabular;
   c. _____ concordant/cylindrical;
   d. _____ discordant/mushroom-shaped;
   e. _____ discordant/irregular.

2. Volcanic rocks can usually be distinguished from plutonic rocks by:
   a. _____ color;
   b. _____ composition;
   c. _____ iron-magnesium content;
   d. _____ size of their mineral grains;
   e. _____ specific gravity.

3. A mafic magma is one having:
   a. _____ 45–52% silica;
   b. _____ 65–80% silica;
   c. _____ 53–65% silica;
   d. _____ less than 45% silica;
   e. _____ more than 80% silica.

4. A magma with a two-stage cooling history commonly has a(an) _____ texture.
   a. _____ porphyritic;
   b. _____ aphanitic;
   c. _____ fragmental;
   d. _____ glassy;
   e. _____ phaneritic.

5. In a continuous branch of Bowen's reaction series, the first plagioclase feldspar to crystallize is rich in the element:
   a. _____ iron;
   b. _____ calcium;
   c. _____ potassium;
   d. _____ aluminum;
   e. _____ magnesium.

6. Assimilation is a process involving:
   a. _____ accumulation of magma in a chamber;
   b. _____ rapid flow of silica-rich lava;
   c. _____ magma reacting with an incorporating preexisting rock;
   d. _____ emplacement of a cylindrical, discordant pluton;
   e. _____ erosion of a volcano and the origin of a volcanic neck.

7. Which of the following igneous rocks have an aphanitic (fine-grained) texture?
   a. _____ tuff-rhyolite-gabbro;
   b. _____ scoria-obsidian-granite;
   c. _____ diorite-basalt-peridotite;
   d. _____ pumice-pegmatite-andesite;
   e. _____ basalt-andesite-rhyolite.

8. Pegmatite has an extremely coarse-grained texture and a composition corresponding closely to that of:
   a. _____ granite;
   b. _____ andesite;
   c. _____ gabbro;
   d. _____ diorite;
   e. _____ basalt.

9. Which one of the following statements is correct?
   a. _____ granite is the most common igneous rock in the ocean basins;
   b. _____ a continuous layer of magma is present beneath Earth's crust;
   c. _____ a discordant pluton is one with boundaries paralleling the layering of the intruded rocks;
   d. _____ quartz is likely to be separated from a mafic magma by crystal settling;
   e. _____ the greater the silica content of a magma, the greater its viscosity.

10. The mafic magma at spreading ridges probably originates by:
    a. _____ heat generated as one plate is subducted beneath another;
    b. _____ partial melting of ultramafic rock of the upper mantle;
    c. _____ assimilation of silica-rich rock of the continental crust;
    d. _____ dewatering and melting of silica-rich sediments at continental margins;
    e. _____ incomplete melting of granite batholiths.

11. Describe the mineral composition and texture of two intrusive and two extrusive igneous rocks.

12. Why is magma originating at subduction zones richer in silica than magma generated at spreading ridges?

13. What are crystal settling and assimilation, and how do they change the composition of a magma?

14. Describe the conditions under which aphanitic (fine-grained), phaneritic (coarse-grained), and glassy textures develop in igneous rocks. Also, give an example of an igneous rock with each texture and specify whether it is an intrusive or extrusive rock.

15. What is viscosity, and what are the primary controls on viscosity?

16. Compare the continuous and discontinuous branches of Bowen's reaction series.

17. What is a batholith? Briefly explain how and where batholiths form.

18. How can partial melting of ultramafic rock of the upper mantle yield mafic magma that is richer in silica?

19. What kind of evidence indicates that assimilation actually takes place?

20. Describe the texture and composition of tuff and explain how it originates.

Points to Ponder

1. In the discontinuous branch of Bowen's reaction series, olivine forms in a specific temperature range, but as the magma continues to cool, it reacts with the remaining melt and changes to pyroxene. Pyroxene in turn changes to amphibole, and amphibole changes to biotite with continued cooling. How is it possible to have any of these minerals other than biotite in an igneous rock?

2. Two rock specimens have the following compositions:

   Specimen 1: 15% biotite, 15% sodium-rich plagioclase, 60% potassium feldspar, and 10% quartz.

   Specimen 2: 10% olivine, 55% pyroxene, 5% hornblende, and 30% calcium-rich plagioclase.

   How would these two rocks differ in color and specific gravity? What was the viscosity of the magmas from which these rocks crystallized? Also, use Figure 4.12 and classify these two rocks.

World Wide Web Activities

▶ **ROB'S GRANITE PAGE**

This site is the work of Robert M. Reed, a Ph.D. candidate in the Department of Geological Sciences, University of Texas at Austin. As you might guess, it is concerned mainly with granite. It has many links to other web sites with information about granite, as well as information about Reed's own research. Check out the various links from his home page.

1. Click on the *Introduction to Granite* heading. It will take you to a page that, among other things, has pictures of granite. Test your knowledge of what the minerals that compose granite look like by pointing your pointer at the various minerals in this image and seeing if you can identify them. Click your mouse to see if you are correct.

2. Click on the *General Stuff* heading. It will take you to a page with links to many other sites. Click on the diamond for *A nice discussion of some California geology (Pt. Reyes), including* **granite rocks** heading. This site contains a discussion of the geology of the Point Reyes Peninsula, including links to specific references about the granitic rocks of this region. What is the age of these granites? What is their origin?

▶ **IGNEOUS ROCKS**

This site contains images and information on ten common igneous rocks. It is part of the Soil Science 223 Rocks and Minerals Reference web site. Click on any of the *rock names* and compare the information and images to the information in this chapter.

For these web site addresses, along with current updates and exercises, log on to

http://www.wadsworth.com/geo

▶ **UNIVERSITY OF TULSA DEPARTMENT OF GEOSCIENCES
IGNEOUS ROCKS AND PROCESSES**

This site contains much information about igneous rocks and processes in general, as well as a few links to other sites.

1. Compare the classifications for igneous rocks presented here with those in this chapter.

2. Click on the *Henry Mountains* site. The Henry Mountains are an excellent example of a laccolith. Compare the images of the Henry Mountains in this site to the diagram of the laccolith shown in Figure 4.18.

▶ **PETE'S BASALT PAGE**

As its name implies, this site is devoted to the igneous rock basalt and the magma from which it crystallizes. Click on *Introductory Basaltology* and learn about the composition and abundance of basaltic magma and see images of the rocks formed from this magma. Also, take the *Virtual Geologic Tour of Kaua'i* and see images and learn about the geology of this Hawaiian Island. While taking the tour, click on *mantle plume* and see how an oceanic hot spot works.

▶ **IGNEOUS ROCKS**

This site is part of the Georgia Science On-Line project. It was created by Dr. Pamela J. W. Gore for Geology 101 at DeKalb College, Clarkston, Georgia. It has good images of all common igneous rocks, along with descriptions of their textures and compositions. It also has a classification chart similar to the one in the text. Scroll down and read about the various igneous rocks shown.

1. How do scoria and vesicular basalt differ?

2. What are the common phenocrysts in andesite porphyry?

3. What is the composition of pumice? What are its uses?

CD-ROM Exploration

Explore the following *In-Terra-Active 2.0* CD-ROM module(s) and increase your understanding of key concepts and processes presented in this chapter.

▷ **SECTION: MATERIALS**
MODULE: IGNEOUS ROCKS: PROCESSES AND STRUCTURE

The crystallization of molten rock can, under different conditions, produce both the finest- and the coarsest-grained rocks known on Earth. At one extreme, volcanic glass (obsidian) is so fine-grained as to be amorphous in structure—that is, not crystalline at all. At the other extreme, pegmatites can have individual crystals that measure 10 meters long. Then there is the odd case of porphyry, a rock that is mixed in texture. **Question: In the case of porphyry, how can you account for the different grain sizes within a single rock?**

CHAPTER

5

Volcanism

Outline

On May 18, 1980, Mount St. Helens in Washington erupted violently. A huge explosion resulted in 63 fatalities and devastation of 600 km² of forest. This 19-km-high steam-and-ash cloud was erupted shortly after the explosion.

Prologue

Thousands of volcanoes have erupted during historic time, but few are as well known as the eruptions of Mount Vesuvius on August 24–25, A.D. 79, that destroyed the cities of Pompeii and Herculaneum. Both were thriving Roman communities along the shore of the Bay of Naples in what is now Italy (Figure 5.1). Fortunately for us, the event was recorded in some detail by Pliny the Younger whose uncle, Pliny the Elder, died while trying to investigate the eruption.

When the eruption began, both Plinys were about 30 km away in the town of Misenum, across the Bay of Naples. At first they were not particularly alarmed when a large cloud appeared over Mount Vesuvius, but soon it became apparent that an eruption was beginning. Pliny the elder, an admiral in the Roman Navy and a naturalist, set sail to investigate and perhaps to evacuate friends from the communities closest to the volcano. When he arrived, volcanic ejecta had already accumulated along the shore, preventing him from landing, so he changed course and made landfall at Stabiae. The next morning while fleeing from noxious fumes emitted by the volcano the Elder Pliny died, probably of a heart attack.

Pliny the Younger's account of Mount Vesuvius's eruption is so vivid that similar eruptions during which large quan-

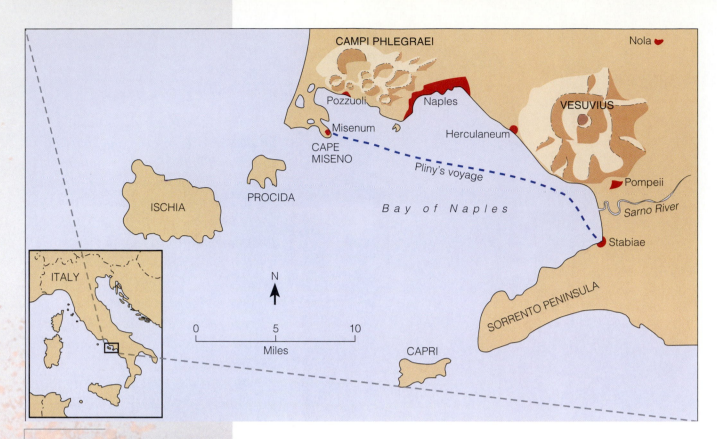

FIGURE 5.1 The region around Mount Vesuvius on the shores of the Bay of Naples in Italy. Eruptions in A.D. 79 destroyed the towns of Stabiae, Pompeii, and Herculaneum. Campi Phlegraei, a few miles northwest of Vesuvius, is also an area of active volcanism.

tities of pumice are blasted into the air are still referred to as plinian. *Pompeii, a city of about 20,000 people about 9 km downwind from the volcano, was buried in nearly 3 m of pyroclastic materials, which covered all but the tallest buildings. Pompeii was so utterly destroyed that, even though its location was known, it was largely forgotten until 1595 when some of the city was rediscovered during the construction of an aqueduct. During the seventeenth and eighteenth centuries, the city was ravaged for artifacts to grace the homes of wealthy Europeans.*

Systematic excavations beginning during the 1800s have exposed much of Pompeii, which is now a popular tourist attraction (Figure 5.2). Probably the most famous attractions are the molds of human bodies that formed when the volcanic debris hardened before the bodies decayed. About 2000 victims have been discovered in the city, including a dog still chained to a post, but what happened to the city's other residents is not known. Some probably escaped by sea or overland, but many may remain buried in the debris beyond the city.

Although Pompeii was the largest city destroyed by Mount Vesuvius, it was not the only one to suffer such a fate. A number of others, such as Stabiae, still remain buried beneath a vast blanket of debris. Herculaneum was just as close to the volcano as Pompeii, but—as opposed to Pompeii, which was buried rather gradually—it was overwhelmed in minutes by surges of incandescent particles and gases in what are known as glowing avalanches. The debris covered the town to a depth of about 20 m. Prior to 1982, only about a dozen skeletons had been discovered, so it was thought that many of Herculaneum's citizens had escaped, but excavations of the city's waterfront revealed hundreds of human skeletons. Many of these were discovered in chambers that probably housed fish-

FIGURE **5.2** The excavated ruins of Pompeii are now a popular tourist attraction. Pompeii, along with Herculaneum and Stabiae, was destroyed during the eruptions of Mount Vesuvius on August 24–25, A.D. 79.

Mount Vesuvius lies just to the south of a 13-km-diameter volcanic depression, or caldera, known as Camp Phlegraei (see Figure 5.1). The last eruption within the caldera took place in 1538, but movement of magma beneath the surface has generated so many earthquakes that about half the population of Pozzuoli, a city within the caldera, has moved away. Furthermore, the city has experienced vertical ground movements also related to moving masses of magma. In Roman times, Pozzuoli stood high, but by the year 1000 it had sunk to about 11 m below sea level, only to rise 12 m between 1000 and 1538. In fact, during a 1538 volcanic outburst from nearby Monte Nuovo, Pozzuoli rose 4 m in just two days! Another episode of uplift began in 1982 when the city rose 1.8 m in about a year and a half.

ing boats, but many more skeletons have since been discovered along the ancient beach.

Before the A.D. 79 eruptions began, Mount Vesuvius was probably not considered a threat, although residents of that area were certainly aware of volcanoes. However, this particular volcano had been quiet for at least 300 years, so it no doubt was of little concern. But it is only one in a chain of volcanoes along Italy's south coast where the African plate is subducted beneath the European plate. Since A.D. 79, Mount Vesuvius has erupted 80 times, most violently in 1631 and 1906; it last erupted in 1944.

Numerous communities on the shores of the Bay of Naples are within easy reach of eruptions from Mount Vesuvius or Campi Phlegraei caldera. Indeed, the city of Naples is situated on the eastern margin of Campi Phlegraei and only a short distance west of Mount Vesuvius (Figure 5.1). Probably no other major population center in the world is in more danger from devastation by volcanoes. The fact that the region remains geologically active was tragically demonstrated in 1980 when an earthquake killed 3000 people in the Naples area.

Introduction

ERUPTING volcanoes are the most impressive manifestations of Earth's dynamic internal processes. During many eruptions, molten rock rises to the surface and flows as incandescent streams or is ejected into the atmosphere in fiery displays that are particularly impressive at night (Figure 5.3). In some parts of the world, volcanic eruptions are commonplace events. The residents of the Philippines, Iceland, Hawaii, and Japan are fully cognizant of volcanoes and their effects (Table 5.1).

Ironically, eruptions of volcanoes are constructive processes when considered in the context of Earth history. The Hawaiian Islands and Iceland owe their existence to volcanism; oceanic crust is continually produced by volcanism at spreading ridges; and volcanic eruptions during Earth's early history released gases that probably formed the atmosphere and surface waters.

FIGURE 5.3   Lava fountains such as these in Hawaii are particularly impressive at night.

Volcanism

VOLCANISM refers to the processes whereby magma and its associated gases rise through the crust and are extruded onto the surface or into the atmosphere. Currently, more than 550 volcanoes are *active*—that is, they have erupted during historic time. Well-known examples of active volcanoes include Mauna Loa and Kilauea on the island of Hawaii, Mount Etna on Sicily, Fujiyama in Japan, Mount St. Helens in Washington, and Mount Pinatubo in the Philippines. Only two other bodies in the solar system are thought to possess active volcanoes: Io, a moon of Jupiter, and perhaps Triton, one of Neptune's moons.

In addition to active volcanoes, numerous *dormant volcanoes* exist that have not erupted recently but may do so again. Mount Vesuvius in Italy had not erupted in human memory until A.D. 79 when it erupted and destroyed the cities of Herculaneum and Pompeii (see the Prologue). Some volcanoes have not erupted during recorded history and show no evidence of doing so again; thousands of these *extinct,* or *inactive,* volcanoes are known.

VOLCANIC GASES

Samples of gases taken from present-day volcanoes indicate that 50 to 80% of all volcanic gases are water vapor. Lesser amounts of carbon dioxide, nitrogen, sulfur gases, especially sulfur dioxide and hydrogen sulfide, and very small amounts of carbon monoxide, hydrogen, and chlorine are also commonly emitted. In areas of recent volcanism, such as Lassen Volcanic National Park in California, one cannot help but notice the rotten-egg odor of hydrogen sulfide gas (Figure 5.4).

When magma rises toward the surface, the pressure is reduced and the contained gases begin to expand. In felsic magmas, which are highly viscous, expansion is inhibited and gas pressure increases. Eventually, the pressure may become great enough to cause an explosion and produce pyroclastic materials such as ash. In contrast, low-viscosity mafic magmas allow gases to expand and escape easily. Accordingly, mafic magmas generally erupt rather quietly.

Most volcanic gases quickly dissipate in the atmosphere and pose little danger to humans, but on several occasions

TABLE 5.1

Some Notable Volcanic Eruptions

DATE	VOLCANO	DEATHS
Aug. 24, 79	Mt. Vesuvius, Italy	3360 killed in Pompeii and Herculaneum.
1586	Kelut, Java	Mudflows kill 10,000.
Dec. 16, 1631	Mt. Vesuvius, Italy	3500 killed.
Aug. 4, 1672	Merapi, Java	3000 killed by mudflows and pyroclastic flows.
Dec. 10, 1711	Awu, Indonesia	3000 killed by pyroclastic flows.
Sept. 22, 1760	Makian, Indonesia	Eruption kills 2000; island evacuated for seven years.
June 8, 1783	Lakagigar, Iceland	Largest historic lava flows: 12 km^3; 9350 die.
July 26, 1783	Asama, Japan	Pyroclastic flows and floods kill 1200+.
May 21, 1792	Unzen, Japan	14,500 die in debris avalanche and tsunami.
Apr. 10, 1815	Tambora, Indonesia	92,000 killed; another 80,000 reported to have died from famine and disease.
Oct. 8, 1822	Galunggung, Java	4011 die in pyroclastic flows and mudflows.
Mar. 2, 1856	Awu, Indonesia	Pyroclastic flows kill 2806.
Aug. 27, 1883	Krakatau, Indonesia	36,417 die; most killed by tsunami.
June 7, 1892	Awu, Indonesia	1532 die in pyroclastic flows.
May 8, 1902	Mt. Pelée, Martinique	St. Pierre destroyed by pyroclastic flow; 28,000 killed.
Oct. 24, 1902	Santa María, Guatemala	5000 killed.
June 6, 1912	Novarupta, Alaska	Largest twentieth-century eruption: about 33 km^3 of pyroclastic materials erupted; no fatalities.
May 19, 1919	Kelut, Java	Mudflows kill 5110, devastate 104 villages.
Jan. 21, 1951	Lamington, New Guinea	2942 killed by pyroclastic flows.
Mar. 17, 1963	Agung, Indonesia	1148 killed.
Aug. 12, 1976	Soufrière, Guadeloupe	74,000 residents evacuated.
May 18, 1980	Mt. St. Helens, Washington	63 killed; 600 km^2 of forest devastated.
Mar. 28, 1982	El Chichón, Mexico	Pyroclastic flows kill 1877.
Nov. 13, 1985	Nevado del Ruiz, Colombia	Mudflows kill 23,000.
Aug. 21, 1986	Oku volcanic field, Cameroon	1746 asphyxiated by cloud of CO_2 released from Lake Nyos.
June 1991	Unzen, Japan	43 killed; at least 8500 fled.
June 1991	Mt. Pinatubo, Philippines	~281 killed during initial eruption; 83 killed by later mudflows; 358 died of illness; 200,000 evacuated.
Feb. 2, 1993	Mt. Mayon, Philippines	At least 70 killed; 60,000 evacuated.
Nov. 22, 1994	Mt. Morapis, Indonesia	Pyroclastic flows kill 60; more than 6000 evacuated.
June 25, 1997	Soufriere Hills, Montserrat	Pyroclastic flows kill 10; 4000 of the island's 11,000 residents evacuated.

SOURCE: American Geological Institute Data Sheets, except for last five entries.

these gases have caused numerous fatalities. In 1783, toxic gases, probably sulfur dioxide, erupted from Laki fissure in Iceland had devastating effects. About 75% of the nation's livestock died, and the haze resulting from the gas caused lower temperatures and crop failures; about 24% of Iceland's population died as a result of the ensuing Blue Haze Famine. The country suffered its coldest winter in 225 years in 1783–1784, with temperatures 4.8°C below the long-term average. The eruption also produced what Benjamin Franklin called a "dry fog" that was responsible for dimming the intensity of sunlight in Europe. The severe winter of 1783–1784 in Europe and eastern North America is attributed to the presence of this "dry fog" in the upper atmosphere.

The particularly cold spring and summer of 1816 are attributed to the 1815 eruption of Tambora in Indonesia, the largest and most deadly eruption during historic time. The eruption of Mayon volcano in the Philippines during the previous year may have contributed to the cool spring and summer of 1816 as well. Another large historic eruption that had widespread climatic effects was the eruption of Krakatau in 1883.

More recently, in 1986, in the African nation of Cameroon 1746 people died when a cloud of carbon dioxide engulfed them. The gas accumulated in the waters of Lake Nyos, which occupies a volcanic crater. No agreement exists on what caused the gas to suddenly burst forth from the lake, but once it did, it flowed downhill along the

FIGURE 5.4 Gases being emitted at the Sulfur Works in Lassen Volcanic National Park, California.

(a)

(b)

FIGURE 5.5 (a) Pahoehoe erupted from the Pu'u O'o vent on Kilauea volcano's southeastern flank. (b) An aa flow in the east rift zone of Kilauea volcano, Hawaii, in 1983. The flow front is about 2.5 m high.

surface because it was denser than air. In fact, the density and velocity of the gas cloud were great enough to flatten vegetation, including trees, a few kilometers from the lake. Unfortunately, thousands of animals and many people, some as far as 23 km from the lake, were asphyxiated.

Residents of the island of Hawaii have coined the term *vog* for volcanic smog. Kilauea volcano has been erupting continuously since 1983, releasing small amounts of lava, copious quantities of carbon dioxide, and about 1000 tons of sulfur dioxide per day. Carbon dioxide is no problem, but sulfur dioxide produces a haze and the unpleasant odor of sulfur. Vog probably poses little or no problem for tourists, but a long-term threat exists for residents of the west side of the island where vog is most common.

LAVA FLOWS AND PYROCLASTIC MATERIALS

Lava flows are frequently portrayed in movies and on television as fiery streams of incandescent rock material posing a great danger to humans. Actually, lava flows are the least dangerous manifestation of volcanism, although they may destroy buildings and cover agricultural land. Most lava flows do not move particularly fast, and because they are fluid, they follow existing low areas. So once a flow erupts from a volcano, determining the path it will take is fairly easy, and anyone in areas likely to be affected can be evacuated.

Two types of lava flows, both of which were named for Hawaiian flows, are generally recognized. A **pahoehoe** (pronounced pah-hoy-hoy) flow has a ropy surface much like taffy (Figure 5.5a). The surface of an **aa** (pronounced ah-ah) flow is characterized by rough, jagged angular blocks and fragments (Figure 5.5b). Pahoehoe flows are less viscous than aa flows; indeed, the latter are viscous enough to break up into blocks and move forward as a wall of rubble.

Columnar joints are common in many lava flows, especially mafic flows, but they also occur in other kinds of flows and in some intrusive igneous rocks (Figure 5.6). Once a lava flow ceases moving, it cools and produces forces that cause fractures called *joints* to open. On the surface of a flow, these joints form polygonal (often six-sided) cracks. These cracks extend downward into the flow, forming parallel columns with their long axes perpendicular to the principal cooling surface. Excellent examples of columnar joints can be seen at Devil's Postpile National Monument in California (Figure 5.6), Devil's Tower National Monument in Wyoming (see Chapter 4 Prologue), the Giant's Causeway in Ireland, and many other areas.

FIGURE **5.7** These bulbous masses of pillow lava form when magma is erupted under water.

(a)

(b)

FIGURE **5.6** (a) Columnar joints in a lava flow at Devil's Postpile National Monument, California. (b) Surface view of the same columnar joints showing their polygonal pattern. The straight lines and polish resulted from glacial ice moving over this surface.

Much of the igneous rock in the upper part of the oceanic crust is of a distinctive type; it consists of bulbous masses of basalt resembling pillows, hence the name **pillow lava**. It was long recognized that pillow lava forms when lava is rapidly chilled beneath water, but its formation was not observed until 1971. Divers near Hawaii saw pillows form when a blob of lava broke through the crust of an underwater lava flow and cooled almost instantly, forming a pillow-shaped structure with a glassy exterior. The remaining fluid inside then broke through the crust of the pillow, resulting in an accumulation of interconnected pillows (Figure 5.7).

Much pyroclastic material is erupted as **ash**, a designation for pyroclastic particles measuring less than 2.0 mm (Figure 5.8). Ash may be erupted in two ways: an ash fall or an ash flow. During an ash fall, ash is ejected into the atmosphere and settles to the surface over a wide area. In 1947, ash erupted from Mount Hekla in Iceland fell 3800 km away on Helsinki, Finland. Ash is also erupted in ash flows, which are coherent clouds of ash and gas that commonly flow along or close to the land surface. These flows can move at more than 100 km per hour, and some of them cover vast areas.

Pyroclastic materials measuring 2 to 64 mm are known as *lapilli,* and any particle larger than 64 mm is called a *bomb* or *block* depending on its shape. Bombs have twisted, streamlined shapes that indicate they were erupted as globs of magma that cooled and solidified during their flight through the air (Figure 5.9). Blocks are angular pieces of rock ripped from a volcanic conduit or pieces of

FIGURE **5.8** Pyroclastic materials: volcanic ash being erupted from Mount Ngauruhoe, New Zealand, during January 1974.

FIGURE **5.9** Pyroclastic materials. The large object on the left is a volcanic bomb; it is about 20 cm long. The streamlined shape of bombs indicates they were erupted as globs of magma that cooled and solidified as they descended. The granular objects in the upper right are pyroclastic materials known as lapilli. The pile of gray material on the lower right is ash. *(Photo courtesy of Sue Monroe.)*

a solidified crust of a magma. Because of their large size, volcanic bomb and block accumulations are not nearly as widespread as ash deposits; instead, they are confined to the immediate area of eruption.

VOLCANOES

Probably no landform is more familiar to the general public than **volcanoes**, which are conical mountains formed around a vent where lava and pyroclastic materials are erupted. The term *volcano* comes from Vulcan, the Roman deity of fire. Because of their danger and obvious connection to Earth's interior, volcanoes have been held in awe by many cultures. Probably no other geologic process, except perhaps earthquakes, has as much lore associated with it. In Hawaiian legends, Pele, the volcano goddess, resides within the crater of Kilauea on the island on Hawaii. In one of her frequent rages, Pele causes earthquakes and lava flows, and she may hurl flaming boulders at those who offend her.

Volcanoes come in many shapes and sizes, but geologists recognize several major categories, each of which has a distinctive eruptive style. One must realize, however, that each volcano has a unique overall history of eruptions and development.

Most volcanoes have a circular depression, or **crater**, at their summit. Craters form as a result of the extrusion of gases and lava from a volcano and are connected via a conduit to a magma chamber below the surface. It is not unusual, though, for magma to erupt from vents on the flanks of large volcanoes where smaller, parasitic cones develop. Mount Etna on Sicily has some 200 smaller vents on its flanks.

Some volcanoes are characterized by a **caldera** rather than a crater. Craters are generally less than 1 km in diameter, whereas calderas greatly exceed this dimension and have steep sides; the Toba caldera in Sumatra measures 100 km long and 30 km wide. One of the best-known calderas in the United States is the misnamed Crater Lake in Oregon—Crater Lake is actually a caldera (Figure 5.10). It formed about 6600 years ago after voluminous eruptions partially drained the magma chamber. This drainage left the summit of the mountain, Mount Mazama, unsupported, and it collapsed into the magma chamber, forming a caldera more than 1200 m deep and measuring 9.7 by 6.5 km. Many calderas probably formed when a summit collapsed during particularly large, explosive eruptions as in the case of Crater Lake, but a few apparently formed when the top of the original volcano was blasted away.

SHIELD VOLCANOES **Shield volcanoes** resemble the outer surface of a shield lying on the ground with the convex side up (Figure 5.11). They have low, rounded profiles with gentle slopes ranging from about 2 to 10 degrees. Their low slopes reflect the fact that they are composed mostly of low-viscosity mafic flows, so the flows spread out and formed thin layers. Shield volcanoes have a summit

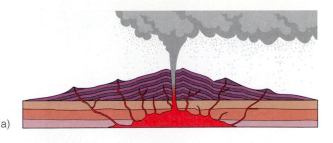
(a)

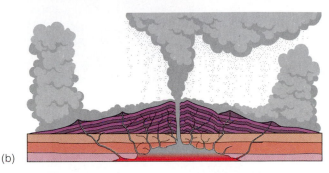
(b)

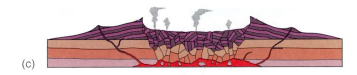
(c)

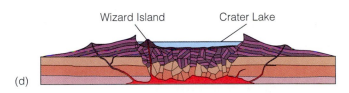
Wizard Island Crater Lake
(d)

(e)

FIGURE **5.10** The sequence of events leading to the origin of Crater Lake, Oregon. (a–b) Ash clouds and ash flows partly drain the magma chamber beneath Mount Mazama. (c) The summit collapses, forming the caldera. (d) Post-caldera eruptions partly cover the caldera floor, and the small volcano known as Wizard Island forms. (e) View from the rim of Crater Lake showing Wizard Island.

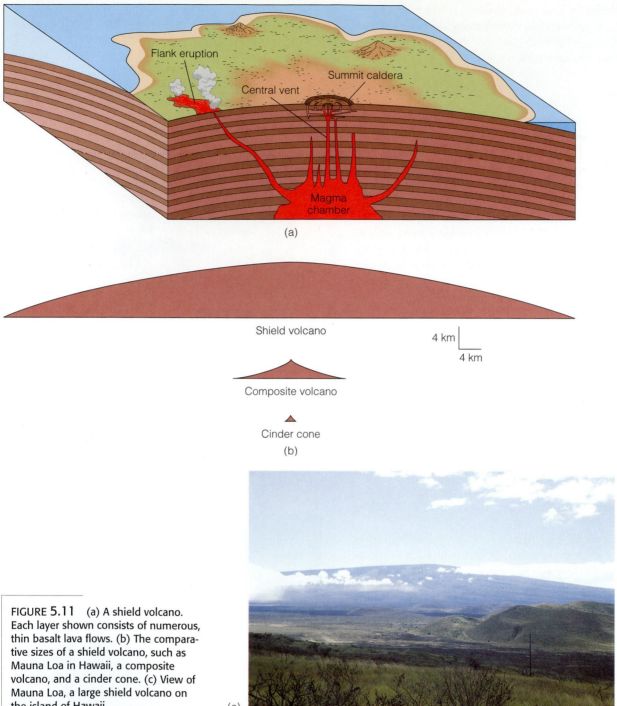

FIGURE **5.11** (a) A shield volcano. Each layer shown consists of numerous, thin basalt lava flows. (b) The comparative sizes of a shield volcano, such as Mauna Loa in Hawaii, a composite volcano, and a cinder cone. (c) View of Mauna Loa, a large shield volcano on the island of Hawaii.

Flank eruption

Summit caldera

Central vent

Magma chamber

(a)

Shield volcano

4 km

4 km

Composite volcano

Cinder cone

(b)

(c)

crater or caldera and a number of smaller cones on their flanks through which lava is erupted (Figure 5.11). A vent opened on the flank of Kilauea and grew to more than 250 m high between June 1983 and September 1986.

Eruptions from shield volcanoes, sometimes called *Hawaiian-type volcanoes,* are quiet compared to those of volcanoes such as Mount St. Helens; lavas most commonly rise to the surface with little explosive activity, so they usually pose little danger to humans. Lava fountains, some up to 400 m high, contribute some pyroclastic

materials to shield volcanoes (Figure 5.3), but otherwise they are composed largely of basalt lava flows; flows comprise more than 99% of the Hawaiian volcanoes above sea level.

Shield volcanoes are most common in oceanic areas, such as the Hawaiian Islands and Iceland, but some are also present on the continents—for example, in East Africa. The island of Hawaii consists of five huge shield volcanoes, two of which, Kilauea and Mauna Loa, are active much of the time. These Hawaiian volcanoes are the largest in the world. Mauna Loa is nearly 100 km across at the base and stands more than 9.5 km above the surrounding seafloor. Its volume is estimated at about 50,000 km³. By contrast, the largest volcano in the continental United States, Mount Shasta in northern California, has a volume of only about 205 km³.

CINDER CONES Volcanic peaks composed of pyroclastic materials resembling cinders are known as **cinder cones** (Figure 5.12). They form when pyroclastic materials are ejected into the atmosphere and fall back to the surface to accumulate around the vent, thus forming a small, steep-sided cone. The slope angle may be as much as 33 degrees, depending on the angle that can be maintained by the irregularly shaped pyroclastic materials. Cinder cones are rarely more than 400 m high, and many have a large, bowl-shaped crater.

Many cinder cones form on the flanks or within the calderas of larger volcanic mountains and appear to represent the final stages of activity, particularly in areas formerly characterized by basalt lava flows. Wizard Island in Crater Lake, Oregon, is a small cinder cone that formed after the summit of Mount Mazama collapsed to form a caldera (Figure 5.10). Cinder cones are common in the southern Rocky Mountain states, particularly New Mexico and Arizona, and many others are found in California, Oregon, and Washington.

Eruptive activity at cinder cones is rather short-lived. For instance, on February 20, 1943, a farmer in Mexico noticed fumes emanating from a crack in his cornfield, and a few minutes later ash and cinders were erupted. Within a month a cinder cone 300 m high had formed. Lava flows broke through the flanks of the new volcano, which was later named Paricutin, and covered two nearby towns. Activity ceased in 1952.

In 1973, on the Icelandic island of Heimaey, the town of Vestmannaeyjar was threatened by a new cinder cone. The initial eruption began on January 23, and within two days a cinder cone, later named Eldfell, rose to about 100 m above the surrounding area (Figure 5.12b). Pyroclastic materials from the volcano buried parts of the town, and by February a massive aa lava flow was advancing toward the town. The flow's leading edge ranged from 10 to 20 m thick, and its central part was as much as 100 m thick. By spraying the leading edge of the flow with seawater, which caused it to cool and solidify, the residents of Vestmannaeyjar successfully diverted the flow before it did much damage to the town.

COMPOSITE VOLCANOES Composite volcanoes, also called **stratovolcanoes**, are composed of both pyroclastic layers and lava flows (Figure 5.13). Typically, both materials have an intermediate composition, and the flows cool to form andesite. Recall that lava of intermediate composition is more viscous than mafic lava. Besides lava flows and pyroclastic layers, a significant proportion of a composite volcano is made up of volcanic mudflows known as **lahars**. Some lahars form when rain falls on layers of loose pyroclastic materials and creates a muddy slurry that moves downslope. On November 13, 1985, lahars resulting from a rather minor eruption of Nevado del Ruiz in Colombia killed about 23,000 people. In the Philippines, 83 of the

(a)

FIGURE **5.12** (a) This small volcanic mountain known as Sunset Crater, Arizona, is a cinder cone. (b) The town of Vestmannaeyjar in Iceland was threatened by lava flows from Eldfell, a cinder cone that formed in 1973. Within two days of the initial eruption on January 23, the new volcano had grown to about 100 m high. Another cinder cone called Helgafell is also visible.

(b)

Pyroclastic layers

Lava flows

(a)

FIGURE 5.13 (a) Composite volcanoes are the typical, large volcanic mountains on continents. They are composed of lava flows, pyroclastic layers, and volcanic mudflows. (b) View of Mount St. Helens, Washington, from the north before its May 1980 eruption. Spirit Lake is in the foreground.

(b)

722 victims of the June 1991 eruptions of Mount Pinatubo were killed by lahars (Table 5.1).

Composite volcanoes are steep sided near their summits, perhaps as much as 30 degrees, but the slope decreases toward the base where it is generally less than 5 degrees. Their concave slopes rise ever steeper to the summit with its central vent through which lava and pyroclastic materials are periodically erupted. Mount St. Helens, Washington, is a good example (Figure 5.13b).

Composite volcanoes are the typical large volcanoes of the continents and island arcs. Familiar examples include Fujiyama in Japan and Mount Vesuvius in Italy, as well as Mount St. Helens and many of the other volcanic peaks in the Cascade Range of western North America (see Perspective 5.1). On June 15, 1991, Mount Pinatubo in the Philippines discharged an estimated 3 to 5 km³ of pyroclastic materials, mostly ash, making it the largest eruption in more than 50 years (Figure 5.14). Fortunately, warnings of an impending eruption were heeded, and some 200,000 people were evacuated from areas around the volcano, yet the eruption still caused 722 deaths (Table 5.1). Another

composite volcano in the Philippines, Mayon volcano, erupted for the twelfth time this century in February 1993.

LAVA DOMES   If the upward pressure in a volcanic conduit is great enough, the most viscous magmas move upward and form bulbous, steep-sided **lava domes** (Figure 5.15). Lava domes are generally composed of felsic lavas, although some are of intermediate composition. Because felsic magma is so viscous, it moves upward very slowly; the lava dome that formed in Santa María volcano in Guatemala in 1922 took two years to grow to 500 m high and 1200 m across. Lava domes contribute significantly to many composite volcanoes. Beginning in 1980 a number of lava domes were emplaced in the crater of Mount St. Helens; most of these were destroyed during subsequent eruptions. Since 1983, Mount St. Helens has been characterized by sporadic dome growth.

In June 1991 a dome in Japan's Unzen volcano collapsed, causing a flow of debris and hot ash that killed 43 people in a nearby town. Lava domes are also often responsible for extremely explosive eruptions. In 1902

(a)

(b)

FIGURE **5.14** (a) Mount Pinatubo in the Philippines is one of many volcanoes in a belt nearly encircling the Pacific Ocean basin. It is shown here erupting on June 12, 1991. A huge, thick cloud of ash and steam rises above Clark Air Force Base, from which about 15,000 people had already been evacuated to Subic Bay Naval Base. Following this eruption, the remaining 900 people at the base were also evacuated. (b) Homes partly buried by a volcanic mudflow (lahar) on June 15, 1991. Note that the roof at the far right is still partly covered by pyroclastic materials.

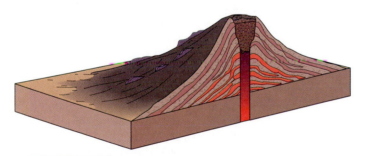

FIGURE **5.15** A cross section showing the internal structure of a lava dome. Lava domes form when a viscous mass of magma, generally of felsic composition, is forced up through a volcanic conduit.

FIGURE **5.16** St. Pierre, Martinique, after it was destroyed by a nuée ardente erupted from Mount Pelée in 1902. Only 2 of the city's 28,000 inhabitants survived.

viscous magma accumulated beneath the summit of Mount Pelée on the island of Martinique. Eventually, the pressure within the mountain increased to the point that it could no longer be contained, and the side of the mountain blew out in a tremendous explosion. When this occurred, a mobile, dense cloud of pyroclastic materials and gases called a **nuée ardente** (French for "glowing cloud") was ejected and raced downhill at about 100 km/hr, engulfing the city of St. Pierre (Figure 5.16).

A tremendous blast hit St. Pierre and leveled buildings; hurled boulders, trees, and pieces of masonry down the streets; and moved a three-ton statue 16 m. Accompanying the blast was a swirling cloud of incandescent ash and gases with an internal temperature of 700°C. Of the 28,000

Volcanism 101

Eruptions of Cascade Range Volcanoes

During the summer of 1914, Mount Lassen in northern California began erupting without warning and culminated with the "Great Hot Blast," a huge steam explosion on May 22, 1915. Mount Lassen is one of 15 large volcanoes in the Cascade Range of northern California, Oregon, Washington, and southern British Columbia, Canada. After Mount Lassen's eruptions, the Cascade volcanoes remained quiet for 63 years. Then, on March 16, 1980, following an inactive period of 123 years, Mount St. Helens in southern Washington (see Figure 5.13b) showed signs of renewed activity, and on May 18 it erupted violently, causing the worst volcanic disaster in U.S. history.

The awakening of Mount St. Helens came as no surprise to geologists of the U.S. Geological Survey (USGS) who warned in 1978 that it was an especially dangerous volcano. Although no one could predict precisely when Mount St. Helens would erupt, the USGS report included maps showing areas where damage from an eruption could be expected. Forewarned with such data, local officials were better prepared to formulate policies when the eruption did occur.

On March 27, 1980, Mount St. Helens began erupting steam and ash and continued to do so during the rest of March and most of April. By late March, a visible bulge had developed on its north face as molten rock was injected into the mountain, and the bulge continued to expand at about 1.5 m per day. On May 18, an earthquake shook the area, the unstable bulge collapsed, and the pent-up volcanic gases below expanded rapidly, creating a tremendous northward-directed lateral blast that blew out the north side of the mountain (Figure 1). The lateral blast accelerated to more than 1000 km/hr, obliterating virtually everything in its path. Some 600 km^2 of forest were completely destroyed; trees were snapped off at their bases and strewn about the countryside, and trees as far as

Figure 1 *The eruption of Mount St. Helens on May 18, 1980. The lateral blast occurred when a bulge on the north face of the mountain collapsed and reduced the pressure on the molten rock within the mountain.*

30 km from the bulge were seared by the intense heat. Tens of thousands of animals were killed; roads, bridges, and buildings were destroyed; and 63 people perished.

Shortly after the lateral blast, volcanic ash and steam formed a 19-km-high cloud above the volcano (see chapter-opening photo). The ash cloud drifted east-northeast, and the resulting ash fall at Yakima, Washington, 130 km to the east, caused almost total darkness at midday. Detectable amounts of ash were deposited over a huge area. Flows of hot gases and volcanic ash raced down the north flank of the mountain, causing steam explosions when they encountered bodies of water or moist ground. Steam explosions continued for weeks, and at least one occurred a year later.

Snow and glacial ice on the upper slopes of Mount St. Helens melted and mixed with ash and other surface debris to form thick, pasty volcanic mudflows. The largest and most destructive mudflow surged down the valley of the North Fork of the Toutle River. Ash and mudflows displaced water in lakes and streams and flooded downstream areas. Ash and other particles carried by the flood waters were deposited in stream channels; many kilometers from Mount St. Helens, the navigation channel of the

Columbia River was reduced from 12 m to less than 4 m as a result of such deposition.

Although the damage resulting from the eruption of Mount St. Helens was significant and the deaths were tragic, it was not a particularly large or deadly eruption compared with some historic eruptions. For example, the 1902 eruption of Mount Pelée on the island of Martinique killed 28,000 people, and the 1815 eruption of Tambora in Indonesia resulted in an estimated 172,000 deaths (Table 5.1). The Tambora eruption produced at least 80 times more ash than the 0.9 km^3 that spewed forth from Mount St. Helens.

Several other currently dormant Cascade Range volcanoes also pose a threat to populated areas. Eruptions of Mount Shasta in northern California would cause damage and perhaps fatalities in several nearby communities, and Mount Hood, Oregon, lies less than 65 km from the densely populated Portland area. But the most dangerous is probably Mount Rainier, Washington (Figure 2).

Rather than lava flows, ash falls, or even a colossal explosion as in the case of Mount St. Helens, the greatest threat from Mount Rainier is volcanic mudflows, or lahars. At least 60 such flows have occurred during the last 10,000 years. The largest flow consisted of nearly 4 km³ of debris, and it covered an area now occupied by more than 120,000 people. No one can predict when another mudflow will take place, but at least one community has taken the threat seriously enough to formulate an emergency evacuation plan. Unfortunately, they would have only 1 or 2 hours to carry out an evacuation.

Figure 2 Mount Rainier as seen from the waterfront of Tacoma, Washington. Its summit elevation of 4392 m makes Mount Rainier the highest peak in the Cascade Range. The summit is only about 80 km from where this picture was taken.

residents of St. Pierre, only 2 survived! One survivor was on the extreme outer edge of the area covered by the nuée ardente, but he was terribly burned. The other survivor, a stevedore incarcerated the night before for disorderly conduct, was in a windowless cell partly below ground level. He remained in his cell badly burned for four days after the eruption until rescue workers heard his cries for help.

MONITORING VOLCANOES AND FORECASTING ERUPTIONS

According to the U.S. Geological Survey, nearly 500 million people live near volcanoes on the margins of Earth's tectonic plates. Many of these volcanoes have erupted explosively during historic time and have the potential to do so again. As a matter of fact, volcanic eruptions are not as unusual as you might think; between 1975 and 1985, 376 separate outbursts occurred. Fortunately, none of these compared to the 1815 eruption of Tambora; nevertheless, fatalities occurred in several instances, the worst being in 1985 in Colombia where about 23,000 perished in mudflows generated by an eruption (Table 5.1). Only a few of these potentially dangerous volcanoes are monitored, including some in Italy, Japan, New Zealand, Russia, and the Cascade Range.

Many of the methods for monitoring active volcanoes were developed at the Hawaiian Volcano Observatory. These methods involve recording and analyzing various changes in both the physical and chemical attributes of volcanoes. *Tiltmeters* are used to detect changes in the slopes of a volcano when it inflates as magma is injected into it, whereas a *geodimeter* uses a laser beam to measure horizontal distances, which also change when a volcano inflates (Figure 5.17). Geologists also monitor gas emissions and changes in the local magnetic and electrical fields of volcanoes.

Of critical importance in volcano monitoring and eruption forecasting are a sudden increase in earthquake activity and the detection of harmonic tremor. *Harmonic tremor* is continuous ground motion, as opposed to the sudden jolts produced by earthquakes. It precedes all eruptions of Hawaiian volcanoes and also preceded the eruption of Mount St. Helens. Such activity indicates that magma is moving below the surface.

The analysis of data gathered during monitoring is not by itself sufficient to forecast eruptions; the past history of a particular volcano must also be known. To determine the eruptive history of a volcano, the record of previous eruptions as preserved in rocks must be studied and analyzed. Indeed, prior to 1980, Mount St. Helens was considered one of the most likely Cascade volcanoes to erupt because detailed studies indicated that it has had a record of explosive activity for the past 4500 years.

For the better monitored volcanoes, such as those in Hawaii, it is now possible to make accurate short-term forecasts of eruptions. In 1960 the warning signs of an eruption of Kilauea were recognized soon enough to evac-

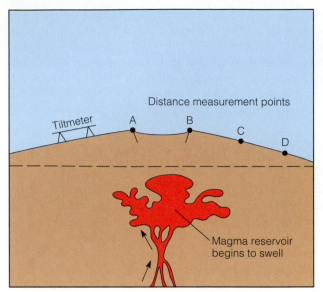

(a) Stage 1

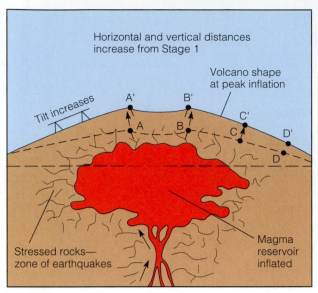

(b) Stage 2

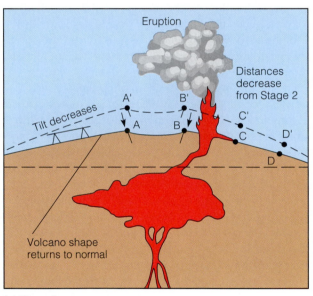

(c) Stage 3

FIGURE 5.17 Volcano monitoring. These diagrams show three stages in a typical eruption of a Hawaiian volcano: (a) The volcano begins to inflate; (b) inflation reaches its peak; (c) the volcano erupts and then deflates, returning to its normal shape.

uate the residents of a small village that was subsequently buried by lava flows.

For some volcanoes, little or no information is available for making predictions. On January 14, 1993, for example, Colombia's Galeras volcano erupted without warning, killing six of ten volcanologists on a field trip and three Colombian tourists. Ironically, the volcanologists were attending a conference on improving methods for predicting volcanic eruptions.

FISSURE ERUPTIONS

Between about 17 million and 5 million years ago, a vast area of 164,000 km² in eastern Washington and parts of Oregon and Idaho were covered by overlapping basalt lava flows. This thick accumulation of lava flows, known as the Columbia River basalts, is now well exposed in the walls of the canyons eroded by the Snake and Columbia Rivers (Figure 5.18). These flows were erupted from long fissures during what are known as **fissure eruptions**, and the fluid lava spread over large areas, forming **basalt plateaus**.

The Columbia River basalt flows have an aggregate thickness of about 1000 m, and some individual flows cover huge areas—the Roza flow, which is 30 m thick, advanced along a front about 100 km wide and covered 40,000 km².

Fissure eruptions and basalt plateaus are not common, although several large areas with these features are known. Currently, this type of activity is known only in Iceland. A number of volcanic mountains are present in Iceland, but the bulk of the island is composed of basalt flows erupted

FIGURE 5.18 The Columbia River basalts are overlapping lava flows with an aggregate thickness of
about 1000 meters.

FIGURE 5.19 Pyroclastic flow deposits in Crater
Lake National Park, Oregon.

from fissures. Two large fissure eruptions, one in A.D. 930
and the other in 1783, account for about half of the magma
erupted in Iceland during historic time. The 1783 eruption
issued from the Laki fissure, which is more than 30 km
long; lava flowed several tens of kilometers from the
fissure, covering more than 560 km², and in one place
filled a valley to a depth of about 200 m.

PYROCLASTIC SHEET DEPOSITS

More than 100 years ago, geologists were aware of vast areas covered by felsic volcanic rocks a few meters to hundreds of meters thick. It seemed improbable that these
could have formed as vast lava flows, but it also seemed unlikely that they were ash fall deposits. Based on observations of historic pyroclastic flows, such as the nuée ardente
erupted by Mount Pelée in 1902, it now seems probable
that these ancient rocks originated as pyroclastic flows,
hence the name **pyroclastic sheet deposits**. They cover far
greater areas than any observed during historic time and
apparently erupted from long fissures rather than from a
central vent. The pyroclastic materials of many of these
flows were so hot they fused together to form *welded tuff*.

It now appears that huge pyroclastic flows issue from
fissures formed during the origin of calderas. For instance,
pyroclastic flows were erupted during the formation of a
large caldera in the area of present-day Crater Lake, Oregon (Figure 5.19). Similarly, the Bishop Tuff of eastern California appears to have been erupted shortly before the
formation of the Long Valley caldera. Interestingly, earthquake activity in the Long Valley caldera and nearby areas

beginning in 1978 may indicate that magma is moving upward beneath part of the caldera. Thus, the possibility of future eruptions in that area cannot be discounted.

Distribution of Volcanoes

RATHER than being distributed randomly, volcanoes are in well-defined zones or belts. More than 60% of all active volcanoes are in the **circum-Pacific belt** that nearly encircles the margins of the Pacific Ocean basin (Figure 5.20). Included in this belt are the volcanoes along the west coast of South America, those in Central America, Mexico, and the Cascade Range, and the Alaskan volcanoes in the Aleutian Island arc. The belt continues on the western side of the Pacific Ocean basin where it extends through Japan, the Philippines, Indonesia, and New Zealand. The circum-Pacific belt includes the southernmost active volcanoes, Mount Erebus in Antarctica, and a large caldera at Deception Island that erupted during 1970.

About 20% of all active volcanoes are in the **Mediterranean belt** (Figure 5.20). This belt includes the famous

Italian volcanoes such as Mount Etna, Stromboli, and Mount Vesuvius.

Most of the rest of the active volcanoes are at or near the mid-oceanic ridges (Figure 5.20). The longest of these ridges is the Mid-Atlantic Ridge, which is near the middle of the Atlantic Ocean basin and curves around the southern tip of Africa where it continues as the Indian Ridge. Branches of the Indian Ridge extend into the Red Sea and East Africa. Mount Kilimanjaro in Africa is on this latter branch (Figure 5.20).

Plate Tectonics, Volcanoes, and Plutons

IN Chapter 4 we discussed the origin and evolution of magma and concluded that (1) mafic magmas are generated beneath spreading ridges and (2) intermediate and felsic magmas form where an oceanic plate is subducted beneath another oceanic plate or a continental plate. Accordingly, most of Earth's volcanism and emplacement of plutons takes place at divergent and convergent plate boundaries.

Much of the mafic magma that forms beneath spreading ridges is simply emplaced at depth as vertical dikes and gabbro plutons (Figure 5.21). Some of this magma rises to the surface where it usually forms submarine lava flows and pillow lavas (Figure 5.7). Indeed, the oceanic crust is

FIGURE 5.20 Most volcanoes are at or near plate boundaries. Two major volcano belts are recognized: the circum-Pacific belt contains about 60% of all active volcanoes, about 20% are in the Mediterranean belt, and most of the rest are located along mid-oceanic ridges.

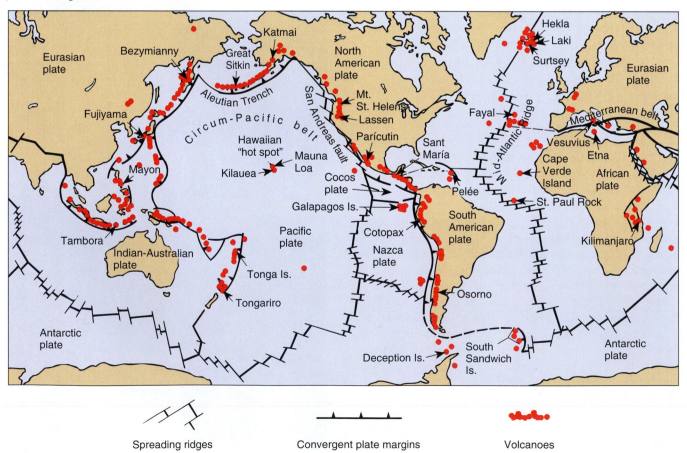

composed largely of gabbro and basalt. Much of this submarine volcanism goes undetected, but researchers in submersible craft have observed the results of these eruptions.

Pyroclastic materials are not common in this environment because mafic lava is very fluid, allowing gases to easily escape, and at great depth water pressure prevents gases from expanding. Should an eruptive center along a ridge build above sea level, however, pyroclastic materials are commonly erupted at lava fountains, but most of the magma issues forth as fluid lava flows that form shield volcanoes.

Excellent examples of divergent plate boundary volcanism are found along the Mid-Atlantic Ridge, particularly where it is taking place above sea level, as in Iceland. In November 1963 a new volcanic island, later named Surtsey, rose from the sea just south of Iceland. The East Pacific rise and the Indian Ridge are also areas of similar volcanism. Not all divergent plate boundaries are beneath sea level as in the previous examples. For instance, divergence and igneous activity is taking place in Africa at the East African Rift system.

The circum-Pacific and Mediterranean belts are made up of composite volcanoes near the leading edges of overriding plates at convergent plate boundaries (see Figure 4.4). The overriding plate may be oceanic as in the case of the Aleutian islands, or it may be continental as, for instance, the South American plate with its chain of volcanoes along its western margin. As already noted, these volcanoes consist largely of pyroclastic materials and lava flows of intermediate to felsic composition. Recall that partial melting of mafic oceanic crust of a subducted oceanic plate generates the magmas, some of which are emplaced along the plate margins as plutons, especially batholiths, and some is erupted to build up composite volcanoes. Some of the more viscous magmas, generally of felsic composition, are emplaced as lava domes within existing volcanoes, thus accounting for the explosive eruptions that characterize convergent plate boundaries.

In previous sections of this chapter, we have alluded to several eruptions at convergent plate boundaries. Good ex-amples are the explosive eruptions of Mount Pinatubo and Mayon volcano in the Philippines, both of which are situated near a plate boundary beneath which an oceanic plate is subducted. Mount St. Helens, Washington (See Perspective 5.1), is similarly situated but is on a continental rather than an oceanic plate. Mount Vesuvius in Italy is one of several active volcanoes in that region that lie on a plate beneath which the northern margin of the African plate is subducted.

Mauna Loa and Kilauea on the island of Hawaii and Loihi just 32 km to the south are within the interior of rigid plate far from any spreading ridge or subduction zone (Figure 5.20). Thus, they are unrelated to divergence or convergence. It is postulated that a **mangle plume** creates a local "hot spot" beneath Hawaii. However, the magma is derived from the upper mantle, as it is at spreading ridges, and accordingly is mafic—so it builds up shield volcanoes.

Loihi is particularly interesting because it represents a stage in the origin of a new Hawaiian island. It is a submarine volcano that rises more than 3000 m above the adjacent seafloor, but its summit is still about 940 below sea level.

Even though the Hawaiian volcanoes are unrelated to spreading ridges or subduction zones, their evolution is related to plate movements. Notice in Figure 2.21 that the ages of the rocks composing the various Hawaiian islands increase toward the northwest; Kauai formed 3.8 to 5.6 million years ago, whereas Hawaii began forming less than 1 million years, and Loihi began forming even more recently. Continuous movement of the Pacific plate over the hot spot, now beneath Hawaii and Loihi, has formed the islands in succession.

Mantle plumes and hot spots have also been proposed to explain volcanism in a few other areas. A mantle plume may exist beneath Yellowstone National Park in Wyoming, for instance. Some source of heat at depth is responsible for the present-day hot springs and geysers such as Old Faithful, but many geologists think the heat source is a body of intruded magma that has not yet completely cooled, rather than a mantle plume.

FIGURE 5.21 Intrusive and extrusive igneous activity at a spreading ridge. The oceanic crust is composed largely of vertical dikes of basaltic composition and gabbro that appears to have crystallized in the upper part of a magma chamber. The upper part of the oceanic crust consists of submarine lavas, especially pillow lavas.

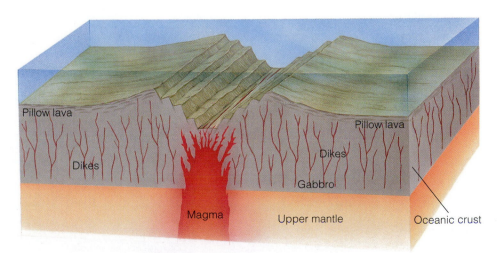

1. Volcanism is the process whereby magma and its associated gases erupt at the surface. Some magma erupts as lava flows, and some is ejected explosively as pyroclastic materials.

2. Only a few percent by weight of a magma consists of gases, most of which is water vapor. Sulfur gases emitted during large eruptions can have far-reaching climatic effects.

3. The surface of an aa lava flow consists of rough, angular blocks, whereas a pahoehoe flow has a smoothly wrinkled surface.

4. Columnar joints form in some lava flows when they cool. Pillow lavas form under water and consist of interconnected bulbous masses.

5. Volcanoes are conical mountains built up around a vent where lava flows and/or pyroclastic materials are erupted.

6. Shield volcanoes have low, rounded profiles and are composed mostly of mafic flows that cooled to form basalt. Small, steep-sided cinder cones form where pyroclastic materials that resemble cinders are erupted and accumulate. Composite volcanoes are composed of lava flows of intermediate composition, layers of pyroclastic materials, and volcanic mudflows known as lahars.

7. Viscous masses of lava, generally of felsic composition, are forced up through the conduits of some volcanoes and form bulbous lava domes. Volcanoes with lava domes are dangerous because they erupt explosively and frequently eject nuée ardentes.

8. The summits of volcanoes are characterized by a circular or oval crater or a much larger caldera. Many calderas form by summit collapse when an underlying magma chamber is partly drained.

9. Fluid mafic lava erupted from long fissures (fissure eruptions) spreads over large areas to form basalt plateaus.

10. Pyroclastic flows erupted from fissures formed during the origin of calderas cover vast areas. Such eruptions of pyroclastic materials form pyroclastic sheet deposits.

11. Most active volcanoes are distributed in linear belts. The circum-Pacific belt and Mediterranean belt contain more than 80% of all active volcanoes.

12. Volcanism and plutonism take place at speading ridges where plates diverge and at convergent plate margins where subduction occurs. Partial melting of a subducted plate generates intermediate and felsic magmas.

13. Magma derived by partial melting of the upper mantle beneath spreading ridges accounts for the mafic plutons and lavas of ocean basins.

14. The two active volcanoes on the island of Hawaii and one just to the south are thought to lie above a hot mantle plume. The Hawaiian Islands developed as a series of volcanoes formed on the Pacific plate as it moved over the mantle plume.

aa

ash

basalt plateau

caldera

cinder cone

circum-Pacific belt

columnar joint

composite volcano (stratovolcano)

crater

fissure eruption

lahar

lava dome

mantle plume

Mediterranean belt

nuée ardente

pahoehoe

pillow lava

pyroclastic sheet deposit

shield volcano

volcanism

volcano

1. The most abundant volcanic gas is:
   a. _____ carbon dioxide;
   b. _____ methane;
   c. _____ hydrogen sulfide;
   d. _____ chlorine;
   e. _____ water vapor.

2. When lava is quickly chilled beneath water, it forms what is known as:
   a. _____ pillow lava;
   b. _____ columnar joints;
   c. _____ spatter cones;
   d. _____ pahoehoe;
   e. _____ pressure ridges.

3. The large depression occupied by Crater Lake, Oregon is actually a:
   a. _____ parasitic cone;
   b. _____ caldera;
   c. _____ lava dome;
   d. _____ lahar;
   e. _____ cinder cone.

4. Basalt plateaus form when:
   a. _____ viscous lava rises to the surface as a large bulge;
   b. _____ explosive volcanism yields large quantities of pyroclastic materials;
   c. _____ volcanic gases react with the atmosphere;
   d. _____ fluid lava flows erupt from fissures;
   e. _____ lava flows outward in all directions from a composite volcano.

5. The only currently active volcanoes in the continental United States are in the:
   a. _____ Rocky Mountains;
   b. _____ Appalachian Mountains;
   c. _____ Cascade range;
   d. _____ Ouachita Mountains;
   e. _____ Teton range.

6. Explosive volcanoes eject ash, lapilli, bombs, and blocks, which are known collectively as:
   a. _____ pyroclastic materials;
   b. _____ phoehoe and aa;
   c. _____ lava domes;
   d. _____ pegmatites;
   e. _____ fissure eruptions.

7. The slopes of shield volcanoes are rarely more than 10 degrees because they are composed mostly of:
   a. _____ ash and lapilli;
   b. _____ felsic magma beneath their summits;
   c. _____ lahars;
   d. _____ pyroclastic sheet deposits;
   e. _____ mafic lava flows.

8. The continuous ground motion that precedes volcanic eruptions is known as:
   a. _____ horizontal displacement;
   b. _____ inflation;
   c. _____ harmonic tremor;
   d. _____ rock stress;
   e. _____ fissure opening.

9. Composite volcanoes are found at:
   a. _____ spreading ridges;
   b. _____ transform plate boundaries;
   c. _____ convergent plate boundaries;
   d. _____ divergent plate boundaries;
   e. _____ submarine ridges.

10. The most commonly erupted type of pyroclastic materials is:
    a. _____ pahoehoe;
    b. _____ cinders;
    c. _____ carbon dioxide;
    d. _____ ash;
    e. _____ lapilli.

11. How do pahoehoe and aa lava flows differ, and what accounts for these differences?

12. Describe how a cinder cone forms and list its characteristics.

13. What are columnar joints and pillow lavas, and how do they form?

14. How do the volcanoes in the circum-Pacific belt and those along spreading ridges differ?

15. Why are lava domes so dangerous? Give an example of a lava dome that has erupted in this century.

16. How do pyroclastic sheet deposits and basalt plateaus differ, and what accounts for the origin of each?

17. Explain how volcanic proclastic materials and volcanic gases can affect climate.

18. Why do shield volcanoes have such low slopes?

19. How does a caldera form? An illustration would be helpful.

20. Describe the events leading to the origin of the Hawaiian Islands. Is there any evidence that the process responsible for their origin is continuing? If so, what?

1. During this century, two Cascade Range volcanoes have erupted. What kinds of evidence would indicate that some of the other volcanoes in this range might erupt in the future?

2. What geologic events would have to occur in order for a chain of volcanoes to form along the east coasts of Canada and the United States?

World Wide **Web** Activities

For these website addresses, along with current updates and excercises, log on to

http://www.wadsworth.com/geo

▶ MICHIGAN TECHNOLOGICAL UNIVERSITY VOLCANOES PAGE

This site provides scientific and educational information about volcanoes to the public. It contains information about current global volcanic activity, research about volcanoes, as well as links to government agencies and research institutes.

1. Click on the *Worldwide Volcanic Reference Map* site. Which regions are experiencing the greatest volcanic activity? Click on the *Recent and Ongoing Volcanic Activity* site. Check out one of the active volcanic sites. What type of volcano is erupting? What is the history of this volcano? Has it been monitored in the past, and is it currently being monitored?

2. Click on the *Remote Sensing of Volcanoes* site. What type of research is being conducted in the area of remote sensing of volcanoes? How is this beneficial to humans?

▶ U.S. GEOLOGICAL SURVEY CASCADES VOLCANO OBSERVATORY

This site contains information about volcanoes and other natural hazards in the western United States and elsewhere in the world. It is an excellent source of information on the hazards of volcanic eruptions and contains specific information about Mount St. Helens, as well as other volcanoes.

1. Click on the *Mt. St. Helens* site under the *Volcanic Information* section. What is the current activity level of Mount St. Helens? What are some of the current research projects on Mount St. Helens?

2. Click on the *Visit a Volcano* site under the *Educational Outreach* section. What are some of the volcanoes you can visit? Report on one of the volcanoes listed on this site.

▶ VOLCANOES AND GLOBAL CLIMATE CHANGE

This is a NASA Facts site and contains information on the relationship between volcanoes and global cooling and ozone depletion. Read about how volcanic eruptions affect global temperatures and how scientists monitor the various components erupted from a volcano.

▶ VOLCANO WORLD

This excellent volcano site is maintained and supported by NASA's Public Use of Earth and Space Science Data Over the Internet program. It contains a wealth of information about volcanoes and volcanic parks and monuments.

1. Click on the *What's Erupting Now* site. It will take you to a map showing all current volcanic activity and a listing of volcanoes. Click on one of the red triangles, which will take you to the volcano shown on the map where you can read about the volcano and see images and videos of its eruption. In what region of the world is current volcanism most active?

2. Click on the *Exploring Earth's Volcanoes* site. It will take you to a map of the world divided into regions. By clicking on a particular region, you can view active volcanoes from that region and learn more about the volcanoes. Compare the volcanoes from the different regions of the world.

3. Click on the *Volcanoes of Other Worlds* site. Check out volcanic activity on the Moon, Mars, and Venus. What kind of volcano(es) are present on Venus? What is the largest known volcano in the solar system and where is it?

4. Click on *How Big Are Volcanic Eruptions?* What is the VEI? What is the largest eruption in the chart, and where did it occur? How often do cataclysmic eruptions take place?

For these web site addresses, along with current updates and excercises, log on to

http://www.wadsworth.com/geo

▶ THE ELECTRONIC VOLCANO

This site is maintained by personnel at the library and Department of Earth Sciences, Dartmouth College, Hanover, New Hampshire. It contains information on active volcanoes, volcanic hazards, catalogs and maps of active volcanoes, current events and research, and a list of journals with articles on active volcanoes. Scroll down to Paricutin volcano and read the account. When and where did Paricutin erupt? Scroll to Benjamin Franklin's "1784 paper." What did he conclude regarding eruptions and climate?

▶ VOLCANO WATCH

This is a weekly newsletter mostly about the volcanoes on the island of Hawaii. It is written for the general public by scientists at the USGS's Hawaiian Volcano Observatory. The site contains information regarding the current status of erupting volcanoes in Hawaii.

▶ TERRESTRIAL VOLCANOES

A site with information on many aspects of volcanism, it includes maps and photos of notable eruptions and descriptions of eruptions and their effects.

1. How much lava was erupted during the 1783 eruption at Laki, Iceland?

2. What are the relationships between lava domes and composite volcanoes?

3. Why does Mount Pinatubo in the Philippines continue to be a major hazard?

▶ GLOBAL VOLCANISM PROGRAM

This database at the Smithsonian Institution contains information on eruptions that have occurred during the last 10,000 years. The site features *Volcanoes of the World, Volcano Net Links,* and the *Bulletin of the Global Volcanism Network,* a weekly newsletter compiled from data submitted by more than 1000 correspondents.

CD-ROM Exploration

Explore the following *In-Terra-Active 2.0* CD-ROM module(s) and increase your understanding of key concepts and processes presented in this chapter.

▶ SECTION: MATERIALS
MODULE: IGNEOUS ROCKS: PROCESSES AND STRUCTURE

The crystallization of molten rock can, under different conditions, produce both the finest- and the coarsest-grained rocks known on Earth. At one extreme, volcanic glass (obsidian) is so fine-grained as to be amorphous in structure—that is, not crystalline at all. At the other extreme, pegmatites can have individual crystals that measure 10 meters long. Then there is the odd case of porphyry, a rock that is mixed in texture. **Question: In the case of porphyry, how can you account for the different grain sizes within a single rock?**

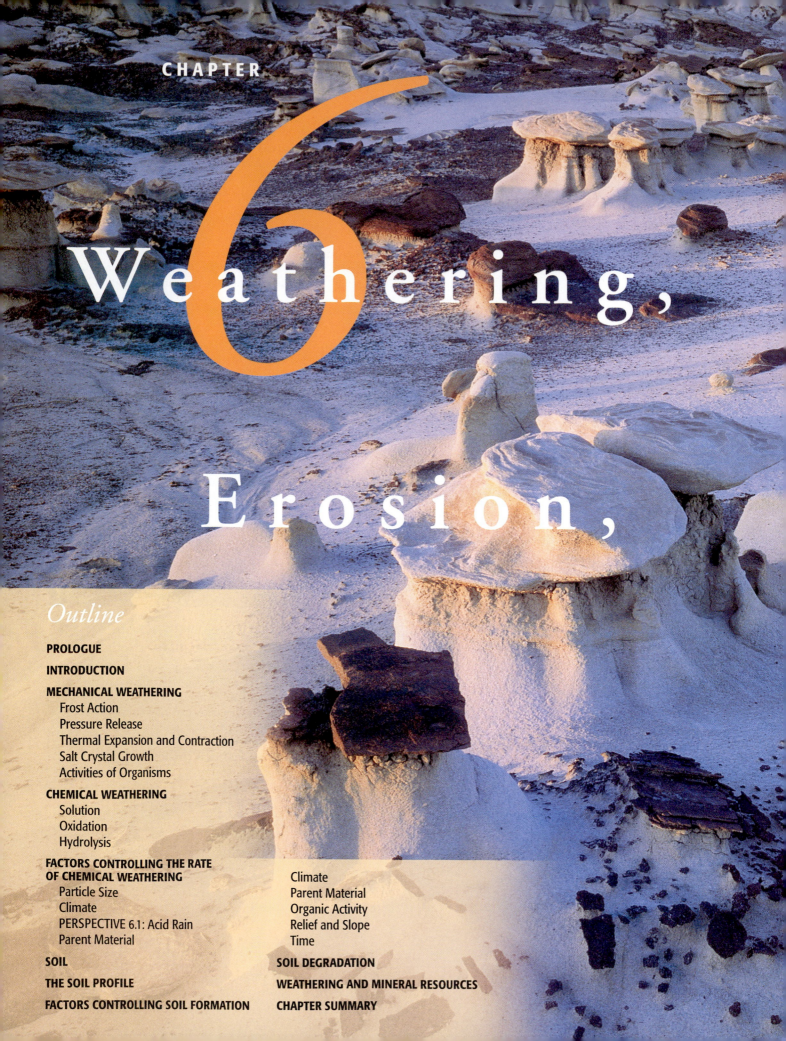

6

Weathering,

Erosion,

PROLOGUE

INTRODUCTION

MECHANICAL WEATHERING
Frost Action
Pressure Release
Thermal Expansion and Contraction
Salt Crystal Growth
Activities of Organisms

CHEMICAL WEATHERING
Solution
Oxidation
Hydrolysis

**FACTORS CONTROLLING THE RATE
OF CHEMICAL WEATHERING**
Particle Size
Climate
PERSPECTIVE 6.1: Acid Rain
Parent Material

Climate
Parent Material
Organic Activity
Relief and Slope
Time

SOIL

THE SOIL PROFILE

FACTORS CONTROLLING SOIL FORMATION

SOIL DEGRADATION

WEATHERING AND MINERAL RESOURCES

CHAPTER SUMMARY

Prologue

North America possesses many areas of exceptional scenery that were shaped by a variety of geologic processes such as volcanism, glaciation, shoreline erosion, and deformation. But certainly some of the most striking examples of scenery yielded by the combined effects of weathering and erosion are found in Bryce Canyon National Park, Utah, and Badlands National Park, South Dakota. In both cases, brilliantly colored rocks have been intricately sculpted to form mazes of interconnected gullies and a variety of oddly shaped features. Scattered localities of similar features are present from Alberta, Canada, to Arizona.

Weathering is a pervasive phenomenon that alters Earth materials by physical and chemical processes. Its effects are not evenly distributed, however, because rocks are not uniform in their resistance to changes induced by weathering. Accordingly, parts of a rock mass may be altered more rapidly than adjacent areas of the same rock mass. Similarly, erosion involving the removal of weathered materials may proceed unevenly, thus yielding irregular surfaces.

Badlands develop in dry areas with sparse vegetation and nearly impermeable yet easily eroded rocks. Rain falling on such unprotected rocks rapidly runs off and intricately dissects the surface by forming numerous closely spaced, small gullies and deep ravines, separated by sharp angular slopes, ridges, and pinnacles.

Weathering and erosion of sedimentary rocks are responsible for the scenery in Bryce Canyon National Park, Utah. *(Photo courtesy of Frank Hanna.)*

and

Soil

(a)   (b)

FIGURE 6.1 Exposures of rocks in Badlands National park, South Dakota. (a) Weathering and erosion of the Brule Formation (on the skyline) yields numerous, closely spaced gullies and sharp, angular slopes, while smooth rounded slopes develop on the Chadron Formation (foreground). (b) Color banding is a distinctive feature of some rocks in the park.

The rocks in Bryce Canyon National Park are commonly described as limy siltstones, meaning they are composed of silt-sized ($^1/_{256}$–$^1/_{16}$ mm) particles, with calcite acting as a cement in the spaces between grains. Some layers, however, are composed mostly of calcite and are thus varieties of limestones. Collectively, the rocks are designated as the Wasatch Formation, which was deposited mostly in a lake 40 to 50 million years ago, but was subsequently uplifted along a large fracture and now forms the Pink Cliffs.

The rocks in Bryce Canyon are some of the most brilliantly colored in the world. The Piute referred to the area as "red rocks standing like men in a bend-shaped canyon." Bryce Canyon is not really a canyon at all; rather, it is the eroded eastern margin of high, fairly flat area known as the Paunsaugunt Plateau. In any case, the rocks respond differently to the effects of weathering and erosion, thus yielding a spectacular example of badlands (see chapter-opening photo). For instance, the limestone layers are harder and more resistant than the silty layers, and weathering and erosion tend to concentrate along numerous, closely spaced frac-

tures. The result is the origin of spires, pillars, monuments, arches, fluted ridges, gullies, and ravines.

Weathering and erosion of the Brule Formation in Badlands National Park have also yielded an excellent example of badlands, whereas the underlying Chadron Formation responds differently to these same processes and forms smoothly rounded slopes (Figure 6.1). All these rocks were originally deposited as sediment in stream channels, their adjacent floodplains, and small lakes. Yet subtle differences in the two formations account for their differing responses to weathering and erosion. The Brule Formation possesses more clay and more chemical cement holding the particles together, so a complex network of small channels is eroded on its surface. In contrast, the Chadron Formation has less clay and cement, water infiltrates more readily, and channeling is uncommon.

South Dakota's badlands are reason enough to visit the park, but it has more to offer; the rocks contain numerous land mammal fossils. The U.S. Park Service has left a number of these fossil mammals exposed but protected for viewing by park visitors. Among the mammals are rodents, doglike carnivores, saber-toothed cats, camels, horses, and extinct hoofed mammals known as titanotheres and oreodonts.

In both of these examples, the combined effects of weathering and erosion are responsible for the spectacular scenery. However, one sometimes

hears that wind erosion also played a role, especially at Bryce Canyon. Wind can be an effective geologic agent in some areas, but its role in modifying Earth's surface is commonly exaggerated; in the case of Bryce Canyon and Badlands National Parks, its impact has been minimal at best.

Introduction

THE physical breakdown (disintegration) and chemical alteration (decomposition) of rocks and minerals at or near Earth's surface is known as **weathering**. It includes processes whereby rocks and minerals are physically and chemically altered so that they are more nearly in equilibrium with a new set of environmental conditions. Many rocks form within the crust where little or no water or oxygen is present and where temperatures, pressures, or both are high. At or near the surface, the rocks are exposed to low temperatures and pressures and are attacked by atmospheric gases, water, acids, and organisms.

Geologists are interested in the phenomenon of weathering because it is an essential part of the rock cycle (see Figure 1.12). The **parent material**, or rock being weathered, is broken down into smaller pieces, and some of its constituent minerals are dissolved or altered and removed from the weathering site. The removal of the weathered materials is known as **erosion**. Running water, wind, or glaciers commonly **transport** the weathered materials elsewhere, where they are deposited as sediment, which may become sedimentary rock. Whether they are eroded or not, weathered rock materials can be further modified to form soil. Thus, weathering provides the raw materials for both sedimentary rocks and soils. Weathering is also important in the origin of some mineral resources such as aluminum ores, and it is responsible for the enrichment of other deposits of economic importance.

Weathering is such a pervasive phenomenon that many people take it for granted or completely overlook it. Nevertheless, it occurs continuously although its rate and impact vary from area to area or even within the same area. Rocks do not weather at the same rate, even in a single rock layer, because of slight differences in composition and structure. Weathering is more intense on fractures than on adjacent areas of unfractured rock, for example. As a result of these variations, **differential weathering** occurs, which means that rocks weather at different rates yielding uneven surfaces and peculiar shapes (Figure 6.2) (see the Prologue).

Two types of weathering are recognized, mechanical and chemical. Both types proceed simultaneously at the weathering site, during erosion and transport, and even in the environments where weathered materials are deposited.

Mechanical Weathering

MECHANICAL **weathering** takes place when physical forces break rock materials into smaller pieces that retain the chemical composition of the parent material. Granite, for instance, may be mechanically weathered to yield smaller pieces of granite, or disintegration may liberate individual mineral grains from it (Figure 6.3). The physical processes responsible for mechanical weathering include frost action, pressure release, thermal expansion and contraction, salt crystal growth, and the activities of organisms.

(a)

(b)

FIGURE 6.2 Differential weathering has yielded these odd shapes and surfaces. (a) Camel Rock near Santa Fe, New Mexico. (b) This intricate, uneven weathering surface at Pebble Beach, California, is an example of honeycomb weathering. *(Photos courtesy of Sue Monroe.)*

FROST ACTION

Frost action involves repeated freezing and thawing of water in cracks and crevices in rocks. When water seeps into a crack and freezes, it expands by about 9% and exerts great force on the walls of the crack, thereby widening and extending it by **frost wedging**. As a result of numerous episodes of freezing and thawing, pieces of rock are even-tually detached from the parent material (Figure 6.4). Frost wedging is particularly effective if the crack is convoluted because, if it is a simple wedge-shaped opening, much of the force of expansion is released up toward the surface. The debris produced by frost wedging in mountains commonly accumulates as large cones of **talus** lying at the bases of slopes (Figure 6.5).

Frost action is most effective in areas where temperatures commonly fluctuate above and below freezing, as in the high mountains of the western United States and Canada. In the tropics and in areas where water is permanently frozen, frost action is of little or no importance.

FIGURE 6.3 Mechanically weathered granite. The sandy material consists of small pieces of granite (rock fragments) and minerals such as quartz and feldspars liberated from the parent material.

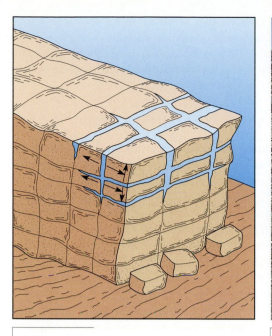

FIGURE 6.4 Frost wedging occurs when water seeps into cracks and expands as it freezes. Repeated freezing and thawing pry loose angular pieces of rock.

FIGURE 6.5 Talus in the Bighorn Mountains, Wyoming.

In the phenomenon known as **frost heaving**, a mass of sediment or soil undergoes freezing, expansion, and actual lifting, followed by thawing, contraction, and lowering of the mass. Frost heaving is particularly evident where water freezes beneath roadways and sidewalks.

PRESSURE RELEASE

The mechanical weathering process called **pressure release** is especially evident in rocks that formed as deeply buried intrusive bodies such as batholiths, but it occurs in other types of rocks as well. When a batholith forms, the magma crystallizes under tremendous pressure (the weight of the overlying rock) and is stable under these pressure conditions. But if the batholith is uplifted and the overlying rock is eroded, the pressure is reduced. However, the rock contains energy that is released by expansion and the formation of **sheet joints**, large fractures that more or less parallel the rock surface (Figure 6.6). Slabs of rock bounded by sheet joints may slip, slide, or spall (break) off of the host rock—a process called **exfoliation**—and accumulate as talus. The large rounded domes of rock resulting from this

FIGURE 6.6 Sheet joints in granite in the Sierra Nevada of California.

process are **exfoliation domes**; examples are found in Yosemite National Park in California and Stone Mountain in Georgia (Figure 6.7).

The fact that solid rock can expand and produce fractures is a well-known phenomenon. In deep mines, masses of rock suddenly detach from the sides of the excavation, often with explosive violence. Spectacular examples of these rock bursts have been recorded in deep mines, where they and related phenomena, such as less violent *popping*, pose a danger to mine workers. In South Africa, about 20 miners are killed by rock bursts every year.

In some quarrying operations,* the removal of surface materials to a depth of only 7 or 8 m has led to the formation of sheet joints in the underlying rock (Figure 6.8). At quarries in Vermont and Tennessee, the excavation of marble exposed rocks that were formerly buried and under great pressure. When the overlying rock was removed, the marble expanded and sheet joints formed.

---

*A quarry is a surface excavation, generally for the extraction of building stone.

FIGURE 6.7 Stone Mountain, Georgia, is a large exfoliation dome.

Some slabs of rock that were bounded by sheet joints burst so violently that quarrying machines weighing more than a ton were thrown from their tracks, and some quarries had to be abandoned because fracturing rendered the stone useless.

THERMAL EXPANSION AND CONTRACTION

During **thermal expansion and contraction**, the volume of solids, such as rocks, changes in response to heating and cooling. In a desert, where the temperature may vary as much as 30°C in one day, rocks expand when heated and contract as they cool. Rock is a poor conductor of heat, so its outside heats up more than its inside, and the surface expands more than the interior, creating stresses that may cause fracturing. Furthermore, dark minerals absorb heat faster than light-colored ones, so differential expansion occurs even between the mineral grains of some rocks.

Experiments in which rocks are heated and cooled repeatedly to simulate years of thermal expansion and contraction indicate that it is not an important agent of mechanical weathering. But thermal expansion and contraction may be a significant mechanical weathering process on the Moon where extreme temperature changes occur quickly.

Daily temperature variation is the most common cause of alternate expansion and contraction, but these changes take place over periods of hours. In contrast, fire can cause very rapid expansion. During a forest fire, rocks may heat very rapidly, especially near the surface, because they conduct heat so poorly. The heated surface layer expands more rapidly than the interior, and thin sheets paralleling the rock surface become detached.

SALT CRYSTAL GROWTH

Under some circumstances, salt crystals forming from solution can cause disaggregation of rocks. Growing crystals exert enough force to widen cracks and crevices or dislodge particles in porous, granular rocks such as sandstones. Even in crystalline rocks such as granite, **salt crystal growth** can pry loose individual mineral grains. To the extent that salt crystal growth produces forces that expand openings in rocks, it is similar to frost wedging. Most salt crystal growth occurs in hot arid areas, although it probably affects rocks in some coastal regions as well.

ACTIVITIES OF ORGANISMS

Animals, plants, and bacteria all participate in the mechanical and chemical alteration of rocks. Burrowing animals, such as worms, reptiles, rodents, and many others, constantly mix soil and sediment particles and bring material from depth to the surface where further weathering may occur. Even materials ingested by worms are further reduced in size, and animal burrows allow gases and water to have easier access to greater depths. The

roots of plants, especially large bushes and trees, wedge themselves into cracks in rocks and further widen them (Figure 6.9).

Chemical Weathering

CHEMICAL weathering is the process whereby rock materials are decomposed by chemical alteration of the parent material. A number of clay minerals, for example, form as the chemically altered products of other minerals. Some minerals are completely decomposed during chemical weathering, but others, which are more resistant, are simply liberated from the parent material. Chemical weathering is accomplished by the action of atmospheric gases, especially oxygen, and water and acids. Organisms also play an important role in chemical weathering. Rocks with lichens (composite organisms consisting of fungi and algae) growing on their surfaces undergo more extensive chemical alteration than lichen-free rocks. Plants remove ions from soil water and reduce the chemical stability of soil minerals, and their roots release organic acids.

SOLUTION

During **solution** the ions of a substance become separated in a liquid, and the solid substance dissolves. Water is a remarkable solvent because its molecules have an asymmetric shape; they consist of one oxygen atom with two hydrogen atoms arranged so that the angle between the two hydrogens is about 104 degrees (Figure 6.10). Because of this asymmetry, the oxygen end of the molecule retains a slight negative electrical charge, whereas the hydrogen end retains a slight positive charge. When a soluble substance such as the mineral halite (NaCl) comes in contact with a water molecule, the positively charged sodium ions are attracted to the negative end of the water molecule, and the negatively charged chloride ions are attracted to the positively charged end of the water molecule (Figure 6.10). Thus, ions are liberated from the crystal structure, and the solid dissolves.

FIGURE **6.9** The contribution of organisms to mechanical weathering. Tree roots enlarge cracks in rocks.

FIGURE 6.10 (a) The structure of a water molecule. The asymmetric arrangement of the hydrogen atoms causes the molecule to have a slight positive electrical charge at its hydrogen end and a slight negative charge at its oxygen end. (b) The dissolution of sodium chloride (NaCl) in water.

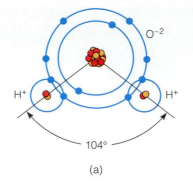

(a)

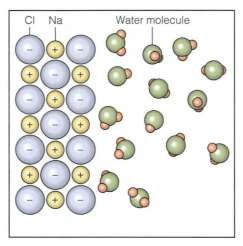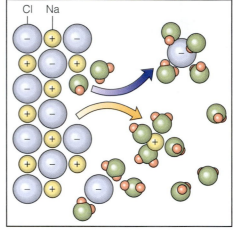

(b)

Most minerals are not very soluble in pure water because the attractive forces of water molecules are not sufficient to overcome the forces between particles in minerals. The mineral calcite ($CaCO_3$), the major constituent of the sedimentary rock limestone and the metamorphic rock marble, is practically insoluble in pure water but rapidly dissolves if a small amount of acid is present. An easy way to make water acidic is by dissociating the ions of carbonic acid as follows:

$$H_2O \; + \; CO_2 \; \rightleftharpoons \; H_2CO_3 \; \rightleftharpoons \; H^+ \; + \; HCO_3^-$$

WATER   CARBON   CARBONIC   HYDROGEN   BICARBONATE
DIOXIDE   ACID   ION   ION

According to this chemical equation, water and carbon dioxide combine to form *carbonic acid,* a small amount of which dissociates to yield hydrogen and bicarbonate ions. The concentration of hydrogen ions determines the acidity of a solution; the more hydrogen ions present, the stronger the acid.

Carbon dioxide from several sources may combine with water and react to form acid solutions. The atmosphere is mostly nitrogen and oxygen, but about 0.03% is carbon dioxide, causing rain to be slightly acidic. Human activities have added materials to the atmosphere that contribute to the problem of acid rain (see Perspective 6.1). Decaying organic matter and the respiration of organisms produces carbon dioxide in soils, so groundwater in humid areas is slightly acidic. Arid regions have sparse vegetation, so groundwater has a limited source of carbon dioxide and tends to be

alkaline—that is, it has a low concentration of hydrogen ions.

Whatever the source of carbon dioxide, once an acidic solution is present, calcite rapidly dissolves according to the following reaction:

$$CaCO_3 \; + \; H_2O \; + \; CO_2 \; \rightleftharpoons \; Ca^{+2} \; + \; 2HCO_3^-$$

CALCITE   WATER   CARBON   CALCIUM   BICARBONATE
DIOXIDE   ION   ION

In many places, the dissolution of the calcite in limestone and marble has had dramatic effects, ranging from small cavities to large caverns such as Mammoth Cave in Kentucky and Carlsbad Caverns in New Mexico.

OXIDATION

Oxidation refers to reactions with oxygen to form oxides or, if water is present, hydroxides. For example, iron rusts when it combines with oxygen to form the iron oxide hematite:

$$4Fe \; + \; 3O_2 \; \rightarrow \; 2Fe_2O_3$$

IRON   OXYGEN   IRON OXIDE
(HEMATITE)

Of course, atmospheric oxygen is abundantly available for oxidation reactions, but oxidation is generally a slow process unless water is present. Most oxidation is carried out by oxygen dissolved in water.

Oxidation is very important in the alteration of ferromagnesian silicates such as olivine, pyroxenes, amphiboles, and biotite. Iron in these minerals combines with oxygen to form the reddish iron oxide hematite (Fe_2O_3) or the yellowish or brown hydroxide limonite [$FeO(OH) \cdot nH_2O$]. The yellow, brown, and red colors of many soils and sedimentary rocks are caused by the presence of small amounts of hematite or limonite.

An oxidation reaction of particular concern in some areas is the oxidation of iron- and sulfur-bearing minerals such as pyrite (FeS_2). Pyrite is commonly associated with coal, so in mine tailings* pyrite oxidizes to form sulfuric acid (H_2SO_4) and iron oxide. Acid soils and waters in coal-mining areas are produced in this manner and present a serious environmental hazard (Figure 6.11).

HYDROLYSIS

Hydrolysis is the chemical reaction between the hydrogen (H^+) ions and hydroxyl (OH^-) ions of water and a mineral's ions. In hydrolysis, hydrogen ions actually replace positive ions in minerals. This replacement changes the composition of minerals by liberating soluble substances and iron that may then be oxidized.

As an illustration of hydrolysis, consider the chemical alteration of feldspars. All feldspars are framework silicates, but when altered they yield compounds in solution and clay minerals, such as kaolinite, which are sheet silicates.

The chemical weathering of potassium feldspar by hydrolysis occurs as follows:

$$2KAlSi_3O_8 \ + \ 2H^+ \ + \ 2HCO_3^- \ + \ H_2O \rightarrow$$

ORTHOCLASE — HYDROGEN ION — BICARBONATE ION — WATER

$$Al_2Si_2O_5(OH)_4 \ + \ 2K^+ \ + \ 2HCO_3^- \ + \ 4SiO_2$$

CLAY (KAOLINITE) — POTASSIUM ION — BICARBONATE ION — SILICA

In this reaction, hydrogen ions attack the ions in the orthoclase structure, and some liberated ions are incorporated in a developing clay mineral, while others simply go into solution. On the right side of the equation is excess silica that would not fit into the crystal structure of the clay mineral.

Factors Controlling the Rate of Chemical Weathering

CHEMICAL weathering processes operate on the surfaces of particles; that is, chemically weathered rocks or minerals are altered from the outside inward. Several factors including particle size, climate, and parent material control the rate of chemical weathering.

PARTICLE SIZE

Because chemical weathering affects particle surfaces, the greater the surface area, the more effective the weathering. It is important to realize that small particles have larger

FIGURE 6.11 The oxidation of pyrite in mine tailings forms acid water as in this small stream. More than 11,000 km of U.S. streams, mostly in the Appalachian region, are contaminated by abandoned coal mines that leak sulfuric acid.

surface areas compared to their volume than do large particles. Notice in Figure 6.12 that a block measuring 1 m on a side has a total surface area of 6 m^2, but when the block is broken into particles measuring 0.5 m on a side, the total surface area increases to 12 m^2. And if these particles are all reduced to 0.25 m on a side, the total surface area increases to 24 m^2. Note that while the surface area in this example increases, the total volume remains the same at 1 m^3.

We can make two important statements regarding the block in Figure 6.12. First, as it is split into a number of smaller blocks, its total surface area increases. Second, the smaller any single block is, the more surface area it has compared to its volume. We can conclude that mechanical weathering, which reduces the size of particles, contributes to chemical weathering by exposing more surface area.

CLIMATE

Chemical processes proceed more rapidly at high temperatures and in the presence of liquids. Accordingly, it is not surprising that chemical weathering is more effective in the tropics than in arid and arctic regions because temperatures and rainfall are high and evaporation rates are low. In

---

Tailings are the rock debris of mining; they are considered too poor for further processing and are left as heaps on the surface.

Acid Rain

One of the consequences of industrialization is atmospheric pollution. Several of the most industrialized nations, such as the United States, Canada, and Russia, have actually reduced their emissions into the atmosphere, but many developing nations continue to increase theirs. Some of the consequences of atmospheric pollution include smog, possible disruption of the ozone layer, global warming, and *acid rain.*

Acidity is an indication of hydrogen ion concentration and is measured on the pH scale (Figure 1). A pH of 7 is neutral, whereas values less than 7 indicate acidic conditions, and values greater than 7 indicate alkaline or basic conditions. Normal rainfall has a pH of about 5.6, making it slightly acidic. Acid rain is defined as rainfall with a pH value of less than 5.0. In addition to acid rain, one may experience acid snow in colder regions and acid fog with pH values as low as 1.7 in some industrialized areas.

Recall that water and carbon dioxide in the atmosphere react to form carbonic acid that dissociates and yields hydrogen ions and bicarbonate ions. The effect of this reaction is that all rainfall is slightly acidic. Thus, acid rain is the direct result of the self-cleansing nature of the atmosphere; that is, many suspended particles or gases in the atmosphere are soluble in water and are removed from the atmosphere during precipitation events.

Several natural processes, including volcanism and the activities of soil bacteria, introduce gases into the atmosphere that cause acid rain. Human activities, however, produce added atmospheric stress. For instance, the burning of fossil fuels (oil, natural gas, and coal) has added carbon dioxide to the atmosphere.

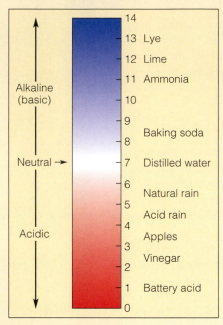

Figure 1 *The pH scale. Values less than 7 are acidic, whereas those greater than 7 are alkaline. This is a logarithmic scale, so a decrease of one unit is a tenfold increase in acidity.*

Nitrogen oxide (NO) from internal combustion engines and nitrogen dioxide (NO_2), which is formed in the atmosphere from NO, react to form nitric acid (HNO_3). Although carbon dioxide and nitrogen gases contribute to acid rain, the greatest culprit is sulfur dioxide (SO_2), which is primarily released by burning coal that contains sulfur. Once in the atmosphere, sulfur dioxide reacts with oxygen to form sulfuric acid (H_2SO_4), the main component of acid rain.

The phenomenon of acid rain was first recognized in England by Robert Angus Smith in 1872, about a century after the beginning of the Industrial Revolution. It was not until 1961, however, that acid rain become a public environmental concern. At that time, it was realized that acid rain is corrosive and irritating, kills vegetation, and has a detrimental effect on surface waters. Since then, the effects of acid rain have been recognized in Europe, especially in Eastern Europe where so much coal is burned, the eastern United States, and southeastern Canada (Figure 2). During the last ten years, the developed countries have made efforts to reduce the impact of acid rain; in the United States, the Clean Air Act of 1990 outlined specific steps to reduce the emissions of pollutants that cause acid rain.

The areas most affected by acid rain invariably lie downwind from coal-burning power plants or other industries that emit sulfur gases. Chemical plants and smelters (plants where metal ores are refined) discharge large quantities of sulfur gases and other substances such as heavy metals. The effect of acid rain in these areas may be modified by the existing geology. If an area is underlain by limestone or alkaline soils, for example, the acid rain tends to be neutralized by the limestone or soil. Areas underlain by granite, on the other hand, are acidic to begin with and have little or no effect on the rain.

The effects of acid rain vary. Small lakes become more acid as they lose the ability to neutralize acid rainfall. As the lakes increase in acidity, various types of organisms disappear, and, in some cases, all life-forms eventually die. Acid rain also causes increased weathering of limestone and marble (recall that both are soluble in weak acids) and, to a lesser degree, sandstone. Such effects are particularly visible on buildings, monuments, and tombstones; a notable example is Gettysburg National Military Park in Pennsylvania, which lies in an area that receives some of the most acidic rain in the country.

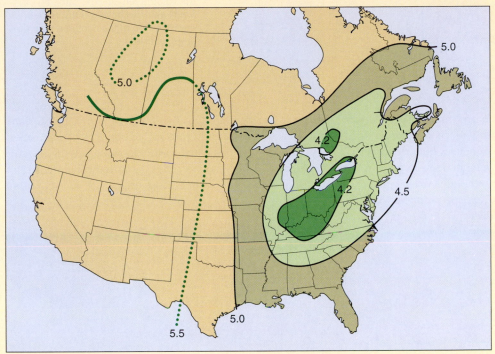

Figure 2 Average pH values for rain and snow in the United States and Canada. The area encompassed by a line has a pH value equal to or less than the value shown on the line.

Although the effects on vegetation in the immediate areas of industries emitting sulfur gas are apparent, some people have questioned whether acid rain has much effect on forests and crops distant from these sources. Nevertheless, many forests in the eastern United States show signs of stress that cannot be attributed to other causes. In Germany's Black Forest, the needles of firs, spruce, and pines are turning yellow and falling off.

Currently, about 20 million tons of sulfur dioxide are released yearly into the atmosphere in the United States, mostly from coal-burning power plants. Power plants built before 1975 have no emission controls and must be addressed if emissions are to be reduced to an ac-

ceptable level. The most effective way to reduce emissions from these older plants is with flue-gas desulfurization (FGD), a process that removes up to 90% of sulfur dioxide from exhaust gases. There are drawbacks to FGD, however. One is that some plants are simply too old to be profitably upgraded; the 85-year-old Phelps Dodge copper smelter in Douglas, Arizona, closed in 1987 for this reason. Other problems with FGD include disposal of sulfur wastes, the lack of control on nitrogen gas emissions, and reduced efficiency of the power plant, which must burn several percent more coal.

Other ways to control emissions include the conservation of electricity; the less electricity used, the lower the emis-

sions of pollutants. Natural gas contains practically no sulfur, but converting to this alternate energy source would require the installation of expensive new furnaces in existing plants.

Acid rain is a global problem that knows no national boundaries. Wind currents may blow pollutants from the source in one country to another where the effects are felt. Developed nations have the economic resources to reduce emissions, but many underdeveloped nations cannot afford to do so. Furthermore, many nations have access to only high-sulfur coal and cannot afford to install FGD devices. Nevertheless, acid rain can be controlled only by the cooperation of all nations contributing to the problem.

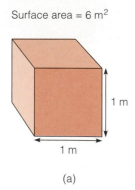

Surface area = 6 m²

1 m

1 m

(a)

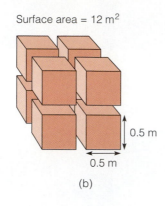

Surface area = 12 m²

0.5 m

0.5 m

(b)

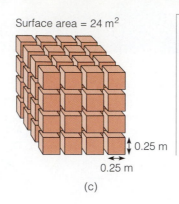

Surface area = 24 m²

0.25 m

0.25 m

(c)

FIGURE **6.12** Particle size and chemical weathering. As a rock is reduced to smaller and smaller particles, its surface area increases, but its volume remains the same. In (a) the surface area is 6 m², in (b) 12 m², and in (c) 24 m², but the volume remains the same at 1 m³. Small particles have more surface area in proportion to their volume than do large particles.

addition, vegetation and animal life are much more abundant in the tropics. Consequently, the effects of weathering extend to depths of several tens of meters, but commonly extend only centimeters to a few meters deep in arid and arctic regions. One should realize, though, that chemical weathering goes on everywhere, except perhaps where Earth materials are permanently frozen.

PARENT MATERIAL

Some rocks are chemically more stable than others and are not altered as rapidly by chemical processes. The metamorphic rock quartzite, composed of quartz, is an extremely stable substance that alters very slowly compared to most other rock types. In contrast, a rock such as basalt, which contains large amounts of calcium-rich plagioclase and pyroxenes, decomposes rapidly because these minerals are chemically unstable. In fact, the stability of common minerals is just the opposite of their order of crystallization in Bowen's reaction series (Table 6.1): The minerals that form last in this series are chemically stable, whereas those that form early are easily altered by chemical processes.

One manifestation of chemical weathering is **spheroidal weathering** (Figure 6.13). In spheroidal weathering, a stone, even one that is rectangular to begin with, weathers to form a spheroidal shape because that is the most stable

shape it can assume. On a rectangular stone, the corners are attacked by weathering processes from three sides, and the edges are attacked from two sides, but the flat surfaces are weathered more or less uniformly (Figure 6.13a). Consequently, the corners and edges are altered more

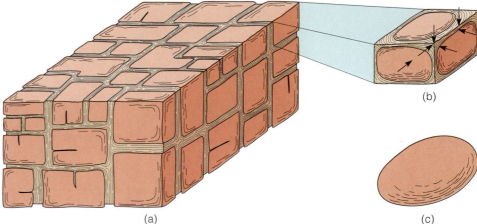

(a)

(b)

(c)

(d)

FIGURE **6.13** Spheroidal weathering. (a) The rectangular blocks outlined by joints are attacked by chemical weathering processes, (b) but the corners and edges are weathered most rapidly. (c) When a block has been weathered so that it is spherical, its entire surface is weathered evenly, and no further change in shape occurs. (d) Spheroidal weathering of granite in Point Reyes National Seashore, California.

TABLE **6.1**	
Stability of Silicate Minerals	
FERROMAGNESIAN SILICATES	NONFERROMAGNESIAN SILICATES
Olivine	Calcium plagioclase
Pyroxene	
Amphibole	Sodium plagioclase
Biotite	Potassium feldspar
	Muscovite
	Quartz

Increasing Stability

(a)

(b)

FIGURE 6.14 (a) Residual soil developed on bedrock near Denver, Colorado. (b) Transported soil developed on a wind-blown dust deposit.

rapidly, the material sloughs off them, and a more spherical shape develops. Once a spherical shape is present, all surfaces are weathered at the same rate.

Spheroidal weathering is often observed in granitic rock bodies cut by fractures. Fluids follow the fracture surfaces and reduce rectangular blocks to a spherical shape (Figure 6.13d).

Soil

IN most places the land surface is covered by a layer of **regolith**, consisting of unconsolidated rock and mineral fragments. Regolith may consist of volcanic ash, sediment deposited by wind, streams, or glaciers; or weathered rock material formed in place as a residue. Some regolith consisting of weathered material, water, air, and organic matter that can support plants is recognized as **soil**.

A good, fertile soil for gardening or farming is about 45% weathered rock material including sand, silt, and clay, but another essential constituent is **humus**. Many soils are dark colored by humus derived by bacterial decay of organic matter. It contains more carbon and less nitrogen than the original material and is resistant to further bacterial decay. Although a fertile soil may contain only a small amount of humus, it is an essential source of plant nutrients and enhances moisture retention.

Some weathered materials in soils are simply sand- and silt-sized mineral grains, especially quartz, but other weathered materials may be present as well. These solid particles are important because they hold soil particles apart, allowing oxygen and water to circulate more freely. Clay minerals are also important constituents of soils and aid in the retention of water as well as supplying nutrients to plants. Soils with excess clay minerals, however, drain poorly and are sticky when wet and hard when dry.

If a body of rock weathers and the weathering residue accumulates over it, the soil so formed is *residual,* meaning that it formed in place (Figure 6.14a). In contrast, *transported soil* develops on weathered material that has been eroded and transported from the weathering site and deposited elsewhere, such as on a stream's floodplain. Many fertile transported soils of the Mississippi River valley and the Pacific Northwest developed on deposits of windblown dust called *loess* (Figure 6.14b).

The Soil Profile

SOIL-FORMING processes begin at the surface and work downward, so the upper layer of soil is more altered from the parent material than the layers below. Observed in vertical cross section, a soil consists of distinct layers, or **soil horizons**, that differ from one another in texture, structure, composition, and color (Figure 6.15). Starting from

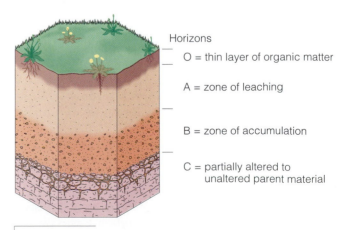

Horizons

O = thin layer of organic matter

A = zone of leaching

B = zone of accumulation

C = partially altered to unaltered parent material

FIGURE 6.15 The soil horizons in a fully developed or mature soil.

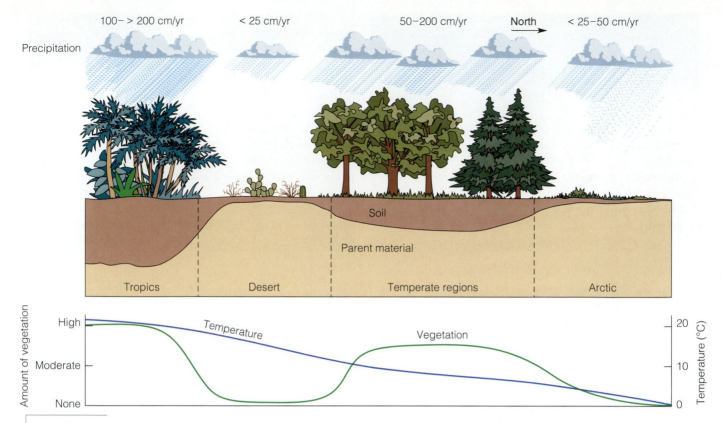

FIGURE 6.16 Schematic representation showing soil formation as a function of the relationship between climate and vegetation, which alters parent material over time. Soil-forming processes operate most vigorously where precipitation and temperatures are high.

the top, the horizons typical of soils are designated O, A, B, and C, but the boundaries between horizons are transitional rather than sharp.

The O horizon, which is generally only a few centimeters thick, consists of organic matter. The remains of plant materials are clearly recognizable in the upper part of the O horizon, but its lower part consists of humus. Horizon O is very thin or absent in soils of arid regions.

Horizon A, called *top soil,* contains more organic matter than horizons B and C, and it is also characterized by intense biological activity because plant roots, bacteria, fungi, and animals such as worms are abundant. Threadlike soil bacteria give freshly plowed soil its earthy aroma. In soils developed over a long period of time, the A horizon consists mostly of clays and chemically stable minerals such as quartz. Water percolating down through horizon A dissolves soluble minerals and carries them away or downward to lower levels in the soil by a process called **leaching** (Figure 6.15).

Horizon B, or *subsoil,* contains fewer organisms and less organic matter than horizon A. It is known as the **zone of accumulation** because soluble minerals leached from horizon A accumulate as irregular masses. If horizon A is stripped away by erosion leaving horizon B exposed, plants do not grow as well, and if horizon B is clayey, it is harder when dry and stickier when wet than other soil horizons.

Horizon C, the lowest soil layer, consists of partially altered parent material grading downward into unaltered parent material (Figure 6.15). In horizons A and B, the composition and texture of the parent material have been so thoroughly altered that the parent material is no longer recognizable. In contrast, rock fragments and mineral grains of the parent material retain their identity in horizon C. Horizon C contains little organic matter.

Factors Controlling Soil Formation

CLIMATE

It has long been acknowledged that climate is the single most important factor in soil origins, but complex interactions among several factors account for soil type, thickness, and fertility (Figure 6.16). Intense chemical weathering in the tropics yields deep soils from which most of the soluble minerals have been removed by leaching. In arctic and desert climates, on the other hand, soils tend to be thin, contain significant quantities of soluble minerals, and are composed mostly of materials derived by mechanical weathering.

A very general classification recognizes three major soil types characteristic of different climatic settings. Soils that develop in humid regions such as the eastern United States and much of Canada are **pedalfers**, a name derived from the Greek word *pedon* meaning "soil" and the chemical symbols for aluminum (Al) and iron (Fe). Because these soils form where abundant moisture is present, most of the soluble minerals have been leached from horizon A. Although it may be gray, horizon A is generally dark colored because of abundant organic matter, and aluminum-rich clays and iron oxides tend to accumulate in horizon B.

Soils found in much of the arid and semiarid western United States, especially the Southwest, are **pedocals**. Pedocal derives its name in part from the first three letters of calcite. These soils contain less organic matter than pedalfers, so horizon A is generally lighter colored and contains more unstable minerals because of less intense chemical weathering. As soil water evaporates, calcium carbonate leached from above commonly precipitates in horizon B where it forms irregular masses of *caliche*. Precipitation of sodium salts in some desert areas where soil water evaporation is intense yields *alkali soils* that cannot support plants.

Laterite is a soil formed in the tropics where chemical weathering is intense and leaching of soluble minerals is complete. These soils are red, commonly extend to depths of several tens of meters, and are composed largely of aluminum hydroxides, iron oxides, and clay minerals; even quartz, a chemically stable mineral, is generally leached out (Figure 6.17).

Although laterites support lush vegetation, they are not very fertile. The native vegetation is sustained by nutrients derived mostly from the surface layer of organic matter. When these soils are cleared of their native vegetation, the surface accumulation of organic matter is rapidly oxidized, and there is little to replace it. Consequently,

when societies practicing slash-and-burn agriculture clear these soils, they can raise crops for only a few years at best. Then the soil is depleted of plant nutrients, the clay-rich laterite bakes brick hard in the tropical sun, and the farmers move on to another area where the process is repeated.

One aspect of laterites is of great economic importance. If the parent material is rich in aluminum, aluminum hydroxides may accumulate in horizon B as *bauxite,* the ore of aluminum. Because such intense chemical weathering currently does not occur in North America, the United States and Canada are dependent on foreign sources for aluminum ores. Some aluminum ores do exist in Arkansas, Alabama, and Georgia, which had a tropical climate about 50 million years ago, but currently it is cheaper to import aluminum ore than to mine these deposits.

PARENT MATERIAL

The same rock type can yield different soils in different climatic regimes, and in the same climatic regime the same soils can develop on different rock types. It is apparent that climate is more important than parent material in determining the type of soil that develops. Nevertheless, rock type does exert some control. For example, the metamorphic rock quartzite will have a thin soil over it because it is chemically stable, whereas an adjacent body of granite will have a much deeper soil.

Soil that develops on basalt will be rich in iron oxides because basalt contains abundant ferromagnesian silicates, but rocks lacking these minerals will not yield an iron oxide–rich soil no matter how thoroughly they are weathered. Also, weathering of a pure quartz sandstone will yield no clay, whereas weathering of clay will yield no sand.

FIGURE **6.17** Laterite, shown here in Madagascar, is a deep, red soil that forms in response to intense chemical weathering in the tropics.

FIGURE **6.18** During the 1930s, the drought-stricken southern Great Plains in the western United States were hard hit by wind erosion. Huge dust storms were common. This one was photographed at Lamar, Colorado, in 1934.

ORGANIC ACTIVITY

Soils depend on organisms for their fertility, and in return they provide a suitable habitat for many organisms. Earthworms—as many as 1 million per acre—ants, sowbugs, termites, centipedes, millipedes, and nematodes, along with various types of fungi, algae, and single-celled animals, make their homes in the soil. All of these contribute to the formation of soils and provide humus when they die and are decomposed by bacterial action.

Much humus in soils is provided by grasses or leaf litter that microorganisms decompose to obtain food. In so doing, they break down organic compounds within plants and release nutrients back into the soil. Additionally, organic acids produced by decaying soil organisms are important in further weathering of parent materials and soil particles.

Burrowing animals constantly churn and mix soils, and their burrows provide avenues for gases and water. Soil organisms, especially some types of bacteria, are extremely important in changing atmospheric nitrogen into a form of soil nitrogen suitable for use by plants.

RELIEF AND SLOPE

Relief is the difference in elevation between high and low points in a region. Because climate changes with elevation, relief affects soil-forming processes largely through elevation. Slope affects soils in two ways. One is simply *slope angle:* the steeper the slope, the less opportunity for soil development because weathered material is eroded faster than soil-forming processes can work. The other slope control is the *direction* a slope faces. In the Northern Hemisphere, north-facing slopes receive less sunlight than south-facing slopes. If a north-facing slope is steep, it may receive no sunlight at all. Consequently, north-facing slopes have soils with cooler internal temperatures, may support different vegetation, and, if in a cold climate, remain frozen longer.

TIME

The properties of a soil are determined by the factors of climate and organisms altering parent material through time; the longer these processes have operated, the more fully developed the soil will be. If a soil is weathered for extended periods of time, however, its fertility decreases as plant nutrients are leached out, unless new materials are delivered. For example, agricultural lands adjacent to major streams such as the Nile River in Egypt have their soils replenished during yearly floods. In areas of active tectonism, uplift and erosion provide fresh materials that are transported to adjacent areas where they contribute to soils.

How much time is needed to develop a centimeter of soil or a fully developed soil a meter or so deep? No definitive answer can be given because weathering proceeds at vastly different rates depending on climate and parent material, but an overall average might be about 2.5 cm per century. However, a lava flow a few centuries old in Hawaii may have a well-developed soil on it, whereas a flow the same age in Iceland will have considerably less soil. Given the same climatic conditions, soil will develop faster on unconsolidated sediment than on *bedrock,* a general term for the rock underlying soil or sediment.

Under optimum conditions, soil-forming processes operate at a rapid rate in the context of geologic time. From the human perspective, though, soil formation is a slow process; consequently, soil is regarded as a nonrenewable resource.

Soil Degradation

Any decrease in soil productivity or loss of soil to erosion is referred to as **soil degradation**. Between 1945 and 1990, the soils of 17% of the world's vegetated land were degraded to some extent as a result of human activities. In North America, 5.3% of the soil has been degraded, and the figures are much higher for all other continents.

Three types of soil degradation are recognized: erosion, chemical deterioration, and physical deterioration. Most soil erosion is caused by the action of wind and water. When the natural vegetation is removed and a soil is pulverized by plowing, the fine particles are easily blown away (Figure 6.18). Falling rain also disrupts soil particles and carries soil with it when it runs off at the surface. This is particularly devastating on steep slopes where the vegetation has been removed by overgrazing, deforestation, or construction.

Soil erosion rates generally increase when rainforest is cleared because the surface is left unprotected (Figure 6.19). Without trees the soil also becomes more compacted and less absorbent, inhibiting the infiltration of water. As a result, surface runoff increases, and the erosion of gullies becomes more common. Flooding is also more frequent because trees with their huge water-holding capacity are no longer present.

Two types of erosion by water are recognized: sheet erosion and rill erosion. **Sheet erosion** is more or less evenly distributed over the surface and removes thin layers of soil. **Rill erosion** takes place when running water scours small channels. If these channels can be eliminated by plowing, they are *rills* (Figure 6.20), but if they are too deep (about 30 cm) to be plowed over, they are *gullies*. Where gullying becomes extensive, croplands can no longer be tilled and must be abandoned.

If soil losses to erosion are minimal, soil-forming processes can keep pace, and the soil remains productive. Should the loss rate exceed the formation rate, however, the most productive upper layer of soil, horizon A, is removed, exposing horizon B. The Soil Conservation Service of the U.S. Department of Agriculture estimates that 25% of the cropland in the United States is eroding faster

FIGURE **6.19** Soil erosion on a bare surface in Madagascar that was once covered by lush forest.

FIGURE **6.20** Rill erosion in a field during a rainstorm. This rill was later plowed over.

than soil-forming processes can replace it. Such losses are problems, of course, but there are additional consequences. The eroded soil is transported elsewhere, perhaps onto neighboring cropland, onto roads, or into channels. Sediment accumulates in canals and irrigation ditches, and agricultural fertilizers and insecticides are carried into streams and lakes.

A soil undergoes chemical deterioration when its nutrients are depleted and its productivity decreases. Loss of soil nutrients is most notable in countries where soils are overused in an attempt to maintain agricultural productivity. Other causes include insufficient use of chemical fertilizers and clearing soils of their natural vegetation. Chemical deterioration of soils occurs on all continents but is most serious in South America where it accounts for 29% of all soil degradation.

Other types of chemical deterioration are pollution and *salinization,* which occurs when the concentration of salts increases in a soil, making it unfit for agriculture. Pollution can be caused by improper disposal of domestic and industrial wastes, oil and chemical spills, and the concentration of insecticides and pesticides in soils. Soil pollution is a particularly serious problem in eastern Europe.

Physical deterioration of soils results when soil particles are compacted under the weight of heavy machinery and livestock, especially cattle. When soils have been compacted, they are more costly to plow, and plants have a more difficult time emerging from the soil. Furthermore, water does not readily infiltrate, so more runoff occurs; this in turn accelerates the rate of water erosion.

In North America, the rich prairie soils of the midwestern United States and the Great Plains of the United States and Canada are suffering significant soil degradation. Nevertheless, this degradation, which is characterized as moderate, is less serious than in many other parts of the world where it is considered severe or extreme. Other areas of concern are the central valleys of California, an area in Washington State, and some parts of Mississippi and Missouri where water erosion rates are high.

Problems experienced during the past have stimulated the development of methods to minimize soil erosion on agricultural lands. Crop rotation, contour plowing, and the construction of terraces have all proved helpful (Figure 6.21). So has no-till planting in which the residue from the harvested crop is left on the ground to protect the surface from the ravages of wind and water.

Weathering and Mineral Resources

In a preceding section, we discussed intense chemical weathering in the tropics and the origin of bauxite, the chief ore of aluminum. Such accumulations of valuable minerals formed by the selective removal of soluble substances are *residual concentrations*. They represent an insoluble residue of chemical weathering. In addition to bauxite, a number of other residual concentrations are economically important, including deposits of clays, nickel, phosphate, tin, diamonds, and gold.

Some limestones contain small amounts of iron carbonate minerals. When the limestone is dissolved during chemical weathering, a residual concentration of insoluble iron oxides accumulates. Residual concentrations of insoluble manganese oxides form in a similar fashion from manganese-rich source rocks. Some of the sedimentary iron deposits (see Chapter 7) of the Lake Superior region were enriched by chemical weathering when the soluble constituents that were originally present were carried away.

Most commercial clay deposits formed by hydrothermal alteration of granitic rocks or by sedimentary processes, but some formed as residual concentrations. A number of kaolinite deposits in the southern United States were formed by the chemical weathering of feldspars in pegmatites and of clay-bearing limestones and dolostones. Kaolinite is a type of clay mineral used in the manufacture of paper and ceramics.

A gossan is a yellow to reddish deposit composed largely of iron hydroxides that formed by the alteration of iron- and sulfur-bearing minerals such as pyrite (FeS_2). The dissolution of these minerals forms sulfuric acid, which causes other metallic minerals to dissolve, and these tend to be carried down toward the water table. Gossans have been used occasionally as sources of iron, but they are far more important as indicators of underlying ore deposits. One of the oldest known underground mines exploited such ores about 3400 years ago in what is now southern Israel.

FIGURE 6.21 Contour plowing, which involves plowing parallel to the contours of the land, can be an effective soil conservation practice. The furrows and ridges are perpendicular to the direction that water would otherwise flow downhill and thus inhibit erosion.

Chapter Summary

1. Mechanical and chemical weathering are processes whereby parent material is disintegrated and decomposed so that it is more nearly in equilibrium with new physical and chemical conditions. The products of weathering include solid particles, soluble compounds, and ions in solution.

2. The residue of weathering can be further modified to form soil, or it can be deposited as sediment, which might become sedimentary rock.

3. Mechanical weathering processes include frost action, pressure release, thermal expansion and contraction, salt crystal growth, and the activities of organisms. Particles liberated by mechanical weathering retain the chemical composition of the parent material.

4. Solution, oxidation, and hydrolysis are chemical weathering processes; they result in a chemical change of the weathered products. Clay minerals, various ions in solution, and soluble compounds are formed during chemical weathering.

5. Chemical weathering proceeds most rapidly in hot, wet environments, but it takes place in all areas, except perhaps where water is permanently frozen.

6. Mechanical weathering aids chemical weathering by breaking parent material into smaller pieces, thereby exposing more surface area.

7. Mechanical and chemical weathering produce regolith, some of which is soil if it consists of solids, air, water, and humus and supports plants.

8. Soils are characterized by horizons that are designated, in descending order, as O, A, B, and C; soil horizons differ from one another in texture, structure, composition, and color.

9. The factors controlling soil formation include climate, parent material, organic activity, relief and slope, and time.

10. Soils called pedalfers develop in humid regions such as the eastern United States and much of Canada. Arid and semiarid regions' soils are pedocals, many of which contain irregular masses of caliche in horizon B.

11. Laterite is a soil resulting from intense chemical weathering as in the tropics. Such soils are deep and red and are sources of aluminum ores if derived from aluminum-rich parent material.

12. Soil degradation is a problem in some areas. Human practices such as construction, agriculture, and deforestation can accelerate soil degradation.

13. Intense chemical weathering is responsible for the origin of residual concentrations, many of which contain valuable minerals such as iron, lead, copper, and clay.

Important Terms

chemical weathering
differential weathering
erosion
exfoliation
exfoliation dome
frost action
frost heaving
frost wedging
humus
hydrolysis
laterite

leaching
mechanical weathering
oxidation
parent material
pedalfer
pedocal
pressure release
regolith
rill erosion
salt crystal growth
sheet erosion

sheet joint
soil
soil degradation
soil horizon
solution
spheroidal weathering
talus
thermal expansion and contraction
transport
weathering
zone of accumulation

1. The layer of weathered rock, sediment, and volcanic ash covering much of Earth's land surface is:
   a. _____ regolith;
   b. _____ pedocal;
   c. _____ humus;
   d. _____ exfoliation;
   e. _____ bauxite.

2. Which one of the following is a mechanical weathering process?
   a. _____ oxidation;
   b. _____ sheet erosion;
   c. _____ pressure release;
   d. _____ solution;
   e. _____ laterization.

3. The mechanical weathering processes of _____ and _____ are similar in that both produce forces that expand openings in rocks.
   a. _____ exfoliation/chemical alteration;
   b. _____ hydrolysis/rill erosion;
   c. _____ erosion/transport;
   d. _____ soil degradation/sheet jointing;
   e. _____ frost wedging/salt crystal growth.

4. The mineral calcite ($CaCO_3$), the primary constituent of limestone, is nearly insoluble in pure water but dissolves rapidly if _____ is present.
   a. _____ carbonic acid;
   b. _____ silicon dioxide;
   c. _____ calcium sulfate;
   d. _____ clay;
   e. _____ residual manganese.

5. Bauxite, the main ore of aluminum, forms where:
   a. _____ mechanical weathering alters volcanic rocks;
   b. _____ thermal expansion and contraction is the dominant weathering process;
   c. _____ chemical weathering is intense;
   d. _____ limestone is altered by salt crystal growth;
   e. _____ thin layers of soil are removed by sheet erosion.

6. An accumulation of valuable minerals by the selective removal of soluble substances is known as a(an):
   a. _____ caliche;
   b. _____ residual concentration;
   c. _____ pedocal;
   d. _____ alkali soil;
   e. _____ organic-rich layer.

7. Hydrolysis is a chemical wathering process during which:
   a. _____ hydrogen and hydroxyl ions of water replace ions in minerals;
   b. _____ pressure release yields large, rounded masses of rock;
   c. _____ soil rich in iron and aluminum forms in the tropics;
   d. _____ angular blocks of rock accumulate at the bases of slopes;
   e. _____ caliche accumulates in horizon B of pedocals.

8. The process whereby soluble materials are removed from a developing soil and carried downward is known as:
   a. _____ frost heaving;
   b. _____ sheet erosion;
   c. _____ leaching;
   d. _____ oxidation;
   e. _____ degradation.

9. The typical soil of the eastern United States and much of Canada is:
   a. _____ pedalfer;
   b. _____ caliche;
   c. _____ pedocal;
   d. _____ laterite;
   e. _____ bauxite.

10. Which one of the following statements is correct?
   a. _____ differential weathering yields smooth surfaces;
   b. _____ the zone of accumulation in a soil is in horizon A;
   c. _____ parent material refers to soil formed in cold climates;
   d. _____ oxidation is a process during which plants release oxygen;
   e. _____ chemical weathering takes place most rapidly in the tropics.

11. Explain why particle size is important in chemical weathering.

12. Describe the chemical alteration of limestone in water containing a small amount of carbon dioxide.

13. How does laterite form, and what are its characteristics?

14. Why is chemical weathering predominant in the tropics, whereas mechanical weathering prevails in dry climates?

15. How does mechanical weathering differ from and contribute to chemical weathering?

16. Describe the phenomenon of pressure release and the landform that might result from this process.

17. Why does chemical weathering of angular blocks of rock commonly yield spherical rock masses? An illustration would be helpful.

18. Explain how parent material, relief, and slope play a role in the origin of soil.

19. Describe the types of soil erosion recognized and list the factors that contribute to soil erosion.

20. How do frost wedging and salt crystal growth disaggregate rocks? Which of these processes is most effective?

Points to Ponder

1. Consider the following: A soil is 1.5 m thick, new soil forms at the rate of 2.5 cm per century, and the erosion rate is 4 mm per year. How much soil will be left after 100 years?

2. How do human practices contribute to soil degradation? What can be done to minimize the impact of such practices on soils?

For this web site address, along with current updates and exercises, log on to

http://www.wadsworth.com/geo

▶ DEVILS MARBLES

This site is maintained by Patrick Jennings and contains various images of Devils Marbles from Australia. Devils Marbles are examples of spheroidal weathering. Click on any of the images on this page. They will take you to larger images and some information about the origin and location of those "marbles."

▶ ROB'S GRANITE PAGE

This site, which is obviously mostly about granite, is maintained by Robert M. Reed, a Ph.D. student in the Department of Geological Sciences, University of Texas at Austin. Click on *Llano Uplift* and then click *Enchanted Rock State Park*. Here you will see images of the Town Mountain Granite. Note especially the well-developed exfoliation domes, which are some of the finest examples seen anywhere. Also click on *Turkey Peak* and see an excellent example of differential weathering and erosion.

▶ WORLD OF CHEMISTRY: THE HOME PAGE OF RALPH LOGAN

This site is maintained by Ralph H. Logon, chemistry instructor at North Lake College in Dallas, Texas. It is mostly devoted to chemistry but also has information of geologic interest. Scroll down to *Places to Visit* and click *Frequently Asked Questions on Chemical Concepts by Category,* and then click *Ecology & Energy Source Questions.* In this section, there are answers to such questions as "What can you tell me about acid rain?" "Why should we be concerned about carbon dioxide in the atmosphere?" and "What is the greenhouse effect?" Answers to questions regarding energy resources in this section will be addressed in the next chapter.

▶ JEFF'S HOMEPAGE & LINK TO THE SUSTAINABLE AGRICULTURE EDUCATION PAGE

This site is maintained by Jeffery M. Dunn, who is majoring in environmental and natural resource policy at Michigan State University in Lansing, Michigan. Click on *Sustainable Agriculture Educational Project Homepage* and see what is meant by sustainable agriculture. Next, click on *Soil Erosion in Agricultural Systems.* What are the positive and negative aspects of no-till planting and leaving crop residue on the field after harvesting?

CD-ROM Exploration

Explore the following *In-Terra-Active 2.0* CD-ROM module(s) and increase your understanding of key concepts and processes presented in this chapter.

▶ SECTION: MATERIALS
MODULE: ROCK CYCLE

During a visit to Thailand, you find a specimen of limestone and another of marble in the same area. **Question: Which rock is older, and how can you tell?**

CHAPTER

7

Sediment and

Outline

These sedimentary rocks making up Checkerboard Mesa in Zion National Park, Utah, belong to the Jurassic-aged Navajo Sandstone, which represents an ancient windblown dune deposit. Vertical fractures intersect inclined layers known as cross-beds, giving this cliff its checkerboard appearance.

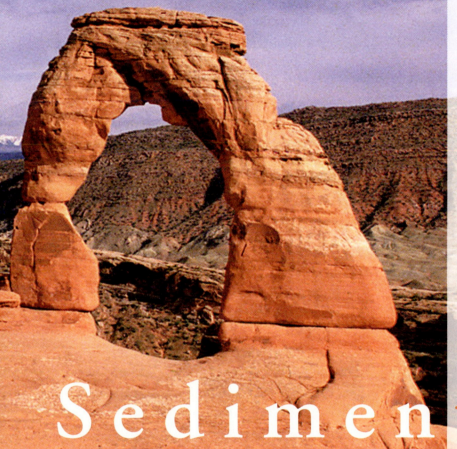

Prologue

About 50 million years ago, two large lakes were present in what are now parts of Wyoming, Utah, and Colorado. Sediments deposited in these lakes became sedimentary rocks known as the Green River Formation. These rocks contain the fossils of millions of fish, plants, and insects. Furthermore, the formation is a potential source of large quantities of oil, combustible gases, and other substances.

Fossil fish skeletons are so common on single surfaces within the Green River Formation that anyone interested in collecting them can do so (Figure 7.1). The abundance of fossil fish indicates that mass deaths took place repeatedly. No one is certain what caused these mass deaths, but some geologists think that blooms of blue-green algae produced toxic substances that killed the fish. Others propose that rapidly changing water temperature or excessive salinity when evaporation was intense

Sedimentary Rocks

was responsible. Whatever the cause, the fish died by the thousands and sank to the lake bottom where little oxygen was present to completely decompose them, thus accounting for the preservation of their skeletons. One area in the formation in Wyoming where fossil plants are particularly abundant has been designated Fossil Butte National Monument.

FIGURE **7.1** Fossil fish from the Green River Formation of Wyoming. *(Photo courtesy of Sue Monroe.)*

FIGURE **7.2** Layers of oil shale of the Green River Formation are exposed along these hillsides.

Fossils in the Green River Formation are interesting, but the formation also contains huge deposits of oil shale, a sedimentary rock consisting of small particles of clay, and an organic substance known as kerogen. By using the appropriate extraction process, liquid oil and combustible gases can be produced from the kerogen of oil shale. To be designated as a true oil shale, the rock must yield at least 10 gallons of oil per ton of rock. A rock that contains so much oil and gas is obviously a potential energy source of great importance. During the Middle Ages (about 1100 to 1450), people in Europe used oil shale as solid fuel for domestic purposes. During the 1850s, small oil shale industries existed in the eastern United States but were discontinued when drilling and pumping oil began in 1859. Oil shale is present on all continents, but the deposits in the Green River Formation are the most extensive (Figure 7.2).

Oil shale yields oil when the rock is heated to nearly 500°C in the absence of oxygen, and the organic matter is driven off as gases and recovered by condensation. Between 25 and 75% of the organic matter in oil shale can be converted to oil and combustible gases during this process. Oil shale of the Green River Formation yields 10 to 140 gallons of oil per ton of rock processed, and the total amount recoverable with present processes is estimated at 80 billion barrels.

As of December 1997, the United States imported more than 9 million barrels of oil per day, so production from the Green River Formation could eventually decrease our reliance on imports. However, no oil is currently being produced from oil shale in the United States because conventional drilling and pumping are less expensive. Nevertheless, the Green River Formation constitutes one of the largest untapped sources of oil in the world. If more effective processes are developed, it could eventually yield even more than the currently estimated 80 billion barrels.

One should realize, though, that at the current and expected consumption rates of oil in the United States, oil production from oil shale will not solve all our energy needs. In addition, the necessary large-scale mining of oil shale would have significant environmental impact. What would be done with the billions of tons of processed rock? Can mining on such a vast scale be done with minimal disruption of wildlife habitats and groundwater systems? Where will the huge volumes of water necessary for processing come from—especially in an area where water is already in short supply? These and other questions are currently being considered by scientists and industry. Perhaps at some future time, the Green River Formation will provide some of our energy needs.

Introduction

SEDIMENTARY rocks, the second major family of rocks, are all composed of materials derived by mechanical and chemical weathering that disintegrate and decompose preexisting rocks (Figure 7.3). As we noted in Chapter 6, various weathering processes yield the raw materials that make up soils and sediment. Recall that weathered materials removed from the weathering site and deposited elsewhere as unconsolidated **sediment** may be further weathered to form soil or it may be transformed into **sedimentary rock**.

Sediment may be *detrital,* meaning it consists of solid particles such as rock fragments or mineral grains liberated during weathering, or it may be *chemical,* consisting of minerals formed from the materials dissolved during chemical weathering. Once derived from parent material, sediment is commonly eroded and transported to another location where it is deposited as an aggregate of

FIGURE 7.3 The derivation of sediment from preexisting rocks. Whether yielded by chemical or mechanical weathering, materials in solution and solid particles are transported and deposited as sediment that, if lithified, becomes sedimentary rock. This illustration simply shows part of the rock cycle in more detail (see Figure 1.12).

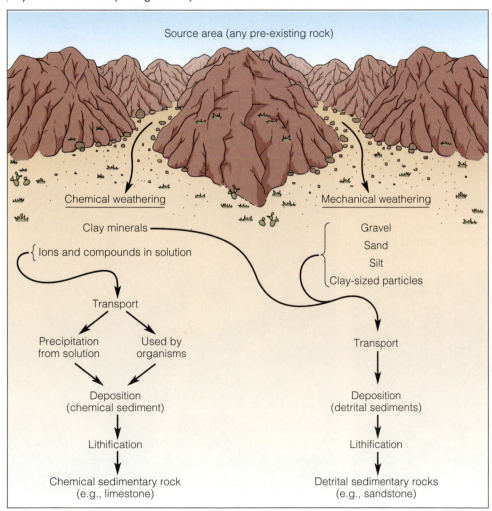

TABLE 7.1

Classification of Sedimentary Particles

SIZE	SEDIMENT NAME	
>2 mm	Gravel	
$1/16$–2 mm	Sand	
$1/256$–$1/16$ mm	Silt	} Mud
<$1/256$ mm	Clay	

loose solids, such as sand on a beach or mud in a lake. In short, the origin or sediment and its subsequent history is simply one part of the rock cycle, which was described in Chapter 1.

Sedimentary rocks are most commonly transformed from sediment by the process known as *lithification*, but a few skipped this unconsolidated sediment stage. For instance, coral reefs form as rock when the reef organisms extract dissolved substances from seawater for their skeletons. Should reef rock be broken apart as during a storm, though, the solid pieces of reef material deposited on the seafloor are sediment.

One important criterion for classifying sedimentary particles is their size (Table 7.1). *Gravel* refers to any sedimentary particle larger than 2 mm, whereas *sand* is any particle, regardless of composition, that measures 1/16 to 2 mm. Gravel- and sand-sized particles are large enough to be observed with the unaided eye or with low-power magnification, but silt- and clay-sized particles are too small to be observed except with very high magnification. Gravel generally consists of rock fragments, whereas sand, silt, and clay particles are mostly individual mineral grains. We should note, though, that *clay* has two meanings: In textural terms, clay refers to sedimentary grains less than 1/256 mm in size; in compositional terms, clay refers to

certain types of sheet silicate minerals (see Figure 3.11). However, most clay-sized particles in sedimentary rocks are in fact clay minerals. Mixtures of silt- and clay-sized particles are generally referred to as *mud*.

Sediment Transport and Deposition

SEDIMENT can be transported by any geologic agent possessing enough energy to move particles of a given size. Glaciers can move particles of any size, whereas wind transports only sand-sized and smaller sediment. Waves and marine currents also transport sediment, but by far the most effective way to erode sediment from the weathering site and transport it elsewhere is by streams.

During sediment transport, *abrasion* reduces the size of particles, and the sharp corners and edges are worn smooth as particles of gravel and sand collide with one another and they become **rounded** (Figure 7.4a). Transport also results in **sorting**, which refers to the size distribution in an aggregate of sediment. If all the particles are about the same size, the sediment is characterized as well sorted, but if a wide range of grain sizes is present, it is poorly sorted (Figure 7.4b). Both rounding and sorting are important properties used to determine the origin of sedimentary rocks; they are discussed more fully in a later section.

Sediment may be transported a considerable distance from its source area, but eventually it is deposited. Some of the sand and mud being deposited at the mouth of the

FIGURE 7.4 Rounding and sorting of sedimentary particles. (a) A deposit consisting of well-sorted and well-rounded gravel. (b) Poorly sorted, angular gravel. *(Photos courtesy of R. V. Dietrich.)*

(a)

(b)

Mississippi River at the present time came from such distant places as Ohio, Minnesota, and Wyoming. Any geographic area in which sediment is deposited is a **depositional environment**. Although no completely satisfactory classification of depositional environments exists, geologists generally recognize three major depositional settings: continental, transitional, and marine, each with several specific depositional environments (Figure 7.5).

Lithification: Sediment to Sedimentary Rock

At present, calcium carbonate mud is accumulating in the shallow waters of Florida Bay, and sand is being deposited in river channels, on beaches, and in sand dunes. These deposits might be compacted and/or cemented and thereby converted into sedimentary rock; the process of transforming sediment into sedimentary rock is **lithification**.

When sediment is deposited, it consists of solid particles and *pore spaces,* which are the voids between particles. If sediment is buried, **compaction**, resulting from the pressure exerted by the weight of overlying sediments, reduces the amount of pore space and thus the volume of the deposit (Figure 7.6). When deposits of mud, which can have as much as 80% water-filled pore space, are

buried and compacted, water is squeezed out, and the volume can be reduced by up to 40%. Sand has up to 50% pore space, although it is generally somewhat less, and it, too, can be compacted so that the sand grains fit more tightly together.

Compaction alone is generally sufficient for lithification of mud, but for sand and gravel deposits **cementation** is necessary to convert the sediment into sedimentary rock (Figure 7.6). Recall from Chapter 6 that calcium carbonate ($CaCO_3$) readily dissolves in water containing a small amount of carbonic acid, and that chemical weathering of feldspars and other silicate minerals yields silica (SiO_2) in solution. These compounds may be precipitated in the pore spaces of sediments, where they act as a cement that effectively binds the sediment together (Figure 7.6).

Calcium carbonate and silica are the most common cements in sedimentary rocks, but iron oxides and hydroxides, such as hematite (Fe_2O_3) and limonite [$FeO(OH)$], respectively, also form a chemical cement in some rocks. Much of the iron oxide cement is derived from the oxidation of iron in ferromagnesian silicates present in the original deposit, although some is carried in by circulating groundwater. The yellow, brown, and red sedimentary rocks exposed in the southwestern United States are colored by small amounts of iron oxide or hydroxide cement.

FIGURE 7.5 Major depositional environments. The environments located along the marine shoreline are transitional from continental to marine. The shallow marine environment corresponds to the continental shelf and can be the site of either sand or carbonate deposition.

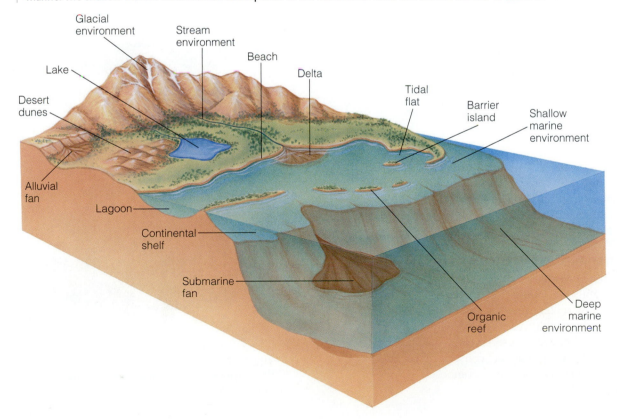

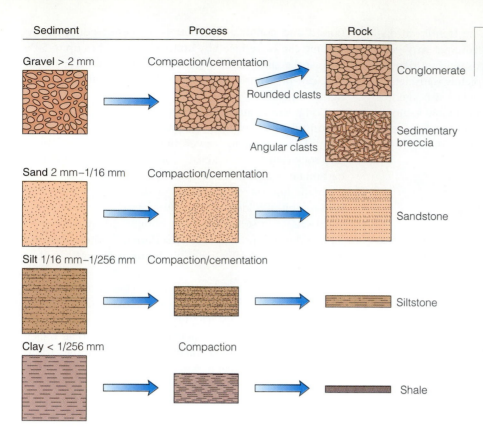

Sediment	Process	Rock
Gravel > 2 mm	Compaction/cementation	Conglomerate (Rounded clasts)
		Sedimentary breccia (Angular clasts)
Sand 2 mm–1/16 mm	Compaction/cementation	Sandstone
Silt 1/16 mm–1/256 mm	Compaction/cementation	Siltstone
Clay < 1/256 mm	Compaction	Shale

FIGURE 7.6 Lithification of detrital sediments by compaction and cementation to form sedimentary rocks.

Sedimentary Rocks

ABOUT 95% of Earth's crust is composed of igneous and metamorphic rocks, but sedimentary rocks are the most common at or near the surface. Approximately 75% of the surface exposures on continents consist of sediments or sedimentary rocks, and they cover most of the seafloor. Sedimentary rocks are generally classified as detrital or chemical; the latter includes a subcategory known as *biochemical* (Table 7.2).

DETRITAL SEDIMENTARY ROCKS

Detrital sedimentary rocks consist of *detritus,* the solid particles of preexisting rocks. They have a *clastic texture,* meaning these rocks are composed of fragments or particles also known as *clasts.* Several varieties of detrital sedimentary rocks are recognized, each of which is characterized by the size of its constituent particles (Table 7.2).

CONGLOMERATE AND SEDIMENTARY BRECCIA Both *conglomerate* and *sedimentary breccia* consist of gravel-sized particles (Table 7.2; Figure 7.7a). The particles commonly measure a few millimeters to a few centimeters, but boulders several meters in diameter are sometimes present. The only difference between these rocks is the shape of their gravel particles; conglomerate consists of rounded gravel, whereas sedimentary breccia is composed of angular gravel called *rubble.*

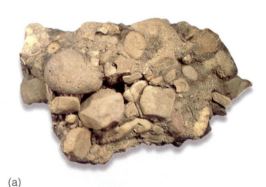

(a)

(b)

FIGURE 7.7 Detrital sedimentary rocks: (a) conglomerate; (b) sandstone. *(Photos courtesy of Sue Monroe.)*

TABLE 7.2

Classification of Sedimentary Rocks

DETRITAL SEDIMENTARY ROCKS

Sediment Name and Size	Description	Rock Name
Gravel (>2 mm)	Rounded gravel particles	Conglomerate
	Angular gravel particles	Sedimentary breccia
Sand (1/16–2mm)	Mostly quartz sand	Quartz sandstone
	Quartz with >25% feldspar	Arkose
Mud (<1/16 mm)	Mostly silt	Siltstone
	Silt and clay	Mudstone* } Mudrocks
	Mostly clay	Claystone*

CHEMICAL SEDIMENTARY ROCKS

Texture	Composition	Rock Name
Varies	Calcite ($CaCO_3$)	Limestone } Carbonates
Varies	Dolomite [$CaMg(CO_3)_2$]	Dolostone
Crystalline	Gypsum ($CaSO_4 \cdot 2H_2O$)	Rock gypsum } Evaporites
Crystalline	Halite (NaCl)	Rock salt

BIOCHEMICAL SEDIMENTARY ROCKS

Texture	Composition	Rock Name
Clastic	Calcium carbonate ($CaCO_3$) shells	Limestone (various types such as chalk and coquina)
Usually crystalline	Altered microscopic shells of silicon dioxide (SiO_2)	Chert
——	Mostly carbon from altered plant remains	Coal

*Mudrocks possessing the property of fissility, meaning they break along closely spaced planes, are commonly called *shale*.

Conglomerate is a fairly common rock type, but sedimentary breccia is rather rare because gravel-sized particles become rounded quickly during transport. So, if a sedimentary breccia is encountered, one can conclude that the augular gravel composing it was not transported very far. High-energy transport agents such as rapidly flowing streams and waves are needed to transport gravel, so gravel tends to be deposited in high-energy environments such as stream channels and beaches.

SANDSTONE   The term *sand* is simply a size designation, so *sandstone* may be composed of grains of any type of mineral or rock fragment. Most sandstones consist primarily of the mineral quartz (Figure 7.7b) with small amounts of a number of other minerals. Geologists recognize several types of sandstones, each characterized by its composition. *Quartz sandstone,* composed mostly of quartz, is the most common, but *arkose,* which contains more than 25% feldspars, is also fairly common (Table 7.2).

MUDROCK   The *mudrocks* include all detrital sedimentary rocks composed of silt- and clay-sized particles. Among the mudrocks, *siltstone,* as the name implies, is composed of silt-sized particles; *mudstone* contains a mixture of silt- and clay-sized particles; and *claystone* is composed mostly of clay (Table 7.2). Some mudstones and claystones are designated as *shale* if they are fissile, which means they break along closely spaced parallel planes.

Mudrocks comprise about 40% of all detrital sedimentary rocks, making them the most common of these rocks. Turbulence in water keeps silt and clay suspended and must therefore be at a minimum if they are to settle. Consequently, deposition occurs in low-energy depositional environments where currents are weak such as in the quiet offshore waters of lakes and in lagoons.

CHEMICAL AND BIOCHEMICAL SEDIMENTARY ROCKS

Chemical sedimentary rocks originate from the materials taken into solution during chemical weathering (Table 7.2). These dissolved materials are transported to lakes and the oceans where they become concentrated. They can be extracted from lake or ocean water to form minerals either by inorganic chemical processes or by the chemical activities of organisms. Some rocks formed by lithification of these minerals have a *crystalline texture,* meaning they consist of a mosaic of interlocking crystals, whereas others have a clastic texture. Rocks formed by the activities of organisms are referred to as **biochemical sedimentary rocks**. In any case, aggregates of minerals accumulate that become lithified by compaction and cementation, just as in detrital sedimentary rocks.

LIMESTONE AND DOLOSTONE The main component of limestone is the calcium carbonate mineral calcite ($CaCO_3$), whereas dolostone is made up of the calcium magnesium carbonate mineral dolomite [$CaMg(CO_3)_2$]. Both calcite and dolomite are carbonate minerals (see Chapter 3), so limestone and dolostone are known as **carbonate rocks**. Recall from Chapter 6 that calcite readily dissolves in water containing a small amount of acid, but the chemical reaction leading to dissolution is reversible, so solid calcite can be precipitated from solution. Accordingly, some limestone, although probably not very much, results from inorganic chemical reactions, such as the limestone known as travertine, which forms in and around hot springs.

Because organisms play such a significant role in their origin, most limestones are conveniently classified as biochemical sedimentary rocks (Figure 7.8; Table 7.2). For example, the limestone known as *coquina* (Figure 7.8b) consists entirely of broken shells cemented by calcium carbonate, and *chalk* is a soft variety of biochemical limestone composed largely of microscopic shells of organisms.

The near-absence of recent dolostone and evidence from chemistry and studies of rocks indicate that most

dolostone was originally limestone that has been changed to dolostone. Many geologists think dolostone originates when magnesium replaces some of the calcium in calcite.

EVAPORITES Evaporites is the collective name for such rocks as *rock salt* and *rock gypsum* that form by inorganic chemical precipitation of minerals from solution (Table 7.2; Figure 7.9). In Chapter 6 we noted that some minerals are dissolved during chemical weathering, but a solution can hold only a certain volume of dissolved mineral matter. If the volume of a solution is reduced by evaporation, the amount of dissolved mineral matter increases in proportion to the volume of the solution and eventually reaches the saturation limit, the point at which precipitation must occur.

Rock salt, composed of the mineral halite (NaCl), is simply sodium chloride that was precipitated from seawater or, more rarely, lake water (Figure 7.9a). Rock gypsum, the most common evaporite rock, is composed of the mineral gypsum ($CaSO_4 \cdot H_2O$), which also precipitates from evaporating water (Figure 7.9b). A number of other evaporite rocks and minerals are known, but most of these are rare.

CHERT *Chert* is a hard rock composed of microscopic crystals of quartz (SiO_2) (Table 7.2; Figure 7.9c). It is found in several varieties including *flint,* which is black because of inclusions of organic matter, and *jasper,* which is red or brown because of iron oxide inclusions. Because chert lacks cleavage and can be shaped to form sharp cutting edges, many cultures have used it for the manufacture of tools, spear points, and arrowheads.

Chert occurs as irregular masses, or *nodules,* in other rocks, especially limestones, and as distinct layers of *bedded chert.* Most nodules in limestones are clearly secondary in origin; that is, they have replaced part of the host rock, apparently by being precipitated from solution.

Bedded chert can be precipitated directly from seawater, but because so little silica is dissolved in seawater, it seems unlikely that most bedded cherts formed this way. It appears that many bedded cherts are biochemical, resulting from accumulations of shells of silica-secreting, single-celled organisms such as radiolarians and diatoms.

COAL *Coal* is a biochemical sedimentary rock composed of the compressed, altered remains of land plants (Table 7.2; Figure 7.10). It forms in swamps and bogs where the water is deficient in oxygen or where organic matter accumulates faster than it decomposes. The bacteria that decompose vegetation in swamps can exist without oxygen, but their wastes must be oxidized, and because no oxygen is present, the wastes accumulate and kill the bacteria. Thus, bacterial decay ceases, and the plants are not completely decomposed. These partly altered plant remains accumulate as layers of organic muck. When buried, this organic muck becomes *peat,* which looks rather like coarse pipe tobacco. Where peat is abundant, as in Ireland and

(a)

(b)

FIGURE 7.8 Two types of limestones. (a) Fossiliferous limestone. (b) Coquina is composed of the broken shells of organisms. *(Photos a and b courtesy of Sue Monroe.)*

FIGURE 7.9 (a) Core of rock salt from a well in Michigan. (b) Rock gypsum. (c) Chert. *(Photos courtesy of Sue Monroe.)*

(a)

(b)

(c)

Scotland, it is burned as a fuel. Peat that is buried more deeply and compressed, especially if it is heated too, is altered to a type of dark brown coal called *lignite,* in which plant remains are still clearly visible. During the change from organic muck to coal, volatile elements of the vegetation such as oxygen, hydrogen, and nitrogen are partly vaporized and driven off, enriching the residue in carbon; lignite contains about 70% carbon as opposed to about 50% in peat.

Bituminous coal, which contains about 80% carbon, is a higher-grade coal than lignite. It is dense and black and has been so thoroughly altered that plant remains can only rarely be seen. The highest grade coal is *anthracite,* which is a metamorphic type of coal (see Chapter 8). It contains up to 98% carbon and, when burned, yields more heat per unit volume than other types of coal.

FIGURE 7.10 Coal is a biochemical sedimentary rock composed of the altered remains of land plants. *(Photo courtesy of Sue Monroe.)*

Reading the Story in Rocks

WHEN geologists investigate sedimentary rocks in the field, they are observing the products of events that occurred during the past. The only record of these events is preserved in the rocks, so geologists must evaluate those aspects of sedimentary rocks that allow inferences to be made about the original processes and the environment of deposition. Sedimentary textures such as sorting and rounding can give clues to the depositional process. Wind-blown dune sands, for example, tend to be well sorted and well rounded. Other aspects of sedimentary rocks that are important in environmental analysis include sedimentary structures and fossils.

SEDIMENTARY STRUCTURES

When sediments are deposited, they contain a variety of features known as **sedimentary structures** that formed as a result of physical and biological processes operating in the depositional environment. One of the most common sedimentary structures is distinct layers known as **strata** or **beds** (Figure 7.11). Beds vary in thickness—from less than a millimeter up to many meters. Individual beds are separated from one another by **bedding planes** and are distinguished by differences in composition, grain size, color, or a combination of features (Figure 7.11). Almost all sedimentary rocks show some kind of bedding; a few, such as limestones that formed as coral reefs, lack this feature, however.

In **graded bedding**, grain size decreases upward within a single bed (Figure 7.12). Most graded bedding appears

FIGURE 7.11 These sedimentary rocks show both horizontal beds and cross-bedding. Cross-beds form when layers are deposited at an angle with respect to the surface upon which they accumulate. The inclination of cross-beds, to the right in this case, indicates ancient current directions.

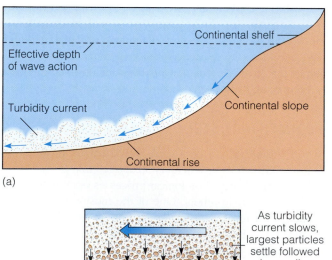

(a)

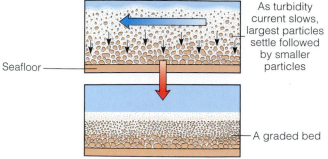

(b)

to have formed from turbidity current deposition, although some forms in stream channels during the waning stages of floods. *Turbidity currents* are underwater flows of sediment-water mixtures that are denser than sediment-free water. These flows move downslope along the bottom of the sea or a lake until they reach the relatively level seafloor or lake floor. There, they rapidly slow down and begin depositing transported sediment—the coarsest first, followed by progressively smaller particles (Figure 7.12).

Many sedimentary rocks are characterized by **cross-bedding**, which consists of layers arranged at an angle to the surface upon which they are deposited (Figure 7.11). Cross-bedding is common in sedimentary rocks that originated in stream channels and shallow marine environments and as desert dunes. Invariably, cross-beds result from transport by wind or water currents and deposition on the downcurrent sides of dunelike structures. Cross-

FIGURE 7.12 (a) Turbidity currents flow downslope along the seafloor (or lake bottom) because of their density. (b) Graded bedding formed by deposition from a turbidity current.

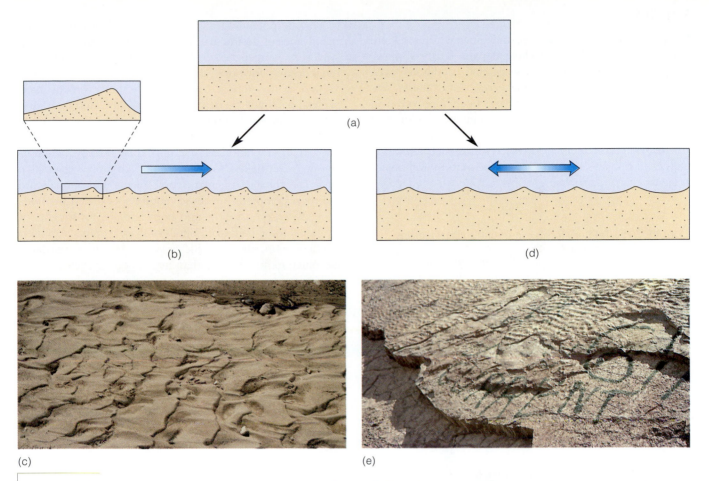

(a)

(b)

(d)

(c)

(e)

FIGURE 7.13 Ripple marks. (a) Undisturbed layer of sand. (b) Current ripple marks form in response to flow in one direction, as in a stream channel. The enlargement of one ripple shows its internal structure. Note that individual layers within the ripple are inclined, showing an example of cross-bedding. (c) Current ripples that formed in a small stream channel. (d) The to-and-fro currents of waves in shallow water deform the surface of the sand layer into wave-formed ripple marks. (e) Wave-formed ripple marks on ancient rocks.

beds are inclined downward, or dip, in the direction of flow. Because their orientation depends on the direction of flow, cross-beds are good indicators of ancient current directions, or *paleocurrents* (Figure 7.11).

In sand deposits one can commonly observe small-scale, ridgelike **ripple marks** on bedding planes. One type of ripple mark is asymmetric in cross section, with a gentle upstream slope and a steep downstream slope. Known as *current ripple marks* (Figure 7.13a), they are formed by currents that move in one direction, as in a stream channel. Like cross-bedding, current ripple marks are good paleocurrent indicators. In contrast, the to-and-fro motion of waves produces ripples that tend to be symmetric in cross section. These *wave-formed ripple marks* (Figure 7.13b) form mostly in the shallow nearshore waters of oceans and lakes.

Mud cracks are found in clay-rich sediment that has dried out (Figure 7.14). When the sediment dries, it shrinks and forms intersecting fractures (mud cracks) that might later be filled with sediment. These features in ancient sedimentary rocks indicate that the sediment was deposited

where periodic drying was possible, as on a river floodplain, near a lake shore, or where muddy deposits are exposed on marine shorelines at low tide.

FIGURE 7.14 Mud cracks form in clay-rich sediments when they dry and shrink.

FOSSILS

Fossils, the remains or traces of ancient organisms (Figure 7.15), are mostly the hard skeletal parts such as shells, bones, and teeth, but under exceptional conditions, even the soft-part anatomy may be preserved. For example, several frozen woolly mammoths have been discovered in Alaska and Siberia with hair, flesh, and internal organs preserved. The remains of organisms are known as *body fossils* to distinguish them from *trace fossils* such as tracks, trails, and burrows (Figure 7.15), which are indications of ancient organic activity.

FIGURE 7.15 (a) Body fossils consist of the actual remains of organisms. Fossil horse teeth are preserved in this rock. (b) Trace fossils are an indication of ancient organic activity. These bird tracks are preserved in mudrock of the Green River Formation of Wyoming. *(Photos courtesy of Sue Monroe.)*

(a)

(b)

For any potential fossil to be preserved, it must escape the ravages of destructive processes such as running water, waves, scavengers, exposure to the atmosphere, and bacterial decay. Obviously, the soft parts of organisms are devoured or decomposed most rapidly, but even the hard skeletal parts will be destroyed unless they are buried and protected in mud, sand, or volcanic ash. Even if buried, skeletal elements may be dissolved by groundwater or destroyed by alteration of the host rock during metamorphism. Nevertheless, fossils are quite common (Perspective 7.1). The remains of microscopic plants and animals are the most common, but these require specialized methods of recovery, preparation, and study and are not sought out by casual fossil collectors. Shells of marine animals are also very common, and even the bones and teeth of dinosaurs are much more common than most people realize.

If it were not for fossils, we would have no knowledge of trilobites, dinosaurs, and other extinct organisms. Thus, fossils constitute our only record of ancient life. They are not simply curiosities, however, but have several practical uses. In many geologic studies, it is necessary to correlate or determine age equivalence of sedimentary rocks in different areas. These correlations are most commonly demonstrated with fossils; we will discuss correlation more fully in Chapter 17. Fossils are also useful in determining environments of deposition.

ENVIRONMENT OF DEPOSITION

Ancient sedimentary rocks acquired most of their properties as a result of the physical, chemical, and biological processes that operated in the original depositional environment. To determine what the depositional environment was, geologists investigate these rock properties, reasoning that the processes responsible for them are the same as those going on at present. For instance, we have every reason to think that wave-formed ripple marks originated by the to-and-fro motion of waves throughout geologic time; thus, we are justified in stating that these structures in an ancient sandstone formed just as they do now. In short, we are simply applying the principle of uniformitarianism (see Chapter 1). Accordingly, geologists with their knowledge of various present-day processes such as sediment transport, wave action, and deposition by streams can make inferences regarding the depositional environment of ancient sediment rocks.

While conducting field studies, geologists commonly make some preliminary interpretations. For example, some sedimentary particles in limestones most commonly form in shallow marine environments where currents are vigorous. Large-scale cross-bedding is typical of but not restricted to desert dunes. Fossils of land plants and animals can be washed into transitional environments, but most of them are preserved in deposits of continental environments. Fossil shells of such marine-dwelling animals as corals obviously indicate marine depositional environments.

Evaluation of rock properties such as sorting of sedimentary particles by size is also useful in environmental interpretation. Sorting results from processes that selectively transport and deposit particles. Windblown dunes are composed of well-sorted sand, because wind cannot transport gravel and it blows silt and clay beyond the areas of sand accumulation. Glaciers and mudflows, on the other hand, are unselective because their energy allows them to transport many different-sized particles, and their deposits tend to be poorly sorted.

Much environmental interpretation is done in the laboratory where the data and rock samples collected during field work can be more fully analyzed. The analyses include microscopic and chemical examination of rock samples, identification of fossils, and graphic representations showing the three-dimensional shapes of rock units and their relationships to other rock units. In addition, the features of sedimentary rocks are compared with those of sediments from present-day depositional environments; once again, the contention is that features in ancient rocks, such as ripple marks, formed during the past in response to the same processes responsible for them now. Finally, when all data have been analyzed, an environmental interpretation is made.

The following example illustrates how environmental interpretations are made. The Navajo Sandstone of the southwestern United States covers a vast area, perhaps as much as 500,000 km². It has an irregular three-dimensional shape, reaches a thickness of about 300 m in the area of Zion National Park, Utah, and consists mostly of well-sorted sand grains measuring about 0.2 to 0.5 mm in diameter. Some of the sandstone beds also possess tracks of dinosaurs and other land-dwelling animals, ruling out the possibility of a marine origin for the rock unit. These features and the fact that the Navajo Sandstone has cross-beds up to 30 m high (see chapter-opening photo) and current ripple marks, both of which appear to have formed in sand dunes, lead to the conclusion that the sandstone represents an ancient desert dune deposit. The cross-beds are generally inclined downward, or dip, to the southwest, indicating that the wind blew mostly from the northeast.

Sediments, Sedimentary Rocks, and Natural Resources

SEDIMENTS and sedimentary rocks or the materials they contain have a variety of uses. Sand and gravel are essential to the construction industry, pure clay deposits are used for ceramics, and limestone is used in the manufacture of cement and in blast furnaces where iron ore is refined to make steel. Rock salt (NaCl) is the source of table salt; sylvite, a potassium chloride (KCl), is used in the manufacture of fertilizers, dyes, and soaps; and gypsum ($CaSO_4 \cdot 2H_2O$) is used to make wallboard. Sand composed mostly of quartz, or what is called silica sand, is used for a variety of purposes including the manufacture of glass, refractory bricks for blast furnaces, and molds for casting iron, aluminum, and copper alloys.

Some valuable sedimentary deposits form when minerals become separated from other transported sediment because of their greater density. These *placer deposits*, as they are called, are surface concentrations that result from density separation in streams or on beaches (Figure 7.16). For instance, heavy minerals may become concentrated as a placer deposit in a depression in a stream channel or at the base of a waterfall. Much of the gold recovered during the California gold rush (1848–1953) came from placer deposits, and placers of a variety of other minerals such as diamonds and tin are important.

The tiny island nation of Nauru, with one of the highest per capita incomes in the world, has an economy based almost entirely on mining and exporting phosphate-bearing sedimentary rock that is used in fertilizers. More than half of Florida's mineral value comes from mining phosphorus from phosphate rock. In addition to fertilizers, phosphorus from phosphate is used in metallurgy, preserving foods, ceramics, and matches.

Dolostones in Missouri are the host rocks for ores of lead and zinc. Diatomite, a lightweight, porous sedimentary rock composed of the microscopic silica (SiO_2)

PAN, CRADLE, LONG-TOM, AND SLUICE WASHING.

FIGURE 7.16 The basic processes of mining placer deposits. In the left foreground, a miner is panning for gold. As stream sediment and water are swirled around in a broad, shallow pan, the heavier gold sinks to the bottom and lighter sand and gravel are washed away. In the right foreground and background, miners are using a Long Tom and a flume sluice, respectively. The miner at the left center is washing sediment in a cradle. Just as in panning, the Long Tom, sluice, and cradle separate heavier gold from other sediment. Also shown is tunneling and winching ore up from a coyote hole that raises sediment from the lowest level in the deposit. Much of the placer mining today is mechanized. *Source: T. H. Watkins,* Gold and Silver into the West *(New York: Bonanza Books, 1971). Picture credit: Bancroft Library.*

The Abundance of Fossils

Only a tiny proportion of all organisms that ever lived are fossilized, but fossils are nevertheless quite common, much more so than most people realize. The reason for this abundance is that so many billions of organisms existed over so many millions of years that even if only 1 in 10,000 was fossilized, the total number of fossils is truly astonishing.

By far the most common types of easily collected fossils are those of marine invertebrates, animals lacking a segmented vertebral column, such as clams, oysters, sea lilies, brachiopods, and corals. Dozens or hundreds of these fossils can be collected in many areas by anyone interested in doing so. In fact, they are so common that, as they weather out of their host rocks, they litter the surface of some localities.

Animals possessing a segmented vertebral column are known as *vertebrates* and include various fish, amphibians, reptiles, birds, and mammals. Their fossils are not as common as those of invertebrates, but even these are better represented by fossils than is generally believed. Thousands of fossilized fish skeletons are found on single surfaces within 50-million-year-old lake deposits in Wyoming, for example (Figure 7.1). And dinosaur bone fragments and complete or partial teeth are easily collected in some places. It is true that entire skeletons of dinosaurs or other land-dwelling animals are rare, but some remarkable concentrations of their fossils have been discovered.

Figure 1 View of some of the 4000 dinosaur bones recovered from Howe Quarry, Wyoming.

Figure 2 Paleontologists excavating rhinoceroses (foreground) and horse (background) skeletons from volcanic ash in northeast Nebraska.

Excavations beginning in 1934 at what is known as the Howe Quarry in Wyoming revealed an extraordinary concentration of Jurassic-age dinosaur fossils (Figure 1). The deposit, which measures only about 18 × 16 m, yielded 4000 bones from at least 20 large dinosaurs. But the most interesting aspect of this discovery was the 12 legs found preserved in an upright position penetrating the muddy sedimentary deposits, indicating that the dinosaurs became mired in mud and perished. The parts of the animals above the surface decomposed and formed a heap of bones, whereas the legs were preserved in the position in which the dinosaurs were trapped.

In 1981 in northwestern Montana, a Cretaceous-age bone bed was discovered that contains an estimated 10,000 duck-billed dinosaurs. Apparently, a vast herd was overcome by ash and gases from a nearby volcano and subsequently buried in volcanic ash. Individual dinosaurs varied from juveniles measuring 3.6 m long to adults 7.6 m long, representing the standing age of the herd at the time of death.

Ten million years ago in what is now northeastern Nebraska, a vast grassland was inhabited by short-legged aquatic rhinoceroses, camels, three-toed horses, saber-toothed deer, land turtles, and many other animals. Thousands of animals perished when a vast cloud of volcanic ash covered the area. Many of the animals died during the initial ash fall, but their remains lay at the surface and partly decomposed before they were buried. In contrast, hundreds of rhinoceroses were killed later when the ash was probably redistributed by wind, and they were buried rapidly as indicated by the large number of complete skeletons (Figure 2).

Scientists known as *paleontologists*, who study the history of life as revealed by fossils, are not often afforded opportunities to find and recover so many well-preserved fossils as in these examples. Ancient calamities provide paleontologists with a unique glimpse of what life was like millions of years ago.

skeletons of single-celled plants, is used in gas purification and to filter a number of fluids such as molasses, fruit juices, water, and sewage. The United States is the world leader in diatomite production, mostly from mines in California, Oregon, and Washington.

Historically, most of the coal mined in the United States has been bituminous coal from the coal fields of the Appalachian coal basin. These coal deposits formed in coastal swamps during the Pennsylvanian Period between 286 and 320 million years ago. Huge lignite and subbituminous coal deposits also exist in the western United States, and these are becoming increasingly important resources.

Anthracite coal is an especially desirable resource because it burns hot with a smokeless flame. Unfortunately, it is the least common type of coal, so most coal used for heating buildings and for generating electrical energy is bituminous (Figure 7.10). Bituminous coal is also used to make *coke*, a hard, gray substance consisting of the fused ash of bituminous coal; coke is prepared by heating the coal and driving off the volatile matter. Coke is used to fire blast furnaces during the production of steel. Synthetic oil and gas and a number of other products are also made from bituminous coal and lignite.

PETROLEUM AND NATURAL GAS

Both petroleum and natural gas are *hydrocarbons*, meaning they are composed of hydrogen and carbon. Hydrocarbons form from the remains of microscopic organisms that exist in the seas and in some large lakes. When these organisms die, their remains settle to the seafloor or lake floor where little oxygen is available to decompose them. They are then buried under layers of sediment. As the depth of burial increases, they are heated and transformed into petroleum and natural gas. The rock in which the hydrocarbons formed is the *source rock*.

For petroleum and natural gas to accumulate in economic quantities, they must migrate from the source rock into some kind of rock, known as *reservoir rock* (Figure 7.17), in which they can be trapped. Effective reservoir rocks contain considerable pore space where appreciable quantities of hydrocarbons can accumulate. Furthermore, the reservoir rocks must possess high *permeability*, or the capacity to transmit fluids; otherwise hydrocarbons cannot be extracted in reasonable quantities. In addition, some kind of impermeable *cap rock* must be present over the reservoir rock to prevent the hydrocarbons from migrating upward (Figure 7.17).

Many hydrocarbon reservoirs consist of nearshore marine sandstones in proximity with fine-grained, organic-rich source rocks. Such oil and gas traps are called *stratigraphic traps* because they owe their existence to variations in the strata (Figure 7.17a). Ancient coral reefs are also good stratigraphic traps. Indeed, some of the oil in the Persian Gulf region is trapped in ancient reefs. *Structural traps* result when rocks are deformed by folding, fracturing, or both (Figure 7.17b).

In the Gulf Coast region, hydrocarbons are commonly found in structures adjacent to salt domes. A vast layer of rock salt was deposited in this region during the Jurassic

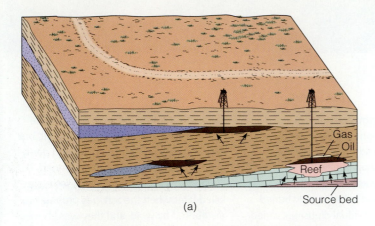

(a)

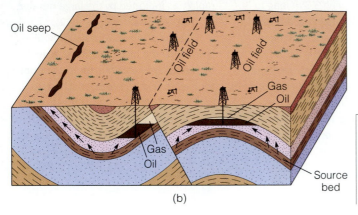

(b)

FIGURE 7.17 Oil and natural gas traps. The arrows in both diagrams indicate the migration of hydrocarbons. In both examples sandstone is the reservoir rock overlain by some kind of impermeable mudrock. (a) Two examples of stratigraphic traps. (b) Two examples of structural traps, one formed by folding, the other by fracturing.

Period as the ancestral Gulf of Mexico formed when North America separated from North Africa. Rock salt is a low-density sedimentary rock, and when deeply buried beneath denser sediments such as sand and mud, it rises toward the surface in pillars known as *salt domes* (see Figure 4.20). As the rock salt rises, it penetrates and deforms the overlying rock layers, forming structures along its margins that may trap petroleum and gas.

Although large concentrations of petroleum are present in many areas, more than 50% of all proven reserves are in the Persian Gulf region! Furthermore, some of the oil fields are gigantic; at least 20 are expected to yield more than 5 billion barrels of oil each, and several have already surpassed this figure. Many nations including the United States are heavily dependent on imports of Persian Gulf oil, a dependence that will increase in the future.

Other sources of petroleum that will probably become increasingly important in the future include *oil shales* and *tar sands*. The United States has about two-thirds of all known oil shales, although South America also has large deposits, and all continents have some oil shale. The richest deposits in the United States are in the Green River Formation of Colorado, Utah, and Wyoming (see the Prologue).

Tar sand is a type of sandstone with viscous, asphaltlike hydrocarbons filling the pore spaces. This substance is the sticky residue of once-liquid petroleum from which the volatile constituents have been lost. Liquid petroleum can be recovered from tar sand, but to do so, large quantities of rock must be mined and processed. Since the United States has few tar sand deposits, it cannot look to this source as a significant future energy resource. The Athabaska tar sands in Alberta, Canada, however, are one of the largest deposits of this type. These deposits are currently being mined, and it is estimated that they contain several hundred billion barrels of recoverable petroleum.

URANIUM

Most of the uranium used in nuclear reactors in North America comes from the complex potassium-, uranium-, vanadium-bearing mineral *carnotite* found in some sedimentary rocks. Some uranium is also derived from *uraninite* (UO_2), a uranium oxide found in granitic rocks and hydrothermal veins. Uraninite is easily oxidized and dissolved in groundwater, transported elsewhere, and chemically reduced and precipitated in the presence of organic matter.

The richest uranium ores in the United States are widespread in the Colorado Plateau area of Colorado and adjoining parts of Wyoming, Utah, Arizona, and New Mexico. These ores, consisting of fairly pure masses and encrustations of carnotite, are associated with plant remains in sandstones that formed in ancient stream channels. Although most of these ores are associated with fragmentary plant remains, some petrified trees also contain large quantities of uranium.

Large reserves of low-grade uranium ore also are found in the Chattanooga Shale. The uranium is finely disseminated in this black, organic-rich mudrock that underlies large parts of several states including Illinois, Indiana, Ohio, Kentucky, and Tennessee. Canada is the world's largest producer and exporter of uranium.

BANDED IRON FORMATION

Banded iron formation is a chemical sedimentary rock of great economic importance. It consists of alternating thin layers of chert and iron minerals, mostly the iron oxides hematite and magnetite (Figure 7.18). Banded iron formations are present on all the continents and account for most of the iron ore mined in the world today. Vast banded iron formations are present in the Lake Superior region of the United States and Canada and in the Labrador trough of eastern Canada.

The origin of banded iron formations is not fully understood, and none are currently forming. Fully 92% of all banded iron formations were deposited in shallow seas during the Proterozoic Eon, between 2.5 and 2.0 billion years ago. Iron is a highly reactive element that in the presence of oxygen combines to form rustlike oxides that are not readily soluble in water. During early Earth history, however, little oxygen was present in the atmosphere, and thus little was dissolved in seawater. Soluble reduced iron (Fe^{+2}) and silica, however, were present in seawater.

Geologic evidence indicates that abundant photosynthesizing organisms were present about 2.5 billion years ago. These organisms, such as bacteria, release oxygen as a by-product of respiration; thus, they released oxygen into seawater and caused large-scale precipitation of iron oxides and silica as banded iron formations.

FIGURE 7.18 Outcrop of banded iron formation in northern Michigan.

1. Sediment consists of mechanically weathered solid particles and minerals extracted from solution by inorganic chemical processes and the activities of organisms.

2. Sedimentary particles are designated in order of decreasing size as gravel, sand, silt, and clay.

3. Sedimentary particles are rounded and sorted during transport although the degree of rounding and sorting depends on particle size, transport distance, and depositional process.

4. Any area where sediment is deposited is a depositional environment. Major depositional settings are continental, transitional, and marine, each of which includes several specific depositional environments.

5. Compaction and cementation are the processes of lithification in which sediment is converted into sedimentary rock. Silica and calcium carbonate are the most common chemical cements, but iron oxide and iron hydroxide cements are important in some rocks.

6. Sedimentary rocks are generally classified as detrital or chemical:

   a. Detrital sedimentary rocks consist of solid particles such as sand or gravel derived from preexisting rocks. They include conglomerate, sedimentary breccia, sandstone, and mudrocks.

   b. Chemical sedimentary rocks are derived from ions in solution by inorganic chemical processes or the biochemical activities of organisms. A subcategory called biochemical sedimentary rocks is recognized.

7. Chemical sedimentary rocks called carbonates contain minerals with the carbonate ion $(CO_3)^{-2}$, as in limestone and dolostone. Dolostone probably forms when magnesium partly replaces the calcium in limestone.

8. Chemical sedimentary rocks known as evaporites include rock salt and rock gypsum, both of which form by inorganic precipitation of minerals from evaporating water.

9. Coal is a type of biochemical sedimentary rock composed of the altered remains of land plants.

10. Sedimentary structures such as bedding, cross-bedding, and ripple marks commonly form in sediments when or shortly after they are deposited. These features preserved in sedimentary rocks help geologists determine ancient current directions and depositional environments.

11. Sediments and sedimentary rocks are the host materials for most fossils. Fossils provide the only record of prehistoric life and are useful for correlation and environmental interpretations.

12. Depositional environments of ancient sedimentary rocks are determined by studying sedimentary textures and structures, examining fossils, and making comparisons with present-day sediments deposited by known processes.

13. Many sediments and sedimentary rocks including sand, gravel, evaporites, coal, and banded iron formations are important natural resources. Most oil and natural gas are found in sedimentary rocks.

Important Terms

bed
bedding plane
biochemical sedimentary rock
carbonate rock
cementation
chemical sedimentary rock
compaction
cross-bedding

depositional environment
detrital sedimentary rock
evaporite
fossil
graded bedding
lithification
mud crack
ripple mark

rounding
sediment
sedimentary rock
sedimentary structure
sorting
strata

1. An aggregate of sand and gravel with particles of markedly different sizes is characterized as:
   a. _____ completely abraded;
   b. _____ poorly sorted;
   c. _____ well rounded;
   d. _____ cross-bedded;
   e. _____ chemically weathered.

2. Cementation is a process during which:
   a. _____ dissolved mineral matter precipitates in pore spaces and binds sediment together;
   b. _____ limestone is converted to dolostone by the addition of magnesium;
   c. _____ sediment layers are deposited on a sloping surface, thus recording paleocurrent direction;
   d. _____ clay-sized particles settle and accumulate as horizontal layers;
   e. _____ seawater evaporates and dissolved substances began to form minerals.

3. An accumulation of minerals by density, as in a stream channel, is known as a:
   a. _____ biochemical sedimentary rock;
   b. _____ ripple mark;
   c. _____ sedimentary structure;
   d. _____ depositional environment;
   e. _____ placer deposit.

4. Which one of the following is a carbonate rock?
   a. _____ rock gypsum;
   b. _____ dolostone;
   c. _____ shale;
   d. _____ conglomerate;
   e. _____ chert.

5. Although rapidly diminishing flow in a stream can yield graded bedding, most of it forms by deposition from:
   a. _____ turbidity currents;
   b. _____ evaporation of seawater;
   c. _____ chemical alteration of limestone;
   d. _____ windblown sand;
   e. _____ settling of mud in lakes.

6. Most limestones have a large component of calcite ($CaCO_3$), which was extraced from seawater by:
   a. _____ lithification;
   b. _____ chemical weathering;
   c. _____ evaporation;
   d. _____ inorganic chemical reactions;
   e. _____ organisms.

7. The chemical sedimentary rock chert is composed of:
   a. _____ silt and clay;
   b. _____ well-sorted and well-rounded sand grains;
   c. _____ microscorpic crystals of quartz;
   d. _____ gravel;
   e. _____ calcium sulfate.

8. Which one of the following is an example of a trace fossil?
   a. _____ clam shell;
   b. _____ wooly mammoth tusk;
   c. _____ mammal bones;
   d. _____ dinosaur footprint;
   e. _____ bird skeleton.

9. Traps for petroleum and natural gas resulting from folding or fracturing of rock layers are _____ traps.
   a. _____ cap rock;
   b. _____ structural;
   c. _____ salt dome;
   d. _____ stratigraphic;
   e. _____ reservoir.

10. The most common mineral in sandstone is:
    a. _____ muscovite;
    b. _____ calcite;
    c. _____ gypsum;
    d. _____ dolomite;
    e. _____ quartz.

11. What is graded bedding, and how does it form?

12. Describe how a placer deposit forms. Give some examples of valuable resources that might be found in placer deposits.

13. How do evaporite rocks form? What are the two most common evaporites, and what are their chemical compositions?

14. What are the differences and similarities between detrital and chemical sedimentary rocks?

15. Describe the process involved in lithification of detrital sediment such as mud and sand.

16. How do body fossils differ from trace fossils? Of what use are fossils in determining depositional environment?

17. Give a brief explanation of how coal forms.

18. Name three sedimentary structures and explain how they originate.

19. Describe the conditions in which petroleum and natural gas might be trapped.

20. What are the two common carbonate rocks? How do they form?

1. As a field geologist, you encounter rock layers consisting of well-sorted, well-rounded sandstone with 10-m high cross-beds and reptile footprints. On top of these layers is a layer of coal and channel-shaped sandstones with land plant fossils and, finally, an uppermost layer of limestone with fossil clams and sea lilies. What can you infer about the depositional environments and how they changed through time?

2. Earth's crust is estimated to contain 51% feldspars, 24% ferromagnesian silicates (biotite, pyroxenes, amphiboles), 12% quartz, and 13% other minerals. Considering these abundances, how can you explain the fact that quartz is by far the most common mineral in sandstones?

For these web site addresses, along with current updates and exercises, log on to

http://www.wadsworth.com/geo

▶ SEDIMENTARY ROCKS

This site contains images and information on 11 common sedimentary rocks. It is part of the Soil Science 223 Rocks and Minerals Reference web site. Click on any of the *rock names* and compare the information and images to the information in this chapter.

▶ GEOLOGY 110 WEB PAGES OF SEDIMENTARY STRUCTURES

This site contains links to two slide collections of sedimentary structures that were used in the Geology 110 class at Duke University. Click on either the *Sedimentary Structures, Part 1* or *Sedimentary Structures, Part 2* heading to view slides of different sedimentary structures. Use this collection to review the different sedimentary structures discussed in this chapter.

▶ SEDIMENTARY ROCKS

This site, part of the Georgia Science On-Line Project, was created by Dr. Pamela J. W. Gore for Geology 101 at DeKalb College, Clarkston, Georgia. It has good images of all common sedimentary rocks, along with descriptions of their textures and compositions. See the images of various sedimentary rocks.

1. What are the three basic components in terrigenous (also called detrital or clastic) sedimentary rocks?

2. What types of siliceous rocks are listed, and what are they composed of?

3. Give the names of two varieties of organic sedimentary rocks.

▶ CHEVRON LEARNING CENTER

Maintained by Chevron Oil Company, its Learning Center has four listed topics, two of which are of particular interest. Click on *What Is Crude Oil?* and learn about the history of oil production.

1. What created the first large-scale demand for petroleum products? When did this occur?

2. Briefly discuss how geologists think crude oil forms.

3. What are the three essentials for the origin of a crude oil field?

Return to the Chevron Learning Center page and click on *A Prospecting Primer.* What are some of the tools used in the search for crude oil?

▶ THE BURGESS SHALE PROJECT: GEOLOGY

The site by NRC-CNRC and Industry Canada has information on the rock cycle and particularly the origin of sedimentary rocks. Click on any of the stages in the rock-cycle diagram and learn about sediments, erosion, compaction, and cementation. See what happens to various sedimentary rocks as they are subjected to exposure, granitization, and pressure and heat. Other aspects of the rock cycle can also be explored by clicking, for example, *metamorphic rocks.* Also, see the information on the Burgess Shale. Go to the Burgess Shale Home Page for information on the fossils of this rock unit.

For these web site addresses, along with current updates and exercises, log on to

http://www.wadsworth.com/geo

WORLD OF CHEMISTRY: THE HOME PAGE OF RALPH LOGAN

This site, which is maintained by Ralph H. Logan, chemistry instructor at North Lake College in Dallas, Texas, is devoted mostly to chemistry but it also has information of geologic interest. Scroll down to *Places to Visit* and click *Frequently Asked Questions on Chemical Concepts by Category,* and then click *Ecology & Energy Source Questions.* In this section see *What Are the Evolutionary Stages of Coal Development?*

1. What are the stages in coal development?

2. What are some of the disadvantages of using coal?

▶ THE FOSSIL COMPANY— ABOUT FOSSILS

Click on any of the fossil icons to see images and learn about fossil mammals, starfish, ammonites, plants, reptiles, and many others.

CD-ROM Exploration

Explore the following *In-Terra-Active 2.0* CD-ROM module(s) and increase your understanding of key concepts and processes presented in this chapter.

▶ SECTION: MATERIALS
MODULE: SEDIMENTARY ROCKS

The compositions, textures, and structures preserved in sedimentary rocks provide the geologist with very clear clues as to the type of setting in which the sedimentary material was originally deposited. The purpose of the module is to explore the links between starting material, transportation and deposition process, and final sedimentary rock. **Question: What can we conclude about the sedimentary environment in which this sandstone accumulated in northwest Wyoming?**

8

Metamorphism
and
Metamorphic

Slate quarry in Wales. Slate is the result of low-grade metamorphism of shale and has a variety of uses. These high-quality slates were formed by a mountain-building episode that occurred approximately 400 to 440 million years ago in the present-day countries of Iceland, Scotland, Wales, and Norway.

Prologue

R o c k s

Its homogeneity, softness, and various textures have made marble, a metamorphic rock formed from limestone or dolostone, a favorite rock of sculptors throughout history. As the value of authentic marble sculptures has increased through the years, the number of forgeries has also increased. With the price of some marble sculptures in the millions of dollars, private collectors and museums need some means of ensuring the authenticity of the work they are buying. Aside from the monetary considerations, it is important that forgeries not become part of the historical and artistic legacy of human endeavor.

Experts have traditionally relied on the artistic style and weathering characteristics to determine whether a marble sculpture is authentic or a forgery. Because marble is not very resistant to weathering, forgers have resorted to a variety of methods to produce the weathered appearance of an authentic ancient work. Now, however, using new techniques, geologists can distinguish a naturally weathered marble surface from one that has been artificially altered.

Although marbles result when the agents of metamorphism (heat, pressure, and fluid activity) are applied to carbonate rocks, the type of marble formed depends, in part, on the original composition of the parent carbonate rock as well as the type and intensity of metamorphism. Therefore, one way to authenticate a marble sculpture is to determine the origin of the marble itself. During the Preclassical, Greek, and Roman periods, the islands of Naxos, Thasos, and Paros in the Aegean Sea—as well as the Greek mainland, Turkey, and Italy—were all sites of major marble quarries.

To identify the source of the marble in various sculptures, geologists employ a wide variety of analytical techniques. These include hand-specimen and thin-section analysis of the marble, trace element analysis by X-ray fluorescence, stable isotopic ratio analysis for carbon and oxygen, and other more esoteric techniques.

The J. Paul Getty Museum in Malibu, California, employed some of these techniques to help authenticate an ancient Greek kouros (a sculptured figure of a Greek youth) thought to have been carved around 530 B.C. (Figure 8.1). The kouros was offered to the Getty Museum in 1984 for a reported price of $7 million. Some of its stylistic features, however, caused some experts to question its authenticity. Consequently, the museum had a variety of geochemical and mineralogical tests performed in an effort to determine the authenticity of the kouros.

Isotopic analysis of the weathered surface and fresh interior of the kouros confirmed that the marble probably came from the Cape Vathy quarries on the island of Thasos in the Aegean Sea. But these results did not prove the age of the kouros—it might still have been a forgery carved from marble taken from an archaeological site on the island.

The kouros was carved from dolomitic marble, and its surface is covered with a complex thin crust (0.01–0.05 mm thick) consisting mostly of whewellite, a calcium oxalate monohydrate mineral. To ensure that the crust is the result of long-term weathering and not a modern forgery, dolomitic marble samples were subjected to a variety of forgery techniques to try to replicate the surface of the kouros. Samples were soaked or boiled in various mixtures for periods of time ranging from hours to months, and their surfaces were treated and retreated to try and match the appearance of the weathered surface of the kouros. Such tests yielded only a few examples

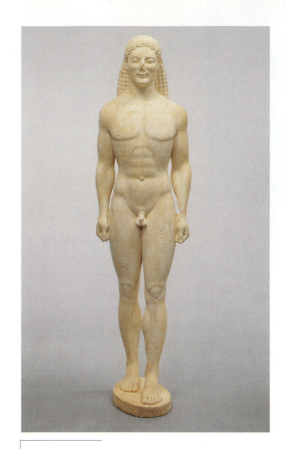

FIGURE 8.1 This Greek kouros, which stands 206 cm tall, is the object of an intensive authentication study by the Getty Museum. Using several geologic tests, it has been determined that the kouros was carved from dolomitic marble that probably came from the Cape Vathy quarries on Thasos.

that appeared similar to the surface of the kouros. Even those samples, however, were different when examined under high magnification or subjected to geochemical analysis. In fact, all of the samples clearly showed that they were the result of recent alteration and not long-term weathering processes.

Though scientific tests have not unequivocally proved authenticity, they have shown that the weathered surface layer of the kouros bears more similarities to naturally occurring weathered surfaces than to known artificially produced surfaces. Furthermore, no evidence indicates that the surface alteration of the kouros is of modern origin.

Despite intensive study by scientists, archaeologists, and art historians, opinion is still divided as to the authenticity of the Getty kouros. Most scientists accept that the kouros was carved sometime around 530 B.C., but most art historians are *doubtful. Pointing to inconsistencies in its style of sculpture for that period, they think that it is a modern forgery. Because of the continuing doubts about the statue's authenticity, the recently opened J. Paul Getty Center has mounted it in an exhibition listing the evidence for and against its authenticity.*

Regardless of the ultimate conclusion on the Getty kouros, geologic testing to authenticate marble sculptures is now an important part of many museums' curatorial functions. In addition, a large body of data about the characteristics and origin of marble is being amassed as more sculptures and quarries are analyzed.

Introduction

METAMORPHIC rocks (from the Greek *meta* meaning "change" and *morpho* meaning "shape") constitute the third major group of rocks. They result from the transformation of other rocks by metamorphic processes that usually occur beneath Earth's surface (see Figure 1.12). During metamorphism, rocks are subjected to sufficient heat, pressure, and fluid activity to change their mineral composition and/or texture, thus forming new rocks. These transformations take place in the solid state, and the type of metamorphic rock formed depends on the original composition and texture of the parent rock, the agents of metamorphism, and the amount of time the parent rock was subjected to the effects of metamorphism.

A large portion of Earth's continental crust is composed of metamorphic and igneous rocks. Together they form the crystalline basement rocks that underlie the sedimentary rocks of a continent's surface. This basement rock is widely exposed in regions of the continents known as *shields,* which have been very stable during the past 600 million years (Figure 8.2). Metamorphic rocks also constitute a sizable portion of the crystalline core of large mountain ranges. Some of the oldest known rocks, dated at 3.96 billion years from the Canadian Shield, are metamorphic, indicating they formed from even older rocks.

Why is it important to study metamorphic rocks? For one thing, they provide information about geologic processes operating within Earth and about the way these processes have varied through time. Furthermore, metamorphic rocks such as marble and slate are used as building materials, and certain metamorphic minerals are economically important. For example, talc is used in cosmetics, in the manufacture of paint, and as a lubricant, while asbestos is used for insulation and fireproofing (see Perspective 8.1).

The Agents of Metamorphism

THE three agents of metamorphism are heat, pressure, and fluid activity. During metamorphism, the original rock undergoes change to achieve equilibrium with its new environment. The changes may result in the formation of new minerals and/or a change in the texture of the rock by the reorientation of the original minerals. In some instances the change is minor, and features of the parent rock can still be recognized. In other cases the rock changes so much that the identity of the parent rock can be determined only with great difficulty, if at all.

Besides heat, pressure, and fluid activity, time is also important to the metamorphic process. Chemical reactions proceed at different rates and thus require different amounts of time to complete. Reactions involving silicate compounds are particularly slow, and because most metamorphic rocks are composed of silicate minerals, it is thought that metamorphism is a slow geologic process.

Asbestos

Asbestos (from the Latin, meaning "unquenchable") is a general term applied to any silicate mineral that easily separates into flexible fibers (Figure 1).

The combination of such features as noncombustibility and flexibility makes asbestos an important industrial material of considerable value. In fact, asbestos has more than 3000 known uses, including brake linings, fireproof fabrics, and heat insulators.

Asbestos can be divided into two broad groups, serpentine and amphibole asbestos. *Chrysotile,* which is a hydrous magnesium silicate with the chemical formula $Mg_3Si_2O_5(OH)_4$, is the fibrous form of serpentine asbestos; it is the most valuable type and constitutes the bulk of all commercial asbestos. Chrysotile's strong, silky fibers are easily spun and can withstand temperatures up to 2750°C.

The vast majority of chrysotile asbestos occurs in serpentine, a type of rock formed by the alteration of ultramafic igneous rocks such as peridotite under low- and medium-grade metamorphic conditions. Serpentine is thought to form from the alteration of olivine by hot, chemically active, residual fluids emanating from cooling magma. The chrysotile asbestos forms veinlets of fiber within the serpentine and may comprise up to 20% of the rock.

At least five varieties of amphibole asbestos are known, but *crocidolite,* a sodium-iron amphibole with the chemical formula $Na_2(Fe^{+3})_2(Fe^{+2})_3Si_8O_{22}$ $(OH)_2$, is the most common. Crocidolite, which is also known as blue asbestos, is a long, coarse, spinning fiber that is stronger but more brittle than chrysotile and also less resistant to heat. The other varieties of amphibole asbestos have fewer uses and are used primarily for insulation.

Crocidolite is found in such metamorphic rocks as slates and schists. It is thought that crocidolite forms by the solid-state alteration of other minerals within the high-temperature and high-pressure environment that results from deep burial.

Despite its widespread use, the federal Environmental Protection Agency (EPA) has instituted a gradual ban on all new asbestos products. The ban was imposed because some forms of asbestos can cause lung cancer and scarring of the lungs if its fibers are inhaled. Because the EPA apparently paid little attention to the issue of risks versus benefits when it enacted this rule, the U.S. Fifth Circuit Court of Appeals overturned the EPA ban on asbestos in 1991.

The threat of lung cancer has also resulted in legislation mandating the removal of asbestos already in place in all public buildings, including all public and private schools. Recently, however, important questions have been raised concerning the threat posed by asbestos and the additional potential hazards that may arise from its improper removal.

The current policy (1993) of the EPA mandates that all forms of asbestos be treated as identical hazards. Yet studies indicate that only the amphibole forms constitute a known health hazard. Chrysotile, whose fibers tend to be curly, does not become lodged in the lungs. Furthermore, its fibers are generally soluble and disappear in tissue. In contrast, crocidolite has long, straight, thin fibers that penetrate the lungs and stay there. These fibers irritate the lung tissue and, over a long period of time, can lead to lung cancer. Thus, crocidolite, and not chrysotile, is overwhelmingly responsible for asbestos-related lung cancer. Because about 95% of the asbestos in place in the United States is chrysotile, many people are questioning whether the dangers from asbestos have been somewhat exaggerated.

Removing asbestos from buildings where it has been installed might cost as much as $100 billion, and some recent studies have indicated that the air in buildings containing asbestos has essentially the same amount of airborne asbestos fibers as the air outdoors. In fact, unless the material containing the asbestos is disturbed, asbestos does not shed fibers. Furthermore, improper removal of asbestos can lead to contamination. In most cases of improper removal, the concentration of airborne asbestos fibers is far higher than if the asbestos had been left in place.

The problem of asbestos contamination is a good example of how geology affects our lives and why a basic knowledge of science is important.

Figure 1 Hand specimen of chrysotile from Thetford, Quebec, Canada. Chrysotile is the fibrous form of serpentine asbestos.

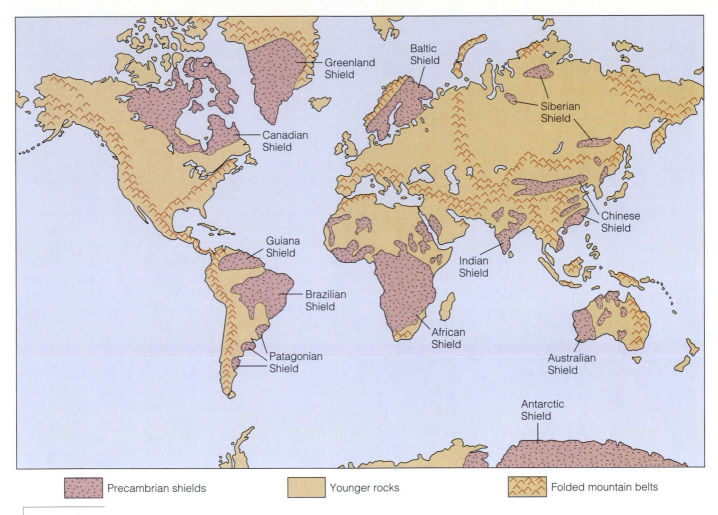

FIGURE 8.2 Shields of the world. Shields are the exposed portion of the crystalline basement rocks that underlie each continent; these areas have been very stable during the past 600 million years.

Legend: Precambrian shields | Younger rocks | Folded mountain belts

HEAT

Heat is an important agent of metamorphism because it increases the rate of chemical reactions that may produce minerals different from those in the original rock. The heat may come from intrusive magmas or result from deep burial in the crust such as occurs during subduction along a convergent plate boundary.

When rocks are intruded by bodies of magma, they are subjected to intense heat that affects the surrounding rock; the most intense heating usually occurs adjacent to the magma body and gradually decreases with distance from the intrusion. The zone of metamorphosed rocks that forms in the country rock adjacent to an intrusive igneous body is usually rather distinct and easy to recognize.

PRESSURE

When rocks are buried, they are subjected to increasingly greater **lithostatic pressure**; this pressure, which results from the weight of the overlying rocks, is applied equally in all directions (Figure 8.3). As rocks are subjected to increasing lithostatic pressure with depth, the mineral grains within a rock may become more closely packed. Under these conditions, the minerals may *recrystallize;* that is, they may form smaller and denser minerals.

Along with lithostatic pressure resulting from burial, rocks may also experience **differential pressures** (Figure 8.4). In this case, the pressures are not equal on all sides, and the rock is consequently distorted. Differential pressures typically occur during deformation associated with mountain building and can produce distinctive metamorphic textures and features.

FLUID ACTIVITY

In almost every region of metamorphism, water and carbon dioxide (CO_2) are present in varying amounts along mineral grain boundaries or in the pore spaces of rocks. These fluids, which may contain ions in solution, enhance metamorphism by increasing the rate of chemical reactions. Under dry conditions, most minerals react very slowly, but when even small amounts of fluid are introduced, reaction rates

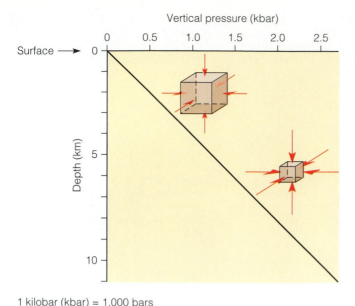

1 kilobar (kbar) = 1,000 bars
Atmospheric pressure at sea level = 1 bar

FIGURE 8.3 Lithostatic pressure is applied equally in all directions in Earth's crust, due to the weight of the overlying rocks. Thus, pressure increases with depth.

increase, mainly because ions can move readily through the fluid and thus enhance chemical reactions and the formation of new minerals.

The following reaction provides a good example of how new minerals can be formed by **fluid activity**. Here, seawater moving through hot basaltic rock of the oceanic crust transforms olivine into the metamorphic mineral serpentine:

$$2Mg_2SiO_4 + 2H_2O \rightarrow Mg_3Si_2O_5(OH)_4 + MgO$$

OLIVINE   WATER   SERPENTINE   CARRIED
AWAY IN
SOLUTION

The chemically active fluids that are part of the metamorphic process come primarily from three sources. The first is water trapped in the pore spaces of sedimentary rocks as they form. A second is the volatile fluid within magma. The third source is the dehydration of water-bearing minerals such as gypsum ($CaSO_4 \cdot 2H_2O$) and some clays.

Types of Metamorphism

THREE major types of metamorphism are recognized: contact metamorphism in which magmatic heat and fluids act to produce change; dynamic metamorphism, which is principally the result of high differential pressures associated with intense deformation; and regional metamorphism, which occurs within a large area and is caused primarily by mountain-building forces. Even though we will discuss each type of metamorphism separately, the boundary between them is not always distinct and depends largely on which of the three metamorphic agents was dominant.

CONTACT METAMORPHISM

Contact metamorphism takes place when a body of magma alters the surrounding country rock. At shallow depths, an intruding magma raises the temperature of the surrounding rock, causing thermal alteration. Furthermore, the release of hot fluids into the country rock by the cooling intrusion can also aid in the formation of new minerals.

Important factors in contact metamorphism are the initial temperature and size of the intrusion as well as the fluid content of the magma and/or country rock. The initial temperature of an intrusion is controlled, in part, by its composition: mafic magmas are hotter than felsic magmas (see Chapter 4) and hence have a greater thermal effect on the rocks directly surrounding them. The size of the intrusion is also important. In the case of small intrusions, such

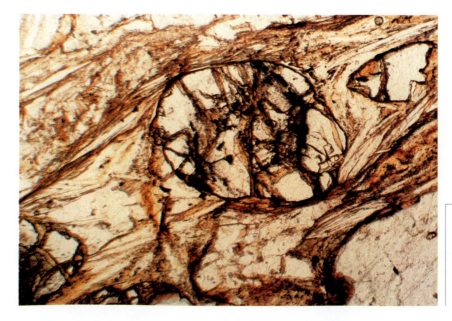

FIGURE 8.4 Differential pressure is pressure that is unequally applied to an object. Rotated garnets are a good example of the effects of differential pressure applied to a rock during metamorphism. These rotated garnets come from a schist in northeast Sardinia. *(Photo courtesy of Eric Johnson.)*

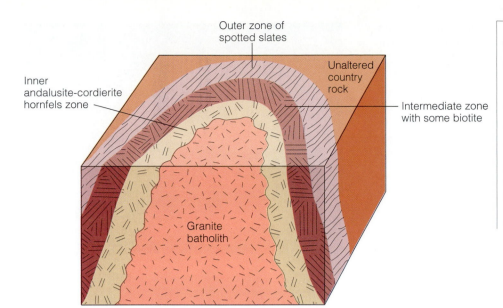

Outer zone of spotted slates

Inner andalusite-cordierite hornfels zone

Unaltered country rock

Intermediate zone with some biotite

Granite batholith

FIGURE 8.5 A metamorphic aureole typically surrounds many igneous intrusions. The metamorphic aureole associated with this idealized granite batholith contains three zones of mineral assemblages reflecting the decreases in temperature with distance from the intrusion. An andalusite-cordierite hornfels forms the inner zone adjacent to the batholith. This is followed by an intermediate zone of extensive recrystallization in which some biotite develops, and farthest from the intrusion is the outer zone, which is characterized by spotted slates.

as dikes and sills, usually only those rocks in immediate contact with the intrusion are affected. Because large intrusions, such as batholiths, take a long time to cool, the increased temperature in the surrounding rock may last long enough for a larger area to be affected.

Fluids also play an important role in contact metamorphism. Many magmas are wet and contain hot, chemically active fluids that may emanate into the surrounding rock. These fluids can react with the rock and aid in the formation of new minerals. In addition, the country rock may contain pore fluids that, when heated by the magma, also increase reaction rates.

Temperatures can reach nearly 900°C adjacent to an intrusion, but they gradually decrease with distance. The effects of such heat and the resulting chemical reactions usually occur in concentric zones known as **aureoles** (Figure 8.5). The boundary between an intrusion and its aureole may be either sharp or transitional (Figure 8.6).

Metamorphic aureoles vary in width depending on the size, temperature, and composition of the intrusion, as well as the composition of the surrounding country rock. Typically, large intrusive bodies have several metamorphic zones, each characterized by distinctive mineral assemblages indicating the decrease in temperature with distance from the intrusion (Figure 8.5). The zone closest to the intrusion, and hence subject to the highest temperatures, may contain high-temperature metamorphic minerals (that is, minerals in equilibrium with the higher-temperature environment) such as sillimanite. The outer zones may be characterized by lower-temperature metamorphic minerals such as chlorite, talc, and epidote.

The formation of new minerals by contact metamorphism depends not only on proximity to the intrusion, but also on the composition of the country rock. Shales, mudstones, impure limestones, and impure dolostones are particularly susceptible to the formation of new minerals by

contact metamorphism, whereas pure sandstones or pure limestones typically are not.

Two types of contact metamorphic rocks are generally recognized: those resulting from baking of country rock and those altered by hot solutions. Many of the rocks resulting from contact metamorphism have the texture of porcelain; that is, they are hard and fine grained. This is particularly true for rocks with a high clay content, such as shale. Such texture results because the clay minerals in the rock are baked, just as a clay pot is baked when fired in a kiln.

During the final stages of cooling when an intruding magma begins to crystallize, large amounts of hot, watery solutions are often released. These solutions may react

FIGURE 8.6 A sharp and clearly defined boundary occurs between the intruding light-colored igneous rock on the left and the dark-colored metamorphosed country rock on the right. The intrusion is part of the Peninsular Ranges Batholith, east of San Diego, California. (Photo courtesy of David J. Matty.)

with the country rock and produce new metamorphic minerals. This process, which usually occurs near Earth's surface, is called *hydrothermal alteration,* and may result in valuable mineral deposits. Geologists think that many of the world's ore deposits result from the migration of metallic ions in hydrothermal solutions. Examples include copper, gold, iron ores, tin, and zinc in various localities including Australia, Canada, China, Cyprus, Finland, Russia, and the western United States.

DYNAMIC METAMORPHISM

Most **dynamic metamorphism** is associated with fault (fractures along which movement has occurred) zones where rocks are subjected to high differential pressures. The metamorphic rocks resulting from pure dynamic metamorphism are called *mylonites* and are typically restricted to narrow zones adjacent to faults. Mylonites are hard, dense, fine-grained rocks, many of which are characterized by thin laminations (Figure 8.7). Tectonic settings where mylonites occur include the Moine Thrust Zone in northwest Scotland and portions of the San Andreas fault in California.

REGIONAL METAMORPHISM

Most metamorphic rocks result from **regional metamorphism**, which occurs over a large area and is usually caused by tremendous temperatures, pressures, and deformation within the deeper portions of the crust. Regional metamorphism is most obvious along convergent plate margins where rocks are intensely deformed and recrystallized during convergence and subduction. Within these metamorphic rocks, there is usually a gradation of meta-

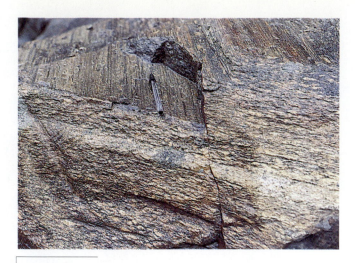

FIGURE 8.7 Mylonite from the Adirondack Highlands, New York. *(Photo courtesy of Eric Johnson.)*

morphic intensity from areas that were subjected to the most intense pressures and/or highest temperatures to areas of lower pressures and temperatures. Such a gradation in metamorphism can be recognized by the metamorphic minerals that are present.

Regional metamorphism is not just confined to convergent margins. It also occurs in areas where plates diverge, though usually at much shallower depths because of the high geothermal gradient associated with these areas.

From field studies and laboratory experiments, certain minerals are known to form only within specific temperature and pressure ranges. Such minerals are known as **index minerals** because their presence allows geologists to recognize low-, intermediate-, and high-grade meta-

TABLE 8.1

Metamorphic Zones and Their Mineral Assemblages for Different Country Rock Types

METAMORPHIC GRADE		METAMORPHIC ZONE FOR CLAY-RICH ROCKS	MINERAL ASSEMBLAGE PRODUCED FOR DIFFERENT COUNTRY ROCKS		
			Mudrocks	*Limestones*	*Mafic Igneous Rocks*
Increasing	Low	Chlorite	Chlorite,* quartz, muscovite, plagioclase	Chlorite,* calcite or dolomite, plagioclase	Chlorite,* plagioclase
		Biotite	Biotite,* quartz, plagioclase		
	Medium	Garnet	Garnet,* mica, quartz, plagioclase	Garnet,* epidote, hornblende, calcite	Garnet,* chlorite, epidote, plagioclase
		Staurolite	Staurolite,* mica, garnet, quartz, plagioclase	Garnet, hornblende,* plagioclase	
	High	Kyanite	Kyanite,* mica, garnet, quartz, plagioclase		
metamorphism					Hornblende,* plagioclase
		Sillimanite	Sillimanite,* garnet, mica, quartz, plagioclase	Garnet, augite,* plagioclase	

*Index mineral.

morphic zones (Table 8.1). A typical progression of index minerals forming primarily in rocks that were originally clay rich involves the sequential formation of the following minerals:

CHLORITE → BIOTITE → AMPHIBOLE → STAUROLITE → SILLIMANITE.

Different rock compositions develop different index minerals. When sandy dolomites are metamorphosed, they produce an entirely different set of index minerals. Therefore, a specific set of index minerals commonly forms in specified rock types as metamorphism progresses.

Classification of Metamorphic Rocks

FOR purposes of classification, metamorphic rocks are commonly divided into two groups: those exhibiting a foliated texture and those with a nonfoliated texture (Table 8.2).

FOLIATED METAMORPHIC ROCKS

Rocks subjected to heat and differential pressure during metamorphism typically have minerals arranged in a parallel fashion that gives them a **foliated texture** (Figure 8.8). The size and shape of the mineral grains determine whether the foliation is fine or coarse. If the foliation is such that the individual grains cannot be recognized without magnification, the rock is said to be slate (Figure 8.9). A coarse foliation results when granular minerals such as quartz and feldspar are segregated into roughly parallel and streaky zones that differ in composition and color as in a gneiss. Foliated metamorphic rocks can be arranged in order of increasingly coarse grain size and perfection of foliation.

Slate is a very fine-grained metamorphic rock that commonly exhibits *slaty cleavage* (Figure 8.9b). Slate is the result of low-grade regional metamorphism of shale or, more rarely, volcanic ash. Because it can easily be split along cleavage planes into flat pieces, slate is an excellent rock for roofing and floor tiles, billiard and pool table tops, and

TABLE 8.2

Classification of Common Metamorphic Rocks

TEXTURE	METAMORPHIC ROCK	TYPICAL MINERALS	METAMORPHIC GRADE	CHARACTERISTICS OF ROCKS	PARENT ROCK
Foliated	Slate	Clays, micas, chlorite	Low	Fine-grained, splits easily into flat pieces	Mudrocks, claystones, volcanic ash
	Phyllite	Fine-grained quartz, micas, chlorite	Low to medium	Fine-grained, glossy or lustrous sheen	Mudrocks
	Schist	Micas, chlorite, quartz, talc, hornblende, garnet, staurolite, graphite	Low to high	Distinct foliation, minerals visible	Mudrocks, carbonates, mafic igneous rocks
	Gneiss	Quartz, feldspars, hornblende, micas	High	Segregated light and dark bands visible	Mudrocks, sandstones, felsic igneous rocks
	Amphibolite	Hornblende, plagioclase	Medium to high	Dark-colored, weakly foliated	Mafic igneous rocks
	Migmatite	Quartz, feldspars, hornblende, micas	High	Streaks or lenses of granite intermixed with gneiss	Felsic igneous rocks mixed with sedimentary rocks
Nonfoliated	Marble	Calcite, dolomite	Low to high	Interlocking grains of calcite or dolomite, reacts with HCl	Limestone or dolostone
	Quartzite	Quartz	Medium to high	Interlocking quartz grains, hard, dense	Quartz sandstone
	Greenstone	Chlorite, epidote, hornblende	Low to high	Fine-grained, green color	Mafic igneous rocks
	Hornfels	Micas, garnets, andalusite, cordierite, quartz	Low to medium	Fine-grained, equidimensional grains, hard, dense	Mudrocks
	Anthracite	Carbon	High	Black, lustrous, subconcoidal fracture	Coal

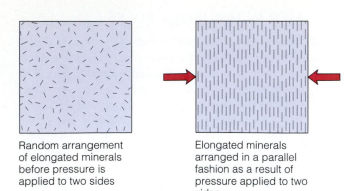

Random arrangement of elongated minerals before pressure is applied to two sides

Elongated minerals arranged in a parallel fashion as a result of pressure applied to two sides

(a)

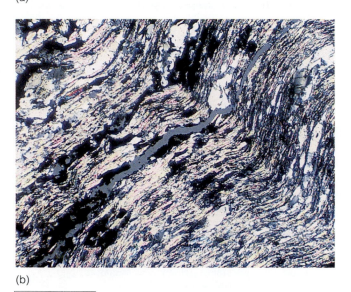

(b)

FIGURE 8.8 (a) When rocks are subjected to differential pressure, the mineral grains are typically arranged in a parallel fashion, producing a foliated texture. (b) Photomicrograph of a metamorphic rock with a foliated texture showing the parallel arrangement of mineral grains.

(a)

(b)

FIGURE 8.9 (a) Hand specimen of slate. (b) This panel of Arvonia Slate from Albemarne Slate Quarry, Virginia, shows bedding (upper right to lower left) at an angle to the slaty cleavage. *(Photo (a) courtesy of Sue Monroe; photo (b) courtesy of R. V. Dietrich.)*

blackboards. The different colors of most slates are caused by minute amounts of graphite (black), iron oxide (red and purple), and/or chlorite (green).

Phyllite is similar in composition to slate but is coarser grained. The minerals, however, are still too small to be identified without magnification. Phyllite can be distinguished from slate by its glossy or lustrous sheen. It represents an intermediate grain size between slate and schist.

Schist is most commonly produced by regional metamorphism. The type of schist formed depends on the intensity of metamorphism and the character of the parent rock (Figure 8.10). Metamorphism of many rock types can yield schist, but most schist appears to have formed from clay-rich sedimentary rocks (Table 8.2).

All schists contain more than 50% platy and elongated minerals, all of which are large enough to be clearly visible. Their mineral composition imparts a *schistosity* or *schistose foliation* to the rock that usually produces a wavy type of parting when split. Schistosity is common in low- to

high-grade metamorphic environments, and each type of schist is known by its most conspicuous mineral or minerals, such as mica schist, chlorite schist, or talc schist.

Gneiss is a metamorphic rock that is streaked or has

FIGURE 8.10 Garnet-mica schist. *(Photo courtesy of Sue Monroe.)*

FIGURE 8.11 Gneiss is characterized by segregated bands of light and dark minerals. This folded gneiss is exposed near Wawa, Ontario, Canada.

segregated bands of light and dark minerals (Figure 8.11). Gneisses are composed mostly of granular minerals such as quartz and/or feldspar with lesser percentages of platy or elongated minerals such as micas or amphiboles. Quartz and feldspar are the principal light-colored minerals, whereas biotite and hornblende are the typical dark-colored minerals. Gneiss typically breaks in an irregular manner, much like coarsely crystalline nonfoliated rocks.

Most gneiss probably results from recrystallization of clay-rich sedimentary rocks during regional metamorphism (Table 8.2). Gneiss also can form from igneous rocks such as granite or older metamorphic rocks.

Another fairly common foliated metamorphic rock is *amphibolite*. A dark-colored rock, it is composed mainly of hornblende and plagioclase. The alignment of the hornblende crystals produces a slightly foliated texture. Many amphibolites result from medium- to high-grade metamorphism of such ferromagnesian silicate-rich igneous rocks as basalt.

In some areas of regional metamorphism, exposures of "mixed rocks" having both igneous and high-grade metamorphic characteristics are present. In these rocks, called *migmatites,* streaks or lenses of granite are usually intermixed with high-grade ferromagnesian-rich metamorphic rocks, imparting a wavy appearance to the rock (Figure 8.12).

Most migmatites are thought to be the product of extremely high-grade metamorphism, and several models for their origin have been proposed. Part of the problem in determining the origin of migmatites is explaining how the granitic component formed. According to one model, the granitic magma formed in place by the partial melting of rock during intense metamorphism. Such an origin is possible providing that the host rocks contained quartz and feldspars and that water was present. Another possibility is that the granitic components formed by the redistribution of minerals by recrystallization in the solid state, that is, by pure metamorphism.

NONFOLIATED METAMORPHIC ROCKS

In some metamorphic rocks, the mineral grains do not show a discernible preferred orientation. Instead, these rocks consist of a mosaic of roughly equidimensional minerals and are characterized as having a **nonfoliated texture** (Figure 8.13). Most nonfoliated metamorphic rocks result from contact or regional metamorphism of rocks in which no platy or elongate minerals are present. Frequently, the only indication that a granular rock has been metamorphosed is the large grain size resulting from recrystallization. Nonfoliated metamorphic rocks are generally of two types: those composed mainly of only one mineral, for example, marble or quartzite; and those in which the different mineral grains are too small to be seen without magnification, such as greenstone and hornfels.

Marble is a well-known metamorphic rock composed predominantly of calcite or dolomite; its grain size ranges from fine to coarsely granular (Figure 8.14). Marble results from either contact or regional metamorphism of limestones or dolostones (Table 8.2). Pure marble is snowy white or bluish, but varieties of all colors exist because of

FIGURE 8.12 Migmatites consist of high-grade metamorphic rock intermixed with streaks or lenses of granite. This Precambrian(?) migmatite is exposed at Thirty Thousand Islands of Georgian Bay, Lake Huron, Ontario, Canada. *(Photo by Ed Bartram, courtesy of R. V. Dietrich.)*

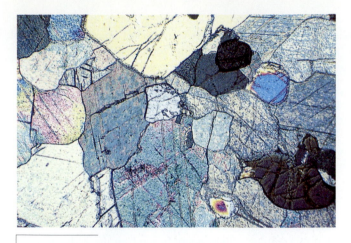

FIGURE 8.13 Nonfoliated textures are characterized by a mosaic of roughly equidimensional minerals as in this photomicrograph of marble.

the presence of mineral impurities in the parent sedimentary rock. The softness of marble, its uniform texture, and its various colors have made it the favorite rock of builders and sculptors throughout history (see the Prologue).

Quartzite is a hard, compact rock formed from quartz sandstone under medium-to-high-grade metamorphic conditions during contact or regional metamorphism (Figure 8.15). Because recrystallization is so complete, metamorphic quartzite is of uniform strength and therefore usually breaks across the component quartz grains rather than around them when it is struck. Pure quartzite is white, but

iron and other impurities commonly impart a reddish or other color to it. Quartzite is commonly used as foundation material for road and railway beds.

The name *greenstone* is applied to any compact, dark-green, altered, mafic igneous rock that formed under low-to-high-grade metamorphic conditions. The green color results from the presence of chlorite, epidote, and hornblende.

Hornfels, a fine-grained, nonfoliated metamorphic rock resulting from contact metamorphism, is composed of various equidimensional mineral grains. The composition of hornfels directly depends on the composition of the original rock, and many compositional varieties are known. The majority of hornfels, however, are apparently derived from contact metamorphism of clay-rich sedimentary rocks or impure dolostones.

Anthracite is a black, lustrous, hard coal that contains a high percentage of fixed carbon and a low percentage of volatile matter. It usually forms from the metamorphism of lower-grade coals by heat and pressure and is thus considered by many geologists to be a metamorphic rock.

Metamorphic Zones

THE first systematic study of metamorphic zones was conducted during the late 1800s by George Barrow and other British geologists working in the Dalradian schists of the southwestern Scottish Highlands. Here, clay-rich sedimentary rocks have been subjected to regional metamorphism, and the resulting metamorphic rocks can be divided into different zones based on the presence of distinctive silicate mineral assemblages. These mineral assemblages, each recognized by the presence of one or more index minerals,

FIGURE 8.14 Marble results from the metamorphism of the sedimentary rock limestone and dolostone. *(Photos courtesy of Sue Monroe.)*

Metamorphism

FIGURE 8.15 Quartzite results from the metamorphism of quartz sandstone. *(Photos courtesy of Sue Monroe.)*

Metamorphism

indicate different degrees of metamorphism. The index minerals Barrow and his associates chose to represent increasing metamorphic intensity were chlorite, biotite, garnet, staurolite, kyanite, and sillimanite (Table 8.1). Note that these are the metamorphic minerals produced from clay-rich sedimentary rocks. Other mineral assemblages and index minerals are produced from rocks with different original compositions (Table 8.1).

The successive appearance of metamorphic index minerals indicates gradually increasing or decreasing intensity of metamorphism. Going from lower- toward higher-grade zones, the first appearance of a particular index mineral indicates the location of the minimum temperature and pressure conditions needed for the formation of that mineral. When the locations of the first appearances of that index mineral are connected on a map, the result is a line of equal metamorphic intensity, or an *isograd*. The region between isograds is known as a **metamorphic zone**. By noting the occurrence of metamorphic index minerals, geologists can construct a map showing the metamorphic zones of an entire area (Figure 8.16).

Numerous studies of different metamorphic rocks have demonstrated that while the texture and composition of any rock may be altered by metamorphism, the overall chemical composition may be little changed. Thus, the different mineral assemblages found in increasingly higher-grade metamorphic rocks derived from the same parent rock result from changes in temperature and pressure (Table 8.1).

Metamorphism and Plate Tectonics

ALTHOUGH metamorphism is associated with all three types of plate boundaries (see Figure 1.11), it is most common along convergent plate margins. Metamorphic rocks form at convergent plate boundaries because temperature and pressure increase as a result of plate collisions.

Figure 8.17 illustrates the various temperature-pressure regimes produced along an oceanic-continental convergent plate boundary. When an oceanic plate collides with a continental plate, tremendous pressure is generated as the oceanic plate is subducted. Because rock is a poor heat conductor, the cold descending oceanic plate heats slowly, and metamorphism is caused mostly by increasing pressure with depth. As subduction continues, both temperature and pressure increase with depth and can result in high-grade metamorphic rocks. Eventually, the descending plate begins to melt and generates a magma that moves upward. This rising magma may alter the surrounding rock by contact metamorphism, producing migmatites in the deeper portions of the crust and hornfels at shallower depths. Such an environment is characterized by high temperatures and low to medium pressures.

While metamorphism is most common along convergent plate margins, many divergent plate boundaries are characterized by contact metamorphism. Rising magma at mid-oceanic ridges heats the adjacent rocks, producing contact metamorphic minerals and textures. Besides contact metamorphism, fluids emanating from the rising

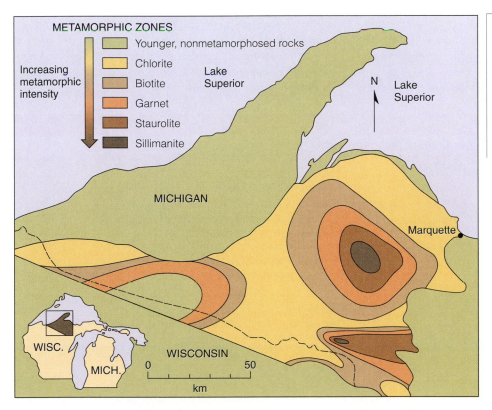

FIGURE 8.16 Metamorphic zones in the Upper Peninsula of Michigan. The zones in this region are based on the presence of distinctive silicate mineral assemblages resulting from the metamorphism of sedimentary rocks during an interval of mountain building and minor granitic intrusion during the Proterozoic Eon, about 1.5 billion years ago.

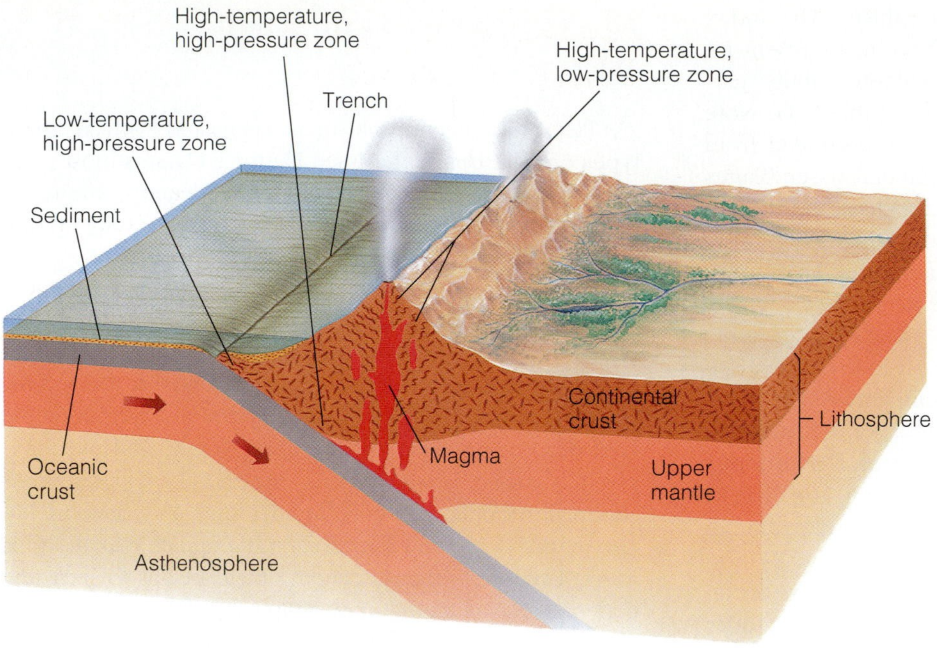

FIGURE 8.17 Various temperature-pressure conditions are produced along an oceanic-continental convergent plate boundary.

magma—and from the reaction of the magma and seawater—very commonly produce hydrothermal solutions that may precipitate minerals of economic value.

Metamorphism and Natural Resources

MANY metamorphic rocks and minerals are valuable natural resources. While these resources include various types of ore deposits, the two most familiar and widely used metamorphic rocks, as such, are marble and slate, which, as previously discussed, have been used for centuries in a variety of ways.

Many ore deposits result from contact metamorphism during which hot, ion-rich fluids migrate from igneous intrusions into the surrounding rock, thereby producing rich ore deposits. The most common sulfide ore minerals associated with contact metamorphism are bornite, chalcopyrite, galena, pyrite, and sphalerite, and two common oxide ore minerals are hematite and magnetite. Tin and tungsten are also important ores associated with contact metamorphism (Table 8.3).

Other economically important metamorphic minerals include talc for talcum powder; graphite for pencils and dry lubricants; garnets and corundum, which are used as abrasives or gemstones, depending on their quality; and andalusite, kyanite, and sillimanite, all of which are used in the manufacture of high-temperature porcelains and temperature-resistant minerals for products such as sparkplugs and the linings of furnaces.

TABLE 8.3			
The Main Ore Deposits Resulting from Contact Metamorphism			
ORE DEPOSIT	MAJOR MINERAL	FORMULA	USE
Copper	Bornite Chalcopyrite	Cu_5FeS_4 $CuFeS_2$	Important sources of copper, which is used in various aspects of manufacturing, transportation, communications, and construction
Iron	Hematite Magnetite	Fe_2O_3 Fe_3O_4	Major sources of iron for manufacture of steel, which is used in nearly every form of construction, manufacturing, transportation, and communications
Lead	Galena	PbS	Chief source of lead, which is used in batteries, pipes, solder, and elsewhere where resistance to corrosion is required
Tin	Cassiterite	SnO_2	Principal source of tin, which is used for tin plating, solder, alloys, and chemicals
Tungsten	Scheelite Wolframite	$CaWO_4$ $(Fe,Mn)WO_4$	Chief sources of tungsten, which is used in hardening metals and manufacturing carbides
Zinc	Sphalerite	$(Zn,Fe)S$	Major source of zinc, which is used in batteries and in galvanizing iron and making brass

1. Metamorphic rocks result from the transformation of other rocks, usually beneath Earth's surface, as a consequence of one or a combination of three agents: heat, pressure, and fluid activity.

2. Heat for metamorphism comes from intrusive magmas or deep burial. Pressure is either lithostatic or differential. Fluids trapped in sedimentary rocks or emanating from intruding magmas can enhance chemical changes and the formation of new minerals.

3. The three major types of metamorphism are contact, dynamic, and regional.

4. Metamorphic rocks are classified primarily according to their texture.

In a foliated texture, platy minerals have a preferred orientation. A nonfoliated texture does not exhibit any discernible preferred orientation of the mineral grains.

5. Foliated metamorphic rocks can be arranged in order of grain size and/or perfection of their foliation. Slate is very fine grained, followed by phyllite and schist; gneiss displays segregated bands of minerals. Amphibolite is another fairly common foliated metamorphic rock.

6. Marble, quartzite, greenstone, and hornfels are common nonfoliated metamorphic rocks.

7. Metamorphic rocks can be arranged into metamorphic zones based on the conditions of metamorphism.

8. Metamorphism can occur along all three kinds of plate boundaries but most commonly occurs at convergent plate margins.

9. Metamorphic rocks formed near Earth's surface along an oceanic-continental plate boundary result from low-temperature, high-pressure conditions. As a subducted oceanic plate descends, it is subjected to increasingly higher temperatures and pressures that result in higher-grade metamorphism.

10. Many metamorphic rocks and minerals, such as marble, slate, graphite, talc, and asbestos, are valuable natural resources.

Important Terms

aureole

contact metamorphism

differential pressure

dynamic metamorphism

fluid activity

foliated texture

heat

index mineral

lithostatic pressure

metamorphic rock

metamorphic zone

nonfoliated texture

regional metamorphism

Review Questions

1. A foliated metamorphic rock composed mainly of hornblende and plagioclase is:
   a. _____ hornfels;
   b. _____ amphibolite;
   c. _____ gneiss;
   d. _____ migmatite;
   e. _____ greenstone.

2. From which of the following rock groups can metamorphic rocks form?
   a. _____ plutonic;
   b. _____ sedimentary;
   c. _____ metamorphic;
   d. _____ volcanic;
   e. _____ all of these.

3. Which of the following is not an agent of metamorphism?
   a. _____ foliation;
   b. _____ heat;

   c. _____ pressure;
   d. _____ fluid activity;
   e. _____ none of these.

4. Pressure exerted equally in all directions on an object is:
   a. _____ differential;
   b. _____ directional;
   c. _____ lithostatic;
   d. _____ shear;
   e. _____ none of these.

5. In which type of metamorphism are magmatic heat and fluids the primary agents of change?
   a. _____ contact;
   b. _____ dynamic;
   c. _____ regional;
   d. _____ local;
   e. _____ thermodynamic.

6. Concentric zones surrounding an igneous intrusion are:

   a. _____ metamorphic layers;
   b. _____ thermodynamic rings;
   c. _____ aureoles;
   d. _____ hydrothermal regions;
   e. _____ none of these.

7. Which type of metamorphism produces the majority of metamorphic rocks?
   a. _____ contact;
   b. _____ dynamic;
   c. _____ regional;
   d. _____ lithostatic;
   e. _____ lithospheric.

8. Which of the following metamorphic rocks displays a foliated texture?
   a. _____ marble;
   b. _____ quartzite;
   c. _____ greenstone;
   d. _____ schist;
   e. _____ hornfels.

9. Metamorphic rocks resulting from dynamic metamorphism are:
   a. _____ fault breccias;
   b. _____ quartzites;
   c. _____ greenstones;
   d. _____ mylonites;
   e. _____ hornfels.

10. Metamorphic zones:
   a. _____ are characterized by distinctive mineral assemblages;
   b. _____ are separated from each other by isograds;
   c. _____ reflect a metamorphic grade;
   d. _____ all of these;
   e. _____ none of these.

11. Along what type of plate boundary is metamorphism most common?
   a. _____ convergent;
   b. _____ divergent;
   c. _____ transform;
   d. _____ mantle plume;
   e. _____ static.

12. Metamorphic rocks form a significant proportion of:
   a. _____ shields;
   b. _____ the cores of mountain ranges;
   c. _____ oceanic crust;
   d. _____ answers (a) and (b);
   e. _____ answers (b) and (c).

13. What is the correct metamorphic sequence of increasingly coarser grain size?
   a. _____ phyllite → slate → gneiss → schist;
   b. _____ slate → phyllite → schist → gneiss;
   c. _____ gneiss → phyllite → schist → slate;
   d. _____ schist → gneiss → phyllite → slate;
   e. _____ slate → schist → gneiss → phyllite.

14. Where does contact metamorphism occur, and what type of changes does it produce?

15. What are aureoles? How can they be used to determine the effects of metamorphism?

16. What is regional metamorphism, and under what conditions does it occur?

17. Describe the two types of metamorphic texture, and explain how they may be produced.

18. Name the three common nonfoliated rocks, and describe their characteristics.

Points to Ponder

1. What specific features of foliated metamorphic rocks make them unsuitable as the foundation rock for a dam? Are there any metamorphic rocks that would make a good foundation? Why?

2. If you were in charge of the EPA, how would you formulate a policy that balances the risks versus the benefits of removing asbestos from public buildings? What role would geologists play in this policy?

World Wide Web Activities

For these web site addresses, along with current updates and exercises, log on to

http://www.wadsworth.com/geo

▶ **METAMORPHIC ROCKS**

This site contains images and information on 11 common metamorphic rocks. It is part of the *Soil Science 223 Rocks and Minerals Reference* web site. Click on any of the *rock names* and compare the information and images to the information in this chapter.

▶ **UNIVERSITY OF TULSA—DEPARTMENT OF GEOSCIENCES METAMORPHIC ROCKS AND PROCESSES**

This site contains much information about metamorphic rocks and processes in general. Read over the material about metamorphic rocks and processes at this site.

▶ **THE GEOLOGY OF THE PT. REYES PENINSULA**

This site contains information about the geology of the Point Reyes Peninsula, California, including information about the metamorphic rocks in the area. Click on the *Pre-Cretaceous metamorphic rocks (pKm)* site. What are the different types of metamorphic rocks found in this area?

▶ **GETTY CENTER WEB SITE**

This site contains information about the newly opened J. Paul Getty Center. At the time of this writing (December 1997), there was nothing on the Greek kouros described in the Prologue, but it might be worth checking to see if anything about it has been added.

Explore the following *In-Terra-Active 2.0* CD-ROM module(s) and increase your understanding of key concepts and processes presented in this chapter.

▶ **SECTION: MATERIALS**
MODULE: METAMORPHIC ROCKS

The coarse-grained, garnet-mica schist (shown on the right of your computer screen) has essentially the same chemical composition as the fine-grained mudstone (shown on the left of the screen). In the field, it is clear that these belong to the same stratigraphic unit. **Question: How can this transformation take place in the solid state?**

▶ **VIRTUAL REALITY FIELD TRIP: METEOR CRATER, ARIZONA**

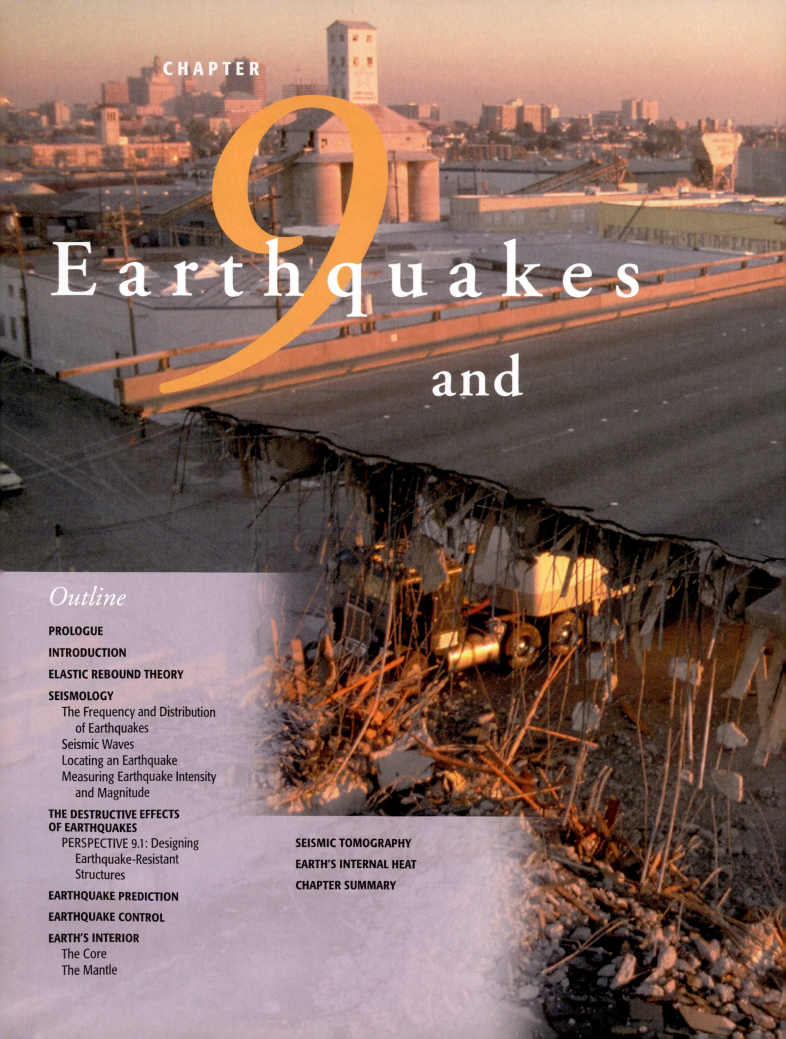

9

Earthquakes

and

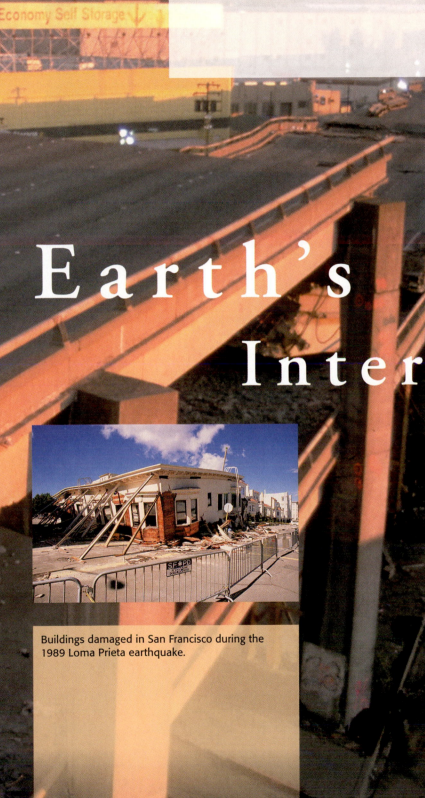

Earth's Interior

In the early morning hours of January 17, 1994, southern California was rocked by a devastating earthquake measuring 6.7 on the Richter Magnitude Scale. The earthquake was centered on the city of Northridge just north of Los Angeles. When the initial shock was over, about 55 people had been killed and thousands were injured. Nine freeways had been severely damaged, thousands of homes and buildings were either destroyed or damaged, and 250 ruptured gas lines ignited numerous fires (Figure 9.1). Thousands of aftershocks followed the main earthquake; many of them added to the damage, which has been estimated at $15 to $30 billion.

Geologists know that southern California is riddled with active faults capable of producing strong earthquakes, but it was a previously unknown fault 15 km beneath Northridge that caused the tragic January 1994 earthquake. Many people are aware that movements on the San Andreas fault and its subsidiary faults have been responsible for many earthquakes in southern California. But only recently have geologists begun to realize that a network of interconnected faults that do not break the surface may be causing much of the earthquake activity in the Los Angeles area. Earthquakes on these hidden

Buildings damaged in San Francisco during the 1989 Loma Prieta earthquake.

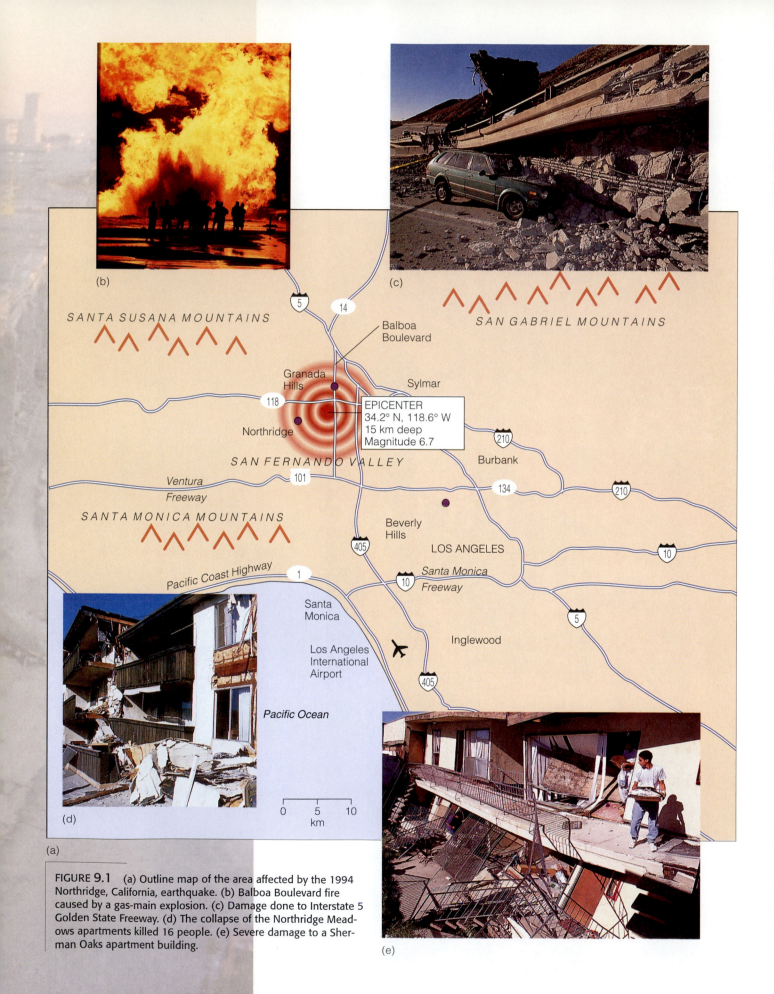

(b)

(c)

SANTA SUSANA MOUNTAINS

SAN GABRIEL MOUNTAINS

5   14

Balboa
Boulevard

Granada
Hills

Sylmar

118

EPICENTER
34.2° N, 118.6° W
15 km deep
Magnitude 6.7

Northridge

210

SAN FERNANDO VALLEY

Burbank

Ventura

101

Freeway

134   210

SANTA MONICA MOUNTAINS

Beverly
Hills

LOS ANGELES

405

10

Pacific Coast Highway

1

Santa Monica
Freeway

10

Santa
Monica

Los Angeles
International
Airport

Inglewood

5

405

Pacific Ocean

(d)

0   5   10
km

(a)

(e)

FIGURE 9.1   (a) Outline map of the area affected by the 1994 Northridge, California, earthquake. (b) Balboa Boulevard fire caused by a gas-main explosion. (c) Damage done to Interstate 5 Golden State Freeway. (d) The collapse of the Northridge Meadows apartments killed 16 people. (e) Severe damage to a Sherman Oaks apartment building.

faults, just as on the obvious ones, are related to movements between the North American and Pacific plates.

How did the Los Angeles area fare during and after this most recent earthquake? Older, unreinforced masonry buildings and more modern wood-frame apartments built over ground-floor garages generally sustained the most damage. Structures built to the stricter building standards in force during the last five years typically escaped unscathed or with only minor damage.

The state transportation department (Caltrans) instituted a program of reinforcing bridges and freeway overpasses soon after the 1971 Sylmar earthquake and began a second phase of reinforcing structures after the 1989 Loma Prieta earthquake. Most of the reinforced structures suffered little or no damage during the Northridge earthquake, but several awaiting reinforcing collapsed, including part of the Santa Monica Freeway, the busiest highway in the world.

Regulations designed to protect utility lines have not yet been implemented, in part because of cost. As a result, numerous power and gas lines were ruptured, and three water aqueducts were severed, cutting off water to at least 40,000 people and power to an estimated 3.1 million residents.

Emergency measures had been well planned and rescue operations went well. Shelters were established, people fed and clothed, and disaster relief offices opened in the area shortly after the earthquake. Based on their experiences in other earthquakes, rescue agencies had invested in better rescue equipment including high-pressure air bags that can lift up to 72 tons, fiber-optic search cameras, and specially trained dogs that can sniff out buried victims. This experience and the up-to-date equipment helped rescue workers locate and extricate victims from the earthquake wreckage.

The Northridge earthquake was tragic, but it was not the "Big One" that Californians have been waiting for. And even though rescue and relief agencies operated efficiently, the earthquake reminds us that much still remains to be done in terms of earthquake prediction and preparedness.

Natural disasters such as earthquakes cannot be prevented, but their impact can be minimized by careful planning. In the case of the Northridge earthquake, emergency services performed very well, thus mitigating the severity of the disaster. In marked contrast, following the Kobe, Japan, earthquake of January 1995, emergency services were hampered by poor planning, and many people suffered needlessly.

Introduction

As one of nature's most frightening and destructive phenomena, earthquakes have always aroused a sense of fear. Even when an earthquake begins, no one can tell how strong the shaking will be or how long it will last. It is estimated that more than 13 million people have died as a result of earthquakes during the past 4000 years, and approximately 1 million of these deaths were during the last century (Table 9.1).

An **earthquake** is defined as Earth vibrations caused by the sudden release of energy beneath the surface, usually as a result of displacement of rocks along fractures known as faults. Following an earthquake, adjustments along a fault commonly generate a series of earthquakes referred to as **aftershocks**. Most of these are smaller than the main shock, but they can cause considerable damage to already weakened structures. Indeed, much of the destruction from the 1755 earthquake in Lisbon, Portugal, was caused by aftershocks. After a small earthquake,

TABLE 9.1

Some Significant Earthquakes

YEAR	LOCATION	MAGNITUDE (ESTIMATED BEFORE 1935)	DEATHS (ESTIMATED)
1556	China (Shanxi Province)	8.0	1,000,000
1755	Portugal (Lisbon)	8.6	70,000
1811–12	USA (New Madrid, Missouri)	7.5	20
1886	USA (Charleston, South Carolina)	7.0	60
1906	USA (San Fernando, California)	8.3	700
1923	Japan (Tokyo)	8.3	143,000
1964	USA (Alaska)	8.6	131
1971	USA (San Fernando, California)	6.6	65
1976	China (Tangshan)	8.0	242,000
1985	Mexico (Mexico City)	8.1	9,500
1988	Armenia	7.0	25,000
1989	USA (Loma Prieta, California)	7.1	63
1990	Iran	7.3	40,000
1992	Turkey	6.8	570
1992	Egypt (Cairo)	5.9	550
1993	India	6.4	30,000
1994	USA (Northridge, California)	6.7	55
1995	Japan (Kobe)	7.2	5,000+
1995	Russia	7.6	2,000+
1996	China (Lijiang)	6.5	304
1997	Iran	5.5	554
1997	Iran	7.3	2,400+
1998	Afghanistan	6.1	5,000+

aftershock activity usually ceases within a few days, but it may persist for months following a large earthquake.

Early humans and cultures explained earthquakes in imaginative and colorful ways, often attributing them to the movements of some organism on which Earth rested. In Japan, the organism was a giant catfish; in Mongolia, a giant frog; in China, an ox; in India, a giant mole; in parts of South America, a whale; and to the Algonquin of North America, an immense tortoise.

The Greek philosopher Aristotle offered what he considered to be a natural explanation for earthquakes. He believed that atmospheric winds were drawn into the interior where they caused fires and swept around various subterranean cavities trying to escape. It was this movement of underground air that caused earthquakes and occasional volcanic eruptions. Today, geologists know that the majority of earthquakes result from faulting associated with plate movements.

Elastic Rebound Theory

BASED on studies conducted after the 1906 San Francisco earthquake, H. F. Reid of Johns Hopkins University proposed the **elastic rebound theory** to explain how energy is released during earthquakes. Reid studied three sets of measurements taken across a portion of the San Andreas fault that had broken during the 1906 earthquake. The measurements revealed that points on opposite sides of the fault had moved 3.2 m during the 50-year period prior to breakage in 1906, with the west side moving northward (Figure 9.2).

According to Reid, rocks on one side of the fault had moved relative to rocks on the other side, and this movement caused the gradual bending of any straight line that crossed the San Andreas fault, such as a fence or road (Figure 9.2). Eventually, the strength of the rocks was exceeded, the rocks on opposite sides of the fault rebounded or "snapped back" to their former undeformed shape, and the energy stored was released as earthquake waves radiating outward from the break. Additional field and laboratory studies conducted by Reid and others have confirmed that elastic rebound is the mechanism that generates earthquakes.

Seismology

SEISMOLOGY, the study of earthquakes, began emerging as a true science around 1880 with the development of *seismographs,* instruments that detect, record, and measure

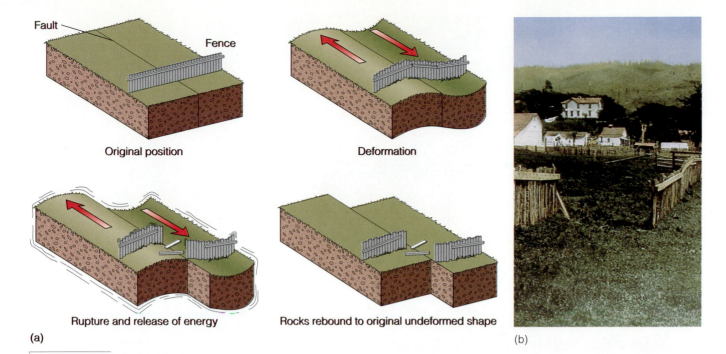

FIGURE 9.2 (a) According to the elastic rebound theory, when rocks are deformed, they store energy and bend. When the inherent strength of the rocks is exceeded, they rupture, releasing the energy in the form of earthquake waves that radiate out in all directions. Upon rupture, the rocks rebound to their former undeformed shape. (b) During the 1906 San Francisco earthquake, this fence in Marin County was displaced 2.5 m.

the various vibrations produced by an earthquake (Figure 9.3). The record made by a seismograph is a *seismogram*.

When an earthquake takes place, energy in the form of *seismic waves* radiates out in all directions from the point of release (Figure 9.4). Most earthquakes result when rocks in the crust rupture along a fault because of the buildup of excessive pressure, which is usually caused by plate movement. Once a rupture begins, it moves along the fault at a velocity of several kilometers per second, for as long as conditions for failure exist. The length of the fault along which rupture occurs can range from a few meters to several hundred kilometers. The longer the rupture, the more time it takes for all the stored energy in the rocks to be released, and therefore the longer the ground will shake.

The location within the crust where rupture initiates, and thus where the energy is released, is referred to as the **focus**, or *hypocenter*. The point on the surface vertically above the focus is the **epicenter**, which is the location that is usually given in news reports on earthquakes (Figure 9.4).

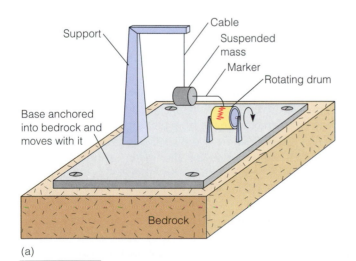

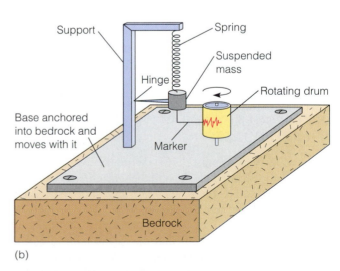

FIGURE 9.3 (a) A horizontal-motion seismograph. Because of its inertia, the heavy mass that contains the marker remains stationary while the rest of the structure moves along with the ground during an earthquake. As long as the length of the arm is not parallel to the direction of ground movement, the marker will record the earthquake waves on the rotating drum. (b) A vertical-motion seismograph. This seismograph operates on the same principle as a horizontal-motion instrument and records vertical ground movement.

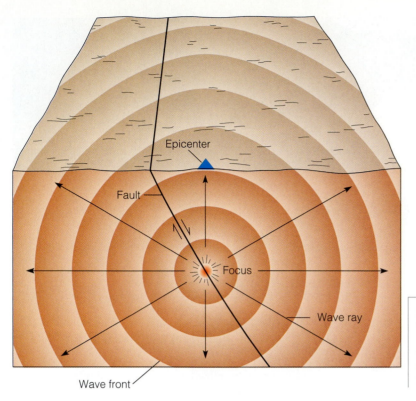

Epicenter

Fault

Focus

Wave ray

Wave front

FIGURE 9.4 The focus of an earthquake is the location where rupture begins and energy is released. The place on the surface vertically above the focus is the epicenter. Seismic wave fronts move out in all directions from their source, the focus of an earthquake. Wave rays are lines drawn perpendicular to wave fronts.

Seismologists recognize three categories of earthquakes based on the depth of their foci. *Shallow-focus* earthquakes have a focal depth of less than 70 km. Earthquakes with foci between 70 and 300 km are referred to as *intermediate focus,* and those with foci greater than 300 km are called *deep focus.* Earthquakes are not evenly distributed among these three categories. Approximately 90% of all earthquake foci are at depths of less than 100 km, whereas only 3% of earthquake foci are deep. Shallow-focus earthquakes are, with few exceptions, the most destructive.

An interesting relationship exists between earthquake foci and plate margins. Earthquakes generated along divergent or transform plate boundaries are always shallow focus, while almost all intermediate- and deep-focus earthquakes occur within the circum-Pacific belt along convergent margins (Figure 9.5).

THE FREQUENCY AND DISTRIBUTION OF EARTHQUAKES

Most earthquakes (almost 95%) take place in seismic belts that correspond to plate boundaries where stresses develop as plates converge, diverge, and slide past each other. Earthquake activity distant from plate margins is minimal but on occasion can be devastating. The relationship between plate margins and the distribution of earthquakes is readily apparent when the locations of earthquake epicenters are superimposed on a map showing the boundaries of Earth's plates (Figure 9.5).

The majority of all earthquakes (approximately 80%) occur in the *circum-Pacific belt,* a zone of seismic activity

nearly encircling the Pacific Ocean basin. Most of these earthquakes result from convergence along plate margins, as in the case of the 1995 Kobe, Japan, earthquake (Figure 9.6). The second major seismic belt is the *Mediterranean-Asiatic belt* where approximately 15% of all earthquakes occur. This belt extends westerly from Indonesia through the Himalayas, across Iran and Turkey, and westerly through the Mediterranean region of Europe. The devastating earthquake that struck Armenia in 1988 killing 25,000 people and the 1990 earthquake in Iran that killed 40,000 are recent examples of the destructive earthquakes that strike this region (Table 9.1).

The remaining 5% of earthquakes occur mostly in the interiors of plates and along oceanic spreading ridge systems. Most of these earthquakes are not very strong, although several major intraplate earthquakes are worthy of mention. For example, the 1811 and 1812 earthquakes near New Madrid, Missouri, killed approximately 20 people and nearly destroyed the town. So strong were these earthquakes that they were felt from the Rocky Mountains to the Atlantic Ocean and from the Canadian border to the Gulf of Mexico. Another major intraplate earthquake struck Charleston, South Carolina, on August 31, 1886, killing 60 people and causing $23 million in property damage (Figure 9.7).

The cause of intraplate earthquakes is not well understood, but geologists think they arise from localized stresses caused by the compression that most plates experience along their margins. A useful analogy might be that of moving a house. Regardless of how careful the movers are, moving something so large without its internal parts shifting slightly is impossible. Likewise, plates are not likely

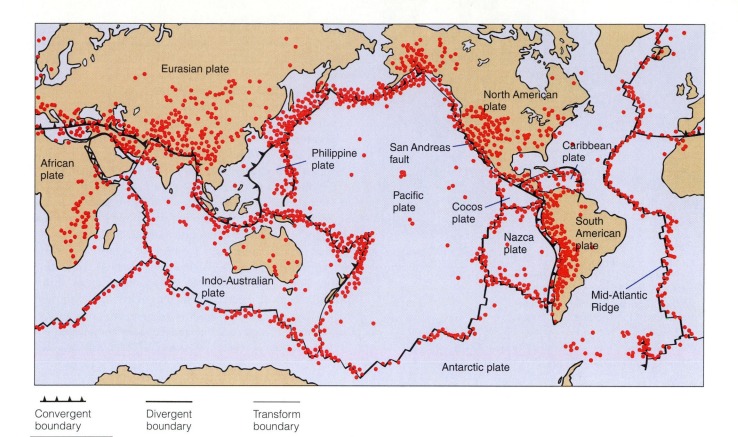

Convergent boundary — Divergent boundary — Transform boundary

FIGURE 9.5 The relationship between earthquake epicenters and plate boundaries. Approximately 80% of earthquakes occur within the circum-Pacific belt, 15% within the Mediterranean-Asiatic belt, and the remaining 5% within plate interiors or along oceanic spreading ridge systems. Each dot represents a single earthquake epicenter.

to move without some internal stresses that occasionally cause earthquakes. Interestingly, many intraplate earthquakes are associated with very ancient and presumed inactive faults that are reactivated at various intervals.

More than 150,000 earthquakes strong enough to be felt are recorded every year by the worldwide network of seismograph stations. In addition, seismologists estimate that about 900,000 earthquakes occur annually that are recorded by seismographs but are too small to be individually cataloged. These small earthquakes result from the energy released as continual adjustments take place between Earth's various plates.

SEISMIC WAVES

The shaking and destruction resulting from earthquakes are caused by two different types of seismic waves: *body waves,* which travel through Earth and are somewhat like sound waves; and *surface waves,* which travel only along the ground surface and are analogous to ocean waves.

An earthquake generates two types of body waves: P-waves and S-waves (Figure 9.8). **P-waves,** or *primary waves,* are the fastest seismic waves and can travel through

FIGURE 9.6 Some of the damage in Kobe, Japan, caused by the January 1995 earthquake in which more than 5000 people died.

FIGURE **9.7** Damage done to Charleston, South Carolina, by the earthquake of August 31, 1886. This earthquake is the largest reported in the eastern United States.

solids, liquids, and gases. P-waves are compressional, or push-pull, waves and are similar to sound waves in that they move material forward and backward along a line in the same direction that the waves themselves are moving (Figure 9.8b). Thus, the material P-waves travel through is expanded and compressed as the wave moves through it and returns to its original size and shape after the wave passes by.

S-waves, or *secondary waves,* are somewhat slower than P-waves and can only travel through solids. S-waves are *shear waves* because they move the material perpendicular to the direction of travel, thereby producing shear stresses in the material they move through (Figure 9.8c). Because liquids (as well as gases) are not rigid, they have no shear strength, and S-waves cannot be transmitted through them.

The velocities of P- and S-waves are determined by the density and elasticity of the materials through which they travel. For example, seismic waves travel more slowly through rocks of greater density but more rapidly through rocks with greater elasticity. *Elasticity* is a property of solids, such as rocks, and means that once they have been

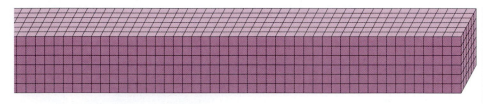

(a) Undisturbed material

FIGURE **9.8** Seismic waves. (a) Undisturbed material. (b) Primary waves (P-waves) compress and expand material in the same direction as the wave movement. (c) Secondary waves (S-waves) move material perpendicular to the direction of wave movement.

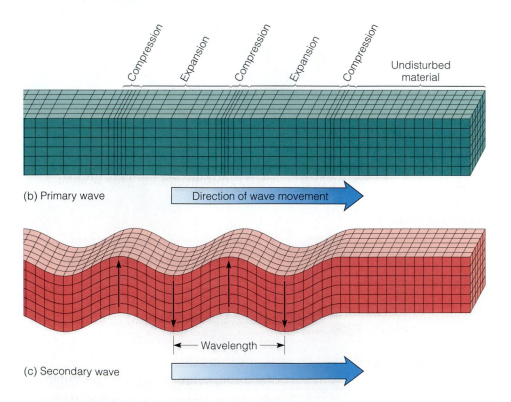

(b) Primary wave

(c) Secondary wave

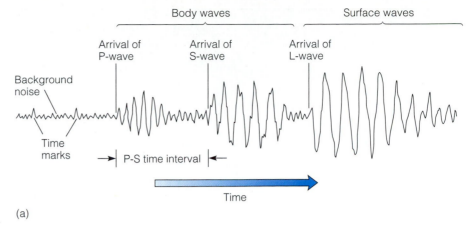

FIGURE 9.9 (a) A schematic seismogram showing the arrival order and pattern produced by P-, S-, and surface waves. When an earthquake occurs, body and surface waves radiate out from the focus at the same time. Because P-waves are the fastest, they arrive at a seismograph first, followed by S-waves, and then by surface waves, which are the slowest. (b) A time-distance graph showing the average travel times for P- and S-waves. The farther away a seismograph station is from the focus of an earthquake, the longer the interval between the arrivals of the P- and S-waves, and hence the greater the distance between the curves on the time-distance graph as indicated by the P-S time interval.

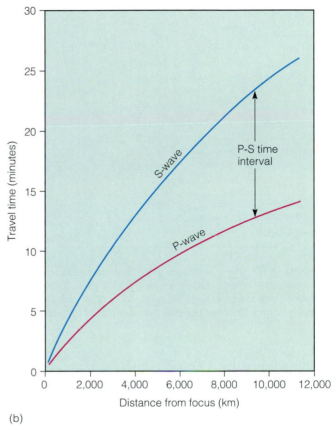

deformed by an applied force, they return to their original shape when the force is no longer present. Because P-wave velocity is greater than S-wave velocity in all materials, P-waves always arrive at seismic stations first.

Surface waves travel along the surface of the ground, or just below it, and are slower than body waves. Unlike the sharp jolting and shaking that body waves cause, surface waves generally produce a rolling or swaying motion, much like the experience of being on a boat.

LOCATING AN EARTHQUAKE

The various seismic waves travel at different speeds and thus arrive at a seismograph at different times. As Figure 9.9 illustrates, the first waves to arrive are P-waves, which travel at nearly twice the velocity of the S-waves that follow. Both the P- and S-waves travel directly from the focus to the seismograph through Earth's interior. Surface waves are the last to arrive because they are the slowest and also travel the longest route along the surface.

By accumulating a tremendous amount of data over the years, seismologists have determined the average travel times of P- and S-waves for any specific distance. These P- and S-wave travel times are published as *time-distance graphs* and illustrate that the difference between the arrival times of P- and S-waves is a function of the seismograph's distance from the focus (Figure 9.9b).

As Figure 9.10 demonstrates, the epicenter of any earthquake can be determined by using a time-distance graph and knowing the arrival times of the P- and S-waves at any three seismograph locations. Subtracting the arrival time of the first P-wave from the arrival time of the first S-wave gives the time interval between the arrivals of the two waves for each seismograph location. Each time interval is then plotted on the time-distance graph, and a line is drawn straight down to the distance axis of the graph, indicating how far away each station is from the focus of the earthquake. Then a circle whose radius equals the distance shown on the time-

distance graph from each of the three seismograph locations is drawn on a map (Figure 9.10). The intersection of the three circles is the location of the earthquake's epicenter. A minimum of three locations is needed because two locations will provide two possible epicenters and one location will provide an infinite number of possible epicenters.

MEASURING EARTHQUAKE INTENSITY AND MAGNITUDE

Geologists measure the strength of an earthquake in two different ways. The first, intensity, is a qualitative assessment of the kinds of damage done by an earthquake. The second, magnitude, is a quantitative measurement of the amount of

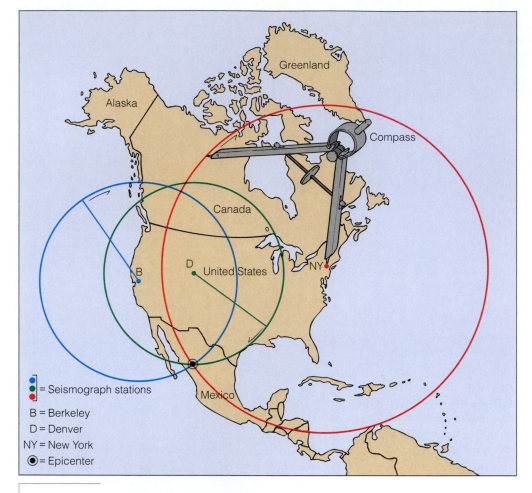

FIGURE 9.10 Three seismograph stations are needed to locate the epicenter of an earthquake. The P-S time interval is plotted on a time-distance graph for each seismograph station to determine the distance that station is from the epicenter. A circle with that radius is drawn from each station, and the intersection of the three circles is the epicenter of the earthquake.

energy released. Each method provides geologists with important data about earthquakes and their effects. This information can then be used to prepare for future earthquakes.

Intensity is a subjective measure of the kind of damage done by an earthquake, as well as people's reaction to it. Since the mid-nineteenth century, geologists have used intensity as a rough approximation of the size and strength of an earthquake. The most common intensity scale used in the United States is the **Modified Mercalli Intensity Scale**, which has values ranging from I to XII (Table 9.2).

While it is generally true that a large earthquake will produce greater intensity values than a small earthquake, many other factors besides the amount of energy released by an earthquake affect its intensity. These include distance from the epicenter, focal depth of the earthquake, population density and local geology of the area, type of building construction employed, and duration of shaking.

If earthquakes are to be compared quantitatively, we must use a scale that measures the amount of energy released and is independent of intensity. Such a scale was developed in 1935 by Charles F. Richter, a seismologist at

the California Institute of Technology. The **Richter Magnitude Scale** measures earthquake **magnitude**, which is the total amount of energy released by an earthquake at its source. It is an open-ended scale with values beginning at 1. The largest magnitude recorded has been 8.6, and though values greater than 9 are theoretically possible, they are highly improbable because rocks are not able to store the energy necessary to generate earthquakes of this magnitude.

To avoid large numbers, Richter used a conventional base-10 logarithmic scale to convert the amplitude of the largest recorded seismic wave to a numerical magnitude value. Therefore, each integer increase in magnitude represents a 10-fold increase in wave amplitude. For example, the amplitude of the largest seismic wave for an earthquake of magnitude 6 is 10 times that produced by an earthquake of magnitude 5, 100 times as large as a magnitude 4 earthquake, and 1000 times that of an earthquake of magnitude 3 ($10 \times 10 \times 10 = 1,000$).

While each increase in magnitude represents a 10-fold increase in wave amplitude, each magnitude increase cor-

TABLE 9.2

Modified Mercalli Intensity Scale

I Not felt except by a very few under especially favorable circumstances.

II Felt only by a few people at rest, especially on upper floors of buildings.

III Felt quite noticeably indoors, especially on upper floors of buildings, but many people do not recognize it as an earthquake. Standing automobiles may rock slightly.

IV During the day felt indoors by many, outdoors by few. At night some awakened. Sensation like heavy truck striking building, standing automobiles rocked noticeably.

V Felt by nearly everyone, many awakened. Some dishes, windows, etc. broken, a few instances of cracked plaster. Disturbance of trees, poles, and other tall objects sometimes noticed.

VI Felt by all, many frightened and run outdoors. Some heavy furniture moved, a few instances of fallen plaster or damaged chimneys. Damage slight.

VII Everybody runs outdoors. Damage negligible in buildings of good design and construction; slight to moderate in well-built ordinary structures; considerable in poorly built or badly designed structures; some chimneys broken. Noticed by people driving automobiles.

VIII Damage slight in specially designed structures; considerable in normally constructed buildings with possible partial collapse; great in poorly built structures. Fall of chimneys, monuments, walls. Heavy furniture overturned. Sand and mud ejected in small amounts.

IX Damage considerable in specially designed structures. Buildings shifted off foundations. Ground noticeably cracked. Underground pipes broken.

X Some well-built wooden structures destroyed; most masonry and frame structures with foundations destroyed; ground badly cracked. Rails bent. Landslides considerable from river banks and steep slopes. Water splashed over river banks.

XI Few if any (masonry) structures remain standing. Bridges destroyed. Broad fissures in ground. Underground pipelines completely out of service.

XII Damage total. Waves seen on ground surfaces. Objects thrown upward into the air.

SOURCE: U.S. Geological Survey.

responds to a roughly 30-fold increase in the amount of energy released. Thus, the 1964 Alaska earthquake with a magnitude of 8.6 released about 900 times the energy of the 1971 San Fernando Valley, California, earthquake of magnitude 6.6.

We have already mentioned that more than 900,000 earthquakes are recorded around the world each year. These figures can be placed in better perspective by reference to Table 9.3, which shows that the vast majority of earthquakes have a Richter magnitude of less than 2.5 and that great earthquakes (those with a magnitude greater than 8.0) occur, on average, only once every five years.

The Richter Magnitude Scale underestimates the energy of very large earthquakes because it measures the highest peak recorded on a seismograph, which records only an instant during an earthquake. For a great earthquake, though, the energy might be released over several minutes and along hundreds of kilometers of a fault. A modification of the Richter Magnitude Scale, known as the *Seismic-Moment Magnitude Scale,* is now used for these earthquakes. On this scale, large earthquakes can exceed magnitude 9.

The Destructive Effects of Earthquakes

THE destructive effects of earthquakes include such phenomena as ground shaking, fire, seismic sea waves, and landslides, as well as disruption of vital services, panic, and psychological shock. The amount of property damage, loss

TABLE 9.3

Average Number of Earthquakes of Various Magnitudes per Year Worldwide

MAGNITUDE	EFFECTS	AVERAGE NUMBER PER YEAR
<2.5	Typically not felt but recorded	900,000
2.5–6.0	Usually felt; minor to moderate damage to structures	31,000
6.1–6.9	Potentially destructive, especially in populated areas	100
7.0–7.9	Major earthquakes; serious damage results	20
>8.0	Great earthquakes; usually result in total destruction	1 every 5 years

SOURCE: Data from *Earthquake Information Bulletin* and Gutenberg and Richter (1949).

of life, and injury depends on the time of day an earthquake occurs, its magnitude, distance from the epicenter, geology of the area, type of construction of various structures, population density, and the duration of shaking. Generally speaking, earthquakes occurring during working and school hours in densely populated urban areas are the most destructive and cause most fatalities and injuries.

Ground shaking usually causes more damage and results in more loss of life and injuries than any other earthquake hazard. Structures built on bedrock generally suffer less damage than those built on poorly consolidated material such as water-saturated sediments or artificial fill. Structures on poorly consolidated or water-saturated material are subjected to ground shaking of longer duration and greater S-wave amplitude than those on bedrock. In addition, fill and water-saturated sediments tend to liquefy, or behave as a fluid, a process known as *liquefaction*. When shaken, the individual grains lose cohesion and the ground flows. Dramatic examples of damage resulting from liquefaction include Niigata, Japan, where large apartment buildings were tipped to their sides after the water-saturated soil of the hillside collapsed in 1964.

During the Loma Prieta earthquake that caused the third game of the 1989 World Series to be postponed, those districts in the San Francisco–Oakland Bay area built on artificial fill or reclaimed bay mud suffered the most damage. In the Marina district of San Francisco, numerous buildings were destroyed, and a fire, fed by broken gas lines, lit up the night sky. The failure of the columns supporting a portion of the two-tiered Interstate 880 freeway in Oakland sent the upper tier crashing down onto the lower one, killing 42 unfortunate motorists (Figure 9.11). The shaking lasted less than 15 seconds but resulted in 63 deaths, 3800 injuries, and $6 billion in property damage and left at least 12,000 people homeless.

In addition to earthquake magnitude and regional geology, the material used and the type of construction also affect the amount of damage done (Perspective 9.1). Adobe and mud-walled structures are the weakest of all and almost always collapse during an earthquake. Unreinforced brick structures and poorly built concrete structures are also particularly susceptible to collapse. The 6.4 magnitude earthquake that struck India in 1993 killed about 30,000 people, whereas the 6.7 magnitude Northridge, California, earthquake in 1994 resulted in only 55 deaths. Both earthquakes were in densely populated regions, but in India the brick and stone buildings could not withstand ground shaking; most collapsed, entombing their occupants (Figure 9.12).

In many earthquakes, particularly in urban areas, fire is a major hazard (Figure 9.13). Nearly 90% of the damage done in the 1906 San Francisco earthquake was caused by fire. The shaking severed many of the electrical and gas lines, which touched off flames and started numerous fires all over the city. Because water mains were ruptured by the earthquake, there was no effective way to fight the fires. Hence, they raged out of control for three days, destroying much of the city.

FIGURE **9.11** Interstate 880 in Oakland, California, was damaged during the 1989 Loma Prieta earthquake. The columns supporting the upper deck of the two-tiered highway failed, and the upper deck fell onto the lower deck, killing 42 motorists. Only 1 of the 51 double-deck spans did not collapse.

Tsunami are destructive seismic sea waves that are usually produced by earthquakes but can also be caused by submarine landslides or volcanic eruptions (Figure 9.14). Tsunami are popularly called tidal waves, although they are not caused by or related to tides. Instead, most tsunami result from the sudden movement of the seafloor, which sets up waves within the water that travel outward, much like ripples that form when a stone is thrown into a pond.

Tsunami travel at speeds of several hundred kilometers per hour and are commonly not noticed in the open ocean because their wave height is usually less than 1 m and the distance between wave crests is typically several hundred kilometers. When tsunami approach shorelines, however, the waves slow down and water piles up to heights of up to 65 m.

FIGURE **9.12** In 1993, India experienced its worst earthquake in more than 50 years. Thousands of brick and stone houses collapsed, killing at least 30,000 people.

FIGURE 9.13 San Francisco Marina district fire caused by broken gas lines during the 1989 Loma Prieta earthquake.

Following a 1946 tsunami that killed 159 people and caused $25 million in property damage in Hawaii, the U.S. Coast and Geodetic Survey established a Tsunami Early Warning System in Honolulu, Hawaii, in an attempt to minimize tsunami devastation. This system combines seismographs and instruments that can detect earthquake-generated sea waves. Whenever a strong earthquake takes place anywhere within the Pacific basin, its location is determined, and instruments are checked to see if a tsunami has been generated. If it has, a warning is sent out to evacuate people from low-lying areas that may be affected.

Earthquake-triggered landslides are particularly dangerous in mountainous regions and have been responsible for tremendous amounts of damage and many deaths. For example, a 1970 Peru earthquake caused an avalanche that destroyed the town of Yungay, resulting in 25,000 deaths.

FIGURE 9.14 As a tsunami crashes into the street behind them, residents of Hilo, Hawaii, run for their lives. This 1946 tsunami was generated by an earthquake in the Aleutian Islands and resulted in considerable property damage to Hilo and the deaths of 159 people.

The Destructive Effects of Earthquakes 187

Designing Earthquake-Resistant Structures

One way to reduce property damage, injuries, and loss of life is to design and build structures as earthquake-resistant as possible. Many things can be done to improve the safety of current structures and of new buildings.

California's Uniform Building Code sets minimum standards for building earthquake-resistant structures and is used as a model around the world. The California code is far more stringent than federal earthquake building codes and requires that structures be able to withstand a 25-second main shock. Unfortunately, many earthquakes are of far longer duration. For instance, the main shock of the 1964 Alaskan earthquake lasted approximately three minutes and was followed by numerous aftershocks. While many of the extensively damaged buildings in this earthquake had been built according to the California code, they were not designed to withstand

shaking of such long duration. Nevertheless, in California and elsewhere, structures built since the California code went into effect have fared much better during moderate to major earthquakes than those built before its implementation.

To design earthquake-resistant structures, engineers must understand the dynamics and mechanics of earthquakes, including the type and duration of ground motion and how rapidly the ground accelerates. An understanding of the area's geology is also important because certain materials such as water-saturated sediments or landfill can lose their strength and cohesiveness during an earthquake. Finally, engineers must be aware of how different structures behave under different earthquake conditions.

With the level of technology currently available, a well-designed, properly constructed building should be able to withstand small, short-duration earthquakes of less than 5.5 magnitude with little or

no damage. In moderate earthquakes (5.5–7.0 magnitude), the damage suffered should not be serious and should be repairable. In a large earthquake of greater than 7.0 magnitude, buildings should not collapse, although they may later have to be demolished.

Many factors enter into the design of an earthquake-resistant structure, but the most important is that the building be tied together; that is, the foundation, walls, floors, and roof should all be joined together to create a structure that can withstand both horizontal and vertical shaking (Figure 1). Almost all the structural failures resulting from earthquake ground movement occur at weak connections, where the various parts of a structure were not securely tied together (Figure 2).

The size and shape of a building can also affect its resistance to earthquakes (Figure 2). Rectangular box-shaped buildings are inherently stronger than those of irregular size or shape because different

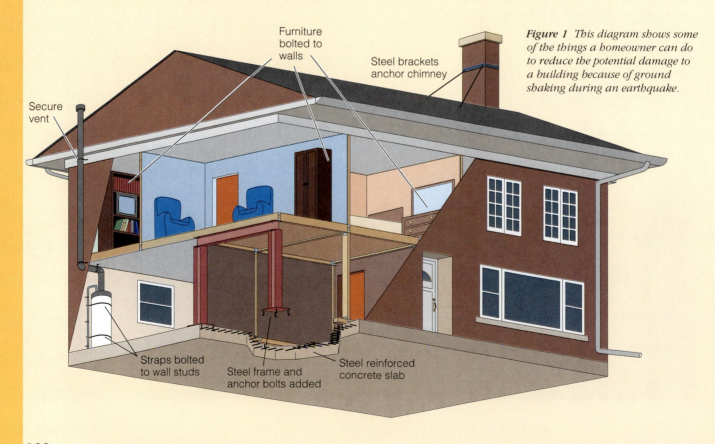

Furniture bolted to walls

Steel brackets anchor chimney

Secure vent

Straps bolted to wall studs

Steel frame and anchor bolts added

Steel reinforced concrete slab

Figure 1 This diagram shows some of the things a homeowner can do to reduce the potential damage to a building because of ground shaking during an earthquake.

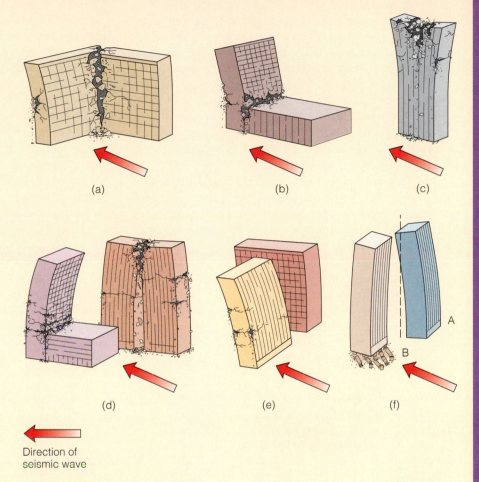

Figure 2 The effects of ground shaking on various tall buildings of differing shapes. (a) Damage will occur if two wings of a building are joined at right angles and experience different motions. (b) Buildings of different heights will sway differently, leading to damage at the point of connection. (c) Shaking increases with height and is greatest at the top of a building. (d) Closely spaced buildings may crash into each other due to swaying. (e) A building whose long axis is parallel to the direction of the seismic waves will sway less than a building whose axis is perpendicular. (f) Two buildings of different design will behave differently even when subjected to the same shaking conditions. Building A sways as a unit and remains standing, while building B, whose first story is composed of tall columns, collapses because most of the swaying takes place in the "soft" first story.

Direction of seismic wave

parts of an irregular building may sway at different rates, increasing the stress and likelihood of structural failure (Figure 2b). Buildings with open or unsupported first stories are particularly susceptible to damage. Some reinforcement must be done or collapse is a distinct possibility.

Tall buildings, such as skyscrapers, must be designed so that a certain amount of swaying or flexing can occur, but not so much that they touch neighboring buildings during swaying (Figure 2d). If a building is brittle and does not give, it will crack and fail. In addition to designed flexibility, engineers must make sure that a building does not vibrate at the same frequency as the ground does during an earthquake. When that happens, the force applied by the seismic waves at ground level is multiplied several times by the time they reach the top of the building (Figure 2c). This condition is particularly troublesome in areas of poorly consolidated sediment (Figure 3). Fortunately, buildings can be designed so that they will sway at a different frequency than the ground.

What about structures built many years ago? Almost every city and town has older single and multistory structures, constructed of unreinforced brick

masonry, poor-quality concrete, and rotting or decaying wood. Just as in new buildings, the most important thing that can be done to increase the stability and safety of older structures is to tie the different components of each building together. This can be done by adding a steel frame to unreinforced parts of a

building such as a garage, bolting the walls to the foundation, adding reinforced beams to the exterior, and using beam and joist connectors whenever possible. Although such modifications are expensive, they are usually cheaper than having to replace a building that was destroyed by an earthquake.

Figure 3 This 15-story reinforced concrete building collapsed due to ground shaking during the 1985 Mexico City earthquake. The soft lake-bed sediments on which Mexico City is built amplified the seismic waves as they passed through.

Earthquake Prediction

CAN earthquakes be predicted? A successful prediction must include a time frame for the occurrence of an earthquake, its location, and its strength. Despite the tremendous amount of information geologists have gathered about the cause of earthquakes, successful predictions are still quite rare. Nevertheless, if reliable predictions can be made, they can greatly reduce the number of deaths and injuries.

From an analysis of historic records and the distribution of known faults, *seismic risk maps* can be constructed that indicate the likelihood and potential severity of future earthquakes based on the intensity of past earthquakes (Figure 9.15). Although such maps cannot predict when an earthquake will occur, they are useful in helping people plan for future earthquakes.

One long-range prediction technique used in seismically active areas involves plotting the location of large earthquakes and their aftershocks to detect areas that have had a history of earthquakes but are currently inactive. Such regions are locked and not releasing energy, making these *seismic gaps* prime locations for future earthquakes. Several seismic gaps along the San Andreas fault have the potential for future major earthquakes.

Changes in elevation and tilting of the land surface have frequently preceded earthquakes and may be warnings of impending quakes. Extremely slight changes in the angle of the ground surface can be measured by *tiltmeters*. Tilt-meters have been placed on both sides of the San Andreas fault to measure tilting of the ground surface that is thought to result from increasing pressure in the rocks. Data from measurements in central California indicate significant tilting occurred immediately preceding small earthquakes. Furthermore, extensive tiltmeter work performed in Japan prior to the 1964 Niigata earthquake clearly showed a relationship between increased tilting and the main shock. While more research is needed, these changes appear to be useful in making short-term earthquake predictions.

Other phenomena that precede earthquakes, known as *precursors,* include fluctuations in the water level of wells and changes in the magnetic field, and the electrical resistance of the ground. These fluctuations are thought to result from changes in the amount of pore space in rocks, due to increasing pressure. A change in animal behavior prior to earthquakes also is frequently mentioned. It may be that animals are sensing small and subtle changes prior to a quake that humans simply do not sense.

Currently, only four nations—the United States, Japan, Russia, and China—have government-sponsored earthquake prediction programs. These programs include laboratory and field studies of rock behavior before, during, and after large earthquakes, as well as monitoring activity along known active faults. Most earthquake prediction work in the United States is done by the United States Geological Survey (USGS) and involves a variety of research into all aspects of earthquake-related phenomena.

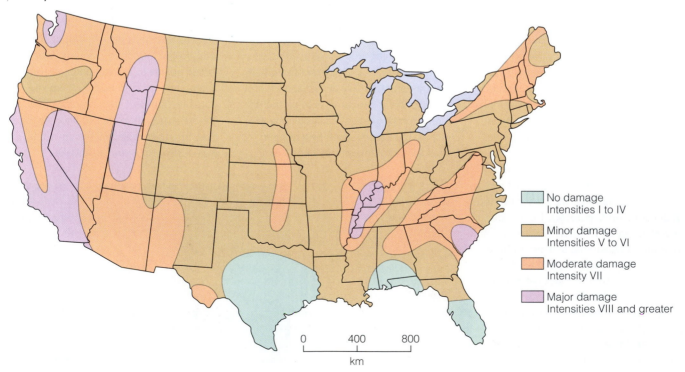

FIGURE **9.15**   A 1969 seismic risk map for the United States, based on intensity data collected by the U.S. Coast and Geodetic Survey.

No damage
Intensities I to IV

Minor damage
Intensities V to VI

Moderate damage
Intensity VII

Major damage
Intensities VIII and greater

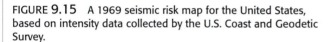

0   400   800
km

FIGURE 9.16 Many of the approximately 242,000 people who died in the 1976 earthquake in Tangshan, China, were killed by collapsing structures. Many of the buildings were constructed of unreinforced brick, which has no flexibility, and quickly fell down during the earthquake.

The Chinese have perhaps the most ambitious earthquake prediction programs anywhere in the world, which is understandable considering their long history of destructive earthquakes. The Chinese earthquake prediction program was initiated soon after two large earthquakes occurred at Xingtai (300 km southwest of Beijing) in 1966. This program includes extensive study and monitoring of all possible earthquake precursors. In addition, the Chinese also emphasize changes in phenomena that can be observed by seeing and hearing without the use of sophisticated instruments. They did successfully predict a 7.5 magnitude earthquake in 1975 soon enough to evacuate the city of Haicheng, which undoubtedly saved thousands of lives. The prediction was based on an unusual sequence of foreshocks and has not resulted in a successful prediction method. Furthermore, they failed to predict the devastating 1976 Tangshan earthquake that killed about 242,000 people (Figure 9.16).

Great strides are being made toward dependable, accurate earthquake predictions, and studies are underway to assess public reactions to long-, medium-, and short-term earthquake warnings. Unless short-term warnings are actually followed by an earthquake, most people will probably ignore warnings as they frequently do now for hurricanes, tornadoes, and tsunami.

Earthquake Control

IF earthquake prediction is still in the future, can anything be done to control earthquakes? Because of the tremendous forces involved, humans are certainly not going to be able to prevent earthquakes. However, there may be ways to dissipate the destructive energy of major earthquakes by releasing it in small amounts that will not cause extensive damage.

During the early to mid-1960s, Denver, Colorado, experienced numerous small earthquakes. This was surprising because Denver had not been prone to earthquakes in the past. In 1962, geologist David M. Evans suggested that Denver's earthquakes were directly related to the injection of contaminated waste water into a disposal well 3674 m deep at the Rocky Mountain Arsenal, northeast of Denver. The U.S. Army initially denied that there was any connection, but a USGS study concluded that the pumping of waste fluids into fractured rocks beneath the disposal well decreased the friction on opposite sides of fractures and, in effect, lubricated them so that movement occurred, causing the earthquakes.

Experiments conducted in 1969 at an abandoned oil field near Rangely, Colorado, confirmed the arsenal hypothesis. Water was pumped in and out of abandoned oil wells, the pore-water pressure in these wells was measured, and seismographs were installed in the area to measure any seismic activity. Monitoring showed that small earthquakes were occurring in the area when fluid was injected and that earthquake activity declined when the fluids were pumped out. What the geologists were doing was starting and stopping earthquakes at will, and the relationship between pore-water pressures and earthquakes was established.

Based on these results, some geologists have proposed that fluids be pumped into the locked segments of active faults to cause small- to moderate-sized earthquakes. They think that this would relieve the pressure on the fault and prevent a major earthquake from occurring. While this plan is intriguing, it also has many potential problems. For instance, no one can guarantee that only a small earthquake might result. Instead, a major earthquake might occur, causing tremendous property damage and loss of life. Who would be responsible? Certainly, a great deal more research is needed before such an experiment is performed, even in an area of low population density.

It appears that until such time as earthquakes can be accurately predicted or controlled, the best means of defense is careful planning and preparation.

Earth's Interior

EARTH'S interior has always been an inaccessible, mysterious realm. During most of historic time, it was perceived as an underground world of vast caverns, heat, and sulfur gases, populated by demons. By the 1860s, scientists knew what Earth's average density was and that pressure and temperature increase with depth. And even though Earth's interior is hidden from direct observation, scientists have a reasonably good idea of its internal structure and composition.

Scientists have known for more than 200 years that planet Earth is not homogeneous throughout. Indeed, Sir Isaac Newton (1642–1727) noted in a study of the planets that Earth's average density—that is, its ratio of mass to volume—is 5.0 to 6.0 g/cm³ (water has a density of 1 g/cm³). In 1797, Henry Cavendish calculated a density value very close to the 5.5 g/cm³ now accepted. To accurately determine Earth's density, its mass and volume must be known. The volume of the nearly spherical Earth can be calculated by knowing its radius, whereas mass can be determined only indirectly by comparing, for example, the gravitational attraction of the Moon and Earth to that of metal spheres of known mass. When these calculations are carried out, a value of 5.52 g/cm³ is derived. Considering that Earth's overall density of 5.5 g/cm³ is much greater than that of surface rocks, most of which range from 2.5 to 3.0 g/cm³, much of Earth's interior must consist of materials with a density greater than the average.

Earth is generally depicted as consisting of concentric layers that differ in composition and density, with each separated from adjacent layers by rather distinct boundaries (see Figure 1.8). Recall that the outermost layer, or the **crust**, is Earth's thin skin. Below the crust and extending about halfway to Earth's center is the **mantle**, which comprises more than 80% of Earth's volume. The central part consists of a **core**, which is divided into a solid inner core and a liquid outer part.

The behavior and travel times of P- and S-waves provide geologists with much information about Earth's internal structure. Seismic waves travel outward as wave fronts from their source areas, although it is most convenient to depict them as *wave rays,* which are lines showing the direction of movement of small parts of wave fronts (Figure 9.4). Any disturbance, such as a passing train or construction equipment, can cause seismic waves, but only those generated by large earthquakes, explosive volcanism, asteroid impacts, and nuclear explosions can travel completely through Earth.

As we noted previously, the velocities of P- and S-waves are determined by the density and elasticity of the materials they travel through, both of which increase with depth. Wave velocity is slowed by increasing density but increases in materials with greater elasticity. Because elasticity increases with depth faster than density, a general increase in the velocity of seismic waves takes place as they penetrate to greater depths. P-waves travel faster than S-waves under all circumstances, but unlike P-waves, S-waves cannot be transmitted through a liquid because liquids have no shear strength (rigidity)—they simply flow in response to shear stress.

As a seismic wave travels from one material into another of different density and elasticity, its velocity and direction of travel change. That is, the wave is bent—a phenomenon known as **refraction**—in much the same way light waves are refracted as they pass from air into a more dense medium such as water (Figure 9.17). Because seismic waves pass through materials of differing density and elasticity, they are continually refracted so that their paths are curved; the only exception is that wave rays are not refracted if their direction of travel is perpendicular to a boundary (Figure 9.17). In that case they travel in a straight line.

In addition to refraction, seismic rays are also **reflected**, much as light is reflected from a mirror. Some of the energy of seismic rays that encounter a boundary separating materials of different density or elasticity is *reflected* back to the surface (Figure 9.17). If we know the wave velocity and the time required for it to travel from its source to the boundary and back to the surface, we can calculate the depth of the reflecting boundary. Such information is useful in determining not only the depths of the various layers but also the depths of sedimentary rocks that may contain petroleum.

Although the velocity of seismic waves changes continuously with depth, their velocity may suddenly increase or decrease. For instance, P-wave velocity increases abruptly at the base of the crust and decreases rapidly at a depth of about 2900 km. These marked changes in seismic wave velocity indicate a boundary known as a **discontinuity** across which a significant change in Earth materials or their properties takes place. These discontinuities are the basis for subdividing Earth's interior into concentric layers.

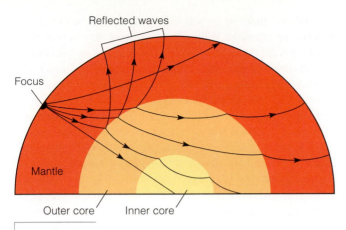

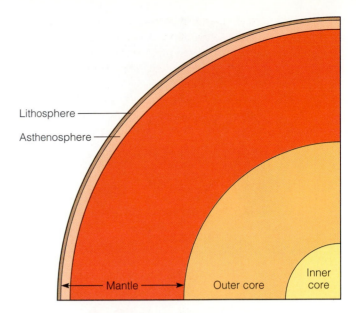

FIGURE **9.17** *Refraction and reflection of P-waves. When seismic waves pass through a boundary separating Earth materials of different density or elasticity, they are refracted, and some of their energy is reflected back to the surface.*

THE CORE

In 1906 R. D. Oldham of the Geological Survey of India postulated the existence of a core that transmits seismic waves at a slower rate than shallower Earth materials. We now know that P-wave velocity decreases markedly at a depth of 2900 km, indicating a major discontinuity now recognized as the core-mantle boundary (Figure 9.18).

The sudden decrease in P-wave velocity at the core-mantle boundary causes P-waves entering the core to be refracted in such a way that very little P-wave energy reaches the surface in the area between 103° and 143° from an earthquake focus (Figure 9.19a). This area in which little P-wave energy is recorded by seismographs is a **P-wave shadow zone.**

In 1926 British physicist Harold Jeffreys realized that S-waves were not simply slowed by the core but were completely blocked by it. So besides a P-wave shadow zone, a much larger and more complete **S-wave shadow zone** exists (Figure 9.19b). At locations greater than 103° from an earthquake focus, no S-waves are recorded, indicating that S-waves cannot be transmitted through the core. S-waves will not pass through a liquid, so it seems that the outer core must be liquid or behave as a liquid. The inner core, however, is thought to be solid because P-wave velocity increases at what is interpreted to be the outer core–inner core boundary (Figure 9.18).

We can estimate the core's density and composition by using seismic evidence and laboratory experiments. For instance, a device known as a *diamond-anvil pressure cell* has been developed in which small samples are studied as they are subjected to pressures and temperatures similar to those in the core. Furthermore, meteorites, which are thought to represent remnants of the material from which the solar system formed, can be used to make estimates of density and composition. For example, meteorites composed of iron and nickel alloys may represent the differen-

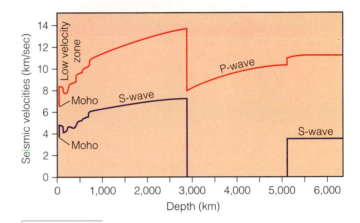

FIGURE **9.18** *Profiles showing seismic wave velocities versus depth. Several discontinuities are shown across which seismic wave velocities change rapidly.*

tiated interiors of large asteroids and approximate the density and composition of Earth's core. The density of the outer core varies from 9.9 to 13.0 g/cm^3. At Earth's center, the pressure is equivalent to about 3.5 million times normal atmospheric pressure.

The core cannot be composed of the minerals most common at the surface because even under the tremendous pressures at great depth, they would still not be dense enough to yield an average density of 5.5 g/cm^3 for the planet. Both the outer and inner core are thought to be composed largely of iron, but pure iron is too dense to be the sole constituent of the outer core. Thus, it must be "diluted" with elements of lesser density. Laboratory experiments and comparisons with iron meteorites indicate that about 12% of the outer core may consist of sulfur, and perhaps some silicon and small amounts of nickel and potassium.

In contrast, pure iron is not dense enough to account for the estimated density of the inner core. Many geologists

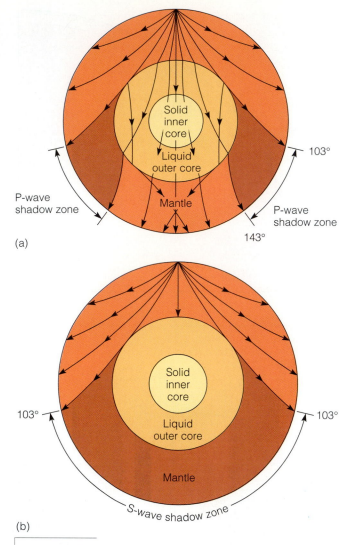

(a)

(b)

FIGURE 9.19 (a) P-waves are refracted so that little P-wave energy reaches the surface in the P-wave shadow zone. (b) The presence of an S-wave shadow zone indicates that S-waves are being blocked within Earth.

think that perhaps 10 to 20% of the inner core also consists of nickel. These metals form an iron-nickel alloy that under the pressure at that depth is thought to be sufficiently dense to account for the inner core's density. When the core formed during early Earth history, it was probably molten and has since cooled to the point that its interior has crystallized. Indeed, the inner core continues to grow as Earth slowly cools and liquid of the outer core crystallizes as iron. Recent evidence also indicates that the inner core rotates faster than the outer core, moving about 20 km/yr relative to the outer core.

THE MANTLE

Another significant discovery about Earth's interior was made in 1909 when the Yugoslavian seismologist Andrija Mohorovičić detected a discontinuity at a depth of about 30

km. While studying arrival times of seismic waves from Balkan earthquakes, Mohorovičić noticed that two distinct sets of P- and S-waves were recorded at seismic stations a few hundred kilometers from an earthquake's epicenter. He reasoned that one set of waves traveled directly from the epicenter to the seismic station, whereas the other waves had penetrated a deeper layer where they were refracted.

From his observations Mohorovičić concluded that a sharp boundary separating rocks with different properties exists at a depth of about 30 km. He postulated that P-waves below this boundary travel at 8 km/sec, whereas those above the boundary travel at 6.75 km/sec. When an earthquake occurs, some waves travel directly from the focus to a seismic station, while others travel through the deeper layer and some of their energy is refracted back to the surface. Waves traveling through the deeper layer travel farther to a seismic station, but they do so more rapidly than those in the shallower layer. The boundary identified by Mohorovičić separates the crust from the mantle and is now called the **Mohorovičić discontinuity**, or simply the **Moho**. It is present everywhere except beneath spreading ridges, but its depth varies: beneath the continents it averages 35 km, but ranges from 20 to 90 km; beneath the seafloor it is 5 to 10 km deep (Figure 9.18).

Although seismic wave velocity in the mantle generally increases with depth, several discontinuities also exist. Between depths of 100 and 250 km, both P- and S-wave velocities decrease markedly. This 100- to 250-km-deep layer is the **low-velocity zone**; it corresponds closely to the *asthenosphere,* a layer in which the rocks are close to their melting point and thus are less elastic, accounting for the observed decrease in seismic wave velocity. The asthenosphere is an important zone because it may be where some magmas are generated. Furthermore, it lacks strength and flows plastically and is thought to be the layer over which the outer, rigid *lithosphere* moves.

Although the mantle's density, which varies from 3.3 to 5.7 g/cm^3, can be inferred rather accurately from seismic waves, its composition is less certain. The igneous rock *peridotite,* which contains mostly ferromagnesian silicates, is considered the most likely component. Laboratory experiments indicate that it possesses physical properties that would account for the mantle's density and observed rates of seismic wave transmissions. Peridotite also forms the lower parts of igneous rock sequences thought to be fragments of the oceanic crust and upper mantle emplaced on land. In addition, peridotite is found as inclusions in volcanic rock bodies such as *kimberlite pipes* that are known to have come from great depths. These inclusions appear to be pieces of the mantle.

Seismic Tomography

THE model of Earth's interior consisting of an iron-rich core and a rocky mantle is probably accurate but is also rather imprecise. Recently, geophysicists have developed a technique called *seismic tomography* that allows them to develop three-dimensional models of Earth's interior. In seismic tomography, numerous crossing seismic waves are analyzed in much the same way radiologists analyze CAT (computerized axial tomography) scans. In CAT scans, X rays penetrate the body, and a two-dimensional image of the inside of a patient is formed. Repeated CAT scans, each from a slightly different angle, are computer analyzed and stacked to produce a three-dimensional picture.

In a similar fashion geophysicists use seismic waves to probe Earth's interior. From its time of arrival and distance traveled, the velocity of a seismic ray is computed at a seismic station. Only average velocity is determined rather than variations in velocity. In seismic tomography, numerous wave rays are analyzed so that "slow" and "fast" areas of wave travel can be detected (Figure 9.20). Recall that seismic wave velocity is controlled partly by elasticity; cold rocks have greater elasticity and therefore transmit seismic waves faster than hot rocks.

Using this technique, geophysicists have detected areas within the mantle at a depth of about 150 km where seismic velocities are slower than expected. These anomalously hot regions lie beneath volcanic areas and beneath mid-oceanic ridges, where convection cells of rising hot mantle rock are thought to exist. In contrast, beneath the older interior parts of continents, where tectonic activity ceased hundreds of millions or billions of years ago, anomalously cold spots are recognized. In effect, tomographic maps and three-dimensional diagrams show heat variations within Earth.

Seismic tomography has also yielded additional and sometimes surprising information about the core. For example, the core-mantle boundary is not a smooth surface but has broad depressions and rises extending several kilometers into the mantle. Of course, the base of the mantle possesses the same features in reverse; geophysicists have termed these features *anticontinents* and *antimountains*. It appears that the surface of the core is continually deformed by sinking and rising masses of mantle material.

As a result of seismic tomography, a much clearer picture of Earth's interior is emerging. It has already given us a better understanding of complex convection within the mantle, including upwelling convection currents thought to be responsible for the movement of lithospheric plates.

Earth's Internal Heat

DURING the nineteenth century, scientists realized that the temperature in deep mines increases with depth. More recently, the same trend has been observed in deep drill holes, but even in these we can measure temperatures directly down to a depth of only a few kilometers. This temperature increase with depth, or **geothermal gradient**, near the surface is about 25°C/km, although it varies from area to area. In areas of active or recently active volcanism, the geothermal gradient is greater than in adjacent nonvolcanic areas, and temperature rises faster beneath spreading ridges than elsewhere beneath the seafloor.

Most of Earth's internal heat is generated by radioactive decay, especially the decay of isotopes of uranium and thorium and to a lesser degree of potassium 40. When these isotopes decay, they emit energetic particles and gamma rays, which heat surrounding rocks. And because rock is such a poor conductor of heat, it takes little radioactive decay to build up considerable heat, given enough time.

Unfortunately, the geothermal gradient is not useful for estimating temperatures at great depth. If we were simply to extrapolate from the surface downward, the temperature at 100 km would be so high that, despite the great pressure, all known rocks would melt. Yet except for pockets of magma, it appears that the mantle is solid rather than liquid because it transmits S-waves. Accordingly, the geothermal gradient must decrease markedly.

Current estimates of the temperature at the base of the crust are 800° to 1200°C. The latter figure seems to be an upper limit: if it were any higher, melting would be expected. Furthermore, fragments of mantle rock in kimberlite pipes, thought to have come from depths of 100 to 300 km, appear to have reached equilibrium at these depths and at a temperature of about 1200°C. At the core-mantle boundary, the temperature is probably between 3500° and 5000°C; the wide spread of values indicates the uncertainties of such estimates. If these figures are reasonably accurate, the geothermal gradient in the mantle is only about 1°C/km.

Considering that the core is so remote and so many uncertainties exist regarding its composition, only very general estimates of its temperature can be made. The maximum temperature at the center of the core is thought to be about 6500°C, very close to the estimated temperature for the surface of the Sun!

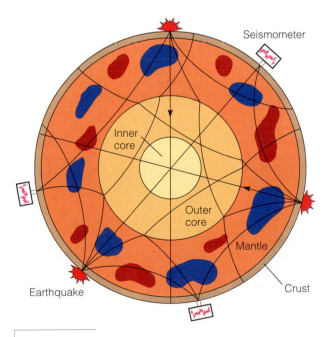

FIGURE 9.20 Numerous earthquake waves are analyzed to detect areas within Earth that transmit seismic waves faster or slower than adjacent areas. Areas of fast wave travel correspond to "cold" regions (blue), whereas "hot" regions (red) transmit seismic waves more slowly.

Chapter Summary

1. Earthquakes are Earth vibrations caused by the sudden release of energy, usually along a fault.

2. According to the elastic rebound theory, pressure builds in rocks on opposite sides of a fault until the strength of the rocks is exceeded and rupture occurs. When the rocks rupture, stored energy is released as they snap back to their original position.

3. Seismology is the study of earthquakes. Earthquakes are recorded on seismographs, and the record of an earthquake is a seismogram.

4. The focus of an earthquake is the point where energy is released. Vertically above the focus on the surface is the epicenter.

5. Approximately 80% of all earthquakes occur in the circum-Pacific belt, 15% within the Mediterranean-Asiatic belt, and the remaining 5% mostly in the interior of the plates or along oceanic spreading ridge systems.

6. The two types of body waves are P-waves and S-waves. P-waves travel through all materials, whereas S-waves do not travel through liquids. P-waves are the fastest waves and are compressional, while S-waves are shear. Surface waves travel along or just below the surface.

7. The distance to the epicenter of an earthquake can be determined by the use of a time-distance graph of the P- and S-waves. Three seismographs are needed to locate an earthquake's epicenter.

8. Intensity is a measure of the kind of damage done by an earthquake and is expressed by values from I to XII in the Modified Mercalli Intensity Scale.

9. Magnitude measures the amount of energy released by an earthquake and is expressed in the Richter Magnitude Scale. Each increase in the magnitude number represents about a 30-fold increase in energy released. The Seismic-Moment Magnitude Scale more accurately estimates the energy released during very large earthquakes.

10. Ground shaking is the most destructive of all earthquake hazards. The amount of damage done by an earthquake depends upon the geology of the area, type of building construction, magnitude of the earthquake, and duration of shaking. Tsunami are seismic sea waves, most of which are produced by earthquakes. They can do a tremendous damage to coastlines.

11. Earth is concentrically layered into an iron-rich core with a solid inner core and a liquid outer part, a rocky mantle, and an oceanic crust and continental crust.

12. Much of the information about Earth's interior has been derived from studies of P- and S-waves. Laboratory experiments, comparisons with meteorites, and studies of inclusions in volcanic rocks provide additional information.

13. Density and elasticity of Earth materials determine the velocity of seismic waves. Seismic waves are refracted when their direction of travel changes. Wave reflection occurs at boundaries across which the properties of rocks change.

14. The behavior of P- and S-waves within Earth and the presence of P- and S-wave shadow zones allow geologists to estimate the density and composition of the interior and to estimate the size and depth of the core and mantle.

15. Earth's inner core is thought to be composed of iron and nickel, whereas the outer core is probably composed mostly of iron with 10 to 20% sulfur and other substances in lesser quantities. Peridotite is the most likely component of the mantle.

16. The geothermal gradient of 25°C/km cannot continue to great depths, otherwise most of Earth would be molten. The geothermal gradient for the mantle and core is probably about 1°C/km. The temperature at Earth's center is estimated to be 6500°C.

Important Terms

aftershock
core
crust
discontinuity
earthquake
elastic rebound theory
epicenter
focus
geothermal gradient

intensity
low-velocity zone
magnitude
mantle
Modified Mercalli Intensity Scale
Mohorovičić discontinuity (Moho)
P-wave
P-wave shadow zone
reflection

refraction
Richter Magnitude Scale
seismology
surface wave
S-wave
S-wave shadow zone
tsunami

Review Questions

1. A magnitude 5 earthquake releases about _____ times as much energy as a magnitude 2 earthquake.
   a. _____ 2.5;
   b. _____ 3;
   c. _____ 30;
   d. _____ 1000;
   e. _____ 27,000.

2. An earthquake's focus is:
   a. _____ the point on the surface where damage is greatest;
   b. _____ the location where rupture begins;
   c. _____ a segment of the Mediterranean-Asiatic belt characterized by deep earthquakes;
   d. _____ the distance traveled by a seismic wave;
   e. _____ a measure of the amplitude of a P-wave on a seismograph.

3. In which one of the following areas would you most likely experience an earthquake?
   a. _____ Japan;
   b. _____ Kansas;
   c. _____ England;
   d. _____ Florida;
   e. _____ Germany.

4. Which type of seismic wave travels fastest, and which type will not travel through a fluid?
   a. _____ refracted/surface;
   b. _____ P-wave/S-wave;
   c. _____ S-wave/reflected wave;
   d. _____ P-wave/surface wave;
   e. _____ surface wave/S-wave.

5. Which one of the following statement is correct?
   a. _____ most large earthquakes take place within plates;
   b. _____ Earth's average density is 2.65 g/cm^3;
   c. _____ oceanic ridges are the sites of deep earthquakes;

   d. _____ seismic wave velocity increases in materials with greater elasticity;
   e. _____ the core-mantle boundary is at a depth of 1000 km.

6. A boundary across which seismic wave velocity changes markedly is known as a(an):
   a. _____ discontinuity;
   b. _____ seismic gap;
   c. _____ aftershock;
   d. _____ low-velocity zone;
   e. _____ S-wave shadow zone.

7. The geothermal gradient is Earth's:
   a. _____ capacity to reflect and refract seismic waves;
   b. _____ most destructive aspect of earthquakes;
   c. _____ temperature increase with depth;
   d. _____ average rate of seismic wave velocity in the mantle;
   e. _____ elastic rebound potential.

8. The intensity of an earthquake is a measure of the:
   a. _____ amount of energy released;
   b. _____ types of damage and people's reactions;
   c. _____ depth and location;
   d. _____ distance from a plate boundary;
   e. _____ time difference between arrivals of P- and S-waves at a seismic station.

9. Earth's mantle is likely composed of the rock:
   a. _____ basalt;
   b. _____ gneiss;
   c. _____ peridotite;
   d. _____ arkose;
   e. _____ kimberlite.

10. Most earthquake activity occurs in the:
    a. _____ mid-oceanic ridge belt;
    b. _____ Mediterranean-Asiatic belt;
    c. _____ divergent plate boundary belt;
    d. _____ circum-Pacific belt;
    e. _____ kimberlite pipe belt.

11. Describe fully how an earthquake's epicenter is located.

12. How does the density and elasticity of Earth materials affect the velocity and direction of travel of seismic waves?

13. Why have geologist developed the Seismic-Moment Magnitude Scale?

14. Explain what the S-wave shadow zone is and what implications it has about Earth's internal structure.

15. How does the elastic rebound theory account for the energy released by an earthquake?

16. Buildings sitting on bedrock are usually damaged less during an earthquake than those sited on unconsolidated sediment. Why?

17. How do P-waves differ from S-waves?

18. Explain why shallow-, intermediate-, and deep-focus earthquakes occur at convergent plate boundaries, whereas divergent plate boundaries have shallow-focus earthquakes.

19. Describe the composition, density, and depth of Earth's mantle and core.

20. Explain how a seismograph works.

Points to Ponder

1. If Earth were completely solid and had the same composition and density throughout, how would P- and S-waves behave as they traveled through Earth?

2. What factors account for higher-than-average heat flow values at spreading ridges? How is heat flow related to the age of crustal rocks?

For these web site addresses, along with current updates and exercises, log on to

http://www.wadsworth.com/geo

▶ **CALIFORNIA INSTITUTE OF TECHNOLOGY SEISMOLOGICAL LABORATORY**

This site provides much information about earthquakes and some of the projects going on in the various labs at the California Institute of Technology.

1. Click on the *Record of the Day* site. Where was that earthquake? What was its magnitude?

2. Click on the *Southern California Earthquake Center Data Center* site. Check out the current activity and location of earthquakes in the southern California area. Where has most of the activity been located for the past week?

▶ **UNITED STATES GEOLOGICAL SURVEY PASADENA FIELD OFFICE**

This site contains many links to other web sites containing information about earthquake activity in the southern California region as well as elsewhere in the world. Click on the *Earthquake Information* site under the *Earthquakes, Monitoring & Publications* section. This site contains a listing on earthquake information for southern California and other areas. Click on any of the sites, and check out the latest earthquake activity in that region.

▶ **U.S. GEOLOGICAL SURVEY NATIONAL EARTHQUAKE INFORMATION CENTER**

This site provides current and general seismicity information as well as links to other sites concerned with earthquake activity. Click on the *Current Seismicity Information* site. Click on one of the *Current Seismicity* map sites such as *World* to see the distribution of earthquakes, their depth, and magnitude. In the *World* site, where is the greatest amount of seismic activity taking place? When and where did the largest magnitude earthquake most recently occur? Based on the information in this chapter, does the current distribution of earthquakes reflect where you would expect most earthquakes to take place?

▶ **TSUNAMI**

This site, hosted by the Department of Geophysics at the University of Washington, is maintained by Catherine Petroff, professor of civil engineering. It was originally developed by Benjamin Cook in 1995 while he was a master's student. It contains data, maps, simulations of tsunamis, and information on the tsunami warning system.

1. What does the term *tsunami* mean?

2. Click on *Physics of Tsunamis*. Explain how a tsunami differs from other water waves.

3. Click on *A Survey of Great Tsunamis* and read about the 1975 tsunami in Hawaii.

▶ **THE GREAT EARTHQUAKE AT NEW MADRID**

This site is maintained by the University of Missouri–Columbia. It contains considerable information on one of the great North American earthquakes that struck the continental interior. Read the history of this earthquake and answer the following:

1. The New Madrid fault zone is in which states?

2. How large were the 1811–1812 earthquakes?

3. What are the chances of another large earthquake in this area?

▶ **CANADIAN NATIONAL EARTHQUAKE HAZARDS PROGRAM**

This site is maintained by the Geological Survey of Canada. It contains information on recent Canadian earthquakes, the Canadian Earthquake Program, and the Canadian National Seismograph Network. Click on any of the subject headings to learn more about the distribution, frequency, and effects of earthquakes in Canada.

▶ SAN ANDREAS FAULT ZONE

This site contains data such as length, slip rate, and type of fault for the San Andreas fault zone and numerous associated faults. It has an alphabetical listing of many faults and maps showing their distribution. Scroll down and click on the *Alphabetical Fault Index*. Check out the San Andreas fault. What is its length and slip rate?

▶ EARTH'S INTERIOR AND PLATE TECTONICS

This site by Rosanna L. Hamilton has descriptions and illustrations of Earth's interior. It also has a summary of depth, volume, composition, and properties of each of Earth's interior divisions. Scroll down and click on *The Earth's Interior*.

1. What divisions of Earth's interior are recognized?

2. How much of Earth's mass is made up of the inner core?

3. Why is the inner core solid and what is it composed of?

▶ ABAG EARTHQUAKE MAPS AND INFORMATION

The Association of Bay Area Government maintains this site which is devoted largely to earthquake activity in the San Francisco Bay area. However, it has information on many aspects of earthquakes. For example, it has a What's New section that was last updated October 15, 1997, and sections on Riding Out Future Quakes, and information on other earthquake hazards. Click on *Fact? or Fiction? A quiz to test your earthquake sense*. Use your textbook and see if you can answer the questions.

CD-ROM Exploration

Explore the following *In-Terra-Active 2.0* CD-ROM module(s) and increase your understanding of key concepts and processes presented in this chapter.

▶ SECTION: INTERIOR
MODULE: INTERNAL STRUCTURE

In 1976, a magnitude 8.0 earthquake rocked the Tangshan area of China, killing more than 240,000 people. During the hour immediately following the quake, seismographs in much of North America were quiet and didn't record arrivals of P or S waves from this killer quake. **Question: How might you account for the lack of P or S wave arrivals at distances of 7,500 to 10,000 miles from an epicenter?**

▶ SECTION: INTERIOR
MODULE: EARTHQUAKE LOCATION

You're visiting a seismograph station in Arizona when earthquake waves begin to register on the instruments. This seismogram shows three clear peaks in energy, separated by periods of reduced activity. Your geology insturctor tells you that the reading comes from a single earthquake reported nearby. **Question: How can one earthquake yield a sesmogram documenting different periods of high energy?**

10

Deformation and

Mountain

View of Mount Everest, Earth's highest
mountain peak.

Prologue

The forces necessary to form lofty mountains are difficult to comprehend, yet when compared with Earth's size even the highest mountains are very small features. In fact, the greatest difference in elevation on Earth is about 20 km, which if depicted to scale on a 1-m diameter globe would be less than 2 mm. Nevertheless, from the human perspective, mountains are indeed impressive features. For instance, Mount Whitney in California, the highest mountain peak in the continental United States, towers more than 4400 m (14,495 ft) above sea level. Alaska's Mount McKinley at 6193 m (20,320 ft) compares with peaks in the Andes of South America, though about 50 Andean peaks exceed 6000 m high! At one time, it was thought that Earth's highest peak was in the Andes, but in 1852 British surveyors in India determined that Mount Everest in the Himalayas stood 8840 m (29,002 ft) high (see chapter-opening photo).

Earth statistics such as the length of the longest river, the greatest oceanic depth, and the coldest temperature are fascinating to many people. For some of these statistics, the figures given are very accurate; for a mountain peak's elevation above sea level, however, the figure is a close approximation because measuring a mountain's height is not as easy as it might seem. First, one has to

Building

be clear on what is mean by highest. *Mount Everest is the highest above sea level, the usual designation for highest; but because of Earth's equatorial bulge, the summit of Chimborazo in the Andes of Ecuador is most distant from Earth's center. It stands 6269 m (20,561 ft) above sea level, but its peak is about 2 km more distance from Earth's center than Mount Everest's. Highest might also be measured from a mountain's base, in which case Mauna Kea on the island of Hawaii would be the record holder. It rises more than 10,203 m (33,476 ft) above its base on the seafloor, but nearly 60% of it is beneath sea level.*

In the case of Mount Everest, its height was originally determined by using an instrument known as a theodolite *to measure vertical angles between the instrument and the peak from two locations of known elevation. More precise surveys done by Indian and Chinese groups during the 1950s and 1960s gave values within 1 foot of one another and an elevation of 8848 m (29,028 ft) is now accepted. However, even this figure may have to be adjusted slightly because sea level is not the same everywhere; it simply conforms to the intensity of Earth's gravitational attraction, which differs from area to area. To account for variations in gravitational attraction, scientists refer to the* geoid, *an imaginary sea-level surface extending continuously through the continents. The problem is the geoid is poorly known beneath mountains.*

Occasionally one hears that K2 (Mount Godwin-Austen) on the India-Pakistan border is higher than Mount Everest. This challenge arose a decade ago when satellite data were briefly interpreted to indicate that K2 was higher than Everest. However, more accurate studies revealed that even though it rises an impressive 8613 m (28,250 ft), it is still in second place.

Introduction

MANY rocks show the effects of *deformation*, meaning that dynamic forces within Earth caused fracturing, contortion of rock layers, or both (Figure 10.1). Seismic activity is a manifestation of the continuing nature of these dynamic forces, as is the uplift of the Teton Range in Wyoming, the ongoing evolution of the Himalayas of Asia, and the present geologic activity in the Andes of South America. In short, Earth is an active planet with a variety of processes being driven by internal heat, particularly the movement of lithospheric plates. Indeed, most of Earth's present-day seismic activity and rock deformation takes place at divergent, convergent, and transform plate boundaries.

Mountains can form in a variety of ways, some of which involve little or no deformation, but in most mountains the rocks have been complexly deformed by

FIGURE 10.1   Rock layers deformed by folding and fracturing.

compressive forces at convergent plate boundaries. The Alps of Europe, the Appalachians of North America, the Himalayas of Asia, and many others owe their existence, and in some cases continuing evolution, to plate convergence. In short, deformation and mountain building are closely related phenomena.

A large part of this chapter is devoted to a review of the various types of geologic structures, their descriptive terminology, and the forces responsible for them. But the study of deformed rocks also has several applications. For instance, geologic structures, such as folds and fractures resulting from deformation, provide a record of the kinds and intensities of forces that operated during the past. By interpreting these structures, geologists can make inferences about Earth history that enable us to satisfy our curiosity about the past or search more efficiently for various natural resources. Understanding the nature of geologic structures helps geologists find and recover resources such as petroleum and natural gas (see Figure 7.17b). Local geologic structures must also be considered when selecting sites for dams, large bridges, and nuclear power plants, especially if the sites are in areas of active deformation.

Deformation

F<small>RACTURED</small> and contorted rocks are said to be **deformed**; that is, their original shape or volume or both have been altered by *stress,* which is the force applied to a given area of rock. If the intensity of the force is greater than the internal strength of the rock, it will be *strained,* that is, deformed by folding or fracturing.

Perhaps the following example will help clarify the meaning of stress and the distinction between stress and strain. The force, or stress, exerted by a person walking on an ice-covered pond is a function of the person's weight and can be expressed as some number of kilograms per square centimeter or pounds per square inch. In either case, it represents the amount of force applied per unit area. If the ice should began to crack, it is being strained, or deformed. To avoid breaking the ice and a very chilly swim, the person might lie down; this does not reduce the total force but does distribute it over a larger area, thus reducing the stress per unit area.

Three types of deforming forces are recognized: compression, tension, and shear. **Compression** results when rocks are squeezed or compressed by external forces directed toward one another. Rock layers subjected to compression are commonly shortened in the direction of stress by folding or faulting (Figure 10.2a). **Tension** results from forces acting in opposite directions along the same line and tends to lengthen rocks or pull them apart (Figure 10.2b). In **shear**, forces act parallel to one another but in opposite directions, resulting in deformation by displacement of adjacent layers along closely spaced planes (Figure 10.2c).

Deformation is characterized as **elastic** if a deformed object returns to its original shape when the forces are relaxed. Squeezing a tennis ball causes it to deform, but once the ball is released, it returns to its original shape. Most rocks show only a very limited amount of elastic behavior. If forces of sufficient intensity are applied to rocks, they are deformed beyond their elastic limit and cannot recover their original shape. Under these conditions, rocks exhibit **plastic deformation**, as when they are folded, or they behave as brittle solids and **fracture**.

The type of deformation depends on the kind of force applied, pressure, temperature, rock type, and length of time the rock is subjected to the stress. A small stress applied over a long period of time will cause plastic deformation. By contrast, a large stress applied rapidly to the same object, as when it is struck by a hammer, will probably result in fracture. Rock type is important because not all rocks respond to stress in the same way. Some rocks are

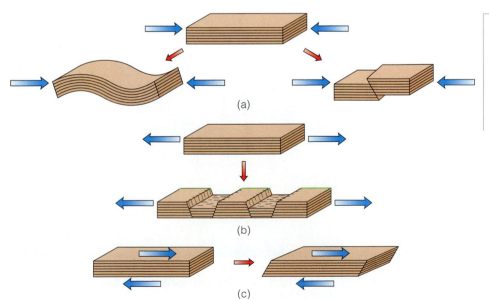

(a)

(b)

(c)

FIGURE 10.2 Stress and possible types of resulting deformation. (a) Compression causes shortening of rock layers by folding or faulting. (b) Tension lengthens rock layers and causes faulting. (c) Shear causes deformation by displacement along closely spaced planes.

characterized as *ductile* because they exhibit a great amount of plastic deformation, whereas brittle rocks show little or none and most commonly deform by fracturing.

Many rocks show the effects of plastic deformation that must have taken place deep within the crust where temperature and pressure are high. At or near the surface, most rocks behave as brittle solids, whereas under conditions of high temperature and high pressure, they commonly deform plastically.

STRIKE AND DIP

According to the principle of original horizontality, sediments are deposited in nearly horizontal layers. Thus, steeply inclined sedimentary rock layers must have been tilted following deposition and lithification. Some igneous rocks, especially ash falls and many lava flows, also form nearly horizontal layers. To describe the orientation of deformed rock layers, geologists use the concept of strike and dip.

Strike is the direction of a line formed by the intersection of a horizontal plane with an inclined plane, such as a rock layer. In Figure 10.3, the surface of any of the tilted rock layers constitutes an inclined plane, and the direction of the line formed by the intersection of a horizontal plane, the water surface, with any of these inclined planes is the strike. The strike line's orientation is determined by using a compass to measure its angle with respect to north. **Dip** is a measure of the maximum angular deviation of an inclined plane from horizontal, so it must be measured perpendicular to the strike direction (Figure 10.3).

Geologic maps indicate strike and dip by using a long line oriented in the strike direction and a short line perpendicular to the strike line and pointing in the dip direction (Figure 10.4). The number adjacent to the strike and dip symbol indicates the dip angle.

FOLDS

If you place your hands on a tablecloth and move them toward one another, the tablecloth is deformed by compression into a series of up- and down-arched folds. Similarly, rock layers within the crust commonly respond to compression by folding. Unlike folds in the tablecloth, however, folding in rock layers is permanent; that is, the rocks have been deformed plastically, so that once folded, they stay folded. Most folding probably takes place deep within the crust because rocks at or near the surface are brittle and generally deform by fracturing rather than by folding.

MONOCLINES, ANTICLINES, AND SYNCLINES   A simple bend or flexure in otherwise horizontal or uniformly dipping rock layers is known as a **monocline** (Figure 10.4). An **anticline** is an up-arched fold, whereas a **syncline** is a down-arched fold (Figure 10.5). A line along the center of a fold is its *axis,* and the rocks on each side of the axis are

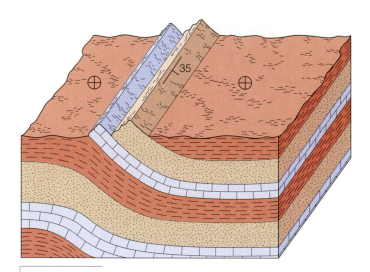

FIGURE 10.4   A monocline. Notice the strike and dip symbols. The other symbols indicate horizontal layers.

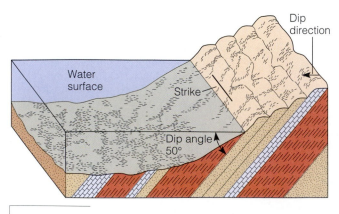

FIGURE 10.3   Strike and dip. Strike is the line formed by the intersection of a horizontal plane (the water surface) with the surface of an inclined plane (the surface of the rock layer). Dip is the maximum angular deviation of the inclined plane from horizontal.

FIGURE 10.5   Anticline and syncline in the Calico Mountains of southeastern California.

limbs. Because folds are most commonly in a series of anticlines alternating with synclines, a limb is generally shared by an anticline and an adjacent syncline (Figure 10.6a).

Even where the exposed view has been deeply eroded, anticlines and synclines can easily be distinguished from each other by strike and dip and by the relative ages of the folded rocks. As Figure 10.6b shows, in an eroded anticline, each limb dips outward or away from the center of the fold, where the oldest rocks are. In eroded synclines, on the other hand, each limb dips inward toward the fold's axis, and the youngest rocks coincide with the center of the fold.

In some folds each limb dips at the same angle, in which case the fold is characterized as *symmetric* (Figure 10.6). Not uncommonly, however, the limbs dip at different angles, and the fold is *asymmetric* (Figure 10.7a). In an *overturned fold,* one limb has been rotated more than 90° from its original position so that both limbs dip in the same direction (Figure 10.7b). Overturned folds are particularly common in mountain ranges that formed by compression at convergent plate boundaries.

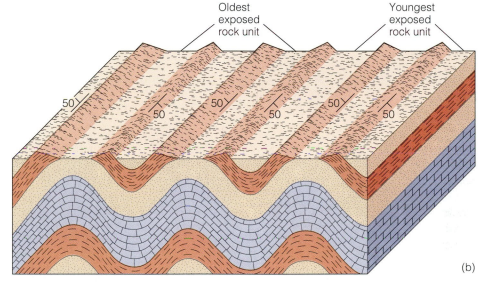

FIGURE **10.6** (a) Syncline and anticline showing the axial plane, axis, and fold limbs. (b) Eroded anticlines and synclines are identified by strike and dip and by the relative ages of the folded layers.

FIGURE **10.7** (a) An asymmetric fold. The axial plane is not vertical, and the fold limbs dip at different angles. (b) Overturned folds. Both fold limbs dip in the same direction, but one limb is inverted.

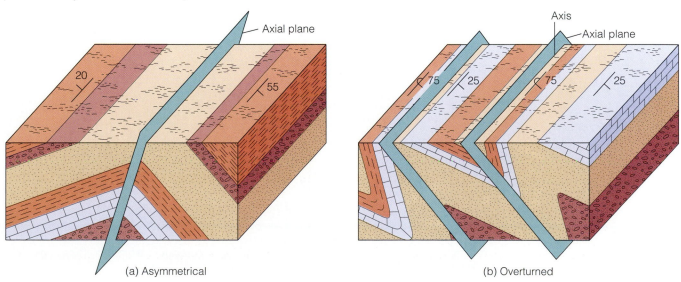

(a) Asymmetrical

(b) Overturned

PLUNGING FOLDS Folds are further characterized as nonplunging or plunging. In *nonplunging folds,* a line known as the axis along the crest of a folded layer is horizontal (Figure 10.6a). **Plunging folds** are much more common; in these, the fold axis is inclined so that it appears to plunge beneath surrounding rocks (Figure 10.8). To differentiate plunging anticlines from plunging synclines, geologists use exactly the same criteria used for nonplunging folds: that is, all rocks dip away from the fold axis in plunging anticlines, whereas in plunging synclines all rocks dip inward toward the axis. The oldest exposed rocks are in the center of an eroded plunging anticline, whereas the youngest exposed rocks are in the center of an eroded plunging syncline (Figure 10.8).

DOMES AND BASINS **Domes** and **basins** are the oval to circular equivalents of anticlines and synclines, which tend to be elongated structures (Figure 10.9). Essentially the same criteria used in recognizing anticlines and synclines are used to identify domes and basins. In an eroded dome, the oldest exposed rocks are in the center, whereas in a basin the opposite is true. All rocks in a dome dip away from a central point, and in a basin they all dip inward toward a central point (Figure 10.9).

Some domes and basins are small structures that are easily recognized by their surface exposure patterns, but many are so large that they can be visualized only on geologic maps or aerial photographs. Many of these large-scale structures formed in the continental interior, not by compression but as a result of vertical uplift of the crust with little additional folding and faulting.

The Black Hills of South Dakota, where a central core of ancient rocks is surrounded by progressively younger rocks, is a good example of an eroded dome. One of the best-known large basins is the Michigan basin. Most of it is buried beneath younger rocks, but strike and dip of ex-

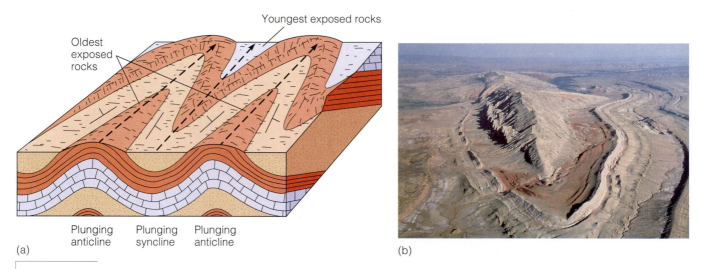

(a)

(b)

FIGURE 10.8 Plunging folds. (a) A block diagram showing surface and cross-sectional views of plunging folds. The long arrow at the center of each fold shows the direction of plunge. (b) Surface view of the eroded, plunging Sheep Mountain anticline in Wyoming. This fold plunges toward the observer.

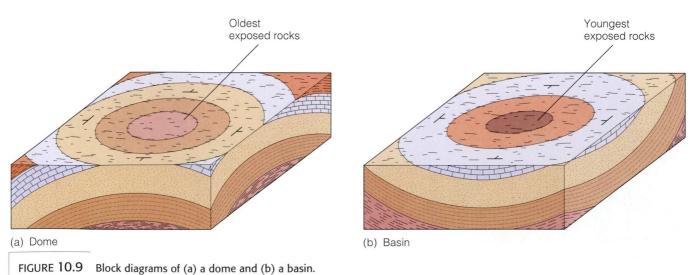

(a) Dome

(b) Basin

FIGURE 10.9 Block diagrams of (a) a dome and (b) a basin.

(b)

(a)

FIGURE 10.10 (a) Erosion along parallel joints in Arches National Park, Utah, yields vertical fins of rock. (b) Intersecting joints forming an "X" pattern at Marquette, Michigan.

posed rocks near the basin margin and thousands of drill holes for oil and natural gas clearly show that the rock layers beneath the surface are deformed into a large basin.

JOINTS

Besides plastic deformation resulting in folding, rocks may also be permanently deformed by fracturing, which yields either joints or faults. **Joints**, which are fractures along which no movement has occurred, are the commonest structures in rocks (Figure 10.10). This lack of movement is what distinguishes joints from faults, which do show movement along fracture surfaces. Coal miners originally used the term *joint* long ago for cracks in rocks that appeared to be surfaces where adjacent blocks were "joined" together.

Joints, which form when brittle rocks deform by fracturing, vary from minute fractures to structures of regional extent that are arranged in parallel or nearly parallel sets. Most joints and joint sets are related to other geologic structures such as faults and large folds. Weathering and erosion of jointed rocks in Utah has produced the spectacular scenery of Arches National Park (see Perspective 10.1).

We have already discussed two other types of joints in earlier chapters: columnar joints and sheet joints. Columnar joints form in some lava flows and in some plutons as cooling magma contracts and fractures (see Figure 5.6). Sheet joints form in response to unloading during mechanical weathering (see Figures 6.6 and 6.8).

FAULTS

A **fault** is a break along which blocks on opposites sides of the fracture move parallel to the fracture surface, which is a *fault plane*. Notice in Figure 10.11 that the rocks adjacent to the fault plane are designated **hanging wall block** and **footwall block**. The hanging wall block is the block of rock overlying the fault, whereas the footwall block lies beneath the fault plane. These two blocks can be recognized on any fault except a vertical one.

To recognize the various types of fault movement, you must identify hanging wall and footwall blocks and understand the concept of relative movement. Geologists refer to relative movement because you usually cannot tell which block actually moved or if both blocks moved. In Figure 10.11, for example, it cannot be determined whether the hanging wall block moved up, the footwall block moved down, or both blocks moved. Nevertheless, the hanging wall block appears to have moved downward relative to the footwall block. Alternatively, one could say that the footwall block appears to have moved upward relative to the hanging wall block.

Like dipping rock layers, fault planes are also incline planes and can be characterized by their strike and dip (Figure 10.11). Two basic types of faults are recognized according to whether the blocks on opposite sides of the fault plane have moved parallel to the direction of dip or along the direction of strike.

Folding, Joints, and Arches

Arches National Park in eastern Utah is noted for its panoramic vistas, which include such landforms as Delicate Arch (Figure 1), Double Arch, Landscape Arch, and many others. Unfortunately, the term *arch* is used for a variety of geologic features of different origin, but here we will restrict the term to mean an opening through a wall of rock that is formed by weathering and erosion.

The arches of Arches National Park continue to form as a result of weathering and erosion of the folded and jointed Entrada Sandstone, the rock underlying much of the park. Accordingly, geologic structures play a significant role in the origin of these arches. Where the Entrada Sandstone was folded into anticlines, it was stretched so that parallel, vertical joints formed. Weathering and erosion occur most vigorously along joints because these processes can attack the exposed rock from both the top and the sides, whereas only the top is attacked in unjointed rocks.

Erosion along joints causes them to enlarge, thereby forming long slender fins of rock between adjacent joints (see Figure 10.10a). Some parts of these fins are more susceptible to weathering and erosion than others, and as the sides are attacked, a recess may form. If it does, eventually pieces of the unsupported rock above the recess will fall away, forming an arch as the original recess is enlarged (Figure 2). Thus, arches are remnants of fins formed by weathering and erosion along joints.

Historical observations show that arches continue to form today. For example, in 1940 Skyline Arch was enlarged when a large block fell from its underside. The park also contains many examples of arches that collapsed during prehistoric time. When arches collapse, they leave isolated pinnacles and spires. Arches National Park is well worth visiting; the pinnacles, spires, and arches are impressive features indeed.

Figure 1 Delicate Arch in Arches National Park, Utah, formed by weathering and erosion of jointed sedimentary rocks. It is 9.7 m wide and 14 m high.

Figure 2 Hole-in-the-wall, or "Baby Arch," shows the early development of an arch in a fin of rock. "Baby Arch" measures 7.6 m wide and 4.5 m high. (Photo courtesy of Sue Monroe.)

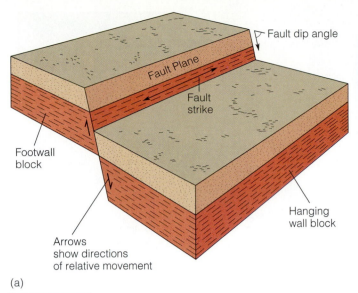

(a)

(b)

FIGURE 10.11 (a) Fault terminology. (b) Fault plane near Klamath Falls, Oregon. *(Photo courtesy of David J. Matty.)*

DIP-SLIP FAULTS In **dip-slip faults**, all movement is parallel to the dip of the fault plane (Figure 10.12a–c); that is, one block moves up or down relative to the block on the opposite side of the fault plane. Depending on the relative movement of the hanging wall and footwall blocks, two types of dip-slip faults are recognized: normal and reverse.

In **normal faults**, the hanging wall block appears to have moved downward relative to the footwall block (Figure 10.12a and Figure 10.13). This type of faulting is caused by tension such as occurs where the crust is stretched and thinned by rifting. The mountain ranges in the Basin and Range Province, a large area in the western United States and northern Mexico, are bounded on one or both sides by large normal faults. A normal fault is also present along the east side of the Sierra Nevada in California where uplift of the block west of the fault has elevated the mountains more than 3000 m above the lowlands to the east. The Teton Range in Wyoming is also bounded by a normal fault.

The second type of dip-slip fault is a **reverse fault** (Figure 10.12b). A reverse fault with a dip of less than 45° is known as a *thrust fault* (Figure 10.12c). Reverse and thrust faults are easily distinguished from normal faults because the hanging wall block moves up relative to the footwall block. Both are caused by compression and are common in mountain ranges that formed at convergent plate boundaries.

STRIKE-SLIP FAULTS Shearing forces are responsible for **strike-slip faulting**, a type of faulting involving horizontal movement in which blocks on opposite sides of a fault plane slide sideways past one another (Figure 10.12d). In other words, all movement is in the direction of the fault plane's strike. One of the best-known strike-slip faults is the San Andreas fault of California. Recall from Chapter 2 that the San Andreas fault is also called a transform fault in plate tectonics terminology.

Strike-slip faults can be characterized as right-lateral or

left-lateral, depending on the apparent direction of offset. In Figure 10.12d, for example, an observer looking at the block on the opposite side of the fault determines whether it appears to have moved to the right or to the left. In this example, movement appears to have been to the left, so the fault is characterized as a *left-lateral strike-slip fault*. Had this been a *right-lateral strike-slip fault*, the block across the fault from the observer would appear to have moved to the right. The San Andreas fault is a right-lateral strike-slip fault.

OBLIQUE-SLIP FAULTS It is possible for movement on a fault to show both dip-slip and strike-slip types of movement. Strike-slip movement may be accompanied by a dip-slip component, giving rise to a combined movement that includes left-lateral and reverse, or right-lateral and normal (Figure 10.12e). Faults showing both dip-slip and strike-slip movement are **oblique-slip faults**.

Mountains

THE term *mountain* refers to any area of land that stands significantly higher than the surrounding country. Some mountains are single, isolated peaks, but much more commonly they are parts of a linear association of peaks and/or ridges called *mountain ranges* that are related in age and origin. A *mountain system* is a mountainous region consisting of several or many mountain ranges such as the Rocky Mountains and Appalachians. Mountain systems are complex linear zones of intense deformation and crustal thickening characterized by many of the geologic structures previously discussed.

Mountain systems are indeed impressive features and represent the effects of dynamic processes operating within Earth. They are large-scale manifestations of tremendous forces that have produced folded, faulted, and thickened parts of the crust. Furthermore, in some mountain systems,

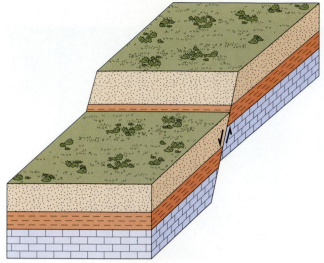

(a) Normal fault

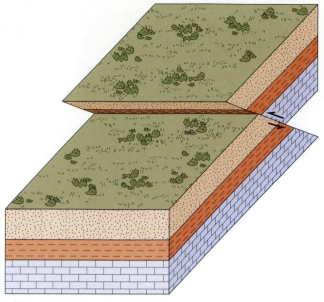

(b) Reverse fault

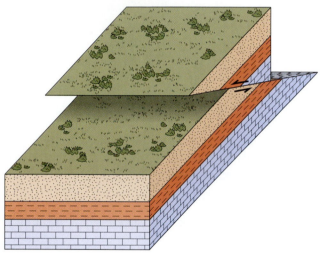

(c) Thrust fault

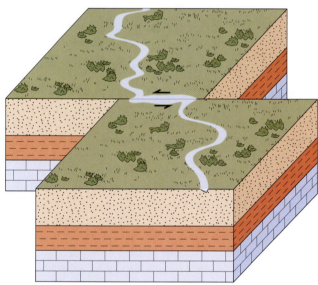

(d) Strike-slip fault

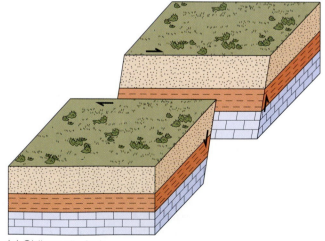

(e) Oblique-slip fault

FIGURE 10.12 Types of faults. (a), (b), and (c) are dip-slip faults. (a) Normal fault—hanging wall block moves down relative to footwall block. (b) and (c) Reverse and thrust faults—hanging wall block moves up relative to footwall block. (d) Strike-slip fault—all movement parallel to the strike of the fault. (e) Oblique-slip fault—combination of dip-slip and strike-slip movements.

such as the Andes of South America and the Himalayas of Asia, the mountain-building processes remain active.

TYPES OF MOUNTAINS

Mountainous topography can develop in a variety of ways, some of which involve little or no deformation. For example, a single volcanic mountain can develop over a hot spot, but more commonly a series of volcanoes forms as a plate moves over the hot spot, as in the case of the Hawaiian Islands.

Mountainous topography also forms where the crust has been intruded by batholiths that are subsequently uplifted and eroded (Figure 10.14). The Sweetgrass Hills of northern Montana consist of resistant plutonic rocks exposed following uplift and erosion of the softer overlying sedimentary rocks.

Block-faulting is yet another way mountains are formed.

FIGURE 10.13   Two small normal faults cutting through layers of volcanic ash in Oregon.

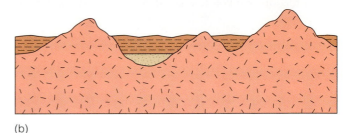

(a)

(b)

FIGURE 10.14   (a) Pluton intruded into sedimentary rocks. (b) Erosion of the softer overlying rocks reveals the pluton and forms small mountains.

Block-faulting involves movement on normal faults so that one or more blocks are elevated relative to adjacent areas. A classic example is the large-scale block-faulting currently occurring in the Basin and Range Province of the western United States, a large area centered on Nevada but extending into several adjacent states and northern Mexico. Here, the crust is being stretched in an east-west direction, and tensional forces produce north-south–oriented, range-bounding faults. Differential movement on these faults has yielded uplifted blocks called *horsts* and down-dropped blocks called *grabens,* which are bounded on both sides by parallel normal faults (Figure 10.15). Erosion of the horsts has yielded the mountainous topography now present, and the grabens have filled with sediments eroded from the horsts.

The processes discussed here can certainly yield mountains. However, the truly large mountain systems of the continents, such as the Alps of Europe and the Appalachians in North America, were produced by compression along convergent plate margins.

Mountain Building: Orogeny

DURING an episode of mountain building, termed an **orogeny,** intense deformation takes place, generally accompanied by metamorphism and the emplacement of plutons, especially batholiths. The processes responsible for orogenies are still not fully understood but are related to plate

movements. In fact, the advent of plate tectonic theory has completely changed the way geologists view the origin of mountains.

Any theory accounting for orogenies must explain the various characteristics of mountain systems, such as their geometry and location; they tend to be long and narrow and to form at or near plate margins. Mountain systems also show intense deformation, especially compression-induced folds and reverse and thrust faults. Furthermore, the deeper, interior parts, or *cores,* of mountain systems are characterized by granitic plutons and regional metamorphism. The presence of deformed shallow and deep marine sedimentary rocks that have been elevated far above sea level is another feature.

Most orogenies occur in response to compressive forces at convergent plate boundaries. Recall from Chapter 2 that three varieties of convergent plate boundaries are recognized: oceanic-oceanic, oceanic-continental, and continental-continental. Although compression at convergent plate boundaries accounts for deformation and uplift of areas

FIGURE 10.15   Block-faulting and the origin of horsts and grabens.

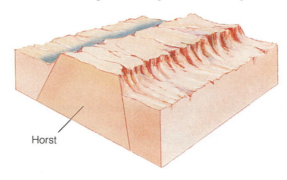

Horst

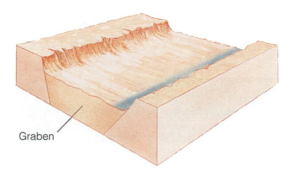

Graben

recognized as mountain systems, erosion is responsible for much of their present-day expression. The Himalayas of Asia, for instance, continue to form as a result of a collision between two plates, but erosion by running water and glaciers is largely responsible for their rugged topography.

OROGENIES AT OCEANIC-OCEANIC PLATE BOUNDARIES

Orogenies occurring where oceanic lithosphere is subducted beneath oceanic lithosphere are characterized by the formation of a volcanic island arc and by deformation,

igneous activity, and metamorphism. The subducted plate forms the outer wall of an oceanic trench, and the inner wall of the trench consists of a subduction complex, or *accretionary wedge*, composed of wedge-shaped slices of highly folded and faulted marine sedimentary rocks and oceanic lithosphere scraped from the descending plate (Figure 10.16). This subduction complex is elevated as a result of uplift along faults as subduction continues. In addition, plate convergence results in low-temperature, high-pressure metamorphism.

Deformation also occurs in the island arc system where it is caused largely by the emplacement of plutons of inter-

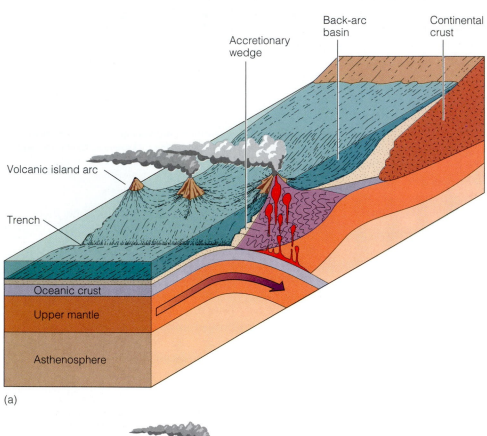

(a)

FIGURE 10.16   (a) Subduction of an oceanic plate beneath an island arc. (b) Continued subduction, emplacement of plutons, and beginning of deformation by thrust faulting and folding of back-arc basin sediments. (c) Thrusting of back-arc basin sediments onto the adjacent continent and suturing of the island arc to the continent.

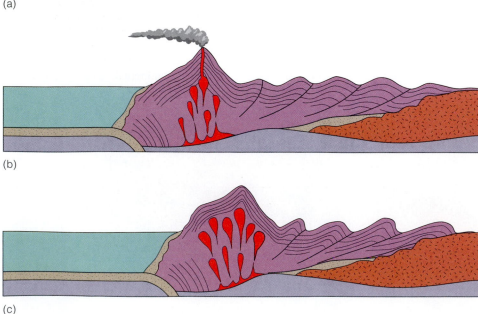

(b)

(c)

mediate and felsic composition, and many rocks show evidence of high-temperature, low-pressure metamorphism. As a result, the overriding oceanic plate is thickened as it is intruded by plutons and becomes more continental. The overall effect of an island arc orogeny is the origin of two more-or-less parallel orogenic belts consisting of a deformed volcanic island arc underlain by batholiths and a seaward belt of deformed trench rocks (Figure 10.16). The Aleutian Islands of Alaska are a good example of this type of deformation.

If a back-arc basin exists between the island arc and a nearby continent, volcanic rocks and sediments derived from the island arc and the adjacent continent are also deformed as the plates continue to converge. The sediments are intensely folded and displaced toward the continent along low-angle thrust faults. Eventually, the entire island arc complex is fused to the edge of the continent, and the back-arc basin sediments are thrust onto the continent and form a thick stack of thrust sheets (Figure 10.16).

OROGENIES AT OCEANIC-CONTINENTAL PLATE BOUNDARIES

The Andes of western South America are perhaps the best example of a continuing orogeny at an oceanic-continental plate boundary where oceanic lithosphere is subducted (Figure 10.17). Among the ranges of the Andes are the highest mountain peaks in the Americas and many active volcanoes; the western part of South America is an extremely active segment of the circum-Pacific earthquake belt. Furthermore, one of Earth's great oceanic trench systems, the Peru-Chile Trench, lies just off the west coast.

Prior to 200 million years ago, the western margin of South America was a passive continental margin, where sediments accumulated on the continental shelf, slope, and rise much as they do now along the east coast of North America (Figure 10.17a). When Pangaea began fragmenting in response to rifting along what is now the Mid-Atlantic Ridge, the South American plate moved westward, and an eastward-moving oceanic plate began subducting beneath the continent (Figure 10.17b). What had been a passive continental margin was now an active one.

As subduction proceeded, rocks of the continental margin and trench were folded and faulted and are now part of an accretionary wedge along the west coast of South America (Figure 10.17c). Subduction also resulted in partial melting of the descending plate, producing an andesitic volcanic arc of composite volcanoes near the edge of the continent. More viscous felsic magmas, mostly of granitic composition, were emplaced as large plutons beneath the volcanic arc (Figure 10.17c). The coastal batholith of Peru, for instance, consists of perhaps 800 individual plutons that were emplaced over several tens of millions of years.

As a result of the events just described, the Andes Mountains consist of a central core of granitic rocks capped by andesitic volcanoes. To the west of this central core along the coast are the deformed rocks of the accretionary wedge. And to the east of the central core are sedimentary rocks that have been intensely folded and thrust eastward onto the continent (Figure 10.17c). Present-day subduction, volcanism, and seismicity along South America's western margin indicate that the Andes Mountains are still actively forming.

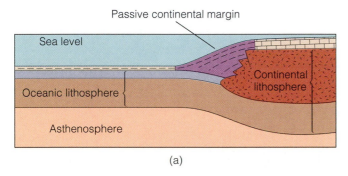

(a)

FIGURE 10.17 Generalized diagrams showing three stages in the development of the Andes of South America. (a) Prior to 200 million years ago, the west coast of South America was a passive continental margin. (b) Orogeny began when the west coast of South America became an active continental margin. (c) Continued deformation, volcanism, and plutonism.

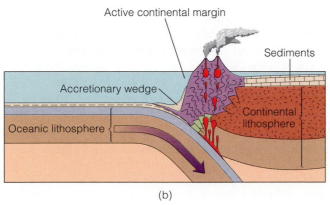

(b)

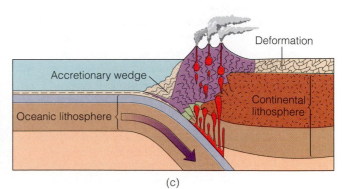

(c)

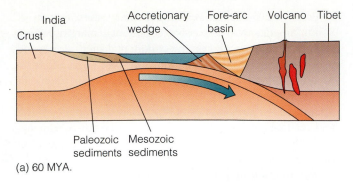

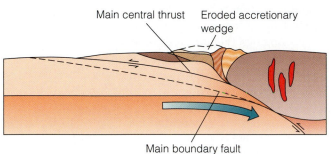

India
Crust
Accretionary wedge
Fore-arc basin
Volcano   Tibet

Paleozoic sediments   Mesozoic sediments

(a) 60 MYA.

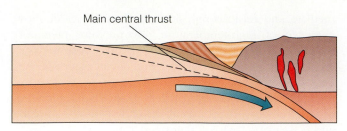

Main central thrust

(b) 40–50 MYA.

Main central thrust   Eroded accretionary wedge

Main boundary fault

(c) 20–40 MYA.

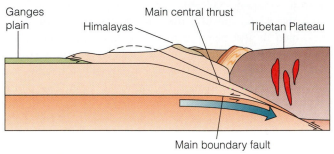

Ganges plain   Himalayas   Main central thrust   Tibetan Plateau

Main boundary fault

(d) 20–0 MYA.

FIGURE 10.18 Simplified cross sections showing the collision of India with Asia and the origin of the Himalayas. (a) The northern margin of India before its collision with Asia. Subduction of oceanic lithosphere beneath southern Tibet as India approached Asia. (b) About 40 to 50 million years ago, India collided with Asia, but because India was too light to be subducted, it was underthrust beneath Asia. (c) Continued convergence accompanied by thrusting of rocks of Asian origin onto the Indian subcontinent. (d) Since about 10 million years ago, India has moved beneath Asia along the main boundary fault. Shallow marine sedimentary rocks that were deposited along India's northern margin now form the higher parts of the Himalayas. Sediment eroded from the Himalayas has been deposited on the Ganges Plain.

OROGENIES AT CONTINENTAL-CONTINENTAL PLATE BOUNDARIES

The best example of orogenesis at a continental-continental plate boundary is provided by the Himalayas of Asia. The Himalayas began forming when India collided with Asia about 40 to 50 million years ago. Prior to that time, India was far south of Asia and separated from it by an ocean basin (Figure 10.18a). As the Indian plate moved northward, a subduction zone formed along the southern margin of Asia where oceanic lithosphere was consumed. Partial melting generated magma, which rose to form a volcanic arc, and large granite plutons were emplaced into what is now Tibet (Figure 10.18a). At this stage, the activity along Asia's southern margin was similar to what is now occurring along the west coast of South America (10-17c).

The ocean separating India from Asia continued to close, and India eventually collided with Asia (Figure 10.18b). As a result, two continental plates became welded, or sutured, together. Thus, the Himalayas are now within a continent rather than along a continental margin. The exact time of India's collision with Asia is uncertain, but between

40 and 50 million years ago, India's rate of northward drift decreased abruptly—from 15 to 20 cm per year to about 5 cm per year. Because continental lithosphere is not dense enough to be subducted, this decrease in rate seems to mark the time of collision and India's resistance to subduction. Consequently, India's leading margin was thrust beneath Asia, causing crustal thickening, thrusting, and uplift. Sedimentary rocks that had been deposited in the sea south of Asia were thrust northward, and two large thrust faults carried rocks of Asian origin onto the Indian plate (Figure 10.18c and d). Rocks deposited in the shallow seas along India's northern margin now form the higher parts of the Himalayas. Since its collision with Asia, India has been underthrust about 2000 km beneath Asia and is still moving north at a rate of about 5 cm per year.

Microplate Tectonics and Mountain Building

OROGENIES at convergent plate boundaries result in material being added to continental margins by a process known as *continental accretion*. Much of the material accreted to continents during these events is simply eroded older continental crust, but a significant amount of new material is added to continents as well—igneous rocks that formed as a result of subduction and partial melting and the suturing of an island arc to a continent, for example. Although subduction is the predominant influence on tectonic history in many regions of orogeny, other processes are also involved in mountain building and continental accretion, especially the accretion of microplates (Figure 10.19).

During the late 1970s and 1980, geologists discovered that parts of many mountain systems are composed of

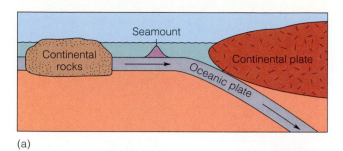

(a)

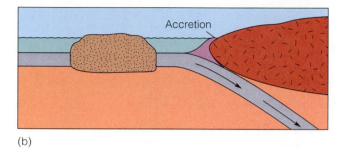

(b)

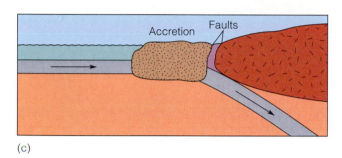

(c)

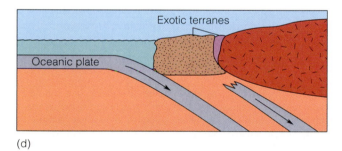

(d)

FIGURE 10.19 Accretion of microplates at a convergent plate boundary. In this example, a seamount and a small block of continental rocks are scraped off a subducting plate and added to the continental margin. Much of the western part of the United States and Canada is composed of accreted microplates.

small accreted lithospheric blocks that are clearly of foreign origin. These **microplates** differ completely from the rocks of the surrounding mountain system. In fact, many microplates are so different from adjacent rocks that most geologists think they formed elsewhere and were carried great distances as parts of other plates until they collided with other microplates or continents.

Geologic evidence indicates that more than 25% of the entire Pacific coast from Alaska to Baja California consists of accreted microplates, some of which were seamounts, volcanic island arcs, oceanic ridges, or, in some cases, simply dis-

placed parts of a continent. It is estimated that more than 100 different-sized microplates have been added to the western margin of North America during the last 200 million years.

Most microplates so far identified are in mountains of the North American Pacific coast region, but a number of others are suspected to be present in different areas as well. They are more difficult to recognize in older mountain systems, such as the Appalachians, because of greater deformation and erosion. Nevertheless, about a dozen microplates have been identified in the Appalachians, although their boundaries are hard to discern.

The basic plate tectonic reconstructions of orogenies and continental accretion remain unchanged, but the details of these reconstructions are decidedly different in view of microplate tectonics. Furthermore, many of these accreted microplates are new additions to a continent rather than reworked older continental material.

Measuring Gravity and the Principle of Isostasy

Sɪʀ Isaac Newton (1642–1727) formulated the law of universal gravitation in which the force of gravitational attraction (F) between two masses (m_1 and m_2) is directly proportional to the products of their mass and inversely proportional to the square of the distance (D) between their centers of mass:

$$F = G \frac{m_1 \times m_2}{D}$$

G in this equation is the universal gravitational constant. This law applies to any two bodies. Consequently, a gravitational attraction exists between Earth and its moon, between the stars, and between an individual and Earth. We generally refer to the gravitational force between an object and Earth as its *weight*.

Gravitational attraction would be the same everywhere on Earth's surface if it were perfectly spherical, homogeneous throughout, and not rotating. As a consequence of rotation, however, a centrifugal force is generated that partly counteracts the force of gravity. An object at the equator weighs slightly less than the same object would at the poles. The force of gravity also varies with distance between the centers of masses, so an object would weigh slightly less above the surface than if it were at sea level.

Geologists use a sensitive instrument called a *gravimeter* to measure variations in the force of gravity. A gravimeter is simple in principle; it contains a weight suspended on a spring that responds to variations in gravity (Figure 10.20). Long ago geologists realized that departures from normal gravitational attraction should exist over buried bodies of ore minerals and salt domes and that geologic structures such as faulted strata could be located by surface gravity surveys (Figure 10.20). Such departures from the expected force of gravity are **gravity anomalies** (Figure 10.20). In other words, the measurement over a body of iron ore indicates an excess of dense material, or simply a

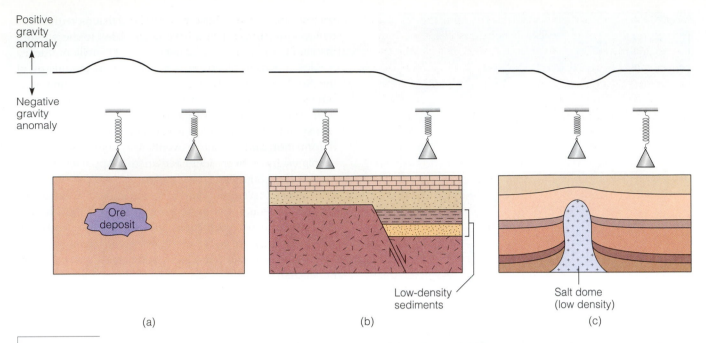

FIGURE 10.20   (a) The mass suspended from a spring in the gravimeter, shown diagrammatically, is pulled downward more over the dense body of ore than it is in adjacent areas, indicating a positive gravity anomaly. (b) A negative gravity anomaly over a buried structure. (c) Rock salt is less dense than most other types of rocks. A gravity survey over a salt dome shows a negative gravity anomaly.

mass excess, between the surface and the center of Earth and is considered to be a *positive gravity anomaly.* A *negative gravity anomaly* indicates a *mass deficiency* exists over low-density materials because the force of gravity is less than the expected average. Large negative gravity anomalies exist over salt domes and at subduction zones, indicating that the crust is not in equilibrium.

More than 150 years ago, British surveyors in India detected a discrepancy of 177 m when they compared the results of two measurements between points 600 km apart. Even though this discrepancy was small, it was an unacceptably large error. The surveyors realized that the gravitational attraction of the nearby Himalaya Mountains probably deflected the plumb line (a cord with a suspended

FIGURE 10.21   (a) A plumb line is normally vertical, pointing to Earth's center of gravity. Near a mountain range, the plumb line should be deflected as shown if the mountains are simply thicker, low-density material resting on denser material, and a gravity survey across the mountains would indicate a positive gravity anomaly. (b) The actual deflection of the plumb line during the survey in India was less than expected. It was explained by postulating that the Himalayas have a low-density root. A gravity survey in this case would show no anomaly because the mass of the mountains above the surface is compensated for at depth by low-density material displacing denser material.

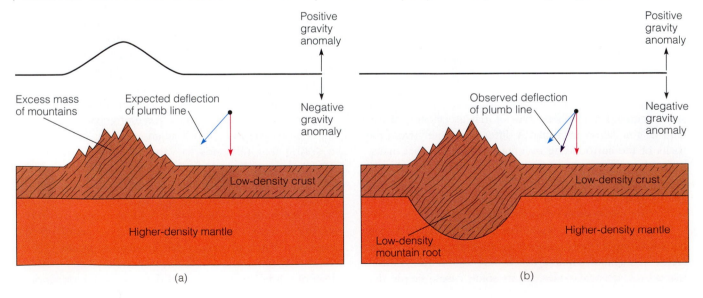

weight) of their surveying instruments from the vertical, thus accounting for the error. Calculations revealed, however, that if the Himalays were simply thicker crust piled on denser material, the error should have been greater than that observed (Figure 10.21)

In 1865 Sir George Airy proposed that, in addition to projecting high above sea level, the Himalayas—and other mountains as well—also project far below the surface and thus have a low-density root (Figure 10.21b). In effect, he was saying that mountains float on denser rock at depth. Their excess mass above sea level is compensated for by a mass deficiency at depth, which would account for the observed deflection of the plumb line during the British survey (Figure 10.21).

Another explanation was proposed by J. H. Pratt who thought that the Himalayas were high because they were composed of rocks of lesser density than those in adjacent regions. Although Airy was correct with respect to the Himalayas, and mountains in general, Pratt was correct in that there are indeed places where the crust's elevation is related to its density. For example, continental crust is thick and less dense than oceanic crust and thus stands high. And the mid-oceanic ridges stand higher than adjacent areas because the crust there is hot and less dense than cooler oceanic crust elsewhere.

Gravity studies have revealed that mountains do indeed have a low-density "root" projecting deep into the mantle. If it were not for this low-density root, a gravity survey across a mountainous area would reveal a huge positive gravity anomaly. The fact that no such anomaly exists indicates that a mass excess is not present, so some of the dense mantle at depth must be displaced by lighter crustal rocks as shown in Figure 10.21. (Seismic wave studies also confirm the existence of low-density roots beneath mountains.)

Both Airy and Pratt agreed that Earth's crust is in floating equilibrium with the more dense mantle below. Now their proposal is known as the **principle of isostasy**. This phenomenon is easy to understand by analogy to an iceberg. Ice is slightly less dense than water, and thus it floats. According to Archimedes' principle of buoyancy, an iceberg will sink in the water until it displaces a volume of water that equals its total weight. When the iceberg has sunk to an equilibrium position, only about 10% of its volume is above water level. If some of the ice above water level should melt, the iceberg will rise in order to maintain the same proportion of ice above and below water.

Earth's crust is similar to the iceberg, that it sinks into the mantle to its equilibrium level. Where the crust is thickest, as beneath mountain ranges, it sinks farther down into the mantle but also rises higher above the equilibrium surface (Figure 10.21). Continental crust being thicker and less dense than oceanic crust stands higher than the ocean basins. Should the crust be loaded, as where widespread glaciers accumulate, it responds by sinking farther into the mantle to maintain equilibrium. In Greenland and Antarctica, the surface of the crust has been depressed below sea

level by the weight of glacial ice. The crust also responds isostatically to widespread erosion and sediment deposition. Notice in Figure 10.22 that isostatic equilibrium is maintained: As erosion proceeds, the mountainous area rises; in areas of widespread sediment deposition, the crust subsides. If the process continues long enough, a mountain system may cease to exist, its former presence being indicated only by a linear zone of plutons and deformed, metamorphosed crustal rocks.

Unloading of the crust causes it to respond by rising upward until equilibrium is again attained. This phenomenon, known as **isostatic rebound**, occurs in areas that are deeply eroded and in areas that were formerly glaciated. Scandinavia, which was covered by a vast ice sheet until about 10,000 years ago, is still rebounding isostatically at a rate of up to 1 m per century. Coastal cities in Scandinavia have been uplifted rapidly enough that docks constructed several centuries ago are now far from shore. Isostatic rebound has

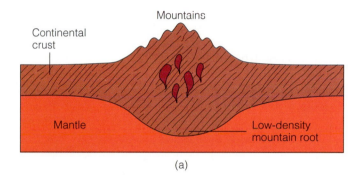

(a)

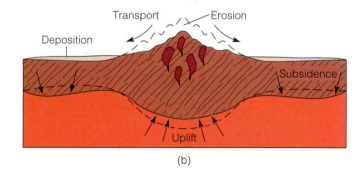

(b)

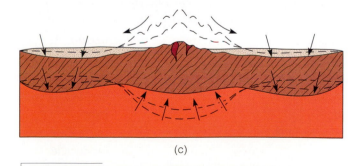

(c)

FIGURE **10.22** A diagrammatic representation showing the isostatic response of the crust to erosion (unloading) and widespread deposition (loading).

also taken place in eastern Canada where the land has risen as much as 100 m during the last 6000 years.

If the principle of isostacy is correct, it implies that the mantle behaves as a liquid. In preceding discussions, however, we said that the mantle must be solid because it transmits S-waves, which will not move through a liquid. How can this apparent paradox be resolved? When considered in terms of the short time necessary for S-waves to pass through it, the mantle is indeed solid. But when subjected to stress over long periods of time, it will yield by flowage, and at these time scales can be considered a viscous liquid. A familiar substance that has the properties of a solid or a liquid, depending on how rapidly deforming forces are applied, is Silly Putty. It will flow under its own weight if given sufficient time, but shatters as a brittle solid if struck a sharp blow.

Chapter Summary

1. Contorted and fractured rocks have been deformed by stress, which is the force applied to a given area of rock.

2. Stress is characterized as compression, tension, or shear. Elastic deformation is not permanent, meaning that when the force is removed, rocks return to their original shape or volume. Plastic deformation and fracture are both permanent types of deformation.

3. The orientation of deformed layers of rock is described by strike and dip.

4. Rock layers that have been buckled into up- and down-arched folds are anticlines and synclines, respectively. They can be identified by strike and dip of the folded rocks and by the relative age of the rocks in the center of eroded folds.

5. Domes and basins are the circular to oval equivalents of anticlines and synclines but are commonly much larger structures.

6. Two types of structures resulting from fracturing are recognized: joints are fractures along which no movement has occurred and faults are fractures along which the blocks on opposite sides of the fracture move parallel to the fracture surface.

7. Joints, which are the commonest geologic structures, form in response to compression, tension, and shear.

8. On dip-slip faults, all movement is in the dip direction of the fault plane. Two varieties of dip-slip faults are recognized: normal faults form in response to tension, whereas reverse faults are caused by compression.

9. On strike-slip faults, all movement is in the direction of the fault plane's strike. They are characterized as right-lateral or left-lateral, depending on the apparent direction of offset of one block relative to the other.

10. Some faults show components of both dip-slip and strike-slip; they are called oblique-slip faults.

11. Mountains can form in a variety of ways, some of which involve little or no folding or faulting. Mountain systems consisting of several mountain ranges result from deformation related to plate movements.

12. A volcanic island arc, deformation, igneous activity, and metamorphism characterize orogenies at oceanic-oceanic plate boundaries. Subduction of oceanic lithosphere at an oceanic-continental plate boundary also results in orogeny.

13. Some mountain systems, such as the Himalayas, are within continents far from a present-day plate boundary. These mountains formed when two continental plates collided and became sutured.

14. According to the principle of isostasy, Earth's crust is floating in equilibrium with the denser mantle below. Continental crust stands higher than oceanic crust because it is thicker and less dense.

Important Terms

anticline	fracture	plastic deformation
basin	gravity anomaly	plunging fold
compression	hanging wall block	principle of isostasy
deformation	isostatic rebound	reverse fault
dip	joint	shear
dip-slip fault	microplate	strike
dome	monocline	strike-slip fault
elastic deformation	normal fault	syncline
fault	oblique-slip fault	tension
footwall block	orogeny	

1. A normal fault is one along which:
   a. _____ all movement is horizontal;
   b. _____ the hanging wall block moves down relative to the footwall block;
   c. _____ the footwall block moves horizontally, whereas the hanging wall block moves vertically;
   d. _____ both dip-slip and strike-slip movement have occurred;
   e. _____ the fault plane is vertical.

2. In an anticline:
   a. _____ fault movement is mainly vertical;
   b. _____ rocks have been deformed by fracture;
   c. _____ a gravity anomaly corresponds to the axis;
   d. _____ all rock layers dip outward from the fold's axis;
   e. _____ rock layers respond to stress by isostatic rebound.

3. Displacement of adjacent rock layers along closely spaced planes results from stress known as:
   a. _____ shear;
   b. _____ elastic;
   c. _____ tension;
   d. _____ compression;
   e. _____ plastic.

4. The strike of a rock layer is defined as:
   a. _____ the kind of deformation occurring when rocks are subjected to both tension and compression;
   b. _____ a circular fold in which all rock layers dip in toward a central point;
   c. _____ the amount of uplift after the crust has been loaded by glacial ice;
   d. _____ a quantity of heat released by rocks as they are deformed;
   e. _____ a line formed by the intersection of a horizontal plane with an inclined plane.

5. Folds in which the axis appears to dip beneath surrounding rocks are referred to as _____ folds.
   a. _____ circular;
   b. _____ reverse;
   c. _____ plunging;
   d. _____ strike-slip;
   e. _____ dome.

6. The fault shown in Figure 10.11a is a _____ fault:
   a. _____ reverse;
   b. _____ oblique;
   c. _____ normal;
   d. _____ thrust;
   e. _____ strike-slip.

7. An episode of intense deformation resulting from compression that yields mountains is termed a(an):
   a. _____ orogeny;
   b. _____ isostasy;
   c. _____ plate separation;
   d. _____ passive continental margin;
   e. _____ gravity anomaly.

8. Which one of the following statements is correct?
   a. _____ an area with a buried body of iron ore possesses a negative gravity anomaly;
   b. _____ mountains have a low-density root projecting deep into the mantle;
   c. _____ mountains forming at convergent plate boundaries result mostly from tension;
   d. _____ a circular fold with the oldest rocks in the center is a monocline;
   e. _____ a reverse fault is one along which the footwall block moves upward relative to the hanging wall block.

9. According to the principle of isostasy:
   a. _____ more heat escapes from oceanic crust than continental crust;
   b. _____ the crust is floating in equilibrium in the denser mantle below;
   c. _____ Earth's mantle behaves as a solid and a liquid;
   d. _____ much of the asthenosphere is molten;
   e. _____ magnetic anomalies result when the crust is loaded by glaciers.

10. Joints are defined as:
    a. _____ anticlines with inclined axes;
    b. _____ areas within orogenic belts where deformation is particularly intense;
    c. _____ areas of crustal thickening resulting from thrust faulting;
    d. _____ strike-slip faults with both horizontal and vertical displacement;
    e. _____ fractures along which no movement has taken place.

11. In Figure 10.11b is the block on the right the footwall or hanging wall block? Explain how you know. Suppose the block on the right moved relatively up. Is this a normal or reverse fault?

12. Explain why two roughly parallel orogenic belts develop where oceanic lithosphere is subducted beneath oceanic lithosphere.

13. Explain in terms of the principle of isostasy why mountains have a "root" extending deep into the mantle and why oceanic crust is lower than continental crust.

14. Domes and basins show the same patterns on geologic maps but differ in two important ways. What are the two criteria for distinguishing one from the other?

15. What types of evidence indicate that deforming forces remain active within Earth?

16. Draw a simple sketch map showing displacement on a right-lateral strike-slip fault.

17. What are the criteria for identifying anticlines and synclines?

18. Define the types of strain recognized. What kinds of stress cause them?

19. How do dip-slip and strike-slip faults differ?

20. Give a brief account of the origin of the Himalayas of Asia. Is there any evidence that orogeny is continuing in this area? If so, what is it?

1. Over 5 million years, rocks are displaced 6000 m along a normal fault. What was the average yearly move- ment on this fault? Is this average likely to represent the actual rate of displacement on this fault? Explain.

2. How is it possible for the same kind of rock to behave both elastically and plastically?

World Wide Web Activities

For these web site addresses, along with current updates and exercises, log on to

http://www.wadsworth.com/geo

▶ STRUCTURAL GEOLOGY

There are not a lot of resources on the WWW devoted only to structural geology. This site, maintained by Steven Henry Schimmrich, contains a listing of data sets, bibliographies, organizations, computer software, and on-line courses in structural geology. Check out the various research projects listed under the Research Information section.

▶ SALT TECTONICS

This site gives an overview of the research being conducted on salt tectonics by Giovanni Guglielmo, at the Applied Geodynamics Laboratory of the Bureau of Economic Geology at the University of Texas at Austin. Much of it will probably not be of interest to the general geology student, but for those planning to major in geology, it shows what is currently being done in the field of salt tectonics.

Click on any of the sites listed under *Examples of My Current Research*. Some of these will require software to run, but it can be downloaded from the links provided.

▶ STRUCTURAL GEOLOGY PHOTO GALLERY GS 326, CORNELL UNIVERSITY

This site, maintained by R. W. Allmendinger, contains a sampling of slides of various geologic structures from field trips and class lectures. Click on any of the designations such as *normal and thrust faults,* then click the thumbnail on the left of the image to see a full-screen photo.

▶ KECK GEOLOGY CONSORTIUM STRUCTURAL GEOLOGY SLIDE SET

This slide set was compiled by H. Robert Burger of Smith College with the support of the W. M. Keck Foundation of Los Angeles, California. The database was created at the Department of Earth and Ocean Sciences, University of British Columbia, Vancouver, BC, Canada. The site has a searchable database. Simply enter a keyword such as *anticline, syncline,* or *normal fault* in the "submit query" space. When the image appears, click to enlarge.

Explore the following *In-Terra-Active 2.0* CD-ROM module(s) and increase your understanding of key concepts and processes presented in this chapter.

▶ **SECTION: INTERIORS**
MODULE: CRUSTAL DEFORMATION

The rock shown was formed in a convergent plate boundary at high temperature and pressure under compresive stress. The deformation occurred at depth in the crust and resulted in the transformation of an existing rock to the rock shown. This detailed information about the rock shown is derived at a glance. **Question: How is it possible to tell so much about a rock simply by examining it for a few seconds?**

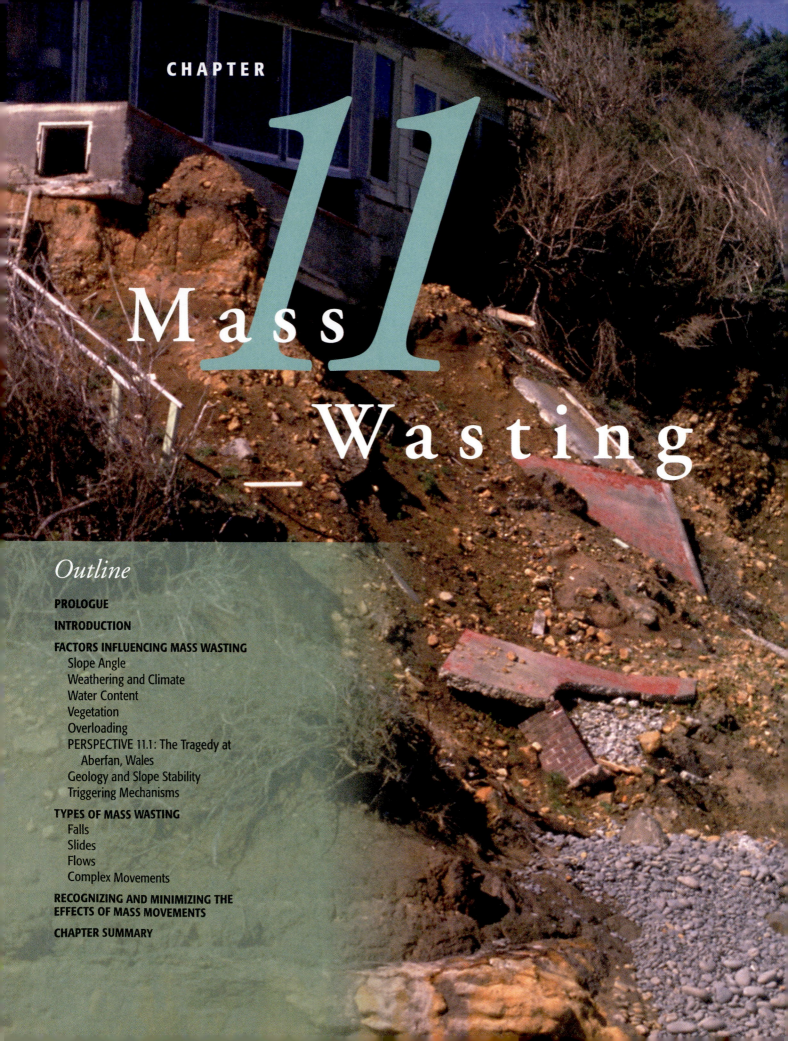

CHAPTER

11

Mass

Wasting

Outline

Prologue

On May 31, 1970, a devastating earthquake occurred about 25 km west of Chimbote, Peru. High in the Peruvian Andes, about 65 km to the east, the violent shaking tore loose a huge block of snow, ice, and rock from the north peak of Nevado Huascarán (6654 m), setting in motion one of this century's worst landslides. Free-falling for about 1000 m, this block of material smashed to the ground, displacing thousands of tons of rock and generating a gigantic debris avalanche (Figure 11.1). Hurtling down the mountain's steep glacial valley at speeds up to 320 km per hour, the avalanche—consisting of more than 50 million m^3 of mud, rock, and water—flowed over ridges 140 m high, obliterating everything in its path.

About 3 km east of the town of Yungay, part of the debris avalanche overrode the valley walls and within seconds buried Yungay, instantly killing more than 20,000 of its residents (Figure 11.1). The main mass of the flow continued down the valley, overwhelming the town of Ranrahirca and several other villages, burying about 5000 more people. When the flow reached the bottom of the valley, its momentum carried it across the Rio Santa and some 60 m up the opposite bank. In a span of roughly four minutes from the time of the initial ground shaking, some 25,000 people died, and most of the area's transportation, power, and communication networks were destroyed.

This sea cliff north of Bodega Bay, California, was undercut by waves during the winter of 1997–1998. As a result, part of the land slid into the ocean, damaging several houses.

FIGURE 11.1 An earthquake 65 km away triggered a landslide on Nevado Huascarán, Peru, that destroyed the towns of Yungay and Ranrahirca and killed more than 25,000 people.

Ironically, the only part of Yungay that was not buried was Cemetery Hill, where 92 people survived by running to its top. As tragic and devastating as this debris avalanche was, it was not the first time a destructive landslide had swept down this valley. In January 1962, another large chunk of snow, ice, and rock broke loose from the main glacier and generated a large debris avalanche that buried several villages and killed about 4000 people.

Introduction

THE topography of land areas are the result of the interaction between Earth's internal processes, type of rocks exposed at the surface, effects of weathering, and the erosional agents of water, ice, and wind. The specific type of landscape developed depends, in part, on which agent of erosion is dominant. Landslides (mass movements), which can be very destructive, are part of the normal adjustment of slopes to changing surface conditions.

Geologists use the term *landslide* in a general sense to cover a wide variety of mass movements that may cause loss of life, property damage, or a general disruption of human activities. In 218 B.C., avalanches in the European Alps buried 18,000 people; an earthquake-generated landslide in Hsian, China, killed an estimated 1 million people in 1556; and 7000 people died when mudflows and avalanches destroyed Huaraz, Peru, in 1941. What makes these mass movements so terrifying, and yet so fascinating, is that they almost always occur with little or no warning and are over in a very short time, leaving behind a legacy of death and destruction (Table 11.1).

Mass wasting (also called *mass movement*) is defined as the downslope movement of material under the direct influence of gravity. Most types of mass wasting are aided by weathering and usually involve surficial material. The material moves at rates ranging from almost imperceptible, as in the case of creep, to extremely fast, as in a rockfall or slide. Though water can play an important role, the relentless pull of gravity is the major force behind mass wasting.

Mass wasting is an important geologic process that can occur at any time and almost any place. While most people associate mass wasting with steep and unstable slopes, it can also occur on near-level land, given the right geologic conditions.

TABLE 11.1

Selected Landslides, Their Cause, and the Number of People Killed

DATE	LOCATION	TYPE	DEATHS
218 B.C.	Alps (European)	Avalanche—destroyed Hannibal's army	18,000
1512	Alps (Biasco)	Landslide—temporary lake burst	>600
1556	China (Hsian)	Landslides—earthquake triggered	1,000,000
1689	Austria (Montaton Valley)	Avalanche	>300
1806	Switzerland (Goldau)	Rock slide	457
1881	Switzerland (Elm)	Rockfall	115
1892	France (Haute-Savoie)	Icefall, mudflow	150
1903	Canada (Frank, Alberta)	Rock slide	70
1920	China (Kansu)	Landslides—earthquake triggered	~200,000
1936	Norway (Loen)	Rockfall into fiord	73
1941	Peru (Huaraz)	Avalanche and mudflow	7000
1959	USA (Madison Canyon, Montana)	Landslide—earthquake triggered	26
1962	Peru (Mt. Huascarán)	Ice avalanche and mudflow	~4000
1963	Italy (Vaiont Dam)	Landslide—subsequent flood	~2000
1966	Brazil (Rio de Janeiro)	Landslides	279
1966	United Kingdom (Aberfan, South Wales)	Debris flow—collapse of mining-waste tip	144
1970	Peru (Mt. Huascarán)	Rockfall and debris avalanche—earthquake triggered	25,000
1971	Canada (St. Jean-Vianney, Quebec)	Quick clays	31
1972	USA (West Virginia)	Landslide and mudflow—collapse of mining-waste tip	400
1974	Peru (Mayunmarca)	Rock slide and debris flow	430
1978	Japan (Myoko Kogen Machi)	Mudflow	12
1979	Indonesia (Sumatra)	Landslide	80
1980	USA (Washington)	Avalanche and mudflow	63
1981	Indonesia (West Irian)	Landslide—earthquake triggered	261
1981	Indonesia (Java)	Mudflow	252
1983	Iran (Northern area)	Landslide and avalanche	90
1987	El Salvador (San Salvador)	Landslide	1000
1988	Chile (Tupungatito area)	Mudflow	41
1989	Tadzhikistan	Mudflow—earthquake triggered	274
1989	Indonesia (West Irian)	Landslide—earthquake triggered	120
1991	Guatemala (Santa Maria)	Landslide	33
1994	Colombia (Paez River Valley)	Avalanche—earthquake triggered	>300
1995	Brazil (Northeastern area)	Mudflow	15
1996	Brazil (Recife)	Mudflow	49
1997	Peru (Ccocha and Pumaranra)	Mudflow	33

SOURCE: Data from J. Whittow, *Disasters: The Anatomy of Environmental Hazards* (Athens, Ga.: University of Georgia Press, 1979); *Geotimes;* and *Earth.*

Furthermore, while the rapid types of mass wasting, such as avalanches and mudflows, typically get the most publicity, the slow, imperceptible types, such as creep, usually do the greatest amount of property damage.

Factors Influencing Mass Wasting

When the gravitational force acting on a slope exceeds its resisting force, slope failure (mass wasting) occurs. The resisting forces helping to maintain slope stability include the slope material's strength and cohesion, the amount of internal friction between grains, and any external support of the slope (Figure 11.2). These factors collectively define a slope's **shear strength**.

Opposing a slope's shear strength is the force of gravity. Gravity operates vertically but has a component acting parallel to the slope, thereby causing instability (Figure 11.2). The greater a slope's angle, the greater the component of force acting parallel to the slope and the greater the chance for mass wasting. The steepest angle that a slope can maintain without collapsing is its *angle of repose*. At this angle, the shear strength of the slope's material exactly counterbalances

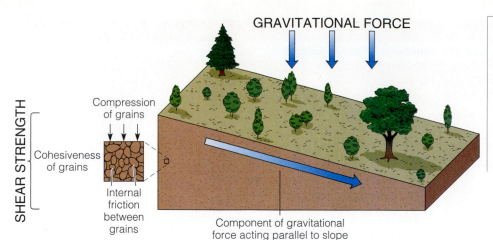

GRAVITATIONAL FORCE

SHEAR STRENGTH

Compression of grains

Cohesiveness of grains

Internal friction between grains

Component of gravitational force acting parallel to slope

FIGURE 11.2 A slope's shear strength depends on the slope material's strength and cohesiveness, the amount of internal friction between grains, and any external support of the slope. These factors promote slope stability. The force of gravity operates vertically but has a component acting parallel to the slope. When this force, which promotes instability, exceeds a slope's shear strength, slope failure occurs.

the force of gravity. For unconsolidated material, the angle of repose normally ranges from 25 to 40 degrees. Slopes steeper than 40 degrees usually consist of unweathered rock.

All slopes are in a state of *dynamic equilibrium,* which means that they are constantly adjusting to new conditions. While we tend to view mass wasting as a disruptive and usually destructive event, it is one of the ways that a slope adjusts to new conditions. Whenever a building or road is constructed on a hillside, the equilibrium of that slope is affected. The slope must then adjust, perhaps by mass wasting, to this new set of conditions.

Many factors can cause mass wasting: a change in slope angle, weakening of material by weathering, increased water content, changes in the vegetation cover, and overloading. Although most of these are interrelated, we will examine them separately for ease of discussion, but we will also show how they individually and collectively affect a slope's equilibrium.

SLOPE ANGLE

Slope angle is probably the major cause of mass wasting. Generally speaking, the steeper the slope, the less stable it is. Therefore, steep slopes are more likely to experience mass wasting than gentle ones.

A number of processes can oversteepen a slope. One of the most common is undercutting by stream or wave action (Figure 11.3). This removes the slope's base, increases the slope angle, and thereby increases the gravitational force acting parallel to the slope. Wave action, especially during storms, often results in mass movements along the shores of oceans or large lakes (see chapter opening photo).

Excavations for road cuts and hillside building sites are another major cause of slope failure (Figure 11.4). Grading the slope too steeply, or cutting into its side, increases the stress in the rock or soil until it is no longer strong enough to remain at the steeper angle, and mass movement ensues. Such action is analogous to undercutting by streams or waves and has the same result, thus explaining why so many mountain roads are plagued by frequent mass movements.

WEATHERING AND CLIMATE

Mass wasting is more likely to occur in loose or poorly consolidated slope material than in bedrock. As soon as solid rock is exposed at Earth's surface, weathering begins to disintegrate and decompose it, reducing its shear strength and increasing its susceptibility to mass wasting. The deeper the weathering zone extends, the greater the likelihood of some type of mass movement.

Recall that some rocks are more susceptible to weathering than others and that climate plays an important role in the rate and type of weathering. In the tropics, where temperatures are high and considerable rainfall occurs, the

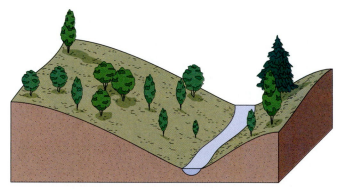

(a)

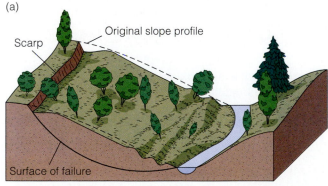

Scarp

Original slope profile

Surface of failure

(b)

FIGURE 11.3 Undercutting by stream erosion (a) removes a slope's base, which increases the slope angle and (b) can lead to slope failure.

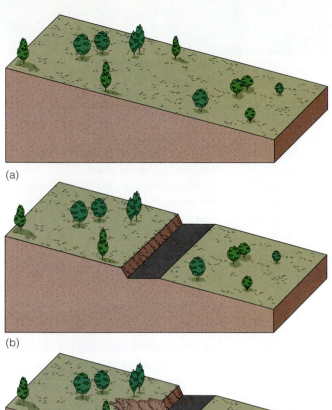

(a)

(b)

(c)

FIGURE 11.4 (a) Highway excavations disturb the equilibrium of a slope by (b) removing a portion of its support as well as oversteepening it at the point of excavation. (c) Such action can result in frequent landslides.

tion between grains, contributing to a loss of cohesion. For example, slopes composed of dry clay are usually quite stable, but when wetted they quickly lose cohesiveness and internal friction and become an unstable slurry. This occurs because clay, which can hold large quantities of water, consists of platy particles that easily slide over each other when wet. For this reason, clay beds are frequently the slippery layer along which overlying rock units slide downslope (see Perspective 11.1).

VEGETATION

Vegetation affects slope stability in several ways. By absorbing the water from a rainstorm, vegetation decreases water saturation of a slope's material that would otherwise lead to a loss of shear strength. Vegetation's root system also helps stabilize a slope by binding soil particles together and holding the soil to bedrock.

The removal of vegetation by either natural or human activity is a major cause of many mass movements. Summer brush and forest fires in southern California frequently leave the hillsides bare of vegetation. Fall rainstorms saturate the ground, causing mudslides that do tremendous damage and cost millions of dollars to clean up (Figure 11.5).

OVERLOADING

Overloading is almost always the result of human activity and typically results from dumping, filling, or piling up of material. Under natural conditions, a material's load is carried by its grain-to-grain contacts, with the friction between the grains maintaining a slope. The additional weight created by overloading increases the water pressure within

effects of weathering extend to depths of several tens of meters, and mass movements most commonly occur in the deep weathering zone. In arid and semiarid regions, the weathering zone is usually considerably shallower. Nevertheless, intense, localized cloudbursts can drop large quantities of water on an area in a short time. With little vegetation to absorb this water, runoff is rapid and frequently results in mudflows.

WATER CONTENT

The amount of water in rock or soil influences slope stability. Large quantities of water from melting snow or heavy storms greatly increase the likelihood of slope failure. The additional weight that water adds to a slope can be enough to cause mass movement. Furthermore, water percolating through a slope's material helps decrease fric-

FIGURE 11.5 A California Highway Patrol officer stands on top of a 2-m-high wall of mud that rolled over a patrol car near the Golden State Freeway on October 23, 1987. Flooding and mudslides also trapped other vehicles and closed the freeway.

The Tragedy at Aberfan, Wales

The debris brought out of underground coal mines in southern Wales typically consists of a wet mixture of various sedimentary rock fragments. This material is usually dumped along the nearest valley slope where it builds up into large waste piles called *tips.* A tip is fairly stable as long as the material composing it is relatively dry and its sides are not too steep.

Between 1918 and 1966, seven large tips composed of mine debris had been built at various elevations on the valley slopes above the small coal-mining village of Aberfan. Shortly after 9:00 A.M. on October 21, 1966, the 250-m-high, rain-soaked Tip No. 7 collapsed, and a black sludge flowed down the valley with the roar of a loud train (Figure 1). Before it came to a halt 800 m from its starting place, the flow had destroyed two farm cottages, crossed a canal, and buried Pantglas Junior School, suffocating virtu-

ally all the children of Aberfan. A total of 144 people died in the flow, among them 116 children who had gathered for morning assembly in the school.

After the disaster, everyone asked, "Why did this tragedy occur, and could it have been prevented?" The subsequent investigation revealed that no stability studies had ever been made on the tips and that repeated warnings about potential failure of the tips, as well as previous slides, had all been ignored.

In 1939, 8 km to the south, a tip constructed under conditions almost identical to those of Tip No. 7 collapsed. Luckily, no one was injured, but unfortunately the failure was soon forgotten and the Aberfan tips continued to grow. In 1944 Tip No. 4 failed, and again no one was injured.

In 1958 Tip No. 7 was sited solely on the basis of available space, with no regard to the area's geology. Despite previ-

ous tip failures and warnings of slope failure by tip workers and others, mine debris was being piled onto Tip No. 7 until the day of the disaster.

What exactly caused Tip No. 7 and the others to fail? The official investigation revealed that the foundation of the tips had become saturated with water from the springs over which they were built. In the case of the collapsed tips, pore pressure from the water exceeded the friction between grains, and the entire mass liquefied like a "quicksand." Behaving like a liquid, the mass quickly moved downhill, spreading out laterally. As it flowed, water escaped from the mass, and the sedimentary particles regained their cohesion.

Following the inquiry, it was recommended that a National Tip Safety Committee be established to assess the dangers of existing tips and advise on the construction of new tip sites.

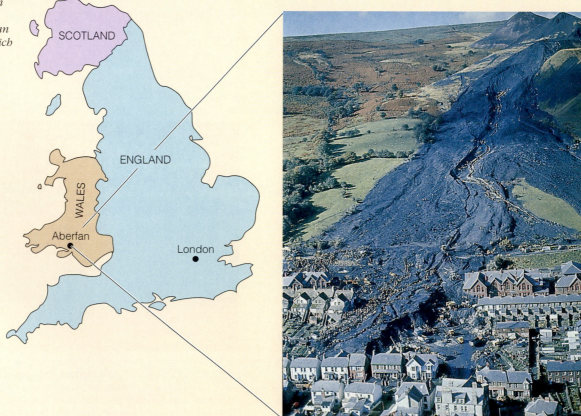

Figure 1 Location map and aerial view of the Aberfan tip disaster in which 144 people died.

SCOTLAND

ENGLAND

WALES

Aberfan

London

the material, which in turn decreases its shear strength, thereby weakening the slope material. If enough material is added, the slope will eventually fail, sometimes with tragic consequences.

GEOLOGY AND SLOPE STABILITY

The relationship between topography and the geology of an area is important in determining slope stability. If the rocks underlying a slope dip in the same direction as the slope, mass wasting is more likely to occur than if the rocks are horizontal or dip in the opposite direction (Figure 11.6). When the rocks dip in the same direction as the slope, water can percolate along the various bedding planes and decrease the cohesiveness and friction between adjacent rock layers (Figure 11.6a). This is particularly true when clay layers are present because clay becomes slippery when wet.

Even if the rocks are horizontal or dip in a direction opposite to that of the slope, joints may dip in the same direction as the slope. Water migrating through them weathers the rock and expands these openings until the weight of the overlying rock causes it to fall (Figure 11.6b).

TRIGGERING MECHANISMS

The factors discussed thus far all contribute to slope instability, but most—though not all—rapid mass movements are triggered by a force that temporarily disturbs slope equilibrium. The most common triggering mechanisms are

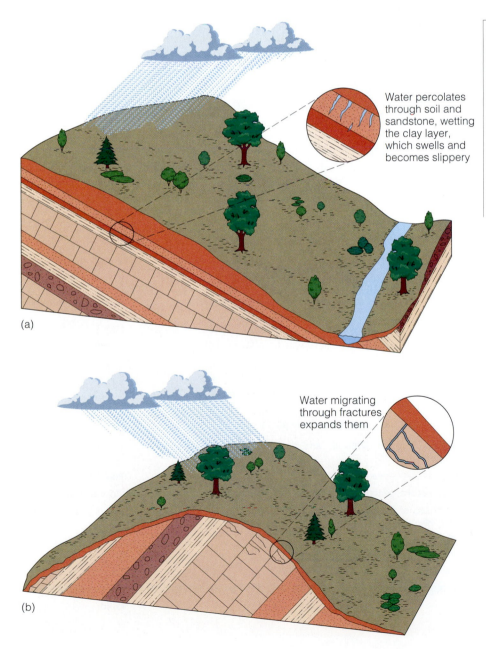

Water percolates through soil and sandstone, wetting the clay layer, which swells and becomes slippery

(a)

Water migrating through fractures expands them

(b)

FIGURE **11.6** (a) Rocks dipping in the same direction as a hill's slope are particularly susceptible to mass wasting. Undercutting of the base of the slope by a stream removes support and steepens the slope at the base. Water percolating through the soil and into the underlying rock increases its weight and, if clay layers are present, wets the clay making them slippery. (b) Fractures dipping in the same directions as a slope are enlarged by chemical weathering, which can remove enough material to cause mass wasting.

strong vibrations from earthquakes and excessive amounts of water from a winter snow melt or a heavy rainstorm.

Volcanic eruptions, explosions, and even loud claps of thunder may also be enough to trigger a landslide if the slope is sufficiently unstable. Many *avalanches,* which are rapid movements of snow and ice down steep mountain slopes, are triggered by the sound of a loud gunshot or, in rare cases, even a person's shout.

Types Of Mass Wasting

GEOLOGISTS recognize a variety of mass movements (Table 11.2). Some are of one distinct type, whereas others are a combination of different types. It is not uncommon for one type of mass movement to change into another along its course. Even though many slope failures are combinations of different materials and movements, classifying them according to their dominant behavior is still convenient.

Mass movements are generally classified on the basis of three major criteria (Table 11.2): (1) rate of movement (rapid or slow); (2) type of movement (primarily falling, sliding, or flowing); and (3) type of material involved (rock, soil, or debris).

Rapid mass movements involve a visible movement of material. Such movements usually occur quite suddenly, and the material moves very quickly downslope. Rapid mass movements are potentially dangerous and frequently result in loss of life and property damage. Most rapid mass movements occur on relatively steep slopes and can involve rock, soil, or debris.

Slow mass movements advance at an imperceptible rate and are usually only detectable by the effects of their movement such as tilted trees and power poles or cracked foundations. Although rapid mass movements are more dramatic, slow mass movements are responsible for the downslope transport of a much greater volume of weathered material.

FALLS

Rockfalls are a common type of extremely rapid mass movement in which rocks of any size fall through the air (Figure 11.7). Rockfalls occur along steep canyons, cliffs, and road cuts and build up accumulations of loose rocks and rock fragments, called *talus,* at their base (see Figure 6.5).

Rockfalls result from failure along joints or bedding planes in the bedrock and are commonly triggered by natural or human undercutting of slopes or by earthquakes. Rockfalls range in size from small rocks falling from a cliff to massive falls involving millions of cubic meters of debris that destroy buildings, bury towns, and block highways.

Rockfalls are a particularly common hazard in mountainous areas where roads have been built by blasting and grading through steep hillsides of bedrock (Figure 11.8). Slopes particularly prone to rockfalls are sometimes covered with wire mesh in an effort to prevent dislodged rocks from falling to the road below. Another tactic is to put up wire mesh fences along the base of the slope to catch or slow down bouncing or rolling rocks.

TABLE 11.2

Classification of Mass Movements and Their Characteristics

TYPE OF MOVEMENT	SUBDIVISION	CHARACTERISTICS	RATE OF MOVEMENT
Falls	Rockfall	Rocks of any size fall through the air from steep cliffs, canyons, and road cuts	Extremely rapid
Slides	Slump	Movement occurs along a curved surface of rupture; most commonly involves unconsolidated or weakly consolidated material	Extremely slow to moderate
	Rock slide	Movement occurs along a generally planar surface	Rapid to very rapid
Flows	Mudflow	Consists of at least 50% silt- and clay-sized particles and up to 30% water	Very rapid
	Debris flow	Contains larger-sized particles and less water than mudflows	Rapid to very rapid
	Earthflow	Thick, viscous, tongue-shaped mass of wet regolith	Slow to moderate
	Quick clays	Composed of fine silt and clay particles saturated with water; when disturbed by a sudden shock, lose their cohesiveness and flow like a liquid	Rapid to very rapid
	Solifluction	Water-saturated surface sediment	Slow
	Creep	Downslope movement of soil and rock	Extremely slow
Complex movements		Combination of different movement types	Slow to extremely rapid

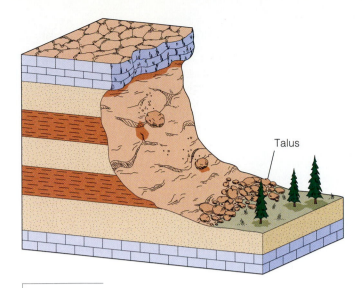

FIGURE 11.7 Rockfalls result from failure along cracks, fractures, or bedding planes in the bedrock and are common features in areas of steep cliffs.

Talus

FIGURE 11.8 Cutting into the hillside to construct this portion of the Pan American Highway in Mexico resulted in a rockfall that completely blocked the road. *(Photo courtesy of R. V. Dietrich.)*

SLIDES

A **slide** involves movement of material along one or more surfaces of failure. The type of material may be soil, rock, or a combination of the two, and it may break apart during movement or remain intact. A slide's rate of movement can vary from extremely slow to very rapid (Table 11.2).

Two types of slides are generally recognized: (1) slumps or rotational slides, in which movement occurs along a curved surface; and (2) rock or block slides, which move along a more-or-less planar surface.

A **slump** involves the downward movement of material along a curved surface of rupture and is characterized by the backward rotation of the slump block (Figure 11.9). Slumps usually occur in unconsolidated or weakly consolidated material and range in size from small individual sets, such as occur along stream banks, to massive, multiple sets that affect large areas and cause considerable damage.

Slumps can be caused by a variety of factors, but the most common is erosion along the base of a slope, which removes support for the overlying material. This local steepening may be caused naturally by stream erosion along its banks (Figure 11.9) or by wave action at the base of a coastal cliff. Slope oversteepening can also be caused by human activity, such as the construction of highways and housing developments. Slumps are particularly prevalent along highway cuts where they are generally the most frequent type of slope failure observed.

While many slumps are merely a nuisance, large-scale slumps involving populated areas and highways can cause extensive damage. Such is the case in coastal southern California where slumping and sliding have been a constant problem. Many areas along the coast are underlain by poorly to weakly consolidated silts, sands, and gravels, interbedded with clay layers, some of which are weathered

ash falls. In addition, southern California is tectonically active so that many of these deposits are cut by faults and joints, which allow the infrequent rains to percolate downward rapidly, wetting and lubricating the clay layers.

Southern California lies in a semiarid climate and is dry most of the year. When it does rain, typically between November and March, large amounts of rain can fall in a short time. Thus, the ground quickly becomes saturated, leading to landslides along steep canyon walls as well as along coastal cliffs (Figure 11.10). Most of the slope failures along the southern California coast are the result of slumping.

A **rock** or *block* **slide** occurs when rocks move downslope along a more-or-less planar surface. Most rock slides take place because the local slopes and rock layers dip in the same direction (Figure 11.11), although they can also occur along fractures parallel to a slope. Rock slides are common occurrences along the southern California coast. At Point Fermin, seaward-dipping rocks with interbedded slippery clay layers are undercut by waves, causing numerous slides (Figure 11.12).

FLOWS

Mass movements in which material flows as a viscous fluid or displays plastic movement are termed *flows*. Their rate of movement ranges from extremely slow to extremely

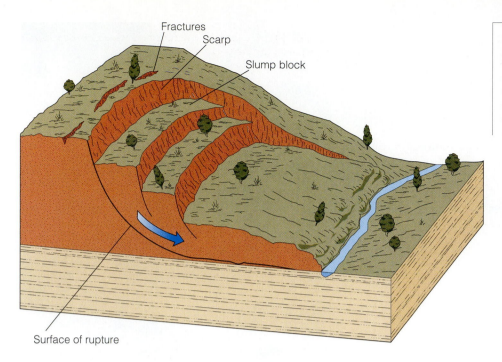

Fractures
Scarp
Slump block

Surface of rupture

FIGURE 11.9 In a slump, material moves downward along a curved surface of a rupture, causing the slump block to rotate backward. Most slumps involve unconsolidated or weakly consolidated material and are typically caused by erosion along the slope's base.

FIGURE 11.10 Undercutting of steep sea cliffs by wave action resulted in massive slumping in the Pacific Palisades area of southern California on March 31 and April 3, 1958. Highway 1 was completely blocked. Note the heavy earthmoving equipment (bottom of photo) for scale.

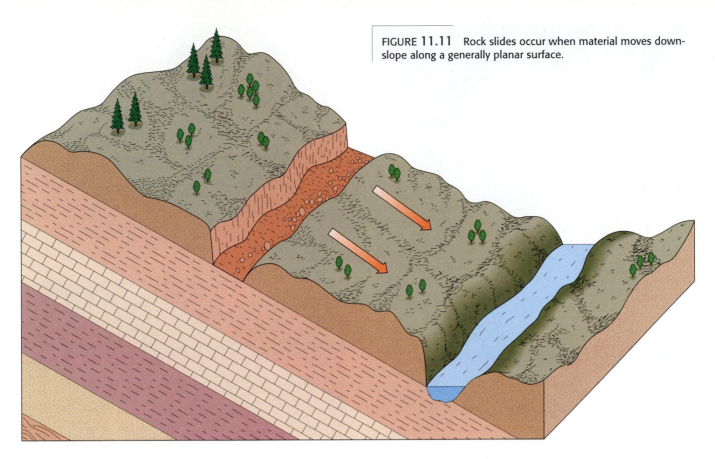

rapid (Table 11.2). In many cases, mass movements begin as falls, slumps, or slides and change into flows further downslope.

Of the major mass movement types, **mudflows** are the most fluid and move most rapidly (at speeds up to 80 km per hour). They consist of at least 50% silt- and clay-sized material combined with a significant amount of water (up to 30%). Mudflows are common in arid and semiarid environments where they are triggered by heavy rainstorms that quickly saturate the regolith, turning it into a raging flow of mud that engulfs everything in its path. Mudflows can also occur in mountain regions (Figure 11.13) and in areas covered by volcanic ash where they can be particularly destructive (see Chapter 5). Because mudflows are so fluid, they generally follow preexisting channels until the slope decreases or the channel widens, at which point they fan out.

Debris flows are composed of larger-sized particles than those in mudflows and do not contain as much water. Consequently, they are usually more viscous than mudflows, typically do not move as rapidly, and rarely are confined to preexisting channels. Debris flows can be just as damaging, though, because they can transport large objects.

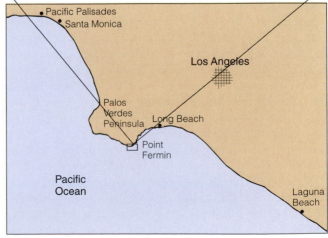

FIGURE 11.12 A combination of interbedded clay beds that become slippery when wet, rocks dipping in the same direction as the slope of the sea cliffs, and undercutting of the sea cliffs by wave action has caused numerous rock slides and slumps at Point Fermin, California. *(Photo courtesy of Eleanora I. Robbins, U.S. Geological Survey.)*

Types of Mass Wasting 233

FIGURE 11.13   A mudflow near Estes Park, Colorado.

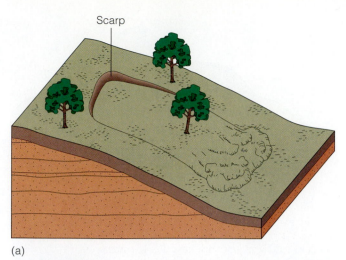

(a)

(b)

FIGURE 11.14   (a) Earthflows form tongue-shaped masses of wet regolith that move slowly downslope. They occur most commonly in humid climates on grassy soil-covered slopes. (b) An earthflow near Baraga, Michigan.

Earthflows move more slowly than either mudflows or debris flows. An earthflow slumps from the upper part of a hillside, leaving a scarp, and flows slowly downslope as a thick, viscous, tongue-shaped mass of wet regolith (Figure 11.14). Like mudflows and debris flows, earthflows can be of any size and are frequently destructive. They occur most commonly in humid climates on grassy soil-covered slopes following heavy rains.

Some clays spontaneously liquefy and flow like water when they are disturbed. Such **quick clays** have caused serious damage and loss of lives in Sweden, Norway, eastern Canada, and Alaska (Table 11.1). Quick clays are composed of fine silt and clay particles made by the grinding action of glaciers. Geologists think these fine sediments were originally deposited in a marine environment where their pore space was filled with salt water. The ions in salt water helped establish strong bonds between the clay particles, thus stabilizing and strengthening the clay. When the clays were subsequently uplifted above sea level, the salt water was flushed out by fresh groundwater, reducing the effectiveness of the ionic bonds between the clay particles and thereby reducing the overall strength and cohesiveness of the clay. Consequently, when the clay is disturbed by a sudden shock or shaking, it essentially turns to a liquid and flows.

An example of the damage that can be done by quick clays occurred in the Turnagain Heights area of Anchorage, Alaska, in 1964 (Figure 11.15). Underlying most of the Anchorage area is the Bootlegger Cove Clay, a massive clay unit of poor permeability. Because the Bootlegger Cove Clay forms a barrier preventing groundwater from flowing through the adjacent glacial deposits to the sea, considerable hydraulic pressure builds up behind the clay. Some of this water has flushed out the salt water in the clay and also has saturated the lenses of sand and silt associated with the clay beds. When the 8.6-magnitude Good Friday earthquake struck on March 27, 1964, the shaking turned parts of the Bootlegger Cove Clay into a quick clay and precipitated a series of massive slides in the coastal bluffs that destroyed most of the homes in the Turnagain Heights subdivision (Figure 11.15b).

Solifluction is the slow downslope movement of water-saturated surface sediment. Solifluction can occur in any climate where the ground becomes saturated with water but is most common in areas of permafrost. *Permafrost,* ground that remains permanently frozen, covers nearly 20% of the world's land surface (Figure 11.16a). During the warmer season when the upper portion of the permafrost thaws, water and surface sediment form a soggy mass that

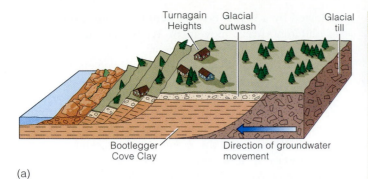

(a)

(b)

FIGURE 11.15 (a) Groundshaking by the 1964 Alaska earthquake turned parts of the Bootlegger Cove Clay into a quick clay, causing numerous slides. (b) Low-altitude photograph of the Turnagain Heights subdivision of Anchorage shows some of the numerous landslide fissures that developed, as well as the extensive damage to buildings in the area. The remains of the Four Seasons apartment building can be seen in the background.

flows by solifluction and produces a characteristic lobate topography (Figure 11.16b).

Construction of the Alaska pipeline from the oil fields in Prudhoe Bay to the ice-free port of Valdez raised numerous concerns over the effect it might have on the permafrost and the potential for solifluction. Some thought that oil flowing through the pipeline would be warm enough to melt the permafrost, causing the pipeline to sink farther into the ground and possibly rupture. After numerous studies were conducted, scientists concluded that the pipeline, completed in 1977, could safely be buried for more than half of its 1280-km length; where melting of the permafrost might cause structural problems to the pipe, it was insulated and installed above ground.

FIGURE 11.16 (a) Distribution of permafrost areas in the Northern Hemisphere. (b) Solifluction flows near Suslositna Creek, Alaska, show the typical lobate topography that is characteristic of solifluction conditions.

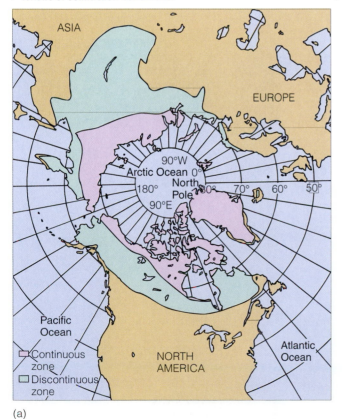

(a)

(b)

Creep, the slowest type of flow, is the most widespread and significant mass wasting process in terms of the total amount of material moved downslope and the monetary damage it does annually. Creep involves extremely slow downhill movement of soil or rock. Although it can occur anywhere and in any climate, it is most effective and significant as a geologic agent in humid regions. In fact, it is the most common form of mass wasting in the southeastern United States and the southern Appalachian Mountains.

Because the rate of movement is essentially imperceptible, we are frequently unaware of creep's existence until we notice its effects: tilted trees and power poles, broken streets and sidewalks, cracked retaining walls or foundations (Figure 11.17). Creep usually involves the whole hillside and probably occurs, to some extent, on any weathered or soil-covered, sloping surface.

Creep is not only difficult to recognize but also to control. Although engineers can sometimes slow or stabilize creep, many times the only course of action is to simply avoid the area if at all possible or, if the zone of creep is relatively thin, design structures that can be anchored into the solid bedrock.

COMPLEX MOVEMENTS

Recall that many mass movements are combinations of different movement types. When one type is dominant, the movement can be classified as one of the movements described thus far. If several types are more-or-less equally involved, it is called a **complex movement**.

The most common type of complex movement is the slide-flow in which there is sliding at the head and then some type of flowage farther along its course. Most slide-flow landslides involve well-defined slumping at the head,

followed by a debris flow or earthflow. Any combination of different mass movement types can, however, be classified as a complex movement.

A *debris avalanche* is a complex movement that often occurs in very steep mountain ranges. Debris avalanches typically start out as rockfalls when large quantities of rock, ice, and snow are dislodged from a mountainside, frequently as a result of an earthquake. The material then slides or flows down the mountainside, picking up additional surface material and increasing in speed. The 1970 Peru earthquake set in motion the debris avalanche that destroyed the town of Yungay (see the Prologue).

Recognizing and Minimizing the Effects of Mass Movements

THE most important factor in eliminating or minimizing the damaging effects of mass wasting is a thorough geologic investigation of the region in question. In this way, former landslides and areas susceptible to mass movements can be identified and perhaps avoided. By assessing the risks of possible mass wasting before construction begins, steps can be taken to eliminate or minimize the effects of such events.

Identifying areas with a high potential for slope failure is important in any hazard-assessment study; these studies include identifying former landslides as well as sites of potential mass movement. Scarps, open fissures, displaced or tilted objects, a hummocky surface, and sudden changes in

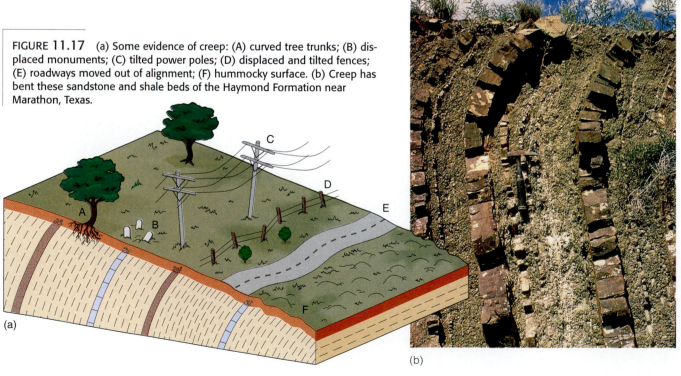

FIGURE 11.17 (a) Some evidence of creep: (A) curved tree trunks; (B) displaced monuments; (C) tilted power poles; (D) displaced and tilted fences; (E) roadways moved out of alignment; (F) hummocky surface. (b) Creep has bent these sandstone and shale beds of the Haymond Formation near Marathon, Texas.

(a)

(b)

vegetation are some of the features indicating former land-slides or an area susceptible to slope failure. The effects of weathering, erosion, and vegetation may, however, obscure the evidence for previous mass wasting.

Soil and bedrock samples are also studied, both in the field and laboratory, to assess such characteristics as composition, susceptibility to weathering, cohesiveness, and ability to transmit fluids. These studies help geologists and engineers predict slope stability under a variety of conditions.

Although most large mass movements usually cannot be prevented, geologists and engineers can employ various methods to minimize the danger and damage resulting from them. Because water plays such an important role in many landslides, one of the most effective and inexpensive ways to reduce the potential for slope failure and to increase existing slope stability is through surface and subsurface drainage of a hillside. Drainage serves two purposes. It reduces the weight of the material likely to slide and increases the shear strength of the slope material by lowering pore pressure.

Surface waters can be drained and diverted by ditches, gutters, or culverts designed to direct water away from slopes. Drainpipes perforated along one surface and driven into a hillside can help remove subsurface water (Figure 11.18). Finally, planting vegetation on hillsides helps stabilize slopes by holding the soil together and reducing the amount of water in the soil.

Another way to help stabilize a hillside is to reduce its slope. Recall that overloading or oversteepening by grading are common causes of slope failure. By reducing the angle

of a hillside, the potential for slope failure is decreased. Two methods are usually employed to reduce a slope's angle. In the *cut-and-fill* method, material is removed from the upper part of the slope and used as fill at the base, thus providing a flat surface for construction and reducing the slope (Figure 11.19a). The second method, which is called *benching,* involves cutting a series of benches or steps into a hillside (Figure 11.19b). This process reduces the overall average slope, and the benches serve as collecting sites for small landslides or rockfalls that might occur. Benching is most commonly used on steep hillsides in conjunction with a system of surface drains to divert runoff.

In some situations, retaining walls can be constructed to provide support for the base of the slope (Figure 11.20). These are usually anchored well into bedrock, backfilled with crushed rock, and provided with drain holes to prevent the buildup of water pressure in the hillside.

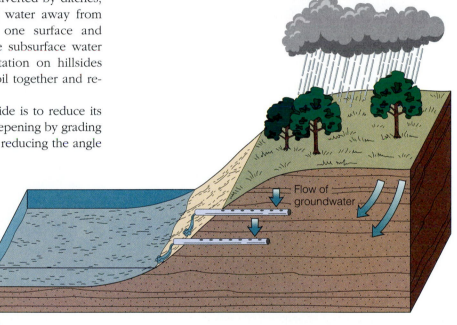

Flow of
groundwater

(a)

FIGURE 11.18 (a) Driving drainpipes that are perforated on one side into a hillside with the perforated side up can remove some subsurface water and help stabilize the hillside. (b) A drainpipe driven into the hillside at Point Fermin, California, helps remove subsurface water and stabilize the slope.

(b)

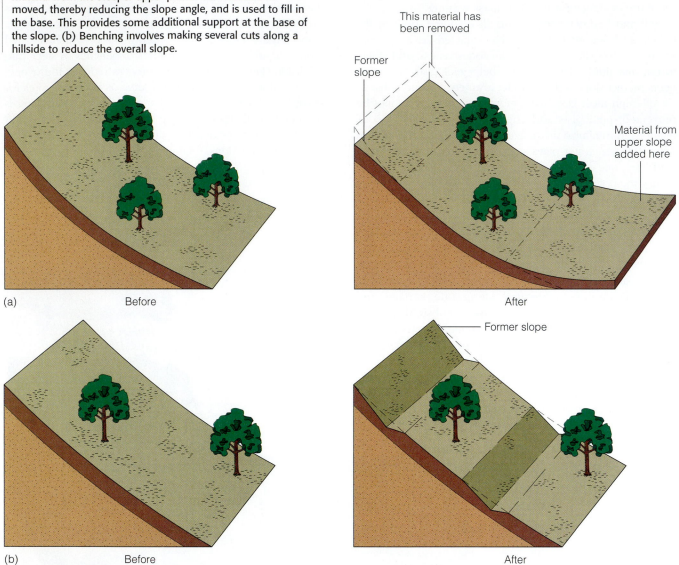

FIGURE **11.19** Two common methods used to help stabilize a hillside and reduce its slope. (a) In the cut-and-fill method, material from the steeper upper part of the hillside is removed, thereby reducing the slope angle, and is used to fill in the base. This provides some additional support at the base of the slope. (b) Benching involves making several cuts along a hillside to reduce the overall slope.

This material has been removed

Former slope

Material from upper slope added here

(a)   Before

After

Former slope

(b)   Before

After

FIGURE **11.20**   (a) Retaining walls anchored into bedrock, backfilled with gravel, and provided with drainpipes can support a slope's base and reduce landslides. (b) Steel retaining wall built to stabilize the slope and keep falling and sliding rocks off of the highway.

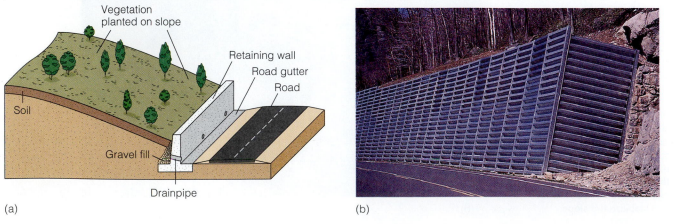

Vegetation planted on slope

Retaining wall

Road gutter

Road

Soil

Gravel fill

Drainpipe

(a)

(b)

1. Mass wasting is the downslope movement of material under the influence of gravity. It occurs when the gravitational force acting parallel to a slope exceeds the slope's strength.

2. Mass wasting frequently results in loss of life, as well as causing millions of dollars in damage annually.

3. Mass wasting can be caused by many factors including slope angle, weathering of slope material, water content, overloading, and removal of vegetation. Usually, several of these factors in combination contribute to slope failure.

4. Mass movements are generally classified on the basis of their rate of movement (rapid versus slow), type of movement (falling, sliding, or flowing), and type of material (rock, soil, or debris).

5. Rockfalls are a common mass movement in which rocks free-fall.

6. Two types of slides are recognized. Slumps are rotational slides involving movement along a curved surface; they are most common in poorly consolidated or unconsolidated material. Rock slides occur when movement takes place along a more-or-less planar surface; they usually involve solid pieces of rock.

7. Several types of flows are recognized on the basis of their rate of movement (rapid vs. slow), type of material (rock, sediment, soil), and amount of water.

8. Mudflows consist of mostly clay- and silt-sized particles and contain more than 30% water. They are most common in semiarid and arid environments and generally follow preexisting channels.

9. Debris flows are composed of larger particles and contain less water than mudflows. They are more viscous and do not flow as rapidly as mudflows.

10. Earthflows move more slowly than either debris flows or mudflows; they move downslope as thick, viscous, tongue-shaped masses of wet regolith.

11. Quick clays are clays that spontaneously liquefy and flow like water when they are disturbed.

12. Solifluction is the slow downslope movement of water-saturated surface material and is most common in areas of permafrost.

13. Creep, the slowest type of flow, is the imperceptible downslope movement of soil or rock. Creep is the most widespread of all types of mass wasting.

14. Complex movements are combinations of different types of mass movements in which one type is not dominant. Most complex movements involve sliding and flowing.

15. The most important factor in reducing or eliminating the damaging effects of mass wasting is a thorough geologic investigation to outline areas susceptible to mass movements.

16. Slopes can be stabilized by retaining walls, draining excess water, regrading slopes, and planting vegetation.

Important Terms

complex movement	mudflow	shear strength
creep	quick clay	slide
debris flow	rapid mass movement	slow mass movement
earthflow	rockfall	slump
mass wasting	rock slide	solifluction

Review Questions

1. Shear strength includes:
   a. _____ the strength and cohesion of material;
   b. _____ the amount of internal friction between grains;
   c. _____ gravity;
   d. _____ all of these;
   e. _____ answers (a) and (b).

2. Which of the following is a factor influencing mass wasting?
   a. _____ gravity;
   b. _____ weathering;
   c. _____ slope angle;
   d. _____ water content;
   e. _____ all of these.

3. Which of the following factors can actually enhance slope stability?
   a. _____ water content;
   b. _____ vegetation;
   c. _____ overloading;
   d. _____ rocks dipping in the same direction as the slope;
   e. _____ none of these.

4. Movement of material along a surface or surfaces of failure is a:
   a. _____ slide;
   b. _____ fall;
   c. _____ flow;
   d. _____ solifluction;
   e. _____ none of these.

5. Downslope movement along an essentially planar surface is a(n):
   a. _____ slump;
   b. _____ rockfall;
   c. _____ earthflow;
   d. _____ landslide;
   e. _____ rock slide.

6. The most widespread and costly type of mass wasting in terms of total material moved and monetary damage is:
   a. _____ creep;
   b. _____ solifluction;
   c. _____ mudflow;
   d. _____ debris flow;
   e. _____ slumping.

7. Which of the following features indicates former landslides or areas susceptible to slope failure?
   a. _____ displaced objects;
   b. _____ scarps;
   c. _____ hummocky surfaces;
   d. _____ open fissures;
   e. _____ all of these.

8. Mass wasting can occur:
   a. _____ on gentle slopes;
   b. _____ on steep slopes;
   c. _____ in flat-lying areas;
   d. _____ all of these;
   e. _____ none of these.

9. A type of mass wasting common in mountainous regions in which talus accumulates is:
   a. _____ creep;
   b. _____ solifluction;
   c. _____ rock falls;
   d. _____ slides;
   e. _____ mudflows.

10. Which of the following helps stabilize a slope?
    a. _____ surface and subsurface drainage;
    b. _____ planting vegetation;
    c. _____ reducing its slope;
    d. _____ construction of a retaining wall;
    e. _____ all of these.

11. What are the forces that help maintain slope stability?

12. What roles do climate and weathering play in mass wasting?

13. Discuss how the relationship between topography and the underlying geology affects slope stability.

14. Where are rockfalls most common, and why?

15. Why are slumps particularly common along road cuts and fills?

16. Why are quick clays so dangerous?

17. What precautions must be taken when building in permafrost areas?

18. Why is creep so prevalent, and why does it do so much damage?

Points to Ponder

1. What features do you look for to determine if the site on which you want to build a house is safe?

2. Do you think it will ever be possible to predict mass wasting events? Explain.

For these web site addresses, along with current updates and exercises, log on to

http://www.wadsworth.com/geo

▶ **U.S. GEOLOGICAL SURVEY EARTH SCIENCE IN THE PUBLIC SERVICE: GEOLOGIC HAZARDS**

The home page of this site contains links to earthquakes, landslides, other geologic hazards, and geomagnetism.

1. Click on the *Landslides* site. Then click on *The National Landslide Information Center (NLIC)* site, which contains links to landslide information, landslide publications, and interesting landslide images and topics. Click on the *Landslide information* site. What is the latest landslide, and what type of damage did it do?

2. Click on the *Interesting landslide topics* site. Under the Landslide Information heading, click on the *What to do and look for* site. What features might be noticed prior to major landsliding?

3. Click on the *Activities of the NLIC* site. What does the NLIC do?

▶ **JCP GEOLOGISTS, INC.**

This site, maintained by a company that provides disclosure reports for the real estate profession in California, contains interesting and useful geologic information and links to other geologic sites. Click on the *Geologic Information* site. Under the Geologic Hazard Information heading, click on the *Landslides* site. Read the information presented. Can landslides be predicted? What are some of the ways landslides might be prevented? What are some of the ways that the buying and selling of property are affected by landslides?

CD-ROM Exploration

Explore the following *In-Terra-Active 2.0* CD-ROM module(s) and increase your understanding of key concepts and processes presented in this chapter.

▶ **SECTION: SURFACE**
MODULE: MASS WASTING

During the winter months of 1998, steady rains up and down the California coast created a record number of rock slides, landslides, and mud flows—different types of mass wasting. In this picture, you can see the devastating effects of mass wasting on an estate in Pacifica, California

One TV meteorologist wished out loud that El Nino would get it all over with at once—delivering a single, huge thunderstorm instead of the light but persistent rains. He argued that many houses would be saved that way, even if the storm sent down in one violent day the same amount of rain that was, in fact, being spread out over several months. **Question: What geologic principles would help support this position?**

Running 12 Water

Outline

Prologue

Tahquamenon Falls in northern Michigan.
(Photo courtesy of R. V. Dietrich.)

According to one report, the flooding in the midwestern United States from late June to August 1993 caused $15 to $20 billion in property damage. Fifty people died as a result of the flooding, 70,000 were left homeless, and more than 200 counties in several states, including every county in Iowa, were declared disaster areas. In addition to Iowa, flooding also occurred in South Dakota, North Dakota, Minnesota, Wisconsin, Nebraska, Missouri, Illinois, and Kansas.

Despite heroic efforts by thousands of volunteers and the National Guard to stabilize levees with sandbags, levees failed everywhere or were simply overtopped by the rising floodwaters. By the end of the first week in July, at least 10 million acres of farmland had been flooded, with more rain expected. On July 16, an additional 17.8 cm of rain fell in North Dakota and Minnesota, and 15.2 cm more fell in South Dakota. By August more than 92,000 km² had been flooded. Grafton, Illinois—at the juncture of the Mississippi and Illinois Rivers—was 80% underwater. And much of St. Charles County, Missouri—near the confluence of the Mississippi and Missouri Rivers—was so extensively flooded that 8000 people were evacuated from this county alone (Figure 12.1).

(a)

FIGURE 12.1 (a) Portage des Sioux, St. Charles County, Missouri, on July 16, 1993. The channel of the Mississippi River is at the far right. Only eight homes in the town were not flooded or were only slightly damaged by flooding. (b) Floodwaters in Portage des Sioux covered the 5.5-m-high pedestal of this statue on the banks of the Mississippi River.

(b)

Obviously, the direct cause of flooding was too much water for the Mississippi and Missouri Rivers and their tributaries to handle. But the reason so much water was present was the unusual weather in the Midwest. Thunderstorms that are normally distributed over a much larger area simply dumped their precipitation in a much smaller region, which received as much as $1\frac{1}{2}$ to 2 times the normal amount of rain.

One important lesson learned from the Flood of '93 is that some floods will occur despite our best efforts at flood control. The 29 dams on the Mississippi and 26 reservoirs on upstream tributaries, as well as about 5800 km of levees, had little success at holding floodwaters in check. No doubt, debate will now focus on the utility of levees in flood control. Levees are effective in protecting many areas during floods, yet in some cases they actually exacerbate the problem by restricting the flow that would otherwise have spread over a floodplain. They are expensive to build and maintain, and their overall effectiveness has been and will continue to be questioned.

The U.S. Army Corps of Engineers has spent about $25 billion in this century to build 500 dams and more than 16,000 km of levees. No one doubts that some of these projects have been successful, at least within the limits of their design. But critics charge that such flood-control projects make the problem of flooding worse, particularly because flood-prone areas tend to be developed once the projects are completed, even though nothing can be done to prevent some floods.

Introduction

AMONG the terrestrial planets, Earth is unique in having abundant liquid water. Fully 71% of Earth's surface is water covered, and its atmosphere contains a small but important quantity of water vapor.

The volume of water on Earth is estimated at 1.36 billion km³, most of which (97.2%) is in the oceans. About 2.15% is frozen in glaciers, and the remaining 0.65% constitutes all the water in streams, lakes, swamps, groundwater, and the atmosphere (Figure 12.2). Only a tiny portion of the total water on Earth is in streams, but running water is nevertheless the most important erosional agent modifying the surface. Even in most desert regions, the effects of running water are manifest, although channels are dry most of the time.

Besides its significance as a geologic agent, running water is important for many other reasons. It is a source of freshwater for industry, domestic use, and agriculture, and about 8% of the electricity used in North America is generated by falling water at hydroelectric stations. Streams have been and continue to be important avenues of commerce. Much of the interior of North America was first explored by following such large streams as the St. Lawrence, Mississippi, and Missouri Rivers.

Much of the following discussion of running water is necessarily descriptive, but one should always be aware that streams are dynamic systems that must continually respond to change. For example, paving in urban areas increases surface runoff to streams, while other human actions such as building dams and impounding reservoirs also alter the dynamics of a stream system. Natural changes, too, affect stream dynamics. When more rain falls in a stream's drainage area due to long-term climatic change, more water flows in the stream's channel, and greater energy is available for erosion and transport of sediments. In short, streams continually adjust to change.

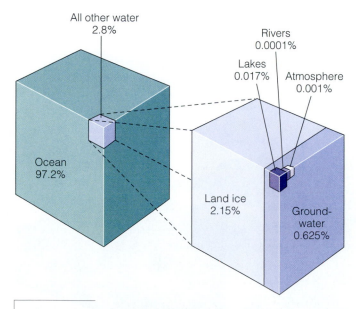

FIGURE 12.2   The relative amounts of water on Earth. Of the estimated 1.36 billion km³ of water, 97.2% is in the oceans; most of the rest, 2.15%, is in glaciers on land.

The Hydrologic Cycle

THE **hydrologic cycle** involves the continuous recycling of water from the oceans, through the atmosphere, and back to the oceans (Figure 12.3). Accordingly, one can think of the oceans as a vast reservoir from which water is continuously withdrawn and returned. The hydrologic cycle, which is powered by solar radiation, is possible because water changes phases easily under Earth surface conditions. Huge quantities of water evaporate from the oceans as the surface waters are heated by solar energy. In fact, about 85% of all water entering the atmosphere is derived from the oceans; the remaining 15% comes from evaporation of water on land.

When water evaporates, the vapor rises into the atmosphere where the complex processes of condensation and cloud formation take place. About 80% of all precipitation falls directly into the oceans, in which case the hydrologic cycle is limited to a three-step process of evaporation, condensation, and precipitation.

About 20% of all precipitation falls on land as rain and snow, and the hydrologic cycle involves more steps: evaporation, condensation, movement of water vapor from the oceans to the continents, precipitation, and runoff and infiltration. Some of the precipitation evaporates as it falls and reenters the hydrologic cycle as vapor; water evaporated from lakes and streams also reenters the cycle as vapor, as does moisture evaporated from plants by *transpiration* (Figure 12.3).

Each year about 36,000 km³ of the precipitation falling on land returns to the oceans by **runoff**, the surface flow of streams. The water returning to the oceans by runoff enters Earth's ultimate reservoir where it begins the hydrologic cycle again. Some of the precipitation falling on land is temporarily stored in lakes, snow fields, and glaciers or

FIGURE 12.3 The hydrologic cycle. During this cycle, water evaporates from the oceans and rises as water vapor to form clouds that release their precipitation either over the oceans or over land. Much of the precipitation falling on land returns to the oceans by surface runoff in streams, thus completing the cycle.

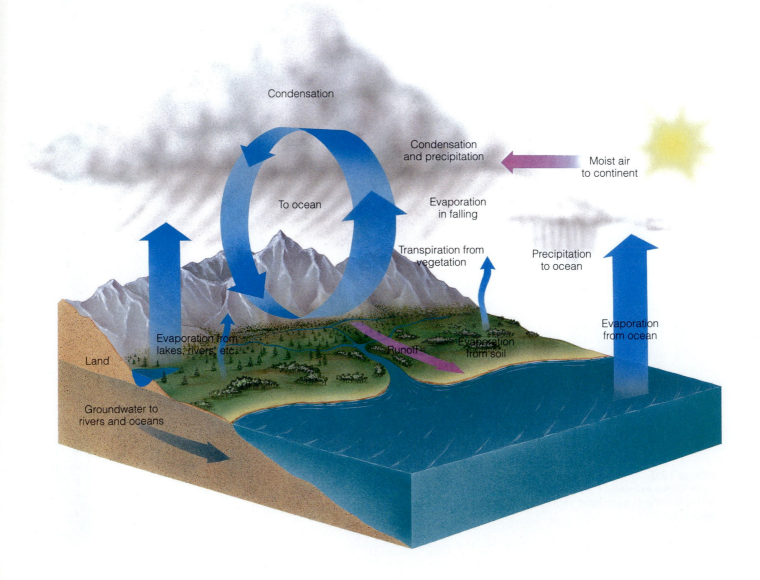

seeps below the surface where it is temporarily stored as groundwater, for up to thousands of years. But eventually, glaciers melt, lakes feed streams, and groundwater flows into streams or directly into the oceans (Figure 12.3). Our concern here is with the comparatively small quantity returning to the oceans as runoff, for the energy of running water is responsible for a great many surface features.

Running Water

THE amount of runoff in any area during a rainstorm depends on **infiltration capacity**, the maximum rate that soil or other surface materials can absorb water. Infiltration capacity depends on several factors, including the intensity and duration of rainfall. Loosely packed, dry soils absorb water faster than tightly packed, wet soils.

If rain is absorbed as fast as it falls, no surface runoff takes place. Should the infiltration capacity be exceeded, or should surface materials become saturated, excess water collects on the surface and, if a slope exists, moves downhill. Even on steep slopes flow is initially slow, and hence causes little or no erosion, but as water continues moving downslope, it accelerates and may move by *sheet flow,* a more-or-less continuous film of water flowing over the surface. Sheet flow is not confined to depressions, and it accounts for *sheet erosion,* a particular problem on some agricultural lands (see Chapter 6).

In *channel flow,* surface runoff is confined to long, troughlike depressions. Channels vary from tiny rills containing a trickling stream of water to the Amazon River of South America, which is 6450 km long and up to 2.4 km wide and 90 m deep. Channelized flow is described by various terms, including rill, brook, creek, stream, and river—most of which are distinguished by size and volume. The term **stream** carries no connotation of size and is used here to refer to all runoff confined to channels regardless of size.

Streams receive water from several sources, including sheet flow and rain falling directly into stream channels. Far more important, though, is the water supplied by soil moisture and groundwater, both of which flow downslope and discharge into streams. In humid areas where groundwater is plentiful, streams may maintain a fairly stable flow year round, even during dry seasons, because they are continuously supplied by groundwater. In contrast, the amount of water in streams of arid and semiarid regions fluctuates widely because these streams depend more on infrequent rainstorms and surface runoff for their water supply.

STREAM GRADIENT, VELOCITY, AND DISCHARGE

Streams flow downhill from a source area to a lower elevation where most of them empty into another stream, a lake, or the sea. Exceptions are streams in arid regions that diminish in a downstream direction by evaporation and

infiltration until they disappear. Streams in regions with numerous caverns may flow into a surface opening and disappear below ground, although they commonly reappear elsewhere (see Chapter 13).

The slope a stream flows over is its **gradient**. If the source (headwaters) of a stream is 1000 m above sea level and the stream flows 500 km to the sea, it drops 1000 m vertically over a horizontal distance of 500 km (Figure 12.4). Its gradient is calculated by dividing the vertical drop by the horizontal distance; in this example, it is 1000 m/500 km = 2 m/km. Gradients vary considerably, even along the course of a single stream. Generally, streams are steeper in their upper reaches where their gradients may be tens of meters per kilometer, but in their lower reaches the gradient may be as little as a few centimeters per kilometer.

Stream velocity and discharge are closely related variables. **Velocity** is simply a measure of the downstream distance traveled per unit of time, and is usually expressed in feet per second (ft/sec) or meters per second (m/sec). Variations in flow velocity occur not only with distance along a stream channel but also across a channel's width. Flow velocity is slower and more turbulent near a stream's banks or bed because of friction than it is farther from these boundaries (Figure 12.5). Other controls on velocity include channel shape and roughness. Broad, shallow channels and narrow, deep channels have proportionally more water in contact with their perimeters than do channels with semicircular cross sections (Figure 12.6). Consequently, the water in semicircular channels flows more rapidly because it encounters less frictional resistance.

Channel roughness is a measure of the frictional resistance within a channel. Frictional resistance to flow is

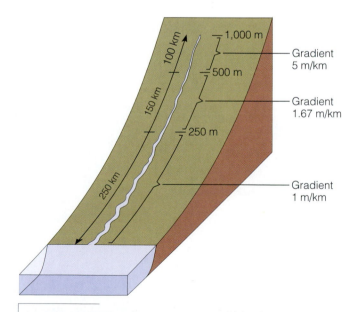

FIGURE 12.4 The average gradient of this stream is 2 m/km. Gradient can be calculated for any segment of a stream as shown in this example. Notice that the gradient is steepest in the headwaters area and decreases downstream.

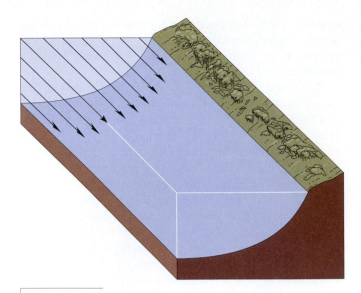

FIGURE 12.5 Flow velocity varies as a result of friction with a stream's banks and bed. The maximum flow velocity is near the center and top in a straight channel. The lengths of the arrows in this illustration are proportional to velocity.

greater in a channel containing large boulders than in one with banks and a bed composed of sand or clay. In channels with abundant vegetation, flow is slower than in barren channels of comparable size.

The most obvious control on velocity is gradient, and one might think that the steeper the gradient, the greater the flow velocity. In fact, the average velocity generally increases in a downstream direction, even though the gradient decreases in the same direction. Three factors contribute to this: First, velocity increases continuously, even as gradient decreases, in response to the acceleration of gravity unless other factors retard flow. Second, in their up-

stream reaches, streams commonly have boulder-strewn, broad, shallow channels, so flow resistance is high and average velocity is correspondingly slower. Downstream, channels generally become more semicircular, and the bed and banks are usually composed of finer-grained materials, reducing the effects of friction. Third, the number of tributary streams joining a larger stream increases in a downstream direction, so the total volume of water (discharge) increases, and increasing discharge results in increased velocity.

Discharge is the total volume of water in a stream moving past a particular point in a given period of time. To determine discharge, one must know the dimensions of the water-filled part of a channel—that is, its cross-sectional area (A)—and its flow velocity (V). Discharge (Q) can then be calculated by the formula $Q = VA$; it is generally expressed in cubic feet per second (ft^3/sec) or cubic meters per second (m^3/sec).

Stream Erosion

EROSION involves the physical removal of dissolved substances and loose particles of soil and rock from a source area. Accordingly, the sediment transported in a stream consists of both dissolved materials and solid particles. Some of the *dissolved load* of a stream is acquired from the streambed and banks where soluble rocks such as limestone and dolostone are exposed. But much of it is carried into streams by sheet flow and groundwater.

The solid sediment carried in streams ranges from clay-sized particles to large boulders. Much of this sediment finds its way into streams by mass wasting (Figure 12.7a), but some is derived directly from the streambed and banks. The power of running water, called **hydraulic action**, is sufficient to set particles in motion.

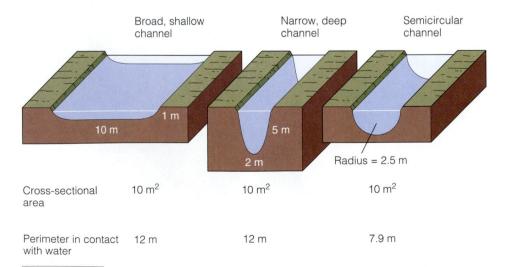

	Broad, shallow channel	Narrow, deep channel	Semicircular channel
Cross-sectional area	10 m²	10 m²	10 m²
Perimeter in contact with water	12 m	12 m	7.9 m

FIGURE 12.6 These three channels have the same cross-sectional area, but each has a different shape. The semicircular channel has the least perimeter in contact with the water and presents the least frictional resistance to flow. If other variables, such as channel roughness, are the same in all of these channels, flow velocity will be greatest in the semicircular channel.

(a)

FIGURE 12.7   (a) Streams such as the Snake River in Idaho receive some of their sediment load by mass wasting processes, frost wedging in this case. (b) Large potholes in a streambed. *(Photos courtesy of R. V. Dietrich.)*

(b)

Another process of erosion in streams is **abrasion**, in which exposed rock is worn and scraped by the impact of solid particles. If running water is transporting sand and gravel, the impact of these particles abrades exposed rock surfaces. One obvious manifestation of abrasion is the occurrence of *potholes* in streambeds (Figure 12.7b). These circular to oval holes form where eddying currents con-

taining sand and gravel swirl around and erode depressions into rock.

Transport of Sediment Load

STREAMS transport both dissolved materials and solid sedimentary particles. Their **dissolved load** consists of ions taken into solution during chemical weathering. This material is not visible but nevertheless is an important part of any stream's sediment load. Among the sedimentary particles transported in a stream, the smallest ones, mostly silt and clay, constitute the **suspended load**. Fluid turbulence keeps these small particles suspended in the water, so they are transported above the streambed. Large sedimentary particles such as sand and gravel make up a stream's **bed load**, which is that part of the sediment load transported along the streambed.

Fluid turbulence is insufficient to keep large sand and gravel particles suspended, so they move along the streambed. However, part of the bed load can be suspended temporarily as when an eddying current swirls across a streambed and lifts sand grains into the water. These particles move forward at approximately the flow velocity, but at the same time they settle toward the streambed where they come to rest, to be moved again later by the same process. This process of intermittent bouncing and skipping is *saltation.*

Particles too large to be suspended even temporarily are transported by rolling or sliding. Obviously, greater flow velocity is required to move particles of these sizes. The maximum-sized particles that a stream can carry define its *competence,* a factor related to flow velocity. *Capacity* is a measure of the total load a stream can carry. It varies as a function of discharge; with greater discharge, more sediment can be carried. A small, swiftly flowing stream may have the competence to move gravel-sized particles but not to transport a large volume of sediment, so it has a low capacity. A large, slow-flowing stream, on the other hand, has a low competence, but may have a very large suspended load, and hence a large capacity.

Stream Deposition

STREAMS can transport sediment a considerable distance from the source area. For instance, some sediments deposited in the Gulf of Mexico by the Mississippi River came from such distant sources as Pennsylvania, Minnesota, and Alberta, Canada. Along the way, sediments may be deposited in a variety of environments, such as stream channels, the floodplains adjacent to channels, and the points where streams flow into lakes or seas or flow from mountain valleys onto adjacent lowlands.

Streams do most of their erosion, sediment transport, and deposition when they flood. Consequently, stream deposits, collectively called **alluvium,** do not represent the continuous day-to-day activity of streams, but rather those periodic, large-scale events of sedimentation associated with flooding.

BRAIDED STREAMS AND THEIR DEPOSITS

Braided streams possess an intricate network of dividing and rejoining channels (Figure 12.8). Braiding develops when a stream is supplied with excessive sediment, which over time is deposited as sand and gravel bars within its channel. During high-water stages, these bars are submerged, but during low-water stages, they are exposed and divide a single channel into multiple channels. Braided streams have broad, shallow channels and are generally characterized as bed load–transport streams. Their deposits are composed mostly of sheets of sand and gravel. Braided streams are common in arid and semiarid regions where little vegetation exists and erosion rates are high. Streams fed by melting glaciers are also commonly braided.

FIGURE **12.8** A braided stream near Sante Fe, New Mexico.

MEANDERING STREAMS AND THEIR DEPOSITS

Meandering streams possess a single, sinuous channel with broadly looping curves known as *meanders* (Figure 12.9). These stream channels are semicircular in cross section along straight reaches, but at meanders they are markedly asymmetric, being deepest near the outer bank, which commonly descends vertically into the channel. The outer bank is called the *cut bank* because flow velocity and turbulence are greatest on that side of the channel where it is eroded. In contrast, flow velocity is at a minimum near the inner bank, which slopes gently into the channel (Figure 12.10a).

As a result of the unequal distribution of flow velocity across meanders, the cut bank is eroded and deposition takes place along the opposite side of the channel. The deposit formed in this manner is a **point bar**, consisting of cross-bedded sand or, in some cases, gravel (Figure 12.10b).

It is not uncommon for meanders to become so sinuous that the thin neck of land separating adjacent meanders is eventually cut off during a flood. The valley floors of meandering streams are commonly marked by crescent-shaped **oxbow lakes**, which are actually cutoff meanders (Figures 12.9 and 12.11). These oxbow lakes may persist as lakes for some time, but eventually they fill with organic matter and fine-grained sediment carried by floods. Once filled, oxbow lakes are called *meander scars*.

FLOODS AND FLOODPLAIN DEPOSITS

Most streams periodically receive more water than their channel can carry, so they spread across low-lying, relatively flat **floodplains** adjacent to their channels (Figure 12.9; see Perspective 12.1). Some floodplains are composed mostly of sand and gravel that were deposited as point bars. When a meandering stream erodes its cut bank and deposits on the opposite bank, it migrates laterally across its floodplain. As deposition in the laterally migrating channel takes place, a succession of point bars develops (Figure 12.12a).

FIGURE 12.9 Aerial view of a meandering stream. The broad, flat area adjacent to the stream channel is the floodplain. Notice the crescent-shaped lakes— these are cutoff meanders, or what are known as oxbow lakes.

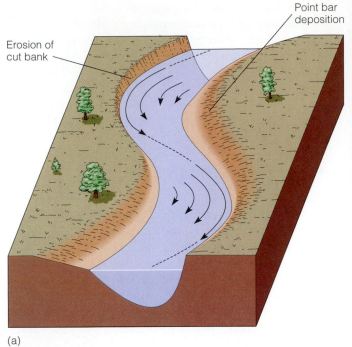

(a)

Point bar deposition

Erosion of cut bank

FIGURE 12.10 (a) The line of maximum velocity (dashed) switches from side to side in a meandering channel. The arrows show relative velocity at various places in the channel. Because of variations in flow velocity, the outer or cut bank is eroded, and a point bar is deposited on the gently sloping side of the meander. (b) Two point bars in a small meandering stream.

(b)

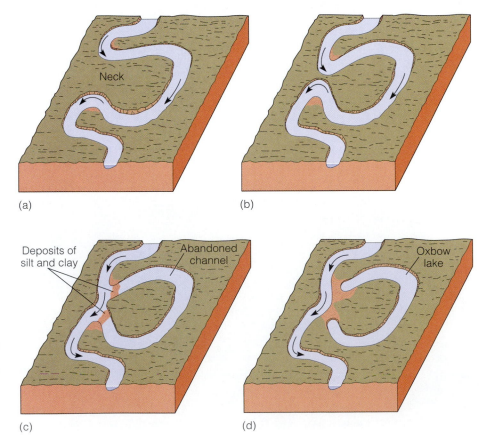

Neck

(a)

(b)

Deposits of silt and clay

Abandoned channel

Oxbow lake

(c)

(d)

FIGURE 12.11 Four stages in the origin of an oxbow lake. In (a) and (b), the meander neck becomes narrower. (c) The meander neck is cut off, and part of the channel is abandoned. (d) When it is completely isolated from the main channel, the abandoned meander is an oxbow lake.

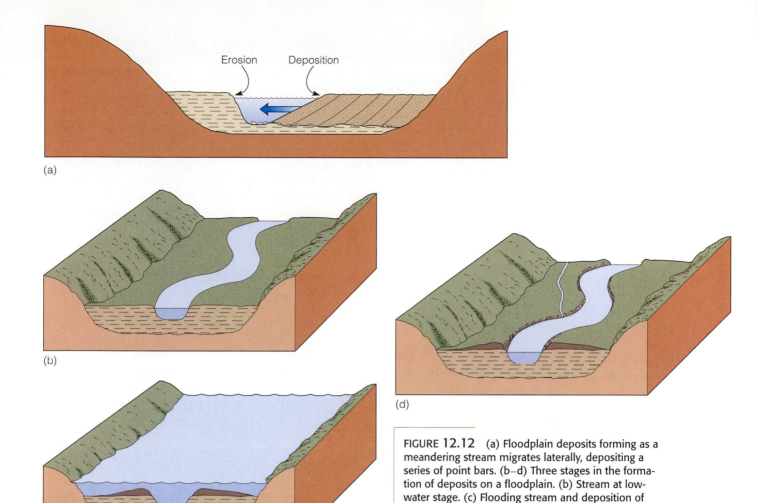

(a)

(b)

(c)

(d)

FIGURE **12.12** (a) Floodplain deposits forming as a meandering stream migrates laterally, depositing a series of point bars. (b–d) Three stages in the formation of deposits on a floodplain. (b) Stream at low-water stage. (c) Flooding stream and deposition of natural levees. The levees form after many such episodes of flooding. (d) After flooding. Notice the tributary stream, which parallels the main stream until it finds a way through the natural levee.

Many floodplains are dominated by fine-grained sediments, mostly mud. When a stream overflows its banks and floods, the velocity of the water spilling onto the floodplain diminishes rapidly because the water encounters greater frictional resistance to flow as it spreads out as a broad, shallow sheet. In response to the diminished velocity, ridges of sandy alluvium known as **natural levees** are deposited along the margins of the stream channel (Figure 12.12b).

Floodwaters spilling from a main channel carry large quantities of mud beyond the natural levees and onto the floodplain. During the waning stages of a flood, floodwaters may flow very slowly or not at all, and the suspended silt and clay eventually settle as layers of mud.

Annual property damage from flooding in the United States exceeds $100 million. And despite the completion of more and more flood-control projects, the amount of property damage is not decreasing. In fact, floodplains are attractive sites for settlement due to the combination of fertile soils, level surfaces for construction, and proximity to water for industry, agriculture, and domestic uses. How-

ever, these human activities generally increase the potential for flooding. Urbanization greatly increases surface runoff because concrete and asphalt compact and cover surface materials, thereby reducing their infiltration capacity. Storm drains in urban areas quickly carry water to nearby streams, many of which flood much more commonly than they did in the past.

DELTAS

When a stream flows into a lake or the sea, its flow velocity decreases rapidly, and it deposits its sediment. As a result, a **delta** forms, causing the local shoreline to build out, or *prograde* (Figure 12.13). The simplest prograding deltas exhibit a characteristic vertical sequence in which *bottomset beds* are successively overlain by *foreset beds* and *topset beds* (Figure 12.13a). This sequence develops when a stream enters another body of water, and the finest sediments are carried some distance beyond the stream's mouth, where they settle from suspension and form bottomset beds. Nearer the stream's mouth, foreset beds form

Predicting and Controlling Floods

When a stream receives more water than its channel can handle, it spills over its banks and occupies part or all of its floodplain. Indeed, floods are so common that, unless they cause considerable property damage or fatalities, they rarely rate more than a passing note in the news. The most extensive recent flooding in the United States was the Flood of '93 in the Midwest (see the Prologue), but since then several other areas have experienced serious flooding. For instance, a series of late December 1996 to early January 1997 storms caused widespread flooding in several western states, resulting in at least 28 deaths and billions of dollars in property damage. The April 1997 spring thaw of a record snowfall caused the Red River to flood, necessitating the evacuation of all 50,000 residents of Grand Forks, North Dakota, and an additional 17,000 people from areas adjacent to the river in Manitoba, Canada. California was hit hard by several January 1998 storms that caused extensive flooding, several deaths, and considerable property damage.

To monitor stream behavior, the U.S. Geological Survey maintains more than 11,000 stream gauging stations, and various state agencies also monitor streams. Data collected at gauging stations can be used to construct a *hydrograph* showing how a stream's discharge varies over time (Figure 1). Hydrographs are useful in planning irrigation and water supply projects, and they give planners a better idea of what to expect during floods.

Stream gauge data are also used to construct *flood-frequency curves* (Figure 2). To construct such a curve, the peak discharges are first arranged in order of volume; the flood with the greatest discharge has a magnitude rank of 1, the second largest is 2, and so on (Table 1). The *recurrence interval*—that is, the time period during which a flood of a given magnitude or larger can be expected over an average of many years—is determined by the equation shown in Table 1. For this stream, floods with magnitude ranks of 1 and 23 have recurrence intervals of 77.00 and 3.35 years, respectively. Once the recurrence interval has been calculated, it is plotted against

discharge, and a line is drawn through the data points (Figure 2).

According to Figure 1, the 10-year flood for the Rio Grande near Lobatos, Colorado, has a discharge of 245 m³/sec. This means that, on average, we can expect one flood of this size or greater to occur within a 10-year interval. One cannot, however, predict that such a flood will take place in any particular year, only that it has a probability of 1 in 10 (1/10) of occurring in any year. Furthermore, 10-year floods are not necessarily separated by 10 years. That is, two such floods could occur in the same year or in successive years, but over a period of centuries their average occurrence would be once every 10 years.

Unfortunately, stream gauge data in the United States have generally been available for only a few decades, and rarely for more than a century. Accordingly, we have a good idea of stream behavior over short periods, the 2-year and 5-year floods, for example, but our knowledge of long-term behavior is limited by the short period of record keeping. Thus, predictions of 50-year or

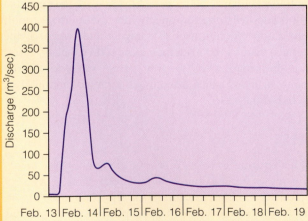

Figure 1 *Hydrographs for Sycamore Creek near Ashland City, Tennessee, for the February 1989 flood. (From U.S. Geological Survey Water-Resources Investigations Report 89–4207.)*

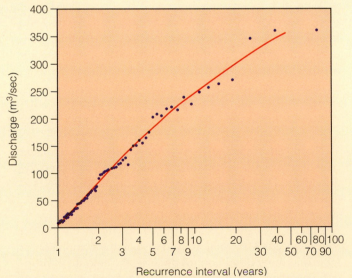

Figure 2 *Flood-frequency curve for the Rio Grande near Lobatos, Colorado. The curve was constructed from the data in Table 1.*

TABLE 1

Some of the Data and Recurrence Intervals for the Rio Grande near Labatos, Colorado

YEAR	DISCHARGE (M³/SEC)	RANK	RECURRENCE INTERVAL
1900	133	23	3.35
1901	103	35	2.20
1902	16	69	1.12
1903	362	2	38.50
1904	22	66	1.17
1905	371	1	77.00
1906	234	10	7.70
1907	249	7	11.00
1908	61	45	1.71
1909	211	13	5.92
.	The greatest yearly discharge is given a magnitude rank (m) ranging from		
.	1 to N ($N = 76$ in this example), and the recurrence interval (R) is		
.	calculated the equation $R = (N + 1)/m$.		
1974	22	64	1.20
1975	68	43	1.79

SOURCE: U.S. Geological Survey Open-File Report 79–681.

100-year floods from Figure 2 are unreliable. In fact, the largest magnitude flood shown in Figure 2 may have been a unique event for this stream that will never be repeated. On the other hand, it may actually turn out to be a magnitude 2 or 3 flood when data for a longer time are available.

Although flood-frequency curves have limited applicability, they are nevertheless helpful in making decisions regarding flood control. Careful mapping of floodplains can identify areas at risk for floods of a given magnitude. For a particular stream, planners must decide what magnitude of flood to protect against because the cost goes up faster than the increasing sizes of floods would indicate.

Federal, state, and local agencies and land-use planners use flood-frequency analyses to develop recommendations and regulations concerning construction on and use of floodplains. Geologists and engineers are interested in the analyses for planning appropriate flood-control projects. They must decide, for example, where dams and basins should be constructed to contain the excess water of floods.

Flood control has been practiced for thousands of years. As noted in the Prologue, common practices are to construct dams that impound reservoirs and to build levees along stream banks. Levees raise the banks of a stream, thereby restricting flow during floods. Unfortunately, deposition within the channel results in raising the streambed, making the levees useless unless they too are raised. Levees along the banks of the Huang He in China caused the streambed to rise more than 20 m above its surrounding floodplain in 4000 years. When the Huang He breached its levees in 1887, more than 1 million people were killed.

Dams and levees alone are insufficient to control large floods (Figure 3), so in many areas floodways are also used. These usually consist of a channel constructed to divert part of the excess water in a stream around populated areas or areas of economic importance. Reforestation of cleared land also helps reduce the potential for flooding because vegetated soil helps prevent runoff by absorbing more water.

When flood-control projects are well planned and constructed, they are functional. What many people fail to realize is that these projects are designed to contain floods of a given size, should larger floods occur, streams spill onto floodplains anyway. Furthermore, dams occasionally collapse, and reservoirs eventually fill with sediment unless dredged. In short, flood-control projects are not only initially expensive but also require constant, costly maintenance. These costs must be weighed against the cost of damage if no control projects were undertaken.

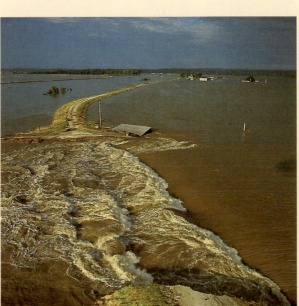

Figure 3 Breached levee in Illinois during the Flood of '93.

(b)

Distributary channel

Lake

Topset beds

Foreset beds

Bottomset beds

(a)

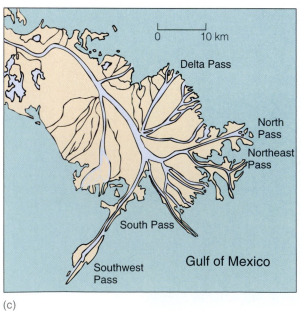

Delta Pass

North Pass

Northeast Pass

South Pass

Southwest Pass

Gulf of Mexico

0   10 km

(c)

FIGURE **12.13**   (a) Internal structure of the simplest type of prograding delta. (b) A small delta, measuring about 20 m across, in which bottomset, foreset, and topset beds are visible. (c) The Mississippi River delta of the U.S. Gulf Coast.

as sand and silt are deposited in gently inclined layers. The topset beds consist of coarse-grained sediments deposited in a network of *distributary channels* traversing the top of the delta. In effect, streams lengthen their channels as they extend across prograding deltas (Figure 12.13).

Many small deltas in lakes have the three-part division described above, but marine deltas are usually much larger, more complex, and more important economically. The Mississippi River delta consists of long fingerlike sand bodies, each deposited in a distributary channel that prograides far seaward (Figure 12.13c). Such deltas are commonly called *bird's-foot deltas* because the projections resemble the toes of a bird.

Progradation of marine deltas is one way that potential reservoirs for oil and gas are formed. Much of the oil and gas production of the Gulf Coast of Texas comes from buried delta deposits, and the present-day deltas of the Niger River in Africa and the Mississippi River are also known to contain reserves of oil and gas. The marshes between distributary channels of deltas are dominated by nonwoody vegetation and are potential areas of coal formation.

ALLUVIAL FANS

Lobate deposits of alluvium on land are known as **alluvial fans** (Figure 12.14). They form best on lowlands adjacent to highlands in arid and semiarid regions where little or no vegetation exists to stabilize surface materials. Following periodic rainstorms, surface materials are quickly saturated and runoff begins. During a particularly heavy rain, all surface flow in a drainage area is funneled into a mountain canyon leading to an adjacent lowland. The stream is confined in the mountain canyon, but when it discharges from the canyon onto the lowland area, it quickly spreads out, its velocity diminishes, and deposition ensues.

The alluvial fans that develop by the process just described are mostly accumulations of sand and gravel, a large proportion of which is deposited by streams. In some cases, the water flowing through a mountain canyon picks up so much sediment that it becomes a viscous mudflow.

FIGURE 12.14 Alluvial fans form where a stream discharges from a mountain canyon onto an adjacent lowland. These alluvial fans are in Death Valley, California.

Consequently, mudflow deposits make up a large part of many alluvial fans.

Drainage Basins and Drainage Patterns

ALL streams consist of a main channel and all smaller *tributary streams* that supply water to it. The Mississippi and all of its tributaries, or any other drainage system for that matter, carry surface runoff from an area known as the **drainage basin**. Individual drainage basins are separated from adjacent ones by topographically higher areas called **divides** (Figure 12.15).

Various **drainage patterns** are recognized based on the regional arrangement of channels in a drainage system. The most common is *dendritic drainage,* consisting of a network of channels resembling tree branching (Figure 12.16a). Dendritic drainage develops on gently sloping surface materials that respond more or less homogeneously to erosion. Areas of flat-lying sedimentary rocks and some terrains of igneous or metamorphic rocks commonly display a dendritic drainage pattern.

Rectangular drainage is characterized by channels with right angle bends and tributaries that join larger streams at right angles (Figure 12.16b). The positions of the channels are strongly controlled by geologic structures, particularly regional joint systems that intersect at right angles.

In some parts of the eastern United States, such as Virginia and Pennsylvania, erosion of folded sedimentary rocks develops a landscape of alternating parallel ridges and valleys. The ridges consist of more resistant rocks, such as sandstone, whereas the valleys overlie less resistant

(a)

FIGURE **12.15** (a) Small drainage basins separated from one another by divides, which are along the crests of the ridges between channels. (b) The drainage basin of the Wabash River, which is one of the tributaries of the Ohio River. All tributary streams within the drainage basin, such as the Vermillion River, have their own smaller drainage basins. Divides are shown by dark red lines.

(b)

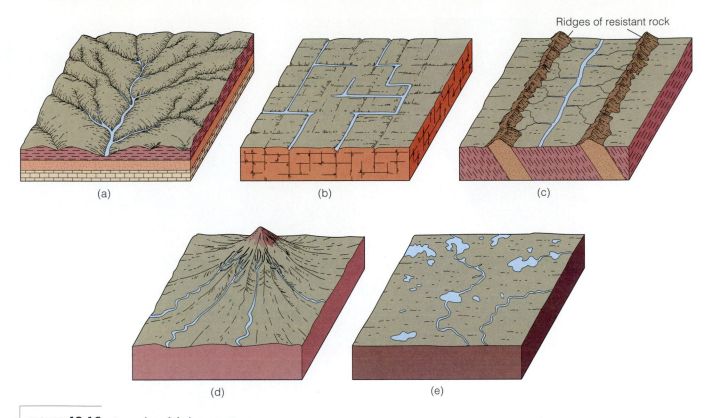

FIGURE **12.16** Examples of drainage patterns.
(a) Dendritic drainage. (b) Rectangular drainage.
(c) Trellis drainage. (d) Radial drainage. (e) Deranged
drainage.

rocks such as shale. Main streams follow the trends of the valleys. Short tributaries flowing from the adjacent ridges join the main stream at nearly right angles, hence the name *trellis drainage* (Figure 12.16c).

In *radial drainage,* streams flow outward in all directions from a central high area (Figure 12.16d). Radial drainage develops on large, isolated volcanic mountains and in areas where the crust has been arched up by the intrusion of plutons such as laccoliths.

In some areas streams flow in and out of swamps and lakes with irregular flow directions. Drainage patterns characterized by such irregularity are called *deranged* (Figure 12.16e). The presence of deranged drainage indicates that it developed recently and has not yet formed an organized drainage system. In areas of Minnesota, Wisconsin, and Michigan that were glaciated until about 10,000 years ago, the previously established drainage systems were obliterated by glacial ice. Following the final retreat of the glaciers, drainage systems became established, but have not yet become fully organized.

Base Level

STREAMS have a lower limit or **base level** to which they can erode (Figure 12.17). Theoretically, a stream could erode its entire valley to very near sea level, so sea level is commonly referred to as *ultimate base level.* In reality,

though, streams never reach ultimate base level because they must have some gradient in order to maintain flow. Streams flowing into depressions below sea level, such as Death Valley, California, have a base level corresponding to the lowest point of the depression and are not limited by sea level.

In addition to ultimate base level, streams have *local* or *temporary base levels.* For instance, a lake or another stream can serve as a local base level for the upstream segment of a stream (Figure 12.17). Likewise, where a stream flows across particularly resistant rock, a waterfall may develop, forming a local base level.

Changes in base level occur when sea level rises or falls with respect to the land, or the land over which a stream flows is uplifted or subsides. During the Pleistocene Epoch when extensive glaciers were present on the Northern Hemisphere continents, sea level was about 130 m lower than at present. Accordingly, streams deepened their valleys by adjusting to a new, lower base level. Rising sea level at the end of the Pleistocene caused base level to rise, and the streams responded by depositing sediments and backfilling previously formed valleys.

Streams adjust to human intervention, but not always in anticipated or desirable ways. Geologists and engineers are well aware that the process of building a dam and impounding a reservoir creates a local base level (Figure 12.18a). Where a stream enters a reservoir, its flow velocity

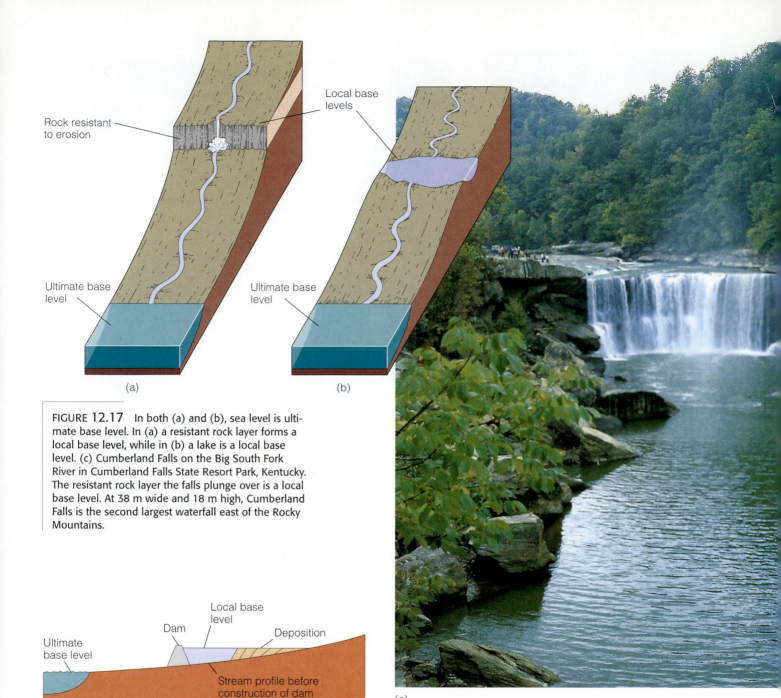

FIGURE 12.17 In both (a) and (b), sea level is ultimate base level. In (a) a resistant rock layer forms a local base level, while in (b) a lake is a local base level. (c) Cumberland Falls on the Big South Fork River in Cumberland Falls State Resort Park, Kentucky. The resistant rock layer the falls plunge over is a local base level. At 38 m wide and 18 m high, Cumberland Falls is the second largest waterfall east of the Rocky Mountains.

FIGURE 12.18 (a) The process of constructing a dam and impounding a reservoir creates a local base level. A stream deposits much of its sediment load where it flows into a reservoir. (b) A stream adjusts to a lower base level when a lake is drained.

diminishes rapidly, and it deposits sediment; consequently, reservoirs eventually fill with sediment unless they are dredged. Another consequence of building a dam is that the water discharged at the dam is largely sediment free, but it still possesses energy to transport sediment. Commonly, such streams simply acquire a new sediment load by vigorously eroding downstream from the dam.

Draining a lake along a stream's course may seem like a small change and well worth the time and expense to expose dry land for agriculture or commercial development. However, draining a lake lowers the base level for that part of the stream above the lake, and the stream will very likely respond by rapid downcutting (Figure 12.18b).

The Graded Stream

A stream's *longitudinal profile* shows the elevations of a channel along its length as viewed in cross section (Figure 12.19). The longitudinal profiles of many streams show a number of irregularities such as lakes and waterfalls, which are local base levels (Figure 12.19a). Over time these irregularities tend to be eliminated; where the gradient is steep, erosion decreases it, and where the gradient is too low to maintain sufficient flow velocity for sediment transport, deposition occurs, steepening the gradient. In short, streams tend to develop a smooth, concave longitudinal profile of equilibrium, meaning that all parts of the system dynamically adjust to one another (Figure 12.19b).

Streams possessing an equilibrium profile are said to be **graded streams**; that is, a delicate balance exists between gradient, discharge, flow velocity, channel characteristics, and sediment load so that neither significant erosion nor deposition takes place within the channel. Such a delicate balance is rarely attained, so the concept of a graded stream is an ideal. Nevertheless, many streams do indeed approximate the graded condition, although not along their entire courses and usually only temporarily.

Even though the concept of a graded stream is an ideal, we can generally anticipate the responses of a graded stream to changes altering its equilibrium. A change in base level, for instance, would cause a stream to adjust as previously discussed. Increased rainfall in a stream's drainage basin would result in greater discharge and flow velocity. In short, the stream would now possess greater energy—energy that must be dissipated within the stream system by, for example, a change in channel shape. A change from a semicircular to a broad, shallow channel would dissipate more energy by friction. On the other hand, the

stream may respond by active downcutting and erode a deeper valley and effectively reduce its gradient until it is once again graded.

Development of Stream Valleys

VALLEYS are common landforms, and with few exceptions they form and evolve mostly as a result of stream processes. The shapes and sizes of valleys vary considerably; some are small, steep-sided *gullies,* others are broad and have gently sloping walls, and some such as the Grand Canyon are steel-walled and deep. Particularly narrow, deep canyons are known as *gorges.* Several processes—in particular downcutting, lateral erosion, sheet wash, headward erosion, and mass wasting—contribute to the origin and evolution of valleys.

If a stream posses more energy than it needs to transport its sediment load, some of its excess energy might be expended by eroding a deeper valley. If downward erosion were the only process operating, valleys would be narrow and steep-sided, but in most cases they are widened as their walls are undercut by streams. Such undermining, termed *lateral erosion,* creates unstable conditions so that part of a stream bank or valley wall moves downslope by mass wasting processes. Furthermore, sheet wash and erosion of rill and gully tributaries carry materials from the valley walls to the main stream.

Besides becoming deeper and wider, stream valleys are commonly lengthened in an upstream direction by *headward erosion* as drainage divides are eroded by entering runoff water (Figure 12.20a). In some cases headward erosion eventually breaches the drainage divide and diverts part of the drainage of another stream by a process called *stream piracy* (Figure 12.20b). Once stream piracy has occurred, both drainage systems must adjust; one now has more water, greater discharge, and greater potential to erode and transport sediment, whereas the other is diminished in all of these aspects.

According to one concept, stream erosion of an area uplifted above sea level yields a distinctive series of landscapes. When erosion begins, streams erode downward; their valleys are deep, narrow, and **V** shaped; and a number of irregularities are present in their profiles (Figure 12.21a). As streams cease eroding downward, they start eroding laterally, thereby establishing a meandering pattern and a broad floodplain (Figure 12.21b). Finally, with continued erosion, a vast, rather featureless plain develops (Figure 12.21c).

Many streams do indeed show an association of features typical of these stages. For instance, the Colorado River flows through the Grand Canyon and closely matches the features in the initial stage of development. Streams in many areas approximate the second stage of development, and certainly the lower Mississippi closely resembles the last stage. Nevertheless, the idea of a sequential develop-

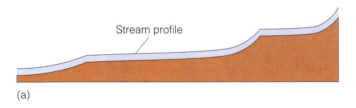

(a)

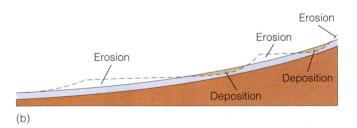

(b)

FIGURE 12.19 (a) An ungraded stream has irregularities in its longitudinal profile. (b) Erosion and deposition along the course of a stream eliminate irregularities and cause it to develop the smooth, concave profile typical of a graded stream.

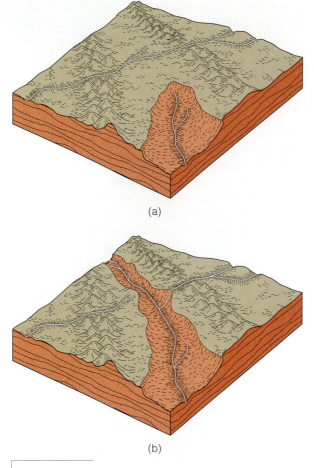

(a)

(b)

FIGURE 12.20 Two stages in stream piracy. (a) In the first stage, the stream at the lower elevation extends its channel by headward erosion. In (b) it has captured some of the drainage of the stream flowing at the higher elevation.

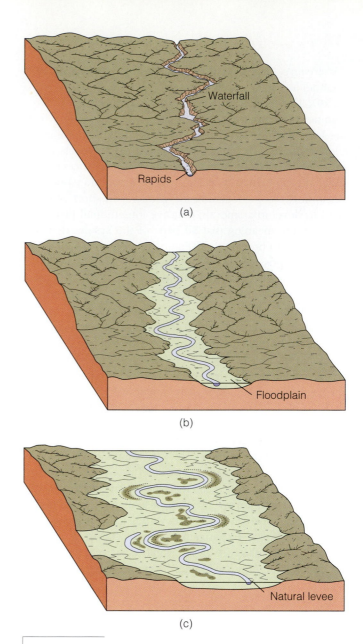

(a)

(b)

(c)

FIGURE 12.21 Idealized stages in the development of a stream and its associated landforms.

ment of stream-eroded landscapes has been largely abandoned because there is no reason to think that streams necessarily follow this idealized cycle. Indeed, a stream on a gently sloping surface near sea level could develop features of the last stage very early in its history. In addition, as long as the rate of uplift exceeds the rate of downcutting, a stream will continue to erode downward and be confined to a narrow canyon.

STREAM TERRACES

Adjacent to many streams are **stream terraces**, which are erosional remnants of floodplains formed when the streams were flowing at a higher level. They consist of a fairly flat upper surface and a steep slope descending to the level of the lower, present-day floodplain (Figure 12.22). In some cases, a stream has several steplike surfaces above its present-day floodplain, indicating that stream terraces formed several times.

Although all stream terraces result from erosion, they are usually preceded by an episode of floodplain formation

and sediment deposition. Subsequent erosion causes the stream to cut downward until it is once again graded, and then it begins eroding laterally and establishes a new floodplain at a lower level. Several such episodes account for the multiple terrace levels seen adjacent to some streams (Figure 12.22). Stream terraces are commonly cut into previously deposited sediment, but some are cut into bedrock where the terrace surface is generally covered by a thin veneer of sediment.

Renewed erosion and the formation of stream terraces are usually attributed to a change in base level. Either uplift of the land over which a stream flows or lowering of sea level yields a steeper gradient and increased flow velocity, thereby initiating an episode of downcutting. When the stream reaches a level at which it is once again graded,

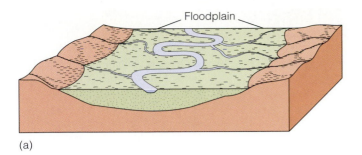

(a)

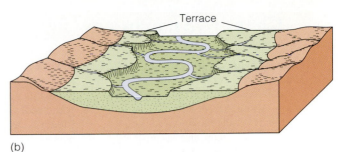

(b)

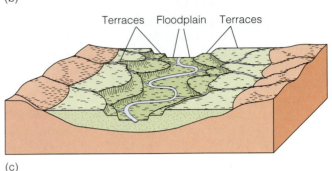

(c)

FIGURE 12.22 Origin of stream terraces. (a) A stream has a broad floodplain adjacent to its channel. (b) The stream erodes downward and establishes a new floodplain at a lower level. Remnants of its old floodplain are stream terraces. (c) Another level of terraces originates as the stream erodes downward again.

downcutting ceases. Although changes in base level no doubt account for many stream terraces, greater runoff in a stream's drainage basin can also result in the formation of terraces.

INCISED MEANDERS

Some streams are restricted to deep, meandering canyons cut into bedrock, where they form features called **incised meanders**. The San Juan River in Utah occupies a meandering canyon more than 390 m deep (Figure 12.23a). These streams, being restricted by rock walls, are generally ineffective in eroding laterally; thus, they lack a floodplain and occupy the entire width of the canyon floor. Some incised meandering streams do erode laterally, thereby cutting off meanders and producing natural bridges (Figure 12.23b).

It is not difficult to understand how a stream can cut downward into solid rock, but forming a meandering pattern in bedrock is another matter. Because lateral erosion is inhibited once downcutting begins, one must infer that the meandering course was established when the stream flowed across an area covered by alluvium. For instance, suppose that a stream near base level has established a meandering pattern. If the land the stream flows over is uplifted, erosion is initiated, and the meanders become incised into the underlying bedrock.

(b)

(a)

FIGURE 12.23 (a) The Goose Necks of the San Juan River in Utah are incised meanders. This meandering canyon is more than 390 m deep. (b) Sipapu Bridge in Natural Bridges National Monument, Utah. This natural bridge formed when an incised meandering stream eroded laterally, cutting through a thin partition of rock between adjacent meanders. It stands 67 m above the canyon floor and has a span of 81.5 m. *(Photo courtesy of Sue Monroe.)*

Development of Stream Valleys 263

Chapter Summary

1. Water is continuously evaporated from the oceans, rises as water vapor, condenses, and falls as precipitation. About 20% of all precipitation falls on land and eventually returns to the oceans, mostly by surface runoff.

2. Runoff can be characterized as either sheet flow or channel flow. Channels of all sizes are called streams.

3. Gradient generally varies from steep to gentle along the course of a stream, being steep in upper reaches and gentle in lower reaches.

4. Flow velocity and discharge are related. A change in one of these parameters causes the other to change as well.

5. A stream and its tributaries carry runoff from its drainage basin. Drainage basins are separated from one another by divides.

6. Streams erode by hydraulic action, abrasion, and dissolution of soluble rocks.

7. The coarser part of a stream's sediment load is transported as bed load, and the finer part as suspended load. Streams also transport a dissolved load of ions in solution.

8. Braided streams are characterized by a complex of dividing and rejoining channels. Braiding occurs when sediment transported by a stream is deposited within channels as sand and gravel bars.

9. Meandering streams have a single, sinuous channel with broad looping curves. Meanders migrate laterally as the cut bank is eroded and point bars form on the inner bank. Oxbow lakes are cutoff meanders in which fine-grained sediments and organic matter accumulate.

10. Rather flat areas paralleling stream channels are floodplains. They may be composed mostly of point bar deposits or mud deposited during floods.

11. Deltas are alluvial deposits at a stream's mouth. Many small deltas in lakes conform to the three-part division of bottomset, foreset, and topset beds, but marine deltas are larger and more complex.

12. Lobate alluvial deposits on land consisting mostly of sand and gravel are alluvial fans. They form best in arid and semiarid regions where erosion rates are high.

13. Sea level is ultimate base level, the lowest level to which streams can erode. Streams also commonly have local base levels such as lakes, other streams, or the points where they flow across particularly resistant rocks.

14. Streams tend to eliminate irregularities in their channels so that they develop a smooth, concave profile of equilibrium. Such streams are graded, meaning that a balance exists between gradient, discharge, flow velocity, channel characteristics, and sediment load so that little or no deposition occurs within its channel.

15. Stream valleys develop by a combination of processes including downcutting, lateral erosion, mass wasting, sheet wash, and headward erosion.

16. Renewed downcutting by a stream possessing a floodplain commonly results in the formation of stream terraces, which are remnants of an older floodplain at a higher level.

17. Incised meanders are generally attributed to renewed downcutting by a meandering stream so that it now occupies a deep, meandering valley.

Important Terms

- abrasion
- alluvial fan
- alluvium
- base level
- bed load
- braided stream
- delta
- discharge
- dissolved load
- divide
- drainage basin
- drainage pattern
- floodplain
- graded stream
- gradient
- hydraulic action
- hydrologic cycle
- incised meander
- infiltration capacity
- meandering stream
- natural levee
- oxbow lake
- point bar
- runoff
- stream
- stream terrace
- suspended load
- velocity

1. The bed load of a stream consists of all material transported by:
   a. _____ suspension and solution;
   b. _____ dissolution of limestone;
   c. _____ saltation and rolling and sliding;
   d. _____ capacity and competence;
   e. _____ infiltration and discharge.

2. A stream having a sinuous channel is referred to as:
   a. _____ graded;
   b. _____ deranged;
   c. _____ meandering;
   d. _____ at base level;
   e. _____ braided.

3. One way a stream erodes is by hydraulic action, which is:
   a. _____ the direct impact of water;
   b. _____ changes in base level;
   c. _____ an increase in gradient;
   d. _____ diminished capacity;
   e. _____ solution activity.

4. The feature separating one drainage basin from another is a(an):
   a. _____ delta;
   b. _____ floodplain;
   c. _____ point bar;
   d. _____ oxbow lake;
   e. _____ divide.

5. A stream's gradient is a measure of its:
   a. _____ capacity to erode bedrock;
   b. _____ rate of infiltration and evaporation;
   c. _____ total sediment load;
   d. _____ vertical drop in a given horizontal distance;
   e. _____ recurrence interval of flooding.

6. The most common drainage pattern resembles the branching of a tree. It is known as:
   a. _____ radial;
   b. _____ incised;
   c. _____ dendritic;
   d. _____ graded;
   e. _____ alluvial.

7. Infiltration capacity is the:
   a. _____ distance a stream flows from its source to the ocean;
   b. _____ maximum rate that surface materials can absorb water;
   c. _____ cross-sectional area of a stream's channel multiplied by its velocity;
   d. _____ total amount of suspended load and bed load carried in a stream;
   e. _____ degree to which a stream has become graded.

8. An alluvial deposit formed where a stream flows into a standing body of water is a(an):
   a. _____ alluvial fan;
   b. _____ stream terrace;
   c. _____ point bar;
   d. _____ natural levee;
   e. _____ delta.

9. Base level is defined as:
   a. _____ the amount of sediment deposited at the base of a mountain slope;
   b. _____ the rate at which stream-transported sediment fills a reservoir;
   c. _____ the amount of headward erosion during a flood;
   d. _____ the lowest level to which a stream can erode;
   e. _____ a profile of equilibrium in which no irregularities exist in a channel.

10. Flow around a meander results in erosion on one bank of the channel and deposition of a(an) _____ on the other bank.
    a. _____ oxbow lake;
    b. _____ point bar;
    c. _____ drainage basin;
    d. _____ alluvial fan;
    e. _____ floodplain.

11. How do solar radiation, the changing phases of water, and runoff cause the recycling of water from the oceans to the continents and back to the oceans?

12. Explain how incised meanders form.

13. Describe an alluvial fan and explain how one develops. Where do alluvial fans develop best?

14. Describe the events leading to the progradation of the simplest type of delta.

15. What is the concept of the graded stream, and why can streams be graded only very temporarily?

16. How are the recurrence intervals of floods determined? Of what use are such determinations?

17. Even though stream gradients generally decrease downstream, flow velocity usually increases. Why?

18. Explain what infiltration capacity is and why it is important in considering runoff.

19. What are point bars and natural levees? How do they form?

20. How can a stream lengthen its channel by erosion and deposition?

1. A stream 2000 m above sea level at its source flows 1500 km to the sea. What is the stream's gradient? Do you think the gradient you calculated will be correct for all segments of this stream? Explain.

2. What long-term changes may occur in the hydrologic cycle? How might human activities bring about such changes?

World Wide Web Activities

For these web site addresses, along with current updates and exercises, log on to

http://www.wadsworth.com/geo

▶ U.S. GEOLOGICAL SURVEY WATER RESOURCES OF THE UNITED STATES

This site contains a wealth of information about streams, flooding, flood forecasting, flood warnings, and groundwater.

1. Click on the *Real-Time Hydrologic Data* site under the USGS Water Resources Sites of Regional and State Offices. At that site, click on your state on the map. What is the current status of the major rivers in your state?

2. Under the Publications heading, click on *On-line Reports*. What are some of the current publications available?

3. Under the Data heading, click on the *Water-Use Data* site. Click on the *Periodic Water Fact* icon for the latest information about a new and interesting fact about water resources.

4. Under the Data heading, click on the *Water-Use Data* site. Click on the *Color maps of water use by State for 1990* icon. Click on any of the *National Water-Use Maps* sites for color maps of water usage in the United States in 1990. What was the domestic, industrial, and agricultural surface water usage for your state?

▶ LINKS TO INFORMATION SOURCES FOR HYDROGEOLOGY, HYDROLOGY, AND ENVIRONMENTAL SCIENCES

This site contains links to information sources for hydrogeology, hydrology, and environmental sciences. The links are listed alphabetically. Check out several sites for information about rivers, flooding, surface water usage, both in the United States and elsewhere in the world.

▶ WW2010 (WEATHER WORLD)

This site has descriptions of all aspects of the hydrologic cycle, including Earth's water budget, evaporation, runoff, and groundwater. It has a good summary and animation of the hydrologic cycle. The site is maintained by the Department of Atmospheric Sciences, University of Illinois.

▶ FLOODPLAIN MANAGEMENT

This site, maintained by the Floodplain Management Association, contains information on all aspects of flooding and floodplain management. Scroll down and click on *Flood Basins,* then go to and read *Basic Facts of Flooding.* Click on *Learning Center* and see Overview of Floods, Cause of Floods, Frequency of Floods, and How Much Risk Is Acceptable.

CD-ROM Exploration

Explore the following *In-Terra-Active 2.0* CD-ROM module(s) and increase your understanding of key concepts and processes presented in this chapter.

▶ SECTION: SURFACE
MODULE: RUNNING WATER

For thousands of years, the Nile River and Delta nurtured the growth of the great Egyptian civilization. During August and September, floods rejuvenated the fields stretching along the river with silt and clay transported from Sudan and Uganda. During the rest of the year, the Nile's sediment load replenished the great delta, compensating for erosion of the Mediterranean Sea. In 1968, the Aswan High Dam was completed in an attempt to gain control over the river. **Question: What changes has the Nile River system undergone since the completion of the Aswan High Dam in 1968?**

▶ VIRTUAL REALITY FIELD TRIP: RIVERS AND STREAMS

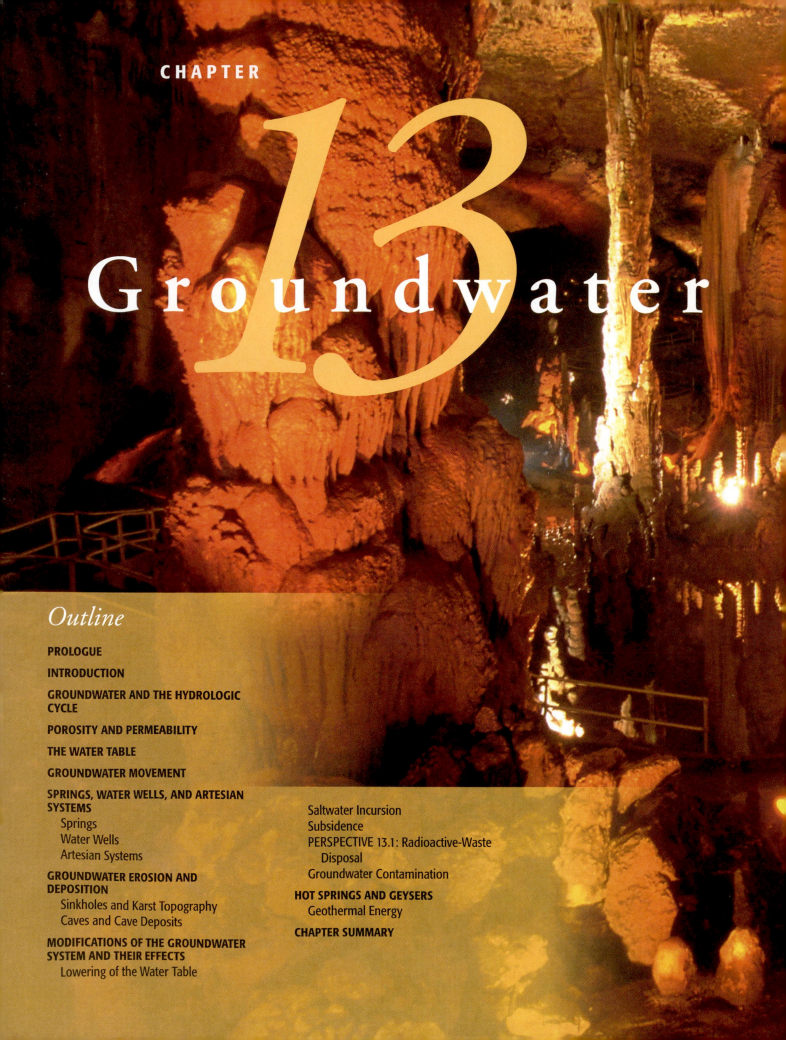

One of the many bathhouses in Bath, England, that were built around hot springs shortly after the Roman Conquest in A.D. 43.

Prologue

Within the limestone region of western Kentucky lies the largest cave system in the world. In 1941 approximately 51,000 acres were set aside and designated as Mammoth Cave National Park. In 1981 it became a World Heritage Site.

From ground level, the topography of the area is unimposing with numerous sinkholes, lakes, valleys, and disappearing streams. Beneath the surface are more than 230 km of interconnecting passageways whose spectacular geologic features have been enjoyed by numerous cave explorers and tourists.

Based on carbon 14 dates from some of the many artifacts found in the cave (such as woven cord and wooden bowls), Mammoth Cave had been explored and used by Native Americans for more than 3000 years prior to its rediscovery in 1799 by a bear hunter named Robert Houchins. During the War of 1812, approximately 180 metric tons of saltpeter (a potassium nitrate mineral), used in the manufacture of gunpowder, were mined from Mammoth Cave. At the end of the war, the saltpeter market collapsed, and Mammoth Cave was developed as a tourist attraction, easily overshadowing the other caves in the area. Over the next 150 years, the discovery of new passageways and caverns helped establish Mammoth Cave as the world's premier cave and the standard against which all others were measured.

Mammoth Cave formed in much the same way as all other caves. Groundwater flowing through the St. Genevieve Limestone eroded a complex network of openings, passageways, and caverns. Flowing through the various caverns is the Echo River, a system of underground streams that eventually joins the Green River at the surface.

The colorful cave deposits are the primary reason millions of tourists have visited Mammoth Cave (Figure 13.1). Other attractions include the Giant's Coffin, a 15-m collapsed block of limestone, and giant rooms such as Mammoth Dome, which is about 58 m high. The cave is also home to more than 200 species of insects and other animals, including about 45 blind species; some of these can be seen on the Echo River Tour, which conveys visitors 5 km along the underground stream.

FIGURE 13.1 Frozen Niagara is a spectacular example of the massive travertine flowstone deposits in Mammoth Cave, Kentucky.

Introduction

GROUNDWATER—the water stored in the open spaces within underground rocks and unconsolidated material—is a valuable natural resource that is essential to the lives of all people. Its importance to humans is not new. Groundwater rights have always been important in North America, and many legal battles have been fought over them. Groundwater also played a crucial role in the development of the U.S. railway system during the nineteenth century when railroads needed a reliable source of water for their steam locomotives. Much of the water used by the locomotives came from groundwater tapped by wells.

Today, the study of groundwater and its movement has become increasingly important as the demand for freshwater by agricultural, industrial, and domestic users has reached an all-time high. More than 65% of the groundwater used in the United States each year goes for irrigation, with industrial use second, followed by domestic needs. These demands have severely depleted the groundwater supply in many areas and led to such problems as ground subsidence and saltwater contamination. In other areas, pollution from landfills, toxic waste, and agriculture has rendered the groundwater supply unsafe.

As the world's population and industrial development expand, the demand for water, particularly groundwater, will increase. Not only is it important to locate new groundwater sources, but, once found, these sources must be protected from pollution and managed properly to ensure that users do not withdraw more water than can be replenished.

Groundwater and the Hydrologic Cycle

GROUNDWATER is one reservoir of the hydrologic cycle (see Figure 12.3), representing approximately 22% (8.4 million km^3) of the world's supply of freshwater. The major source of groundwater is precipitation that infiltrates the ground and moves through the soil and pore spaces of rocks. Other sources include water infiltrating from lakes and streams, recharge ponds, and wastewater-treatment systems. As the groundwater moves through soil, sediment, and rocks, many of its impurities, such as disease-causing microorganisms, are filtered out. Not all soils and rocks are good filters, however, and some serious pollutants are not removed. Groundwater eventually returns to the surface when it enters lakes, streams, or the ocean.

Porosity and Permeability

POROSITY and permeability are important physical properties of Earth materials and are largely responsible for the amount, availability, and movement of groundwater. Water soaks into the ground because the soil, sediment, or rock has open spaces or pores. **Porosity** is the percentage of a material's total volume that is pore space. Porosity most often consists of the spaces between particles in soil, sediments, and sedimentary rocks, but other types of porosity can include cracks, fractures, faults, and vesicles in volcanic rocks (Figure 13.2).

Porosity varies among different rock types and depends on the size, shape, and arrangement of the material composing the rock (Table 13.1). Most igneous and metamorphic rocks, as well as many limestones and dolostones, have very low porosity because they are composed of tightly interlocking crystals. Their porosity can be increased, however, if they have been fractured or weathered by groundwater. This is particularly true for massive limestone and dolostone whose fractures can be enlarged by acidic groundwater.

By contrast, detrital sedimentary rocks composed of well-sorted and well-rounded grains can have very high porosity because any two grains touch only at a single point, leaving relatively large open spaces between the grains (Figure 13.2a). Poorly sorted sedimentary rocks, on the other hand, typically have low porosity because finer grains fill in the space between the larger grains, further reducing porosity (Figure 13.2b). In addition, the amount of cement between grains can also decrease porosity.

Although porosity determines the amount of groundwater a rock can hold, it does not guarantee that the water can be extracted. A material's capacity for transmitting fluids is its **permeability**. Permeability depends on not only porosity but also the size of the pores or fractures and their interconnections. For example, deposits of silt or clay are typically more porous than sand or gravel, but they have low permeability because the pores between the clay particles are very small, and the molecular attraction between the particles and water is great, thereby preventing movement of the water. In contrast, pore spaces between grains in sandstone and conglomerate are much larger, and the molecular attraction on the water is therefore low. Chemical and biochemical sedimentary rocks, such as limestone and dolostone, and many igneous and metamorphic rocks that are highly fractured can also be very permeable provided that the fractures are interconnected.

A permeable layer transporting groundwater is called an **aquifer**, from the Latin *aqua* meaning "water." The most effective aquifers are deposits of well-sorted and well-rounded sand and gravel. Limestones in which fractures and bedding planes have been enlarged by solution are also good aquifers. Shales and many igneous and metamorphic rocks make poor aquifers because they are typically impermeable. Rocks such as these and any other materials that prevent the movement of groundwater are **aquicludes**.

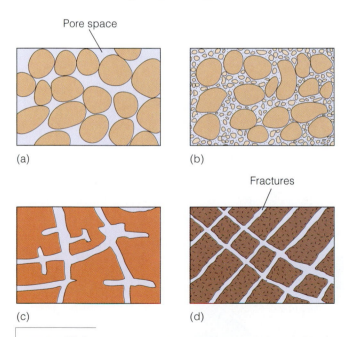

Pore space

Fractures

(a)   (b)

(c)   (d)

FIGURE 13.2 A rock's porosity is dependent on the size, shape, and arrangement of the material composing the rock. (a) A well-sorted sedimentary rock has high porosity while (b) a poorly sorted one has low porosity. (c) In soluble rocks such as limestones, porosity can be increased by solution, whereas (d) crystalline metamorphic and igneous rocks can be rendered porous by fracturing.

TABLE 13.1	
Porosity Values for Different Materials	

MATERIAL	POROSITY (%)
Unconsolidated sediment	
Soil	55
Gravel	20–40
Sand	25–50
Silt	35–50
Clay	50–70
Rocks	
Sandstone	5–30
Shale	0–10
Solution activity in limestone, dolostone	10–30
Fractured basalt	5–40
Fractured granite	10

SOURCE: U.S. Geological Survey, Water Supply Paper 2220 (1983) and others.

The Water Table

WHEN precipitation occurs over land, some of it evaporates, some is carried away by runoff in streams, and the remainder seeps into the ground. As this water moves down from the surface, some of it adheres to the material that it is moving through and halts its downward progress. This region is the **zone of aeration**, and its water is called *suspended water* (Figure 13.3). The pore spaces in this zone contain both water and air.

Beneath the zone of aeration lies the **zone of saturation** where all of the pore spaces are filled with groundwater (Figure 13.3). The base of the zone of saturation varies from place to place but usually extends to a depth where an impermeable layer is encountered or to a depth where confining pressure closes all open space. Extending irregularly upward a few centimeters to several meters from the zone of saturation is the *capillary fringe*. Water moves upward in this region because of surface tension, much as water moves upward through a paper towel.

The surface separating the zone of aeration from the underlying zone of saturation is the **water table** (Figure 13.3). In general, the configuration of the water table is a subdued replica of the overlying land surface; that is, it has its highest elevations beneath hills and its lowest elevations in valleys. In most arid and semiarid regions, however, the water table is quite flat and is below the level of river valleys.

Several factors contribute to the surface configuration of a region's water table. These include regional differences in the amount of rainfall, permeability, and the rate of groundwater movement. During periods of high rainfall, groundwater tends to rise beneath hills because it cannot flow fast enough into the adjacent valleys to maintain a level surface. During droughts, the water table falls and tends to flatten out because it is not being replenished.

Groundwater Movement

GROUNDWATER velocity varies greatly and depends on many factors. Velocities range from 250 m per day in some extremely permeable material to less than a few centimeters per year in nearly impermeable material. In most ordinary aquifers, the average velocity of groundwater is a few centimeters per day.

Gravity provides the energy for the downward movement of groundwater. Water entering the ground moves through the zone of aeration to the zone of saturation (Figure 13.3). When water reaches the water table, it continues to move through the zone of saturation from areas where the water table is high toward areas where it is lower, such as at streams, lakes, or swamps. Only some of the water follows the direct route along the slope of the water table. Most of it takes longer curving paths downward and then enters a stream, lake, or swamp from below, because it moves from areas of high pressure toward areas of lower pressure within the saturated zone.

Springs, Water Wells, and Artesian Systems

ADDING water to the zone of saturation is known as *recharge*, and it causes the water table to rise. Water may be added by natural means, such as rainfall or melting snow, or artificially at recharge basins or wastewater-treatment plants. If groundwater is discharged without sufficient replenishment, the water table drops. Ground-

FIGURE 13.3 The zone of aeration contains both air and water within its open space, whereas all open space in the zone of saturation is filled with groundwater. The water table is the surface separating the zones of aeration and saturation. Within the capillary fringe, water rises upward by surface tension from the zone of saturation into the zone of aeration.

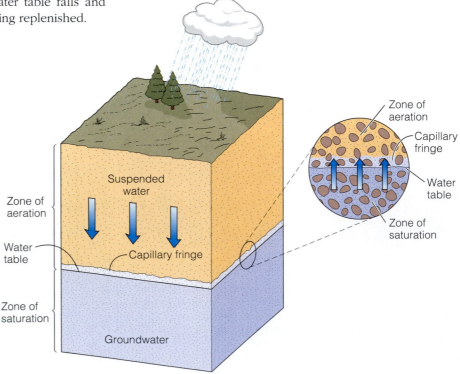

water discharges naturally whenever the water table intersects the ground surface, as at a spring or along a stream, lake, or swamp. Groundwater can also be discharged artificially by pumping water from wells.

SPRINGS

A **spring** is a place where groundwater flows or seeps out of the ground. Springs have always fascinated people because the water flows out of the ground for no apparent reason and from no readily identifiable source. It is not surprising that springs have long been regarded with superstition and revered for their supposed medicinal value and healing powers. Nevertheless, there is nothing mystical or mysterious about springs.

Although springs can occur under a wide variety of geologic conditions, they all form in basically the same way (Figure 13.4). When percolating water reaches the water table or an impermeable layer, it flows laterally, and if this flow intersects the surface, the water discharges as a spring. The Mammoth Cave area in Kentucky is underlain by fractured limestones that have been enlarged into caves by solution activity (see the Prologue). In this geologic environment, springs occur where the fractures and caves intersect the ground surface allowing groundwater to exit onto the surface. Most springs are along valley walls where streams have cut valleys below the regional water table.

Springs can also develop wherever a perched water table intersects the surface (Figure 13.5). A *perched water table* may occur wherever a local aquiclude is present within a larger aquifer, such as a lens of shale within a sandstone. As water migrates through the zone of aeration, it is stopped by the local aquiclude, and a localized zone of

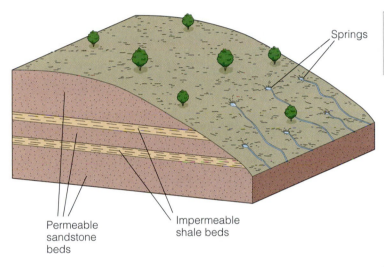

Permeable sandstone beds

Impermeable shale beds

FIGURE 13.4 Springs form wherever laterally moving groundwater intersects Earth's surface. Most commonly, they form when percolating water reaches an impermeable layer and migrates laterally until it seeps out at the surface.

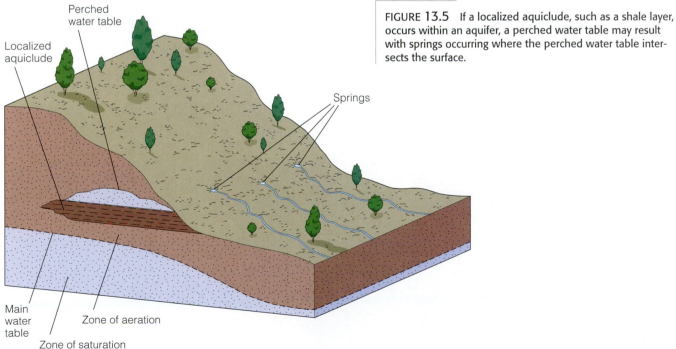

Perched water table

Localized aquiclude

Springs

Main water table

Zone of aeration

Zone of saturation

FIGURE 13.5 If a localized aquiclude, such as a shale layer, occurs within an aquifer, a perched water table may result with springs occurring where the perched water table intersects the surface.

saturation "perched" above the main water table forms. Water moving laterally along the perched water table may intersect the surface to produce a spring.

WATER WELLS

A **water well** is made by digging or drilling into the zone of saturation. Once the zone of saturation is reached, water percolates into the well and fills it to the level of the water table. Most wells must be pumped to bring the groundwater to the surface.

When a well is pumped, the water table in the area around the well is lowered because water is removed from the aquifer faster than it can be replenished. A **cone of depression** thus forms around the well, varying in size according to the rate and amount of water being withdrawn (Figure 13.6). If water is pumped out of a well faster than it can be replaced, the cone of depression grows until the well goes dry. This lowering of the water table normally does not pose a problem for the average domestic well, provided that the well is drilled sufficiently deep into the zone of saturation. The tremendous amounts of water used by industry and irrigation, however, may create a large cone of depression that lowers the water table sufficiently to cause shallow wells in the immediate area to go dry (Figure 13.6). This situation is not uncommon and frequently results in lawsuits by the owners of the shallow dry wells. Furthermore, lowering of the regional water table is becoming a serious problem in many areas, particularly in the southwestern United States where rapid growth has placed tremendous demands on the groundwater system. Unrestricted withdrawal of groundwater cannot continue indefinitely, and the rising costs and decreasing supply of groundwater should soon limit the growth of this region of the United States.

ARTESIAN SYSTEMS

The term **artesian system** can be applied to any system in which groundwater is confined and builds up high hydrostatic (fluid) pressure. Water in such a system is able to rise above the level of the aquifer if a well is drilled through the confining layer, thereby reducing the pressure and forcing the water upward. For an artesian system to develop, three geologic conditions must be present (Figure 13.7): (1) the aquifer must be confined above and below by aquicludes to prevent water from escaping; (2) the rock sequence is usually tilted and exposed at the surface, enabling the aquifer to be recharged; and (3) there is sufficient precipitation in the recharge area to keep the aquifer filled.

The elevation of the water table in the recharge area and the distance of the well from the recharge area determine the height to which artesian water rises in a well. The surface defined by the water table in the recharge area, called the *artesian-pressure surface,* is indicated by the sloping dashed line in Figure 13.7. If there were no friction in the aquifer, well water from an artesian aquifer would rise exactly to the elevation of the artesian-pressure surface. Friction, however, slightly reduces the pressure of the aquifer water and consequently the level to which artesian water rises. This is why the pressure surface slopes.

An artesian well will flow freely at the ground surface only if the wellhead is at an elevation below the artesian-pressure surface. In this situation, the water flows out of the well because it rises toward the artesian-pressure surface, which is at a higher elevation than the wellhead. In a nonflowing artesian well, the wellhead is above the artesian-pressure surface, and the water will rise in the well only as high as the artesian-pressure surface.

In addition to artesian wells, many artesian springs also exist. Such springs can occur if a fault or fracture intersects the confined aquifer, allowing water to rise above the aquifer. Desert oases are commonly artesian springs.

Because the geologic conditions necessary for artesian water can occur in a variety of ways, artesian systems are quite common in many areas of the world underlain by sedimentary rocks. One of the best-known artesian systems in the United States underlies South Dakota and extends south to central Texas. The majority of the artesian water from this system is used for irrigation. The aquifer of this artesian system, the Dakota Sandstone, is recharged where it is exposed along the margins of the Black Hills of South Dakota. The hydrostatic pressure in this system was originally great enough to produce free-flowing wells and to

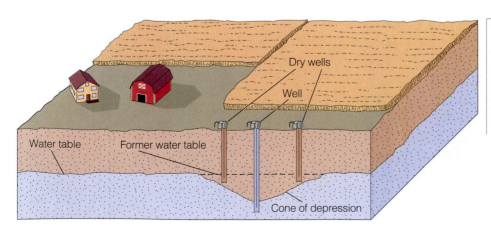

FIGURE **13.6** A cone of depression forms whenever water is withdrawn from a well. If water is withdrawn faster than it can be replenished, the cone of depression will grow in depth and circumference, lowering the water table in the area and causing nearby shallow wells to go dry.

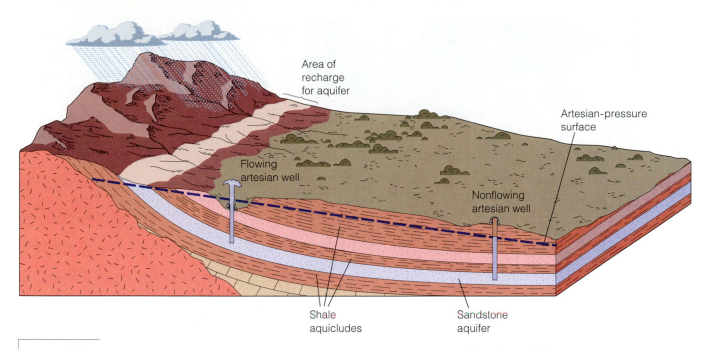

FIGURE 13.7   An artesian system must have an aquifer confined above and below by aquicludes, the aquifer must be exposed at the surface, and there must be sufficient precipitation in the recharge area to keep the aquifer filled. The elevation of the water table in the recharge area, which is indicated by the sloping dashed line (the artesian-pressure surface), defines the highest level to which well water can rise. If the elevation of a wellhead is below the elevation of the artesian-pressure surface, the well will be free-flowing because the water will rise toward the artesian-pressure surface, which is at a higher elevation than the wellhead. If the elevation of a wellhead is at or above that of the artesian-pressure surface, the well will be nonflowing.

operate waterwheels. The extensive use of water for irrigation over the years has reduced the pressure in many of the wells so that they are no longer free-flowing and the water must be pumped.

Groundwater Erosion and Deposition

WHEN rainwater begins seeping into the ground, it immediately starts to react with the minerals it contacts, weathering them chemically. In an area underlain by soluble rock, groundwater is the principal agent of erosion and is responsible for the formation of many major features of the landscape.

Limestone, a common sedimentary rock composed primarily of the mineral calcite ($CaCO_3$), underlies large areas of Earth's surface. Although limestone is practically insoluble in pure water, it readily dissolves if a small amount of acid is present. Carbonic acid (H_2CO_3) is a weak acid that forms when carbon dioxide combines with water ($H_2O + CO_2 \rightarrow H_2CO_3$) (see Chapter 6). Because the atmosphere contains a small amount of carbon dioxide (0.03%) and carbon dioxide is also produced in soil by the decay of organic matter, most groundwater is slightly acidic. When groundwater percolates through the various openings in limestone, the slightly acidic water readily reacts with the calcite to dissolve the rock by forming soluble calcium bicarbonate, which is carried away in solution (see Chapter 6).

SINKHOLES AND KARST TOPOGRAPHY

In regions underlain by soluble rock, the ground surface may be pitted with numerous depressions that vary in size and shape. These depressions, called **sinkholes** or merely *sinks,* mark areas where the underlying rock is soluble (Figure 13.8). Sinkholes usually form in one of two ways. The first is when the soluble rock below the soil is dissolved by seeping water. Natural openings in the rock are enlarged and filled in by the overlying soil. As the groundwater continues to dissolve the rock, the soil is eventually removed, leaving depressions that are typically shallow with gently sloping sides.

Sinkholes also form when a cave's roof collapses, usually producing a steep-sided crater. Sinkholes formed in this way are a serious hazard, particularly in populated areas. In regions prone to sinkhole formation, the depth and extent of underlying cave systems must be mapped before any development to ensure that the underlying rocks are thick enough to support planned structures.

A **karst topography** is one that has developed largely by groundwater erosion (Figure 13.9). The name *karst* is derived from the plateau region of the border area of Slovenia, Croatia, and northeastern Italy where this type of topography is well developed. In the United States, regions of karst topography include large areas of southwestern Illinois, southern Indiana, Kentucky, Tennessee, northern Missouri, Alabama, and central and northern Florida.

Karst topography is characterized by numerous caves, springs, sinkholes, solution valleys, and disappearing

FIGURE **13.8** This sinkhole formed on May 8 and 9, 1981, in Winter Park, Florida, due to a drop in the water table after prior dissolution of the underlying limestone. The sinkhole destroyed a house, numerous cars, and the municipal swimming pool. It has a diameter of 100 m and a depth of 35 m.

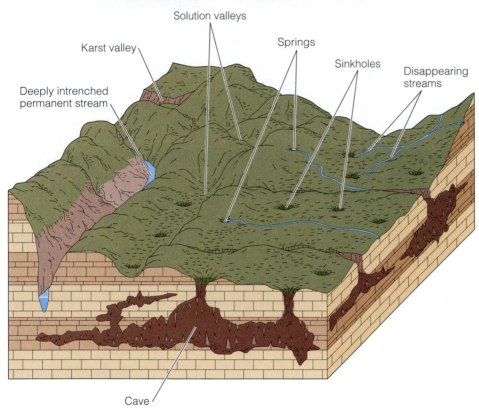

Solution valleys

Karst valley

Springs

Sinkholes

Disappearing streams

Deeply intrenched permanent stream

Cave

FIGURE **13.9** Some of the features of karst topography.

streams (Figure 13.9). When adjacent sinkholes merge, they form a network of larger, irregular, closed depressions called *solution valleys. Disappearing streams,* another feature of areas of karst topography, are so named because they typically flow only a short distance at the surface and then disappear into a sinkhole. The water continues flowing underground through various fractures or caves until it surfaces again at a spring or other stream.

Karst topography can range from the spectacular high-relief landscapes of China to the subdued and pockmarked landforms of Kentucky (Figure 13.10). Common to all karst topography, though, is the presence of thick-bedded, readily soluble rock at the surface or just below the soil and

enough water for solution activity to occur. Karst topography is, therefore, typically restricted to humid and temperate climates.

CAVES AND CAVE DEPOSITS

Caves are perhaps the most spectacular examples of the combined effects of weathering and erosion by groundwater. As groundwater percolates through carbonate rocks, it dissolves and enlarges original fractures and openings to form a complex interconnecting system of crevices, caves, caverns, and underground streams. A **cave** is usually defined as a naturally formed subsurface opening that is

(a)

(b)

FIGURE 13.10 (a) The Stone Forest, 126 km southeast of Kunming, People's Republic of China, is a high-relief karst landscape formed by the dissolution of carbonate rocks. (b) Solution valleys, sinkholes, and sinkhole lakes dominate the subdued karst topography east of Bowling Green, Kentucky.

generally connected to the surface and is large enough for a person to enter. A *cavern* is a very large cave or a system of interconnected caves.

More than 17,000 caves are known in the United States. Most of them are small, but some are quite large and spectacular. Some of the more famous caves in the United States are Mammoth Cave, Kentucky (see the Prologue); Carlsbad Caverns, New Mexico; Lewis and Clark Caverns, Montana; and Meramec Caverns, Missouri, which Jesse James and his outlaw band often used as a hideout. While the United States has many famous caves, the deepest known cave in North America is the 536-m-deep Arctomys Cave in Mount Robson Provincial Park, British Columbia, Canada.

Caves and caverns form as a result of the dissolution of carbonate rocks by weakly acidic groundwater (Figure 13.11). Groundwater percolating through the zone of aeration slowly dissolves the carbonate rock and enlarges its fractures and bedding planes. Upon reaching the water table, the groundwater migrates toward the region's surface streams. As the groundwater moves through the zone of saturation, it continues to dissolve the rock and gradually forms a system of horizontal passageways through which the dissolved rock is carried to the streams. As the surface streams erode deeper valleys, the water table drops in response to the lower elevation of the streams. The water that flowed through the system of horizontal passageways now percolates to the lower water table where a new system of passageways begins to form. The abandoned channelways now form an interconnecting system of caves and caverns. Caves eventually become unstable and collapse, littering their floors with fallen debris.

When most people think of caves, they think of the seemingly endless variety of colorful and bizarre-shaped deposits found in them. Although a great many different types of cave deposits exist, most form in essentially the same manner and are collectively known as **dripstone**. As water seeps through a cave, some of the dissolved carbon dioxide in the water escapes, and a small amount of calcite is precipitated. In this manner, the various dripstone deposits are formed.

Stalactites are icicle-shaped structures hanging from cave ceilings that form as a result of precipitation from dripping water (Figure 13.12). With each drop of water, a thin layer of calcite is deposited over the previous layer, forming a cone-shaped projection that grows down from the ceiling.

The water that drips from a cave's ceiling also precipitates a small amount of calcite when it hits the floor. As additional calcite is deposited, an upward-growing projection called a *stalagmite* forms (Figure 13.12). If a stalactite and stalagmite meet, they form a *column*. Groundwater seeping from a crack in a cave's ceiling may form a vertical sheet of rock called a *drip curtain,* while water flowing across a cave's floor may produce *travertine terraces* (Figure 13.11).

Modifications of the Groundwater System and Their Effects

GROUNDWATER is a valuable natural resource that is rapidly being exploited with little regard to the effects of overuse and misuse. Currently, about 20% of all water used in the United States is groundwater. This percentage is increasing, and unless this resource is used more wisely, sufficient amounts of clean groundwater will not be available in the future. Modifications of the groundwater system may have many consequences including (1) lowering of the water table, causing wells to dry up; (2) loss of hydrostatic pressure, causing once free-flowing wells to require pumping; (3) saltwater encroachment; (4) subsidence; and (5) contamination of the groundwater supply.

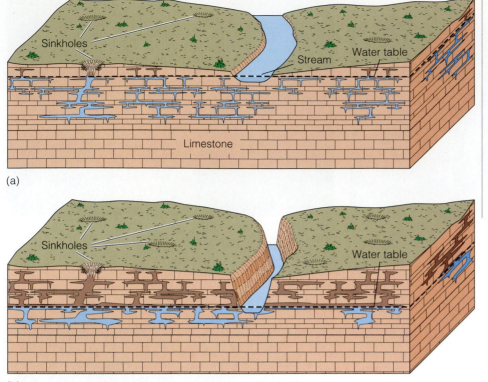

Sinkholes

Stream

Water table

Limestone

(a)

Sinkholes

Water table

(b)

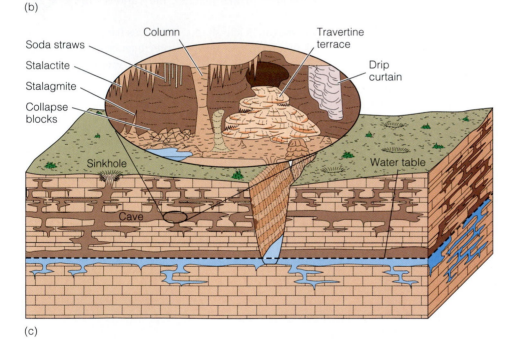

Column

Travertine terrace

Soda straws

Drip curtain

Stalactite

Stalagmite

Collapse blocks

Sinkhole

Water table

Cave

(c)

FIGURE 13.11 The formation of caves. (a) As groundwater percolates through the zone of aeration and flows through the zone of saturation, it dissolves the carbonate rocks and gradually forms a system of passageways. (b) Groundwater moves along the surface of the water table, forming a system of horizontal passageways through which dissolved rock is carried to the surface streams, thus enlarging the passageways. (c) As the surface streams erode deeper valleys, the water table drops, and the abandoned channelways form an interconnecting system of caves and caverns.

LOWERING OF THE WATER TABLE

Withdrawing groundwater at a significantly greater rate than it is replaced by either natural or artificial recharge can have serious effects. For example, the High Plains aquifer is one of the most important aquifers in the United States. Underlying most of Nebraska, large parts of Colorado and Kansas, portions of South Dakota, Wyoming, and New Mexico, as well as the panhandle regions of Oklahoma and Texas, it accounts for approximately 30% of the groundwater used for irrigation in the United States (Figure 13.13). Ir-

rigation from the High Plains aquifer is largely responsible for the region's high agricultural productivity, which includes a significant percentage of the nation's corn, cotton, and wheat and half of U.S. beef cattle. Large areas of land (more than 14 million acres) are currently irrigated with water pumped from the High Plains aquifer. Irrigation is popular because yields from irrigated lands can be triple what they would be without irrigation.

While the High Plains aquifer has contributed to the high productivity of the region, it cannot continue provid-

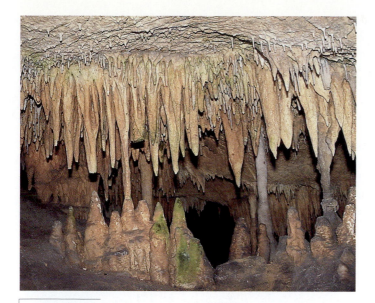

FIGURE 13.12 Stalactites are the icicle-shaped structures seen hanging from the ceiling, whereas the upward-pointing structures on the cave floor are stalagmites. Several columns are present where the stalactites and stalagmites have met in this chamber of Luray Caves, Virginia.

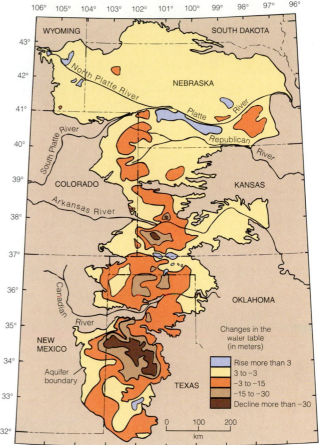

FIGURE 13.13 Areal extent of the High Plains aquifer and changes in the water table, predevelopment to 1980.

ing the quantities of water that it has in the past. In some parts of the High Plains, from 2 to 100 times more water is being pumped annually than is being recharged. Consequently, water is being removed from the aquifer faster than it is being replenished, causing the water table to drop significantly in many areas.

What will happen to this region's economy if long-term withdrawal of water from the High Plains aquifer greatly exceeds its recharge rate so that it can no longer supply the quantities of water necessary for irrigation? Solutions range from going back to farming without irrigation to diverting water from other regions such as the Great Lakes. Farming without irrigation would result in greatly decreased yields and higher costs and prices for agricultural products, while the diversion of water from elsewhere would cost billions of dollars and the price of agricultural products would still rise.

SALTWATER INCURSION

The excessive pumping of groundwater in coastal areas can result in *saltwater incursion* such as occurred on Long Island, New York, during the 1960s. Along coastlines where permeable rocks or sediments are in contact with the ocean, the fresh groundwater, being less dense than seawater, forms a lens-shaped body above the underlying salt water (Figure 13.14a). The weight of the freshwater exerts pressure on the underlying salt water. As long as rates of recharge equal rates of withdrawal, the contact between the fresh groundwater and the seawater will remain the same. If excessive pumping occurs, a deep cone of depression forms in the fresh groundwater (Figure 13.14b). Because some of the pressure from the overlying fresh

water has been removed, salt water forms a *cone of ascension* as it rises to fill the pore space that formerly contained freshwater. When this occurs, wells become contaminated with salt water and remain contaminated until recharge by freshwater restores the former level of the fresh-groundwater water table.

Saltwater incursion is a major problem in many rapidly growing coastal communities. As the population in these areas grows, greater demand for groundwater creates an even greater imbalance between recharge and withdrawal.

To counteract the effects of saltwater incursion, recharge wells are often drilled to pump water back into the groundwater system (Figure 13.14c). Recharge ponds that allow large quantities of fresh surface water to infiltrate the groundwater supply may also be constructed.

SUBSIDENCE

As excessive amounts of groundwater are withdrawn from poorly consolidated sediments and sedimentary rocks, the water pressure between grains is reduced, and the weight of the overlying materials causes the grains to pack closer together, resulting in subsidence of the ground. As more

Modifications of the Groundwater System and Their Effects　279

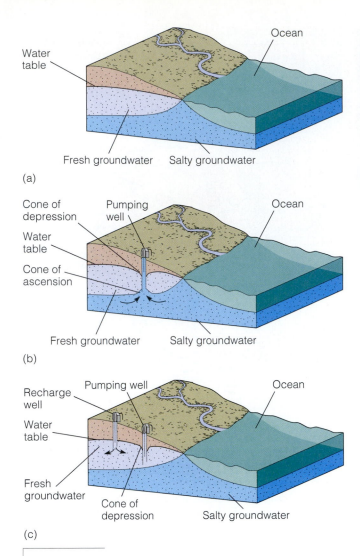

(a)

Water table · Ocean · Fresh groundwater · Salty groundwater

(b)

Cone of depression · Pumping well · Ocean · Water table · Cone of ascension · Fresh groundwater · Salty groundwater

(c)

Recharge well · Pumping well · Ocean · Water table · Fresh groundwater · Cone of depression · Salty groundwater

FIGURE **13.14** Saltwater incursion. (a) Because freshwater is not as dense as salt water, it forms a lens-shaped body above the underlying saltwater. (b) If excessive pumping occurs, a cone of depression develops in the fresh groundwater, and a cone of ascension forms in the underlying salty groundwater that may result in salt water contamination of the well. (c) Pumping water back into the groundwater system through recharge wells can help lower the interface between the fresh groundwater and the salty groundwater and reduce saltwater incursion.

FIGURE **13.15** The Leaning Tower of Pisa, Italy. The tilting is partly the result of subsidence due to removal of groundwater.

and more groundwater is pumped to meet the increasing needs of agriculture, industry, and population growth, subsidence is becoming more prevalent.

The San Joaquin Valley of California is a major agricultural region that relies largely on groundwater for irrigation. Between 1925 and 1975, groundwater withdrawals in parts of the valley caused subsidence of almost 9 m. Other examples of subsidence in the United States include New Orleans, Louisiana, and Houston, Texas, both of which have subsided more than 2 m, and Las Vegas, Nevada, which has subsided 8.5 m.

Elsewhere in the world, the tilt of the Leaning Tower of Pisa is partly due to groundwater withdrawal (Figure 13.15). The tower started tilting soon after construction be-

gan in 1173 because of differential compaction of the foundation. During the 1960s, the city of Pisa withdrew ever larger amounts of groundwater, causing the ground to subside further; as a result, the tilt of the tower increased until it was considered in danger of falling over. Strict control of groundwater withdrawal and stabilization of the foundation have now reduced the amount of tilting to about 1 mm per year, ensuring that the tower should stand for several more centuries.

The extraction of oil can also cause subsidence. Long Beach, California, has subsided 9 m as a result of 34 years of oil production. More than $100 million of damage was done to the pumping, transportation, and harbor facilities in this area because of subsidence and encroachment of the sea (Figure 13.16). Once water was pumped back into the oil reservoir, subsidence virtually stopped.

GROUNDWATER CONTAMINATION

A major problem facing our society is the safe disposal of the numerous pollutant by-products of an industrialized economy. We are becoming increasingly aware that our streams, lakes, and oceans are not unlimited reservoirs for waste and that we must find new safe ways to dispose of pollutants.

FIGURE 13.16 The withdrawal of petroleum from the oil field in Long Beach, California, resulted in 9 m of ground subsidence because of sediment compaction. Not until water was pumped back into the reservoir to replace the petroleum did ground subsidence essentially cease. (2 to 29 feet = 0.6 to 8.8 meters)

The most common sources of groundwater contamination are sewage, landfills, toxic waste–disposal sites (see Perspective 13.1), and agriculture. Once pollutants get into the groundwater system, they will spread wherever groundwater travels, which can make their containment difficult. Furthermore, because groundwater moves so slowly, it takes a very long time to cleanse a groundwater reservoir once it has become contaminated.

In many areas, septic tanks are the most common way of disposing of sewage. A septic tank slowly releases sewage into the ground where it is decomposed by oxidation and microorganisms and filtered by the sediment as it percolates through the zone of aeration. In most situations, by the time the water from the sewage reaches the zone of saturation, it has been cleansed of any impurities and is safe to use (Figure 13.17a). If the water table is close to the surface or if the rocks are very permeable, water entering the zone of saturation may still be contaminated and unfit to use.

Landfills are also potential sources of groundwater contamination (Figure 13.17b). Not only does liquid waste seep into the ground, but rainwater also carries dissolved chemicals and other pollutants down into the groundwater reservoir. Unless the landfill is carefully designed and lined below by an impermeable layer such as clay, many toxic compounds such as paints, solvents, cleansers, pesticides, and battery acid will find their way into the groundwater system.

Toxic-waste sites where dangerous chemicals are either buried or pumped underground are an increasing source of groundwater contamination. The United States alone must dispose of several thousand metric tons of hazardous chemical waste per year. Unfortunately, much of this waste has been, and still is being, improperly dumped and is contaminating the surface water, soil, and groundwater.

Hot Springs and Geysers

THE subsurface rocks in regions of recent volcanic activity usually stay hot for thousands of years. Groundwater percolating through these rocks is heated and, if returned to the surface, forms hot springs or geysers. Yellowstone National Park in the United States, Rotorua, New Zealand, and Iceland are all famous for their hot springs and geysers. All are sites of recent volcanism, and consequently their subsurface rocks and groundwater are very hot. The water in some hot springs, however, is circulated deep into Earth, where it is warmed by the normal increase in temperature, the geothermal gradient (see Chapter 9).

A **hot spring** (also called a *thermal spring* or *warm spring*) is a spring in which the water temperature is warmer than the temperature of the human body (37°C) (Figure 13.18). Some hot springs are much hotter, with temperatures ranging up to the boiling point in many instances. Of the approximately 1100 known hot springs in the United States, more than 1000 are in the Far West, while the rest are in the Black Hills of South Dakota, Georgia, the Ouachita region of Arkansas, and the Appalachian region.

Hot springs are also common in other parts of the world. One of the most famous is at Bath, England, where shortly after the Roman conquest of Britain in A.D. 43, numerous bathhouses and a temple were built around the hot springs (see chapter-opening photo).

Geysers are hot springs that intermittently eject hot water and steam with tremendous force. The word comes from the Icelandic *geysir*, which means "to gush or rush forth." One of the most famous geysers in the world is Old Faithful in Yellowstone National Park in Wyoming (Figure 13.19). With a thunderous roar, it erupts a column of hot

Radioactive-Waste Disposal

One of the problems of the nuclear age is finding safe storage sites for the radioactive waste from nuclear power plants, the manufacture of nuclear weapons, and the radioactive by-products of nuclear medicine. Radioactive waste can be grouped into two categories: low-level and high-level waste. Low-level wastes are low enough in radioactivity that, when properly handled, they do not pose a significant environmental threat. Most low-level wastes can be safely buried in controlled dump sites where the geology and groundwater system are well known and careful monitoring is provided.

High-level radioactive waste, such as the spent uranium fuel assemblies used in nuclear reactors and the material used in nuclear weapons, is extremely dangerous because of high amounts of radioactivity; it therefore presents a major environmental problem. Currently, some 28,000 metric tons of spent uranium fuel are being stored in 70 "temporary" sites in 35 states, while awaiting shipment to a permanent site in Nevada. Furthermore, the Department of Energy (DOE) estimates that by the year 2000 the nation will have produced almost 50,000 metric tons of highly radioactive waste that must be disposed of safely.

In 1986, Congress chose Yucca Mountain as the only candidate to house the nation's ever-increasing amounts of civilian high-level radioactive waste (Figure 1). Congress also authorized the DOE to study the suitability of the site. Such a facility must be able to isolate high-level waste from the environment for at least 10,000 years, which is the minimum time the waste will remain dangerous. The Yucca Mountain site will have a capacity of 70,000 metric tons of waste and will not be completely filled until around the year 2030, at which time its entrance shafts will be sealed and backfilled (Figure 2).

The canisters holding the waste are designed to remain leakproof for at least 300 years, so there is some possibility that leakage could occur over the next 10,000 years. The DOE thinks, however, that the geology of the area will prevent radioactive isotopes from entering the groundwater system. Under an Environmental Protection Agency (EPA) regulation, a radioactive dump site must be located so that the travel time for groundwater from the site to the outside environment is at least 1000 years.

The radioactive waste at the Yucca Mountain repository will be buried in a volcanic tuff at a depth of about 300 m. The water table in the area will be an additional 200 to 420 m below the dump site. Thus, the canisters will be stored in the zone of aeration, which was one of the reasons Yucca Mountain was selected. Only about 15 cm of rain fall in this area per year, and only a small amount of this percolates into the ground. Most of the water that does seep into the ground evaporates before it migrates very far, so the rock at the

Figure 1 The location and aerial view of Nevada's Yucca Mountain.

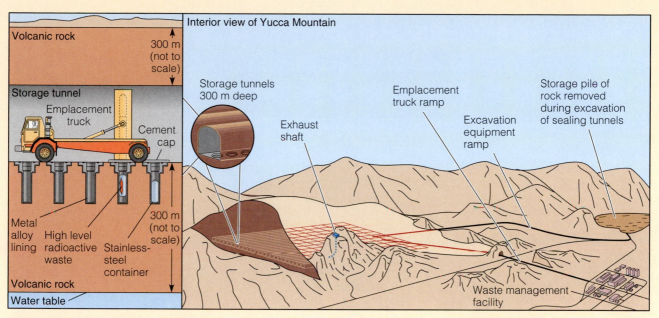

Figure 2 *Schematic diagram of the proposed Yucca Mountain high-level radioactive waste–disposal facility.*

The following labels appear in the diagram:

Volcanic rock
300 m (not to scale)
Interior view of Yucca Mountain
Storage tunnel
Emplacement truck
Cement cap
Storage tunnels 300 m deep
Exhaust shaft
Emplacement truck ramp
Excavation equipment ramp
Storage pile of rock removed during excavation of sealing tunnels
Metal alloy lining
High level radioactive waste
Stainless-steel container
300 m (not to scale)
Volcanic rock
Water table
Waste management facility

depth the canisters are buried will be very dry, helping prolong their lives.

Geologists think that the radioactive waste at Yucca Mountain is most likely to contaminate the environment if it is in liquid form; if liquid, it could seep into the zone of saturation and enter the groundwater supply. But because of the low moisture in the zone of aeration, there is little water to carry the waste downward, and it will take more than 1000 years to reach the zone of saturation. In fact, the DOE estimates that the waste will take longer than 10,000 years to move from the repository to the water table.

Some geologists are concerned that the climate will change during the next 10,000 years. If the region should become more humid, more water will percolate through the zone of aeration. This will increase the corrosion rate of the canisters and could cause the water table to rise, thereby decreasing the

travel time between the repository and the zone of saturation. This area of the country was much more humid during the Ice Age 1.6 million to 10,000 years ago (see Chapter 14).

Another concern is the seismic activity in the area. It is in fact riddled with faults. At least 27 earthquakes with magnitudes greater than 3 occurred in the area between 1852 and 1991. A quake occurred on June 29, 1992, with a magnitude of 5.6 and an epicenter only 32 km from Yucca Mountain. Nevertheless, the DOE is convinced that earthquakes pose little danger to the underground repository itself because the disruptive effects of an earthquake are usually confined to the surface.

Finally, some suggest that the DOE has not thoroughly evaluated the economic potential of the area. Exploration is occurring around the Yucca Mountain site, and some Nevada government

officials think that there is geologic evidence for various metals and possibly oil and gas in the area. Should human intrusion occur during the thousands of years that the site is supposed to be isolated, dangerous radiation could be released into the environment. Others think the economic potential is being sufficiently evaluated and that the area in question has a low economic potential.

While it appears that Yucca Mountain meets all of the requirements for a safe high-level radioactive waste dump, the site is still controversial. In fact, the government has spent $3 billion since 1986 studying Yucca Mountain and will have dug a huge U-shaped tunnel through the site in early 1997 to ensure the subsurface geology is suitable for a high-level radioactive waste–disposal site. Nevertheless, there are those critics of the project who think the site is not safe and should be abandoned.

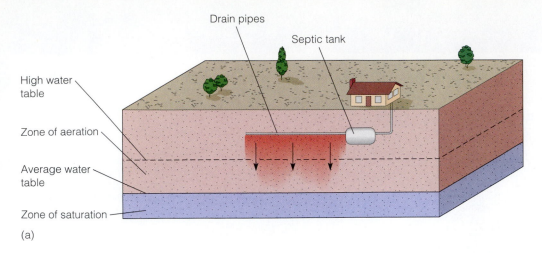

High water table

Drain pipes

Septic tank

Zone of aeration

Average water table

Zone of saturation

(a)

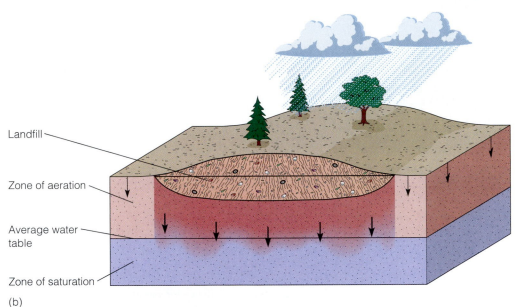

Landfill

Zone of aeration

Average water table

Zone of saturation

(b)

FIGURE 13.17 (a) A septic system slowly releases sewage into the zone of aeration. Oxidation, bacterial degradation, and filtering by the sediments usually remove all natural impurities before they reach the water table. If the rocks are very permeable or the water table is too close to the septic system, contamination of the groundwater can result. (b) Unless there is an impermeable barrier between a landfill and the water table, pollutants can be carried into the zone of saturation and contaminate the groundwater supply.

FIGURE 13.18 Hot springs are springs with a water temperature greater than 37°C. This hot spring is in West Thumb Geyser Basin, Yellowstone National Park, Wyoming.

FIGURE 13.19 Old Faithful Geyser in Yellowstone National Park, Wyoming, is one of the world's most famous geysers, erupting approximately every 30 to 90 minutes.

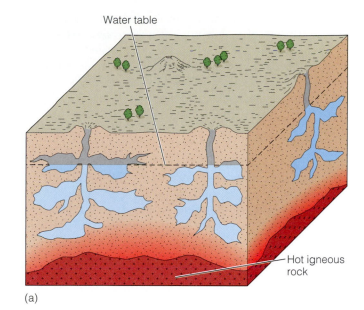

(a)

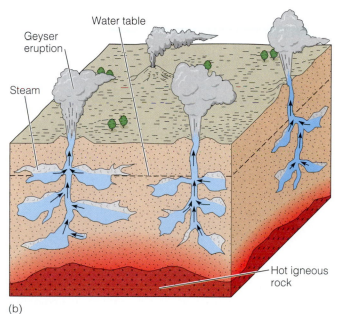

(b)

FIGURE 13.20 The formation of a geyser. (a) Groundwater percolates down into a network of interconnected openings and is heated by hot igneous rocks. The water near the bottom of the fracture system is under greater pressure than that near the top and consequently must be heated to a higher temperature before it will boil. (b) Any rise in temperature of the water above its boiling point or a drop in pressure will cause the water to change to steam, which quickly pushes the water above it up and out of the ground, producing a geyser eruption.

water and steam every 30 to 90 minutes. Other well-known geyser areas are found in Iceland and New Zealand.

Geysers are the surface expression of an extensive underground system of interconnected fractures within hot igneous rocks (Figure 13.20). Groundwater percolating into the network of fractures is heated as it comes into contact with the hot rocks. Because the water near the bottom of the fracture system is under greater pressure than that near the top, it must be heated to a higher temperature before it will boil. Thus, when the deeper water is heated to very near the boiling point, a slight rise in temperature or a drop in pressure, such as from escaping gas, will cause it to instantly change to steam. The expanding steam quickly pushes the water above it out of the ground and into the air, thereby producing a geyser eruption. After the eruption, relatively cool groundwater starts to seep back into the fracture system where it is heated to near its boiling temperature, and the eruption cycle begins again. Such a process explains how geysers can erupt with some regularity.

Hot spring and geyser water typically contains large quantities of dissolved minerals because most minerals dissolve more rapidly in warm water than in cold water. Due to this high-mineral content, the waters of many hot springs are believed by some to have medicinal properties.

Numerous spas and bathhouses have been built throughout the world at hot springs to take advantage of these supposed healing properties.

When the highly mineralized water of hot springs or geysers cools at the surface, some of the material in solution is precipitated, forming various types of deposits. The amount and type of precipitated minerals depend on the solubility and composition of the material that the groundwater flows

Hot Springs and Geysers 285

through. If the groundwater contains dissolved calcium carbonate ($CaCO_3$), then *travertine* or *calcareous tufa* (both of which are varieties of limestone) are precipitated. Spectacular examples of hot spring travertine deposits are found at Mammoth Hot Springs in Yellowstone National Park (Figure 13.21) and at Pamukhale in Turkey.

GEOTHERMAL ENERGY

Energy harnessed from steam and hot water trapped within the crust is **geothermal energy**, a desirable and relatively nonpolluting alternate form of energy. Approximately 1 to 2% of the world's current energy needs could be met by geothermal energy. In those areas where it is plentiful, geothermal energy can supply most, if not all, of the energy needs, sometimes at a fraction of the cost of other types of energy. Some of the countries currently using geothermal energy in one form or another include Iceland, the United States, Mexico, Italy, New Zealand, Japan, the Philippines, and Indonesia.

The city of Rotorua, New Zealand, is world famous for its volcanoes, hot springs, geysers, and geothermal fields. Since the first well was sunk in the 1930s, more than 800 wells have been drilled to tap the hot water and steam. Geothermal energy in Rotorua is used in a variety of ways, including home, commercial, and greenhouse heating.

In the United States, the first commercial geothermal electrical generating plant was built in 1960 at The Geysers, about 120 km north of San Francisco, California. Here, wells were drilled into the numerous near-vertical fractures underlying the region. As pressure on the rising groundwater decreases, the water changes to steam that is piped directly to electricity-generating turbines.

As oil reserves decline, geothermal energy is becoming an attractive alternative, particularly in parts of the western United States, such as the Salton Sea area of southern California, where geothermal exploration and development have begun. While geothermally generated electricity is a generally clean source of power, it can also be expensive because most geothermal waters are acidic and very corrosive. Consequently, the turbines must either be built of expensive corrosion-resistant alloy metals or frequently replaced. Furthermore, geothermal power is not inexhaustible. The steam and hot water removed for geothermal power cannot be easily replaced, and eventually pressure in the wells drops to the point at which the geothermal field must be abandoned.

FIGURE 13.21 Minerva Terrace at Mammoth Hot Springs in Yellowstone National Park, Wyoming, formed when calcium carbonate–rich hot spring water cooled, precipitating travertine deposits.

Chapter Summary

1. The water stored in the pore spaces of subsurface rocks and unconsolidated material is groundwater.

2. Groundwater is part of the hydrologic cycle and represents approximately 22% of the world's supply of freshwater.

3. Porosity is the percentage of a rock, sediment, or soil consisting of pore space. Permeability is the ability of a rock, sediment, or soil to transmit fluids. A material that transmits groundwater is an aquifer, and one that prevents the movement of groundwater is an aquiclude.

4. The water table is the surface separating the zone of aeration (in which pore spaces are filled with both air and water) from the zone of saturation (in which all pore spaces are filled with water).

5. Groundwater moves slowly through the pore spaces in the zone of aeration and moves through the zone of saturation to outlets such as streams, lakes, and swamps.

6. A spring occurs wherever the water table intersects the surface. Some springs are the result of a perched water table—that is, a localized aquiclude within an aquifer and above the regional water table.

7. Water wells are made by digging or drilling into the zone of saturation. When water is pumped out of a well, a cone of depression forms. If water is pumped out faster than it can be recharged, the cone of depression deepens and enlarges and may locally drop to the base of the well, resulting in a dry well.

8. Artesian systems are those in which confined groundwater builds up high hydrostatic pressure. Three conditions must generally be met before an artesian system can form: the aquifer must be confined above and below by aquicludes; the aquifer is usually tilted and exposed at the surface so it can be recharged; and precipitation must be sufficient to keep the aquifer filled.

9. Karst topography results from groundwater weathering and erosion and is characterized by sinkholes, caves, solution valleys, and disappearing streams.

10. Caves form when groundwater in the zone of saturation weathers and erodes soluble rock such as limestone. Cave deposits, called dripstone, result from the precipitation of calcite.

11. Modifications of the groundwater system can cause serious problems. Excessive withdrawal of groundwater can result in dry wells, loss of hydrostatic pressure, saltwater encroachment, and ground subsidence.

12. Groundwater contamination is becoming a serious problem and can result from sewage, landfills, and toxic waste.

13. Hot springs and geysers are found where groundwater is heated by hot subsurface volcanic rocks or by the geothermal gradient. Geysers are hot springs that intermittently eject hot water and steam.

14. Geothermal energy comes from the steam and hot water trapped within the crust. It is a relatively nonpolluting form of energy that is used as a source of heat and for generating electricity.

Important Terms

aquiclude
aquifer
artesian system
cave
cone of depression
dripstone
geothermal energy

geyser
groundwater
hot spring
karst topography
permeability
porosity
sinkhole

spring
water table
water well
zone of aeration
zone of saturation

1. The percentage of a material's total volume of pore space is its:
   a. _____ solubility;
   b. _____ saturation;
   c. _____ permeability;
   d. _____ aeration quotient;
   e. _____ porosity.

2. The water table is a surface separating the:
   a. _____ zone of porosity from the underlying zone of permeability;
   b. _____ capillary fringe from the underlying zone of aeration;
   c. _____ capillary fringe from the underlying zone of saturation;
   d. _____ zone of aeration from the underlying zone of saturation;
   e. _____ zone of saturation from the underlying zone of aeration.

3. Groundwater:
   a. _____ moves slowly through the pore spaces of Earth materials;
   b. _____ moves fastest through the central area of a material's pore space;
   c. _____ can move upward against the force of gravity;
   d. _____ moves from areas of high pressure toward areas of low pressure;
   e. _____ all of these.

4. A perched water table:
   a. _____ occurs wherever there is a localized aquiclude within an aquifer;
   b. _____ is frequently the site of springs;
   c. _____ lacks a zone of aeration;
   d. _____ answers (a) and (b);
   e. _____ answers (b) and (c).

5. An artesian system is one in which:
   a. _____ water is confined;
   b. _____ water can rise above the level of the aquifer when a well is drilled;
   c. _____ there are no aquicludes;
   d. _____ answers (a) and (c);
   e. _____ answers (a) and (b).

6. Which of the following is *not* an example of groundwater deposition?
   a. _____ stalagmite;
   b. _____ dripstone;
   c. _____ karst topography;
   d. _____ stalactite;
   e. _____ column.

7. Rapid withdrawal of groundwater can result in:
   a. _____ a cone of depression;
   b. _____ ground subsidence;
   c. _____ saltwater incursion;
   d. _____ loss of hydrostatic pressure;
   e. _____ all of these.

8. The water in hot springs and geysers:
   a. _____ is believed to have curative properties;
   b. _____ is noncorrosive;
   c. _____ contains large quantities of dissolved minerals;
   d. _____ answers (a) and (b);
   e. _____ answers (a) and (c).

9. Karst topography is characterized by:
   a. _____ caves;
   b. _____ springs;
   c. _____ sinkholes;
   d. _____ solution valleys;
   e. _____ all of these.

10. What is the correct order, from highest to lowest, of groundwater usage in the United States?
    a. _____ agricultural, industrial, domestic;
    b. _____ industrial, domestic, agricultural;
    c. _____ domestic, agricultural, industrial;
    d. _____ agricultural, domestic, industrial;
    e. _____ industrial, agricultural, domestic.

11. What percentage of the world's supply of freshwater is represented by groundwater?
    a. _____ 5;
    b. _____ 18;
    c. _____ 22;
    d. _____ 43;
    e. _____ 50.

12. Discuss the role of groundwater in the hydrologic cycle.

13. What types of materials make good aquifers and aquicludes?

14. Why is the water table a subdued replica of the surface topography?

15. Where are springs likely to occur?

16. What is a cone of depression and why is it so important?

17. Why are some artesian wells free-flowing while others must be pumped?

18. How does groundwater weather and erode?

19. Discuss the various ways that a groundwater system may become contaminated.

Points to Ponder

1. One of the concerns geologists have about using Yucca Mountain as a repository for nuclear waste is that the climate may change during the next 10,000 years and become more humid, thus allowing more water to percolate through the zone of aeration. What would the average rate of groundwater movement have to be during the next 10,000 years to reach the canisters containing radioactive waste buried at a depth of 300 m?

2. Why should we be concerned with how fast the groundwater supply is being depleted in some areas?

For these web site addresses, along with current updates and exercises, log on to

http://www.wadsworth.com/geo

▶ THE VIRTUAL CAVE

This site contains images of various cave features from around the world. In addition, it contains links to and information on other cave sites elsewhere in the world.

1. Click on any of the headings for cave features. View the images and read the text about each image.

2. Click on the site link to learn about caves in your area or near where you may be traveling. What are some of the caves near you?

▶ U.S. GEOLOGICAL SURVEY WATER RESOURCES OF THE UNITED STATES

This site contains a wealth of information about streams, flooding, flood forecasting, flood warnings, and groundwater.

1. Under the Publications heading, click on *On-line Reports*. What are some of the current publications available?

2. Under the Data heading, click on the *Water-Use Data* site. Click on the *Periodic Water Fact* icon for the latest information about new and interesting facts about water resources.

3. Under the Data heading, click on the *Water-Use Data* site. Click on the *3-D graphic of total water use in the U.S.* icon. Where is water withdrawal greatest in the United States? Does this surprise you?

4. Under the Data heading, click on the *Water-Use Data* site. Click on the *Color maps of water use by state for 1990* icon. Click on any of the *National Water-Use Maps* sites for color maps of U.S. groundwater usage in 1990. What was the domestic, industrial, and agricultural groundwater withdrawal for your state?

5. Click on the *California-Nevada Ground Water Atlas* site for information on groundwater in this region, including quality and problems with subsidence. This is the first segment of the Ground Water Atlas of the United States, and it has some great maps.

▶ YUCCA MOUNTAIN PROJECT STUDIES

This site is devoted to the Yucca Mountain Project, which is the only current site being considered for long-term storage of the nation's high-level radioactive waste. It includes information, maps, and images.

1. Click on the *What's News (News and Hot Topics)* site. Check out what has happened at Yucca Mountain in the previous year.

2. Click on the *Frequently Asked Questions* site. Click on the *Why study Yucca Mountain?* site. After reading this information, why is it important we study Yucca Mountain, and why should the average citizen be concerned about this site?

▶ LINKS TO INFORMATION SOURCES FOR HYDROGEOLOGY, HYDROLOGY, AND ENVIRONMENTAL SCIENCES

This site contains links, which are listed alphabetically, to information sources for hydrogeology, hydrology, and environmental sciences. Check out several sites for information about groundwater and groundwater usage, both in the United States and elsewhere in the world.

Explore the following *In-Terra-Active 2.0* CD-ROM module(s) and increase your understanding of key concepts and processes presented in this chapter.

▶ **SECTION: SURFACE**
MODULE: GROUND WATER

This circular depression in a field in Indiana is much like the tip of an iceberg: what you see is only the beginning of the story. Below the surface, both natural and manmade changes in the availability of groundwater can cause far-reaching and sometimes unpredictable effects. **Question: How does this depression relate to subsurface geology, the water table, and movements in the groundwater system?**

▶ **VIRTUAL REALITY FIELD TRIP: RIVERS AND STREAMS**

Glaciers

This view in Glacier National Park, Montana, shows the rugged, angular peaks and ridges and broad valleys typical of mountains eroded by glaciers.

Prologue

and

Glaciation

The Great Ice Age, which ended about 10,000 years ago, was followed by a general warming trend that was periodically interrupted by short relatively cool periods. One cool period, from about 1500 to the mid-to-late 1800s, was characterized by the expansion of small glaciers in mountain valleys and the persistence of sea ice at high latitudes for longer periods than had occurred previously. This interval of nearly four centuries is known as the Little Ice Age.

The climatic changes leading to the Little Ice Age actually began about 1300. During the preceding centuries, Europe had experienced rather mild temperatures, and the North Atlantic Ocean was warmer and more storm-free than it is now. During this time, the Vikings discovered and settled Iceland, and by 1200 about 80,000 people resided there. They also sailed to Greenland and North America and established two colonies on the former and one on the latter. As the climate deteriorated, however, the North Atlantic became stormier, and sea ice was present farther south and persisted longer each year. As a result of poor sea conditions and political problems in Norway, all shipping across the North Atlantic ceased, and the colonies in Greenland and North America eventually disappeared.

During the Little Ice Age, many of the small glaciers in Europe and Iceland expanded and moved far down their valleys, reaching their greatest historic extent by the early 1800s. A small ice cap formed in Iceland where none had existed previously, and glaciers in Alaska and the mountains of the western United States and Canada also expanded to their greatest limits during historic time. Although glaciers caused some problems in Europe where they advanced across roadways and pastures, destroying some villages in Scandinavia and threatening villages elsewhere, their overall impact on humans was minimal. Far more important from the human perspective was that during much of the Little Ice Age the summers in northern latitudes were cooler and wetter.

Although worldwide temperatures were slightly lower during this time, the change in summer conditions rather than cold winters or glaciers caused most of the problems. Particularly hard hit were Iceland and the Scandinavian countries, but at times much of northern Europe was affected (Figure 14.1). Growing seasons were shorter during many years, resulting in food shortages and a number of famines. Iceland's population declined from its high of 80,000 in 1200 to about 40,000 by 1700. Between 1610 and 1870, sea ice was observed near Iceland for as much as three months a year, and each time the sea ice persisted for long periods, poor growing seasons and food shortages followed.

Exactly when the Little Ice Age ended is debatable. Some authorities put the end at 1880, whereas others think it ended as early as 1850. In any case, during the late 1800s, the sea ice was retreating north, glaciers were retreating back up their valleys, and summer weather became more moderate.

FIGURE 14.1   During the Little Ice Age, many European glaciers, such as this one in Switzerland, extended much farther down their valleys than they do at the present. This painting, called *The Unterer Grindelwald,* was painted in 1826 by Samuel Birmann (1793–1847).

Introduction

MOST people have some idea of what a glacier is, but many confuse glaciers with other masses of snow and ice. A **glacier** is a mass of ice composed of compacted and recrystallized snow flowing under its own weight on land. Accordingly, sea ice in the polar regions is not glacial ice, nor are drifting icebergs glaciers even though they may have derived from glaciers that flowed into the sea. Snow fields in high mountains may persist in protected areas for years, but these are also not glaciers because they are not actively moving.

At first glance, glaciers may appear to be static, but like other geologic agents such as streams, glaciers are dynamic systems that are continually adjusting to changes. For instance, just as streams vary their sediment load depending on available energy, increases or decreases in the amount of ice in a glacier alter its ability to erode and transport sediment.

At present, glaciers cover nearly 15 million km², or about one-tenth of Earth's land surface (Table 14.1). Numerous glaciers exist in the mountains of the western United States, especially Alaska, and western Canada and in the Andes in South America, the Alps of Europe, the Himalayas of Asia, and other high mountains. Even Mount Kilimanjaro in Africa, although near the equator, is high enough to support glaciers. In fact, Australia is the only continent lacking glaciers. By far the largest existing glaciers are in Greenland and Antarctica; both areas are almost completely covered by glacial ice (Table 14.1).

Glaciers, as particularly effective agents of erosion and transport, deeply scour the land they move over, thereby producing a number of easily recognized erosional landforms. They also deposit sediment, giving rise to a variety of depositional landforms. Although many landscapes that glaciers continue to modify can be found, most glacial landscapes developed during the Pleistocene Epoch, or what is commonly called the Ice Age (1.6 million to 10,000 years ago). During the Pleistocene, glaciers covered much more extensive areas than they do now, particularly on the Northern Hemisphere continents.

Glaciers and the Hydrologic Cycle

GLACIERS contain about 2.15% of the world's water, which constitutes one reservoir in the hydrologic cycle. However, many glaciers at high latitudes, as in Alaska, flow directly into the sea where they melt or where icebergs break off by a process called *calving* and drift out to sea where they eventually melt. At lower latitudes where they can exist only at high elevations, glaciers flow to lower elevations where they melt and the water returns to the oceans by surface runoff. In areas of low precipitation, as in parts of the western United States, glaciers are important freshwater reservoirs that release water to streams during the dry season.

In addition to melting, glaciers lose water by sublimation, a process in which ice changes directly to water vapor without an intermediate liquid phase. Water vapor so derived rises into the atmosphere where it may condense and fall once again as snow or rain. In any case, it too is eventually returned to the oceans.

The Origin of Glacial Ice

ICE is a mineral in every sense of the word; it has a crystalline structure and possesses characteristic physical and chemical properties. Accordingly, geologists consider glacial ice to be rock, although it is a type of rock that is easily deformed. It forms in a fairly straightforward manner (Figure 14.2). When an area receives more winter snow than can melt during the spring and summer seasons, a net accumulation occurs. Freshly fallen snow consists of about 80% air and 20% solids, but it compacts as it accumulates, partly thaws, and refreezes; in the process, the original

TABLE 14.1	
Present-Day Ice-Covered Areas	
Antarctica	12,653,000 km²
Greenland	1,802,600
Northeast Canada	153,200
Central Asian ranges	124,500
Spitsbergen group	58,000
Other Arctic islands	54,000
Alaska	51,500
South American ranges	25,000
West Canadian ranges	24,900
Iceland	11,800
Scandinavia	5,000
Alps	3,600
Caucasus	2,000
New Zealand	1,000
USA (other than Alaska)	650
Others	about 800
	14,971,550
Total volume of present ice: 28 to 35 million km³	

SOURCE: C. Embleton and C. A. King, *Glacial Geomorphology* (New York: Halsted Press, 1975).

snow layer is converted to a granular type of ice known as **firn**. Deeply buried firn is further compacted and is finally converted to **glacial ice**, consisting of about 90% solids (Figure 14.2). When accumulated snow and ice reach a critical thickness of about 40 m, the pressure on the ice at depth is sufficient to cause deformation and flow, even though it

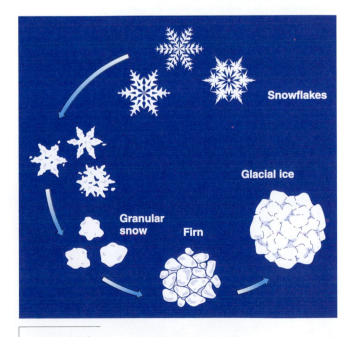

FIGURE 14.2 The conversion of freshly fallen snow to firn and glacial ice.

remains solid. Once the critical thickness is reached and flow begins, the moving mass of ice becomes a glacier.

Plastic flow, which causes permanent deformation, occurs in response to pressure and is the primary way that glaciers move. They may also move by **basal slip,** which takes place when a glacier slides over its underlying surface (Figure 14.3). Basal slip is facilitated by the presence of meltwater that reduces frictional resistance between the underlying surface and the glacier. If a glacier is frozen to the underlying surface, though, it moves only by plastic flow.

Types of Glaciers

GEOLOGISTS generally recognize two basic types of glaciers: valley and continental. A **valley glacier,** as its name implies, is confined to a mountain valley or perhaps to an interconnected system of mountain valleys (Figure 14.4). Large valley glaciers commonly have several smaller tributary glaciers, much as large streams have tributaries. Valley glaciers flow from higher to lower elevations and are invariably small in comparison to continental glaciers. The Bering Glacier in Alaska, for instance, is about 200 km long, whereas the Salmon Glacier in the Canadian Rocky Mountains is nearly 500 m thick.

Continental glaciers, also called *ice sheets,* cover vast areas (at least 50,000 km^2) and are unconfined by topography (Figure 14.5). In contrast to valley glaciers, which flow downhill within the confines of a valley, continental glaciers flow out in all directions from a central area of accu-

FIGURE **14.4** A large valley glacier in Alaska. Notice the tributaries to the large glacier.

mulation. Currently, continental glaciers exist only in two areas, Greenland and Antarctica. In both areas the ice is more than 3000 m thick in the central areas, becomes thinner toward the margins, and covers all but the highest mountains. The continental glacier in Greenland covers more than 1 million km^2, and the East and West Antarctic Glaciers merge to form a continuous ice sheet covering 12.653 million km^2 (Table 14.1). During the Pleistocene Epoch, glaciers covered large parts of the Northern Hemisphere continents. Many of the erosional and depositional landforms in much of Canada and the northern tier of the United States formed as a consequence of continental glaciation during the Pleistocene.

Although valley and continental glaciers are easily differentiated by their size and location, an intermediate variety called an *ice cap* is also recognized. Ice caps are similar to but smaller than continental glaciers and cover less than 50,000 km^2, such as the 6000 km^2 Penny Ice Cap on Baffin Island, Canada. Some ice caps form when valley glaciers grow and overtop the divides and passes between adjacent valleys and coalesce to form a continuous ice cap. They also form on fairly flat terrain, including some of the islands of the Canadian Arctic and Iceland.

The Glacial Budget

JUST as a savings account grows and shrinks as funds are deposited and withdrawn, glaciers expand and contract in response to accumulation and wastage. Their behavior can

FIGURE **14.3** Movement of a glacier by a combination of plastic flow and basal slip. If a glacier is solidly frozen to the underlying surface, it moves only by plastic flow. Notice that the top of the glacier moves farther in a given time than the bottom does.

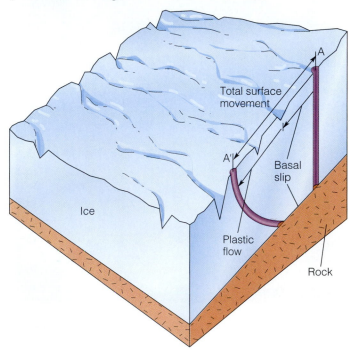

FIGURE 14.5 The Antarctic ice sheet, one of two areas presently covered by continental glaciers.

perennially covered by snow. In contrast, the lower part of the same glacier is a **zone of wastage**, where losses from melting, sublimation, and calving of icebergs exceed the rate of accumulation (Figure 14.6).

At the end of winter, a glacier's surface is usually completely covered with the accumulated seasonal snowfall. During spring and summer, the snow begins to melt, first at lower elevations and then progressively higher up the glacier. The elevation to which snow recedes during a wastage season is called the *firn limit* (Figure 14.6). One can easily identify the zones of accumulation and wastage by noting the position of the firn limit.

Observations of a single glacier reveal that the position of the firn limit usually changes from year to year. If it does not change or shows only minor fluctuations, the glacier is said to have a balanced budget; that is, additions in the zone of accumulation are exactly balanced by losses in the zone of wastage, and the distal end, or *terminus,* of the glacier remains stationary (Figure 14.6a). When the firn limit moves down the glacier, the glacier has a positive budget; its additions exceed its losses, and its terminus advances (Figure 14.6b). If the budget is negative, the glacier recedes, and its terminus retreats up the glacial valley (Figure 14.6).

be described in terms of a **glacial budget**, which is essentially a balance sheet of accumulation and wastage. The upper part of a valley glacier is a **zone of accumulation** where additions exceed losses, and the glacier's surface is

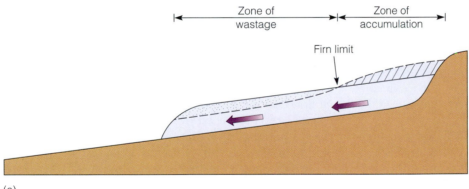

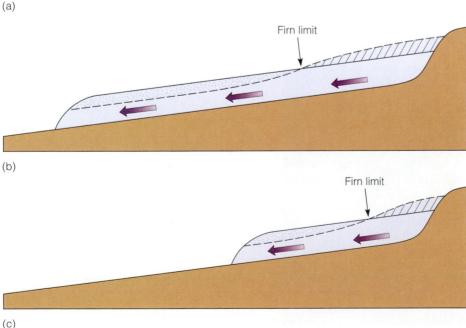

FIGURE 14.6 Response of a hypothetical glacier to changes in its budget. (a) If the losses in the zone of wastage, shown by stippling, equal additions in the zone of accumulation, shown by crosshatching, the terminus of the glacier remains stationary. (b) Gains exceed losses, and the glacier's terminus advances. (c) Losses exceed gains, and the glacier's terminus retreats, although the glacier continues to flow.

The Glacial Budget 297

But even though a glacier's terminus may be receding, the glacial ice continues to move toward the terminus by plastic flow and basal slip. If a negative budget persists long enough, though, a glacier recedes and thins until it no longer flows, thus becoming a *stagnant glacier*.

Although we used a valley glacier as our example, the same budget considerations control the flow of continental glaciers as well. In the case of the Antarctic ice sheet, the entire ice sheet is in the zone of accumulation, but wastage takes place where it flows into the ocean.

Rates of Glacial Movement

In general, valley glaciers move more rapidly than continental glaciers, but the rates for both vary, ranging from centimeters to tens of meters per day. Valley glaciers moving down steep slopes flow more rapidly than glaciers of comparable size on gentle slopes, assuming that all other variables are the same. The main glacier in a valley glacier system contains a greater volume of ice and thus has a greater discharge and flow velocity than its tributaries (Figure 14.4). Temperature exerts a seasonal control on valley glaciers because although plastic flow remains rather constant year-round, basal slip is more important during warmer months when meltwater is more abundant.

Flow rates also vary within the ice itself. Flow velocity generally increases in the zone of accumulation until the firn limit is reached; from that point, the velocity becomes progressively slower toward the glacier's terminus. Valley glaciers are similar to streams, in that the valley walls and floor cause frictional resistance to flow, so the ice in contact with the walls and floor moves more slowly than the ice some distance away (Figure 14.7).

Notice in Figure 14.7 that flow velocity increases upward until the top few tens of meters of ice are reached, but there is little or no additional increase after that point. This upper ice constitutes the rigid part of the glacier that moves as a result of basal slip and plastic flow below. The

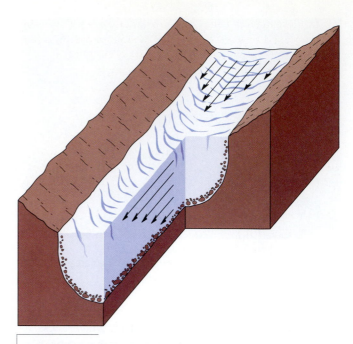

FIGURE 14.7 Flow velocity in a valley glacier varies both horizontally and vertically. Velocity is greatest at the top-center of the glacier because friction with the walls and floor of the trough causes flow to be slower adjacent to these boundaries. The length of the arrows in the figure is proportional to velocity.

fact that this upper 40 m or so of ice behaves as a brittle solid is clearly demonstrated by large fractures known as *crevasses* that develop when a valley glacier flows over a step in its valley floor where the slope increases or where it flows around a corner (Figure 14.8). In either case, the glacial ice is stretched (subjected to tension), and large crevasses develop, but they extend down only to the zone of plastic flow. In some cases, a valley glacier descends over such a steep precipice that crevasses break up the ice into a jumble of blocks and spires, and an *ice fall* develops (Figure 14.8).

FIGURE 14.8 Crevasses and an ice fall in a glacier in Alaska.

The flow rates of valley glaciers are also complicated by *glacial surges,* which are bulges of ice that move through a glacier at a velocity several times faster than the normal flow. Although surges are best documented in valley glaciers, they take place in ice caps and perhaps continental glaciers as well (Perspective 14.1).

One reason continental glaciers move comparatively slowly is that they exist at higher latitudes and are frozen to the underlying surface most of the time, which limits the amount of basal slip. Some basal slip does occur even beneath the Antarctic ice sheet, but most of its movement is by plastic flow. Nevertheless, some parts of continental glaciers manage to achieve extremely high flow rates. Near the margins of the Greenland ice sheet, for instance, the ice is forced between mountains in what are called *outlet glaciers.* In some of these outlets, flow velocities exceeding 100 m per day have been recorded.

In parts of the continental glacier covering West Antarctica, geologists have identified ice streams in which flow rates are much greater than in adjacent glacial ice. Drilling has revealed a 5-m-thick layer of water-saturated sediment beneath these ice streams, which apparently act to facilitate movement of the ice above. Some geologists think that geothermal heat from active volcanism melts the underside of the ice, thus accounting for the layer of water-saturated sediment.

Glacial Erosion and Transport

Gʟᴀᴄɪᴇʀꜱ as moving solids can erode and transport huge quantities of materials, especially unconsolidated sediment and soil. In many areas of Canada and the northern United States, glaciers transported boulders, known as *glacial erratics,* for long distances before depositing them (Figure 14.9).

Important erosional processes associated with glaciers include bulldozing, plucking, and abrasion. Bulldozing, although not a formal geologic term, is fairly self-explanatory: a glacier simply shoves or pushes unconsolidated materials in its path. An observer in Norway aptly described this effective process in 1744 during the Little Ice Age: "When at times [the glacier] pushes forward a great sound is heard, like that of an organ and it pushes in front of it unmeasurable masses of soil, grit and rocks bigger than any house could be, which it then crushes small like sand."* *Plucking,* also called *quarrying,* results when glacial ice freezes in the cracks and crevices of a bedrock projection and eventually pulls it loose.

Bedrock over which sediment-laden glacial ice has moved is effectively eroded by **abrasion** and commonly develops a *glacial polish,* a smooth surface that glistens in reflected light (Figure 14.10a). Abrasion also yields *glacial striations,* consisting of rather straight scratches rarely more than a few millimeters deep on rock surfaces (Figure 14.10b). During abrasion, rocks are thoroughly pulverized so that they yield an aggregate of clay- and silt-sized particles having the consistency of flour, hence the name *rock flour.* Rock flour is so common in streams discharging from glaciers that the water generally has a milky appearance.

Continental glaciers can derive sediment from mountains projecting through them, and windblown dust settles on their surfaces. Otherwise, most of their sediment comes from the surface they move over and is transported in the lower part of the ice sheet. In contrast, valley glaciers carry sediment in all parts of the ice, but it is concentrated at the base and along the margins. Some of the marginal sediment is derived by abrasion and plucking, but much of it is supplied by mass wasting processes.

---

*Quoted in C. Officer and J. Page, *Tales of the Earth* (New York: Oxford University Press, 1993), 99.

FIGURE 14.9 A glacial erratic in Montana.

Surging Glaciers

Normally, the snout (terminus) of a glacier possessing a positive budget advances a few meters to a few tens of meters per year. Under some circumstances, though, a glacier that has shown no anomalous activity for many years suddenly advances as much as several tens of meters per day for a few months or years and then returns to normal. In 1986 the terminus of Hubbard Glacier in Alaska began advancing at about 10 m per day, and more recently, in 1993, Alaska's Bering Glacier advanced more than 1.5 km in just three weeks (Figure 1).

Surging glaciers, constituting a tiny fraction of all glaciers, have been known for more than 100 years but are still not fully understood. None are present in the United States outside Alaska, and the only ones in Canada are in the Yukon Territory and the Queen Elizabeth Islands of the Arctic. Others are present in the Andes of South America, the Himalayas in Asia, Iceland, and some of Russia's Arctic islands. Many of these surging glaciers are difficult to study because of their remote locations, although some of them surge at somewhat regular intervals. In one case, geologists accurately predicted the 1982–1983 surge of the Variegated Glacier of Alaska, based on its past history of surges.

Predicting glacial surges is of more than academic interest. Although few people live in areas directly threatened by an advancing ice front, surges can advance across highways and endanger facilities such as the Alaska pipeline. Geologists are currently monitoring the Black Rapids Glacier in Alaska for just this reason. Furthermore, during surges huge quantities of meltwater are released, which threaten downstream areas with flooding. In some cases, a surging glacier advances across a stream valley, thus forming an ice-dammed lake. Unfortunately, ice dams are unstable and commonly collapse, thereby rapidly releasing water that causes catastrophic flooding.

Besides their direct or indirect threats to humans and their structures, surging glaciers have caused numerous animal deaths and disruptions of habitats. For example, during the 1986 surge of the Hubbard Glacier in Alaska, glacial ice advanced across a shallow marine embayment, isolating it from the open sea (Figure 1). Hundreds of seals and porpoises were trapped in the former bay and died, although environmentalists did manage to save some of them.

The onset of a surge is marked by a noticeable thickening in the upper part of a glacier. As this thickened bulge of ice begins moving at several times the normal velocity toward the glacier's terminus, thousands of crevasses appear in the surface. Commonly, the bulge reaches the terminus, which then advances rapidly—as much as 20 km in two years in one case. Rarely, the bulge stops before reaching the terminus.

Even though some surges take place following a period of unusually heavy snowfall in the zone of accumulation, plastic flow alone cannot account for the accelerated flow rates in most surging glaciers. Accordingly, basal slip must be more important during these events.

Two theories, neither being mutually exclusive, have been proposed to account for accelerated basal slip. One theory holds that thickening in the zone of accumulation is accompanied by thinning in the zone of wastage, thereby increasing the glacier's slope. Eventually, the stress at the bottom of the upper part of a glacier causes channels below the glacier to close, which forces water out under the entire glacier, causing accelerated sliding. According to the other theory, pressure beneath a glacier resting upon soft sediment forces fluids through the sediment layer, which is easily deformed, thus allowing the overlying glacier to slide more effectively.

Figure 1 During a 1986 surge, the terminus of Hubbard Glacier in Alaska advanced across Russell Fiord, the shallow embayment at the right back. Environmentalists saved some of the marine mammals trapped in the former bay but hundreds of seals and porpoises died.

(a)

(b)

FIGURE 14.10 (a) Glacial polish on quartzite near Marquette, Michigan. (b) Glacial striations in basalt at Devil's Postpile National Monument, California.

EROSIONAL LANDFORMS OF VALLEY GLACIERS

Some of the world's most inspiring scenery is produced by valley glaciers. Many mountain ranges are scenic to begin with, but when modified by valley glaciers, they take on a unique aspect of jagged, angular peaks and ridges in the midst of broad valleys (see chapter-opening photo).

U-SHAPED GLACIAL TROUGHS A U-shaped glacial trough is one of the most distinctive features of valley glaciation (Figure 14.11c). Mountain valleys eroded by running water are typically V-shaped in cross section; that is, they have valley walls that descend to a narrow valley bottom (Figure 14.11a). In contrast, valleys scoured by glaciers are deepened, widened, and straightened so that they possess very steep or vertical walls but have broad, rather flat valley floors; thus, they exhibit a U-shaped profile (Figure 14.12). Many glacial troughs contain triangular-shaped *truncated spurs,* which are cutoff or truncated ridges that extend into the preglacial valley (Figure 14.11c).

During the Pleistocene, when glaciers were more extensive, sea level was about 130 m lower than at present, so glaciers flowing into the sea eroded their valleys to much greater depths than they do now. When the glaciers melted at the end of the Pleistocene, sea level rose, and

FIGURE 14.11 Erosional landforms produced by valley glaciers. (a) A mountain area before glaciation. (b) The same area during the maximum extent of the valley glaciers. (c) After glaciation.

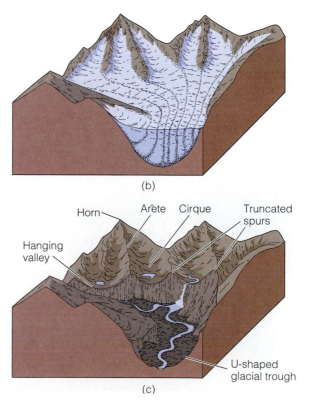

(b)

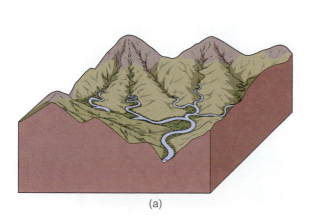

(a)

Horn Arête Cirque Truncated spurs

Hanging valley

U-shaped glacial trough

(c)

FIGURE **14.12** A U-shaped glacial trough in the Beartooth Mountains, Wyoming-Montana.

level than that of the main valley (Figure 14.14). As Figure 14.11 shows, the large glacier in the main valley vigorously erodes, whereas the smaller glaciers in tributary valleys are less capable of large-scale erosion. When the glaciers disappear, the smaller tributary valleys remain as hanging valleys. Accordingly, streams flowing through hanging valleys plunge over vertical or steep precipices.

CIRQUES, ARÊTES, AND HORNS Perhaps the most spectacular erosional landforms in areas of valley glaciation are at the upper ends of glacial troughs and along the divides separating adjacent glacial troughs. Valley glaciers form and move out from steep-walled, bowl-shaped depressions called **cirques** at the upper end of their troughs (Figure 14.11c, 14.15a). Cirques are typically steep-walled on three sides, but one side is open and leads into the glacial trough.

Although the details of cirque origin are not fully understood, they apparently form by erosion of a preexisting depression on a mountain side. As snow and ice accumulate in the depression, frost wedging and plucking enlarge it until it takes on the typical cirque shape. Many cirques have a lip or threshold, indicating that the glacial ice not only moves outward but rotates as well, scouring out a depression rimmed by rock.

Cirques become wider and are cut deeper into mountain sides by headward erosion as a consequence of abrasion, plucking, and several mass wasting processes. The largest cirque known is the Walcott Cirque in Antarctica, which is

the ocean filled the lower ends of the glacial troughs so that now they are long, steep-walled embayments called **fiords** (Figure 14.13).

Lower sea level during the Pleistocene was not responsible for the formation of all fiords. Unlike running water, glaciers can erode a considerable distance below sea level. In fact, a glacier 500 m thick can stay in contact with the seafloor and effectively erode it to a depth of about 450 m before the buoyant effects of water cause the glacial ice to float!

HANGING VALLEYS Although waterfalls can form in several ways, some of the world's highest and most spectacular are found in recently glaciated areas. Bridalveil Falls in Yosemite National Park, California, plunges from a **hanging valley**, which is a tributary valley whose floor is at a higher

FIGURE **14.13** Geirangerfjorden, a fiord in Norway.

FIGURE **14.14** Bridalveil Falls in Yosemite National Park, California, plunges about 190 m from a hanging valley.

(a)

(b)

FIGURE 14.15 (a) Cirques on Mount Whitney in California. (b) This knifelike ridge between glacial troughs in California is an arête.

16 km wide and 3 km deep. The fact that cirques expand laterally and by headward erosion accounts for the origin of two other distinctive erosional features, arêtes and horns. **Arêtes**—narrow, serrated ridges—can form in two ways. In many cases, cirques form on opposite sides of a ridge, and headward erosion reduces the ridge until only a thin partition of rock remains. The same effect is produced when erosion in two parallel glacial troughs reduces the intervening ridge to a thin spine of rock (Figures 14.11c and 14.15b).

The most majestic of all mountain peaks are **horns**; these steep-walled, pyramidal peaks are formed by headward erosion of cirques. For a horn to form, a mountain peak must have at least three cirques on its flanks, all of which erode headward (Figure 14.11c). Excellent examples of horns include Mount Assiniboine in the Canadian Rockies, the Grand Teton in Wyoming, and the most famous of all, the Matterhorn in Switzerland (Figure 14.16).

EROSIONAL LANDFORMS OF CONTINENTAL GLACIERS

Areas eroded by continental glaciers tend to be smooth and rounded because they bevel and abrade high areas that projected into the ice. Rather than yielding the sharp, angular landforms typical of valley glaciation, continental glaciers produce a landscape of rather flat topography interrupted by rounded hills.

In a large part of Canada, particularly the vast Canadian Shield region (see Chapter 8), continental glaciation has stripped off the soil and unconsolidated surface sediment, revealing extensive exposures of striated and polished bedrock. Similar though smaller bedrock exposures are also widespread in the northern United States from Maine through Minnesota.

Another result of erosion by continental glaciers in these areas is the complete disruption of drainage systems that have not yet become fully reestablished. Accordingly, de-

ranged drainage characterizes large parts of these regions (see Figure 12.16e). These *ice-scoured plains,* as they are known, also have numerous lakes and swamps, low relief, extensive bedrock exposures, and little or no soil.

Glacial Deposits

A̲L̲L̲ sediment deposited by glacial ice or meltwater streams discharging from glaciers is **glacial drift**. In several upper midwestern states and parts of southern Canada glacial drift is an important source of groundwater and rich soils, and in many areas it is exploited for sand and gravel.

Geologists recognize two distinct types of glacial drift, till and stratified drift. **Till** consists of sediment deposited directly by glacial ice. It is not sorted nor stratified; that is, its particles are not separated by size or density, and it does

FIGURE 14.16 The Matterhorn in Switzerland is a well-known horn.

not exhibit any layering. Till deposited by valley glaciers looks much like the till of continental glaciers except that the latter's deposits are much more extensive and have generally been transported much farther.

Stratified drift is sorted by size and density and, as its name implies, is layered. In fact, most of the sediments recognized as stratified drift are braided stream deposits; the streams in which they were deposited received their water and sediment load directly from melting glacial ice.

LANDFORMS COMPOSED OF TILL

Landforms composed of till include several types of moraines and elongated hills known as drumlins.

END MORAINES

The terminus of either a valley or a continental glacier may become stabilized in one position for some period of time, perhaps a few years or even decades. Stabilization of the ice front does not mean that the glacier has ceased flowing, only that it has a balanced budget. When an ice front is stationary, flow within the glacier continues, and any sediment transported within or upon the ice is dumped as a pile of rubble at the glacier's terminus. These deposits are **end moraines**, which continue to grow as long as the ice front remains stationary (Figure 14.17). End moraines of valley glaciers are commonly crescent-shaped ridges of till spanning the valley occupied by the glacier. Those of continental glaciers similarly parallel the ice front but are much more extensive.

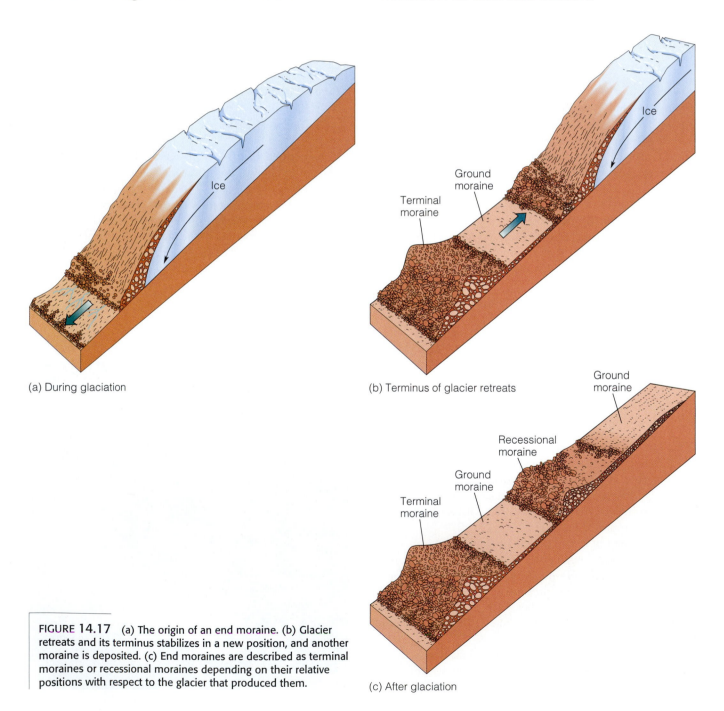

(a) During glaciation

(b) Terminus of glacier retreats

(c) After glaciation

FIGURE 14.17 (a) The origin of an end moraine. (b) Glacier retreats and its terminus stabilizes in a new position, and another moraine is deposited. (c) End moraines are described as terminal moraines or recessional moraines depending on their relative positions with respect to the glacier that produced them.

Following a period of stabilization, a glacier may advance or retreat, depending on changes in its budget. If it advances, the ice front overrides and modifies its former moraine. Should it have a negative budget, though, the ice front retreats toward the zone of accumulation. As the ice front recedes, till is deposited as it is liberated from the melting ice and forms a layer of **ground moraine** (Figure 14.17b). Ground moraine has an irregular, rolling topography, whereas end moraine consists of long ridgelike accumulations of sediment.

After a glacier has retreated for some time, its terminus may once again stabilize, and it will deposit another end moraine. Because the ice front has receded, such moraines are called **recessional moraines** (Figure 14.17c). During the Pleistocene, continental glaciers in the mid-continent region extended as far south as the southern parts of Ohio, Indiana, and Illinois. Their outermost end moraines, marking the greatest extent of the glaciers, go by the special name **terminal moraine** (valley glaciers also deposit terminal moraines). As the glaciers retreated from the positions where their terminal moraines were deposited, they temporarily ceased retreating numerous times and deposited many recessional moraines.

LATERAL AND MEDIAL MORAINES Valley glaciers transport considerable sediment along their margins. Much of this sediment is abraded and plucked from the valley walls, but a significant amount falls or slides onto the glacier's surface by mass wasting processes. In any case, this sediment is deposited as long ridges of till known as **lateral moraines** along the margin of the glacier (Figure 14.18).

Where two lateral moraines merge, as when a tributary glacier flows into a larger glacier, a **medial moraine** forms (Figure 14.18). Although medial moraines are identified by their position on a valley glacier, they are, in fact, formed from the coalescence of two lateral moraines. One can generally determine how many tributaries a valley glacier has by the number of its medial moraines.

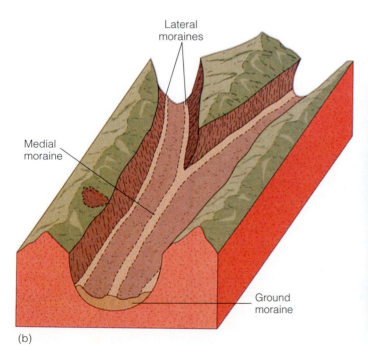

(a)

(b)

FIGURE **14.18** (a) and (b) The material transported and deposited along the margins of a valley glacier is lateral moraine. Where two lateral moraines merge, a medial moraine forms. (c) The two parallel ridges extending from this mountain valley are lateral moraines.

(c)

DRUMLINS In many areas where continental glaciers have deposited till, the till has been reshaped into elongated hills known as **drumlins**. Some drumlins measure as much as 50 m high and 1 km long, but most are much smaller. From the side, a drumlin looks like an inverted spoon with the steep end on the side from which the glacial ice advanced, and the gently sloping end pointing in the direction of ice movement (Figure 14.19). Drumlins can therefore be used to determine the direction of ice movement.

Drumlins are rarely found as single, isolated hills; instead, they occur in *drumlin fields* containing hundreds or thousands of drumlins. Drumlin fields are found in several states and Ontario, Canada, but perhaps the finest example is near Palmyra, New York.

One hypothesis for the origin of drumlins holds that they form in the zone of plastic flow as glacial ice modifies preexisting till into streamlined hills. According to another hypothesis, drumlins form when huge floods of glacial meltwater modify deposits of till.

LANDFORMS COMPOSED OF STRATIFIED DRIFT

Stratified drift is deposited by both valley and continental glaciers, but as one would expect, it is more extensive in areas of continental glaciation.

OUTWASH PLAINS AND VALLEY TRAINS Glaciers discharge meltwater laden with sediment most of the time, except perhaps during the coldest months. This meltwater forms a series of braided streams that radiate out from the front of continental glaciers over a wide region (Figure 14.20). So much sediment is supplied to these streams that much of it is deposited within the channels as sand and gravel bars. The vast blanket of sediments so formed is an **outwash plain** (Figure 14.20).

FIGURE 14.19 These elongated hills in Antrim County, Michigan, are drumlins. *(Photo courtesy of B. M. C. Pape.)*

Occasionally, huge quantities of meltwater are discharged from glaciers when, for instance, a volcano erupts beneath them. A recent example is an eruption beneath Iceland's largest ice cap that melted the base of the glacier. Meltwater accumulated in a caldera, also beneath the ice cap, and on November 5, 1996, suddenly burst forth in Iceland's largest flood in 60 years. Huge blocks of ice estimated to weigh 900 metric tons were ripped from the glacier and deposited nearly 5 km away.

Valley glaciers discharge huge amounts of meltwater and, like continental glaciers, have braided streams extending from them. But these streams are confined to the lower parts of glacial troughs, and their long, narrow deposits of stratified drift are known as **valley trains**.

Outwash plains and valley trains commonly contain numerous circular to oval depressions, many of which are the sites of small lakes. These depressions are *kettles,* which form when a retreating ice sheet or valley glacier leaves a block of ice that is subsequently partly or wholly buried (Figure 14.20). When the ice block eventually melts, it leaves a depression; if the depression extends below the water table, it becomes the site of a small lake. Some outwash plains have so many kettles that they are called *pitted outwash plains*. Although kettles are most common in outwash plains, they can also form in end moraines.

KAMES AND ESKERS Conical hills as much as 50 m high, known as **kames**, are composed of stratified drift (Figure 14.20). Many kames form when a stream deposits sediment in a depression on a glacier's surface; as the ice melts, the deposit is lowered to the surface. They also form in cavities within or beneath stagnant ice.

Long sinuous ridges of stratified drift, many of which meander and have tributaries, are **eskers** (Figure 14.20). Some eskers are quite high, as much as 100 m, and can be traced for more than 100 km. Most eskers are in areas once covered by continental glaciers, but they are also associated with large valley glaciers. The sorting and stratification of the sediments within eskers clearly indicate deposition by running water. The properties of ancient eskers and observations of present-day glaciers indicate that they form in tunnels beneath stagnant ice (Figure 14.20).

GLACIAL LAKES AND THEIR DEPOSITS

During the Pleistocene Epoch, many so-called *pluvial lakes* lakes existed in the western United States. These lakes formed as a result of greater precipitation and overall cooler temperatures that lowered the evaporation rate. Lake Bonneville, which covered 50,000 km², was the largest pluvial lake; the Great Salt Lake in Utah is the remnant of this once much larger lake. Another large pluvial lake existed in Death Valley, California, which is now the hottest, driest place in North America.

In contrast to pluvial lakes, which are far from glaciers, *proglacial lakes* are formed by meltwater accumulating along the margins of glaciers, so at least one shoreline is

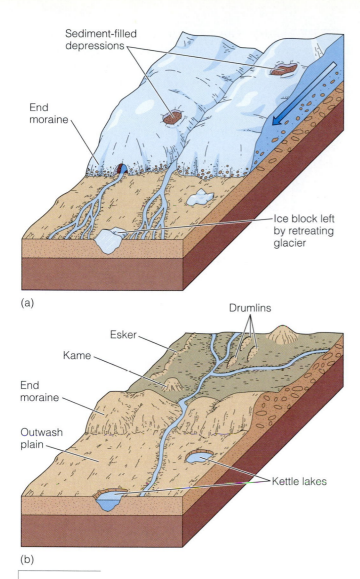

Sediment-filled
depressions

End
moraine

(a)

Ice block left
by retreating
glacier

Drumlins

Esker

Kame

End
moraine

Outwash
plain

Kettle lakes

(b)

FIGURE 14.20 Two stages in the origin of kettles, kames, eskers, and an outwash plain: (a) during glaciation and (b) after glaciation.

the ice itself. Lake Agassiz, named in honor of the French naturalist Louis Agassiz, was a large proglacial lake covering about 250,000 km² of North Dakota and Minnesota, and Manitoba, Saskatchewan, and Ontario, Canada. It persisted until the glacial ice along its northern margin melted; then it drained northward into Hudson Bay.

Numerous proglacial lakes existed during the Pleistocene, but most of them eventually disappeared as they drained or were filled with sediments. The most notable exception is the Great Lakes, all of which first formed as proglacial lakes.

At their greatest extent, the glaciers covered the entire Great Lakes region and extended far to the south. As the ice sheet retreated northward during the late Pleistocene, the ice front periodically stabilized, and numerous recessional moraines were deposited. By about 14,000 years ago, parts of the Lake Michigan and Lake Erie basins were ice-free, and glacial meltwater began forming proglacial

lakes (Figure 14.21). As the retreat of the ice sheet continued—although periodically interrupted by minor readvances of the ice front—the Great Lakes basins were uncovered, and the lakes expanded until they eventually reached their present size and configuration (Figure 14.21). Currently, the Great Lakes contain nearly 23,000 km³ of water, about 18% of the water in all freshwater lakes.

Glacial lakes, like all lakes, are areas of deposition. Sediment may be carried into them and deposited as small deltas, but of special interest are the fine-grained deposits. Mud deposits in glacial lakes are commonly finely laminated (having layers less than 1 cm thick) and consist of alternating light and dark layers. Each light-dark couplet is a *varve* (Figure 14.22), which represents an annual episode of deposition; the light layer formed during the spring and summer and consists of silt and clay; the dark layer formed during the winter when the smallest particles of clay and organic matter settled from suspension when the lake froze over. The number of varves indicates how many years a glacial lake existed.

Dropstones are another distinctive feature of glacial lakes containing varved deposits (Figure 14.22). These are pieces of gravel, some of boulder size, in otherwise very fine-grained deposits. Most of them were probably carried into the lakes by icebergs that eventually melted and released sediment contained in the ice.

Causes of Glaciation

So far we have examined the effects of glaciation but have not addressed the central questions of what causes large-scale glaciation and why there have been so few episodes of widespread glaciation. For more than a century, scientists have been attempting to develop a comprehensive theory explaining all aspects of ice ages, but they have not yet been completely successful. One reason for their lack of success is that the climatic changes responsible for glaciation, the cyclic occurrence of glacial-interglacial episodes, and short-term events such as the Little Ice Age operate on vastly different time scales.

Only a few periods of glaciation are recognized in the geologic record, each separated from the others by long intervals of mild climate. Such long-term climatic changes probably result from slow geographic changes related to plate tectonic activity. Moving plates can carry continents to high latitudes where glaciers can exist, provided that they receive enough precipitation as snow. Plate collisions, the subsequent uplift of vast areas far above sea level, and the changing atmospheric and oceanic circulation patterns caused by the changing shapes and positions of plates also contribute to long-term climatic change.

A theory explaining ice ages must also address the fact that during the Pleistocene Ice Age (1.6 million to 10,000 years ago) several intervals of glacial expansion separated by warmer interglacial periods occurred. At least four major episodes of glaciation have been recognized in North America, and six or seven glacial advances and retreats occurred in Europe. These intermediate-term climatic

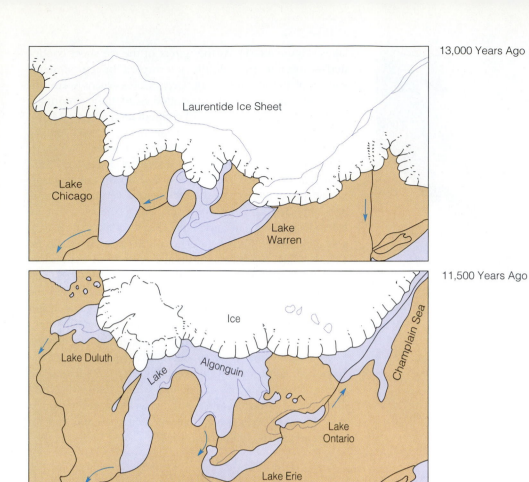

13,000 Years Ago

11,500 Years Ago

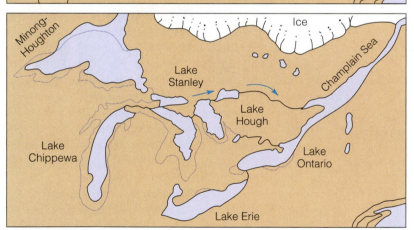

9,500 Years Ago

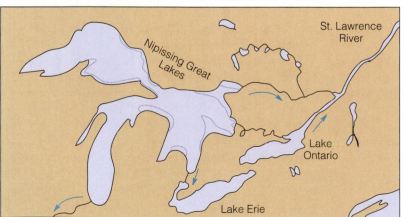

6,000 Years Ago

FIGURE 14.21 Four stages in the evolution of the Great Lakes, which formed originally as proglacial lakes. As the glacial ice retreated northward, the lake basins began filling with meltwater. The dotted lines indicate the present-day shorelines of the lakes.

FIGURE **14.22** Glacial varves with a dropstone. *(Photo courtesy of Canadian Geological Survey.)*

events take place on time scales of tens to hundreds of thousands of years. The cyclic nature of this most recent episode of glaciation has long been a problem in formulating a comprehensive theory of climatic change.

THE MILANKOVITCH THEORY

A particularly interesting hypothesis for intermediate-term climatic events was put forth by the Yugoslavian astronomer Milutin Milankovitch during the 1920s. He proposed that minor irregularities in Earth's rotation and orbit are sufficient to alter the amount of solar radiation received at any given latitude and hence can affect climatic changes. Now called the **Milankovitch theory**, it was initially ignored but has received renewed interest during the last 20 years.

Milankovitch attributed the onset of the Pleistocene Ice Age to variations in three parameters of Earth's orbit (Figure 14.23). The first of these is orbital eccentricity, which is the degree that the orbit departs from a perfect circle. Calculations indicate a roughly 100,000-year cycle between times of maximum eccentricity. This corresponds closely to 20 warm-cold climatic cycles that occurred during the Pleistocene. The second parameter is the angle between Earth's axis and a line perpendicular to the plane of the ecliptic. This angle shifts about 1.5° from its current value of 23.5° during a 41,000-year cycle. The third parameter is the precession of the equinoxes, which causes the position of the equinoxes and solstices to shift slowly around Earth's elliptical orbit in a 23,000-year cycle.

Continuous changes in these three parameters cause the

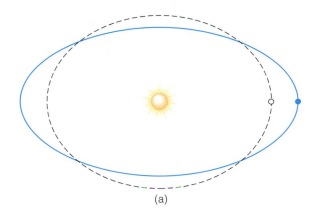

(a)

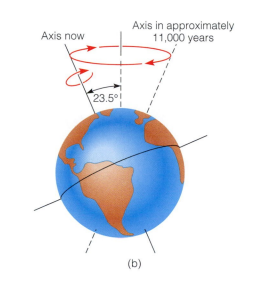

(b)

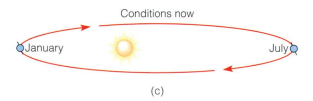

Conditions now

(c)

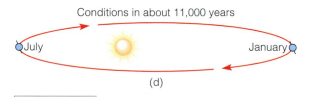

Conditions in about 11,000 years

(d)

FIGURE **14.23** (a) Earth's orbit varies from nearly a circle (dashed line) to an ellipse (solid line) and back again in about 100,000 years. (b) Earth moves around its orbit while spinning about its axis, which is tilted to the plane of the ecliptic at 23.5° and points toward the North Star. Earth's axis of rotation slowly moves and traces out the path of a cone in space. (c) At present Earth is closest to the Sun in January when the Northern Hemisphere experiences winter. (d) In about 11,000 years, as a result of precession, Earth will be closer to the Sun in July, when summer occurs in the Northern Hemisphere.

amount of solar heat received at any latitude to vary slightly over time. The total heat received by the planet, however, remains little changed. Milankovitch proposed, and now many scientists agree, that the interaction of these three parameters provided the triggering mechanism for the glacial-interglacial episodes during the Pleistocene.

SHORT-TERM CLIMATIC EVENTS

Climatic events having durations of several centuries, such as the Little Ice Age, are too short to be accounted for by plate tectonics or Milankovitch cycles. Several hypotheses have been proposed, including variations in solar energy and volcanism.

Variations in solar energy could result from changes within the Sun itself or from anything that would reduce the amount of energy Earth receives from the Sun. The latter could result from the solar system passing through clouds of interstellar dust and gas or from substances in the atmosphere reflecting solar radiation back into space. Records kept over the past 75 years indicate that during this time the amount of solar radiation has varied only slightly. So although variations in solar energy may influence short-term climatic events, such a correlation has not been demonstrated.

During large volcanic eruptions, tremendous amounts of ash and gases are spewed into the atmosphere where they reflect incoming solar radiation and reduce atmospheric temperatures. Recall from Chapter 5 that small droplets of sulfur gases remain in the atmosphere for years and can have a significant effect on climate. Several large-scale volcanic events have been recorded, such as the 1815 eruption of Tambora and 1991 eruption of Pinatubo, and are known to have had climatic effects. However, no relationship between periods of volcanic activity and periods of glaciation has yet been established.

Chapter Summary

1. Glaciers are masses of ice on land that move by plastic flow and basal slip. Valley glaciers are confined to mountain valleys and flow from higher to lower elevations, whereas continental glaciers cover vast areas and flow outward in all directions from a zone of accumulation.

2. A glacier forms when winter snowfall in an area exceeds summer melt and therefore accumulates year after year. Snow is compacted and converted to glacial ice, and when the ice is about 40 m thick, pressure causes it to flow.

3. Glaciers move by plastic flow and basal slip. Plastic flow involves deformation in response to pressure, whereas basal slip takes place when a glacier slides over its underlying surface.

4. The behavior of a glacier depends on its budget, which is the relationship between accumulation and wastage. If a glacier possesses a balanced budget, its terminus remains stationary; a positive or negative budget results in advance or retreat of the terminus, respectively.

5. Glaciers move at varying rates depending on slope, discharge, and season. Valley glaciers tend to flow more rapidly than continental glaciers.

6. Glaciers are powerful agents of erosion and transport because they are solids in motion. They are particularly effective at eroding soil and unconsolidated sediment, and they can transport any size sediment supplied to them.

7. Continental glaciers transport most of their sediment in the lower part of the ice, whereas valley glaciers may carry sediment in all parts of the ice.

8. Erosion of mountains by valley glaciers yields several sharp, angular landforms including cirques, arêtes, and horns. U-shaped glacial troughs, fiords, and hanging valleys are also products of valley glaciation.

9. Continental glaciers abrade and bevel high areas, producing a smooth, rounded landscape known as an ice-scoured plain.

10. Depositional landforms include moraines, which are ridgelike accumulations of till. Several types of moraines are recognized, including terminal, recessional, lateral, and medial moraines.

11. Drumlins are composed of till that was apparently reshaped into streamlined hills by continental glaciers or floods of glacial meltwater.

12. Stratified drift consists of sediments deposited in or by meltwater streams issuing from glaciers; it is found in outwash plains and valley trains. Ridges called eskers and conical hills called kames are also composed of stratified drift.

13. Major glacial intervals separated by tens or hundreds of millions of years probably occur as a result of the changing positions of plates, which in turn cause changes in oceanic and atmospheric circulation patterns.

14. Currently, the Milankovitch theory is widely accepted as the explanation for glacial-interglacial intervals.

15. The reasons for short-term climatic changes, such as the Little Ice Age, are not understood. Two proposed causes for these events are changes in the amount of solar energy received by Earth and volcanism.

abrasion
arête
basal slip
cirque
continental glacier
drumlin
end moraine
esker
fiord
firn
glacial budget

glacial drift
glacial ice
glacier
ground moraine
hanging valley
horn
kame
lateral moraine
medial moraine
Milankovitch theory
outwash plain

plastic flow
recessional moraine
stratified drift
terminal moraine
till
U-shaped glacial trough
valley glacier
valley train
zone of accumulation
zone of wastage

Review Questions

1. Glaciers move by a combination of _____ and _____.
   a. _____ fracture/surging;
   b. _____ creep/outwash;
   c. _____ drift/ice rafting;
   d. _____ plastic flow/basal slip;
   e. _____ saltation/suspension.

2. If the terminus of a glacier retreats:
   a. _____ it must have a negative budget;
   b. _____ the rate of accumulation exceeds the rate of wastage;
   c. _____ plastic flow ceases and the glacier moves only by basal slip;
   d. _____ the ice at depth becomes more brittle and crevasses stop forming;
   e. _____ the firn limit moves to a lower elevation.

3. A boulder deposited by glacial activity far from its source is known as a(an):
   a. _____ esker;
   b. _____ glacial erratic;
   c. _____ striation;
   d. _____ glacial polish;
   e. _____ medial moraine.

4. Two distinctive types of glacial drift are recognized, _____ and _____.
   a. _____ glacial grooves/moraines;
   b. _____ dropstones/glacial erratics;
   c. _____ U-shaped glacial troughs/eskers;
   d. _____ kames/cirques;
   e. _____ till/stratified drift.

5. The sediment deposited along the margin of valley glacier makes up a(an):
   a. _____ lateral moraine;
   b. _____ ice-scoured plain;
   c. _____ drumlin field;
   d. _____ outwash plain;
   e. _____ valley train.

6. One hypothesis for the origin of drumlins is that:
   a. _____ glacial ice flows more rapidly where it moves over a saturated layer of sediment;
   b. _____ they form where two lateral moraines merge;
   c. _____ they form in the zone of plastic flow as ice modifies till into streamlined hills;
   d. _____ circular depressions form when buried blocks of ice melt;
   e. _____ ice-rafted boulders are deposited far from land.

7. A bowl-shaped depression at the upper end of a U-shaped glacial trough is a(an):
   a. _____ fiord;
   b. _____ cirque;
   c. _____ valley train;
   d. _____ zone of wastage;
   e. _____ horn.

8. Which one of the following statements is correct?
   a. _____ ice sheet is another commonly used name for valley glacier;
   b. _____ till consists of layered, sorted deposits of sand and gravel;
   c. _____ erosion of mountains by valley glaciers yields a smooth, rounded landscape;
   d. _____ eskers and kames are glacial deposits;
   e. _____ glaciers flow fastest when solidly frozen to their underlying surfaces.

9. According to the Milankovitch theory:
   a. _____ glacial episodes are triggered by slight irregularities in Earth's rotation and orbit;
   b. _____ glaciers expand and contract depending on changes in their budgets;
   c. _____ glaciers were more widespread during the Pleistocene Epoch than at present;
   d. _____ the Little Ice Age was the last glacial event of the Pleistocene;
   e. _____ glaciers flow fastest when they move by both plastic flow and basal slip.

10. The number of medial moraines on a valley glacier indicates the number of its:
    a. _____ eskers;
    b. _____ valley trains;
    c. _____ tributary glaciers;
    d. _____ striations;
    e. _____ drumlins.

11. What is the relative importance of plastic flow and basal slip for glaciers at high and low latitudes?

12. Explain in terms of the glacial budget how a glacier might first advance and then retreat to the point that it becomes a stagnant glacier.

13. Describe drumlins and explain two ways that they might form.

14. How do glaciers acquire a sediment load?

15. Describe the type of end moraines recognized and explain how each forms.

16. How does the Milankovitch theory explain the onset of glacial-interglacial intervals?

17. Describe how a landscape eroded by valley glaciers differs from one eroded by a continental glacier. What distinctive landforms are present in each of these areas?

18. Use diagrams to show how the rates of movement of a valley glacier vary from bottom to top and across the surface. Why do the rates vary?

19. How do crevasses form in a glacier, and why do they extend downward only about 40 m?

20. Describe the sequence of events leading to the origin of an arête and a hanging valley.

Points to Ponder

1. In a roadside rock exposure, you observe a deposit of alternating light and dark laminated mud containing a few large boulders. Explain the sequence of events responsible for deposition of these sediments.

2. In North America, valley glaciers are common in Alaska and western Canada, and small ones are present in the mountains of the western United States; none, however, occur east of the Rocky Mountains. How can you explain this distribution of glaciers?

World Wide Web Activities

For these web site addresses, along with current updates and exercises, log on to

http://www.wadsworth.com/geo

▶ **GLACIER RESEARCH GROUP**

This site is maintained by the Climate Change Research Center, University of New Hampshire, Institute for the Study of Earth, Oceans, and Space. It contains an overview of what it is and what it does, as well as a listing of its current projects and links to other glacially related sites. Click on the *GRG Photo Gallery* site under The Group heading. What are some of the projects the GRG is involved in? How do these relate to glaciers and glaciation?

▶ **U.S. GEOLOGICAL SURVEY ICE AND CLIMATE PROJECT**

This site contains links to the various Ice and Climate Projects being undertaken by the USGS. Click on *Benchmark Glaciers* site under the Ice and Climate Projects links heading. What are the three benchmark glaciers being studied, and where are they located? What measurements are being taken on the three glaciers? How do these glaciers compare to each other?

For these web site addresses, along with current updates and exercises, log on to

http://www.wadsworth.com/geo

▶ ICE AGES AND GLACIATION

This site is maintained by the Department of Geological and Environmental Sciences at Hartwick College, New York. Take a virtual trip through the Ice Age and see maps and photographs of glacial features.

1. What two ice sheets were present in North America during the Wisconsinan, and where was each centered?

2. See the image of Gorner Glacier in Switzerland. What is the dark material in the center of the glacier? Also notice in this image the Matterhorn in the background and the typical topography developed in areas eroded by valley glaciers.

3. See the image of Bridalveil Falls in Yosemite National Park, California, and explain how the falls formed.

4. How did the Madison Boulder in the White Mountains of New Hampshire get to its present location?

▶ MIDWESTERN U.S. 16,000 YEARS AGO

This site, maintained by the Illinois State Museum, has a wealth of information, maps, and photographs on glaciers, glaciation, and particularly the Late Pleistocene of the Midwest. See the sections on *Late Pleistocene Landscapes, Late Pleistocene Plants and Animals,* and *The End of Pleistocene Extinctions.*

CD-ROM Exploration

Explore the following *In-Terra-Active 2.0* CD-ROM module(s) and increase your understanding of key concepts and processes presented in this chapter.

▷ **SECTION: SURFACE**
MODULE: GLACIERS

A boulder of granite gneiss from Ontario has been sitting on top of Observatory Hill on the University of Wisonsin-Madison campus for thousands of years. Unlike Stonehenge and the Pyramids, there is no evidence that the ancient inhabitants of Wisconsin had the ability to move such stones. **Question: How could this boulder have gotten on this hill?**

CHAPTER

15

The Work of Wind

Outline

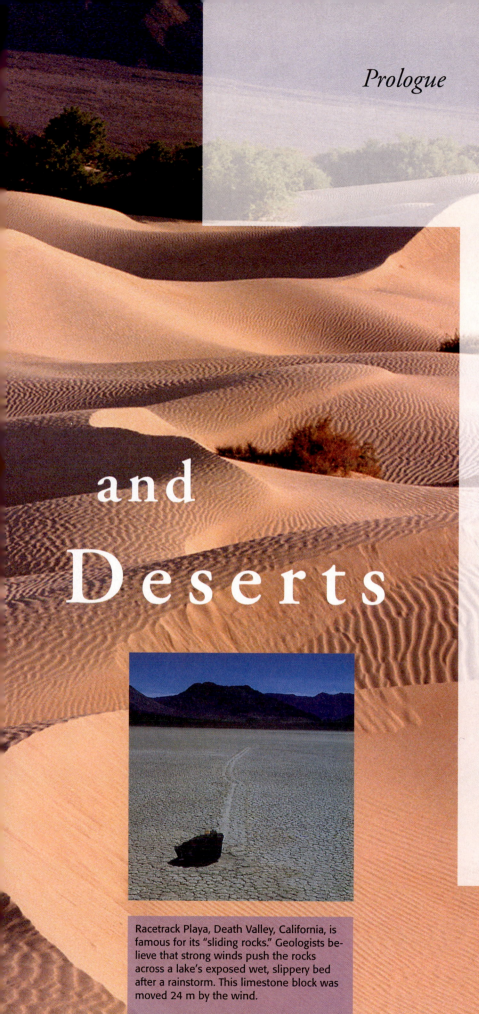

and Deserts

Prologue

During the last few decades, deserts have been advancing across millions of acres of productive land, destroying rangelands, croplands, and even villages. Such expansion, estimated at 70,000 km^2 per year, has exacted a terrible toll in human suffering. Because of the relentless advance of deserts, hundreds of thousands of people have died of starvation or been forced to migrate as "environmental refugees" from their homelands to camps where the majority are severely malnourished. This expansion of deserts into formerly productive lands is **desertification**.

Most regions undergoing desertification lie along the margins of existing deserts. These margins have a delicately balanced ecosystem that serves as a buffer between the desert on one side and a more humid environment on the other. Their potential to adjust to increasing environmental pressures from natural causes or human activity is limited. Ordinarily, desert regions expand and contract gradually in response to natural processes such as climatic change, but much recent desertification has been greatly accelerated by human activities. In many areas, the natural vegetation has been cleared as crop cultivation has expanded into increasingly drier fringes to support the growing population. Because these areas are especially prone

Racetrack Playa, Death Valley, California, is famous for its "sliding rocks." Geologists believe that strong winds push the rocks across a lake's exposed wet, slippery bed after a rainstorm. This limestone block was moved 24 m by the wind.

to droughts, crop failures are common occurrences, leaving the land bare and susceptible to increased wind and water erosion.

Because grasses constitute the dominant natural vegetation in most fringe areas, raising livestock is a common economic activity. Usually, these areas achieve a natural balance between vegetation and livestock as nomadic herders graze their animals on the available grasses. In many fringe areas, livestock numbers have been greatly increasing in recent years, and they now far exceed the land's capacity to support them. As a result, the vegetation cover that protects the soil has diminished, causing the soil to crumble. This leads to further drying of the soil and accelerated soil erosion by wind and water (Figure 15.1).

Drilling water wells also contributes to desertification because human and livestock activity around a well site strips away the vegetation. With its vegetation gone, the topsoil blows away, and the resultant bare areas merge with the surrounding desert. In addition, the water used for irrigation from these wells sometimes contributes to desertification by increasing the salt content of the soil. As the water evaporates, a small amount of salt is deposited in the soil and is not flushed out as it would be in an area that receives more rain.

Over time, the salt concentration becomes so high that plants can no longer grow. Desertification resulting from soil salinization is a major problem in North Africa, the Middle East, southwest Asia, and the western United States.

Collecting firewood for heating and cooking is another major cause of desertification, particularly in many less-developed countries where wood is the major fuel source. In the Sahel of Africa (a belt 300–1100 km wide that lies south of the Sahara), the expanding population has removed all trees and shrubs in the areas surrounding many towns and cities. Journeys of several days on foot to collect firewood are common there. Furthermore, the use of dried animal dung to supplement firewood has exacerbated desertification because important nutrients in the dung are not returned to the soil.

The Sahel averages between 10 and 60 cm of rainfall per year, 90% of which evaporates when it falls. Because drought is common in the Sahel, the region can support only a limited population of livestock and humans. Traditionally, herders and livestock existed in a natural balance with the vegetation, following the rains north during the rainy season and returning south to greener rangeland during the dry seasons. Some areas

FIGURE 15.1 A sharp line marks the boundary between pasture and an encroaching dune in Niger, Africa. As the goats eat the remaining bushes, the dune will continue to advance, and more land will be lost to desertification.

were alternately planted and left fallow to help regenerate the soil. During fallow periods, livestock fed off the stubble of the previous year's planting, and their dung helped fertilize the soil.

With the emergence of new nations and increased foreign aid to the Sahel during the 1950s and 1960s, nomads and their herds were restricted, and large areas of grazing land were converted to cash crops such as peanuts and cotton that have a short growing season. Expanding human and animal populations and more intensive agriculture increased the demands on the land. These factors, combined with the worst drought of the century (1968–1973), brought untold misery to the people of the Sahel. Without rains, the crops failed and the livestock denuded the land of what little vegetation remained. As a result, nearly 250,000 people and 3.5 million cattle died of starvation, and the adjacent Sahara expanded southward as much as 150 km.

The tragedy of the Sahel and prolonged droughts in other desert fringe areas serve to remind us of the delicate equilibrium of ecosystems in such regions. Once the fragile soil cover has been removed by erosion, it takes centuries for new soil to form (see Chapter 6).

Introduction

MOST people associate the work of wind with deserts. Wind is an effective geologic agent in desert regions, but it also plays an important role wherever loose sediment can be eroded, transported, and deposited, such as along shorelines or on the plains. Therefore, we will first consider the work of wind in general and then discuss the distribution, characteristics, and landforms of deserts.

Sediment Transport by Wind

WIND is a turbulent fluid and therefore transports sediment in much the same way as running water. Although wind typically flows at a greater velocity than water, it has a lower density and, thus, can carry only clay- and silt-size particles as *suspended load*. Sand and larger particles are moved along the ground as *bed load*.

BED LOAD

Sediments too large or heavy to be carried in suspension by water or wind are moved as bed load either by *saltation* or by rolling and sliding. As we discussed in Chapter 12, saltation is the process by which a portion of the bed load moves by intermittent bouncing along a streambed. Saltation also occurs on land. Wind starts sand grains rolling and lifts and carries some grains short distances before they fall back to the surface. As the descending sand grains hit the surface, they strike other grains, causing them to bounce along by saltation (Figure 15.2). Wind-tunnel experiments have shown that once sand grains begin moving, they will continue to move, even if the wind drops below the speed necessary to start them moving! This happens because once saltation begins, it sets off a chain reaction of collisions between grains that keeps the sand grains in constant motion.

Saltating sand usually moves near the surface, and even when winds are strong, grains are rarely lifted higher than about a meter. If the winds are very strong, these wind-whipped grains can cause extensive abrasion. A car's paint can be removed by sandblasting in a short time, and its windshield will become completely frosted and translucent from pitting.

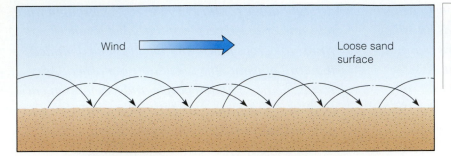

FIGURE 15.2 Most sand is moved near the ground surface by saltation. Sand grains are picked up by the wind and carried a short distance before falling back to the ground where they usually hit other grains, causing them to bounce and move in the direction of the wind.

SUSPENDED LOAD

Silt- and clay-sized particles constitute most of a wind's suspended load. Even though these particles are much smaller and lighter than sand-sized particles, wind usually starts the latter moving first. The reason for this phenomenon is that a very thin layer of motionless air lies next to the ground where the small silt and clay particles remain undisturbed. The larger sand grains, however, stick up into the turbulent air zone where they can be moved. Unless the stationary air layer is disrupted, the silt and clay particles remain on the ground, providing a smooth surface. This phenomenon can be observed on a dirt road on a windy day. Unless a vehicle travels over the road, little dust is raised even though it is windy. When a vehicle moves over the road, it breaks the calm boundary layer of air and disturbs the smooth layer of dust, which is picked up by the wind and forms a dust cloud in the vehicle's wake.

In a similar manner, when a sediment layer is disturbed, silt- and clay-sized particles are easily picked up and carried in suspension by the wind, creating clouds of dust or even dust storms. Once these fine particles are lifted into the atmosphere, they may be carried thousands of kilometers from their source. For example, large quantities of fine dust from the southwestern United States were blown eastward and fell on New England during the Dust Bowl of the 1930s (see Figure 6.18).

Wind Erosion

ALTHOUGH wind action produces many distinctive erosional features and is an extremely efficient sorting agent, running water is still responsible for most erosional landforms in arid regions, even though stream channels are typically dry. Wind erodes material in two ways: abrasion and deflation.

ABRASION

Abrasion involves the impact of saltating sand grains on an object and is analogous to sandblasting. The effects of abrasion are usually minor because sand, the most common agent of abrasion, is rarely carried more than 1 m above the surface. Rather than creating major erosional features, wind abrasion merely modifies existing features by etching, pitting, smoothing, or polishing. Thus, wind abrasion is most effective on soft sedimentary rocks.

Ventifacts are a common product of wind abrasion; these are stones whose surfaces have been polished, pitted, grooved, or faceted by the wind (Figure 15.3). If the wind blows from different directions or if the stone is moved, the ventifact will have multiple facets. Ventifacts are most common in deserts, yet they can also form wherever stones are exposed to saltating sand grains, as on beaches in humid regions and some outwash plains in New England.

FIGURE 15.3 (a) A ventifact forms when windblown particles (1) abrade the surface of a rock (2) forming a flat surface. If the rock is moved, (3) additional flat surfaces are formed. (b) Large ventifacts lying on desert pavement in Death Valley National Monument, California.

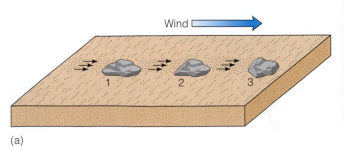

(a)

(b)

Yardangs are larger features than ventifacts and also result from wind erosion (Figure 15.4). They are elongated and streamlined ridges that look like an overturned ship's hull. They are typically found grouped in clusters aligned parallel to the prevailing winds. They probably form by differential erosion in which depressions, parallel to the direction of wind, are carved out of a rock body, leaving sharp, elongated ridges. These ridges may then be further modified by wind abrasion into their characteristic shape.

DEFLATION

Another important mechanism of wind erosion is **deflation**, which is the removal of loose surface sediment by the wind. Among the characteristic features of deflation in many arid and semiarid regions are *deflation hollows* or *blowouts* (Figure 15.5). These shallow depressions of variable dimensions result from differential erosion of surface materials. Ranging in size from several kilometers in diameter and tens of meters deep to small depressions only a few meters wide and less than a meter deep, deflation hollows are common in the southern Great Plains region of the United States.

In many dry regions, the removal of sand-sized and smaller particles by wind leaves a surface of pebbles, cobbles, and boulders. As the wind removes the fine-grained material from the surface, the effects of gravity and occasional floodwaters rearrange the remaining coarse particles into a mosaic of close-fitting rocks called **desert pavement** (Figures 15.3 and 15.6). Once a desert pavement forms, it protects the underlying material from further deflation.

Wind Deposits

ALTHOUGH wind is of minor importance as an erosional agent, it is responsible for impressive deposits, which are primarily of two types. The first, dunes, occur in several

FIGURE 15.4 Profile view of a streamlined yardang in the Roman playa deposits of the Kharga Depression, Egypt. *(Photo courtesy of Marion A. Whitney.)*

distinctive types, all of which consist of sand-sized particles that are usually deposited near their source. The second is loess, which consists of layers of windblown silt and clay that are deposited over large areas downwind and commonly far from their source.

THE FORMATION AND MIGRATION OF DUNES

The most characteristic features associated with sand-covered regions are **dunes**, which are mounds or ridges of wind-deposited sand. Dunes form when wind flows over and around an obstruction. Most dunes have an asymmetric profile, with a gentle windward slope and a steeper downwind, or leeward, slope that is inclined in the direction of the prevailing wind (Figure 15.7a). Sand grains move up the gentle windward slope by saltation and accumulate on the leeward side forming an angle between 30

FIGURE 15.5 A deflation hollow in Death Valley, California.

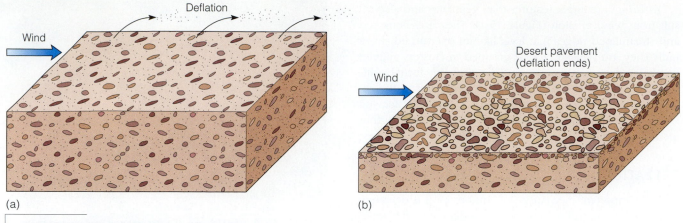

(a)   (b)

FIGURE 15.6 (a) Desert pavement forms when deflation removes fine-grained material from the ground surface, leaving larger-sized particles. (b) As deflation continues and more material is removed, the larger particles are concentrated and form a desert pavement, which protects the underlying material from additional deflation.

and 34 degrees from the horizontal, which is the angle of repose of dry sand. When this angle is exceeded by accumulating sand, the slope collapses, and the sand slides down the leeward slope, coming to rest at its base. As sand moves from a dune's windward side and periodically slides down its leeward slope, the dune slowly migrates in the direction of the prevailing wind (Figure 15.7b).

DUNE TYPES

Four major dune types are generally recognized (barchan, longitudinal, transverse, and parabolic), although intermediate forms between the major types also exist. The size, shape, and arrangement of dunes result from the interaction of such factors as sand supply, the direction and velocity of the prevailing wind, and the amount of vegetation. While dunes are usually found in deserts, they can also occur wherever there is an abundance of sand such as along the upper parts of many beaches.

Barchan dunes are crescent-shaped dunes whose tips point downwind (Figure 15.8). They form in areas having a generally flat, dry surface with little vegetation, a limited supply of sand, and a nearly constant wind direction. Most barchans are small, with the largest reaching about 30 m in height. Barchans are the most mobile of the major dune types, moving at rates that can exceed 10 m per year.

Longitudinal dunes (also called *seif dunes*) are long, parallel ridges of sand aligned generally parallel to the direction of the prevailing winds; they form where the sand supply is somewhat limited (Figure 15.9). Longitudinal dunes result when winds converge from slightly different directions to produce the prevailing wind. They range in size from about 3 m to more than 100 m high, and some stretch for more than 100 km. These dunes are especially well developed in central Australia, where they cover nearly one-fourth of the continent. They also cover extensive areas in Saudi Arabia, Egypt, and Iran.

Transverse dunes form long ridges perpendicular to the prevailing wind direction in areas where abundant sand is available and little or no vegetation exists (Figure 15.10). When viewed from the air, transverse dunes have a wave-like appearance, and the areas they cover are therefore sometimes called *sand seas*. The crests of transverse dunes can be as high as 200 m, and the dunes may be as much as 3 km wide.

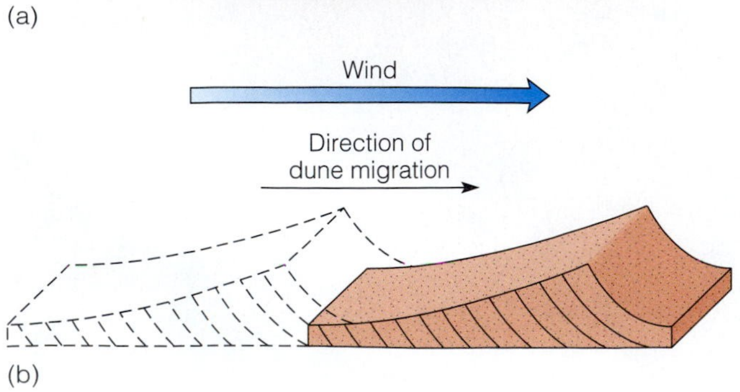

(a)

(b)

FIGURE 15.7 (a) Profile view of a sand dune. (b) Dunes migrate when sand moves up the windward side and slides down the leeward slope. Such movement of the sand grains produces a series of crossbeds that slope in the direction of wind movement.

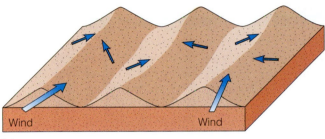

(a)

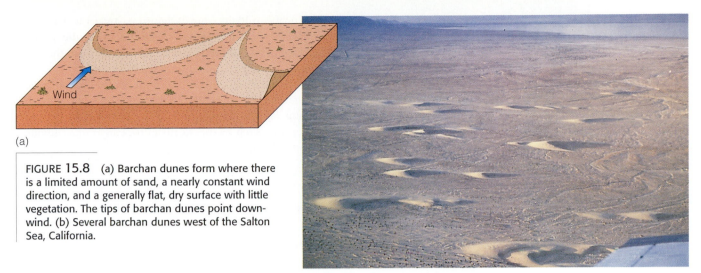

(b)

FIGURE 15.8 (a) Barchan dunes form where there is a limited amount of sand, a nearly constant wind direction, and a generally flat, dry surface with little vegetation. The tips of barchan dunes point downwind. (b) Several barchan dunes west of the Salton Sea, California.

(a)

(b)

FIGURE 15.9 (a) Longitudinal dunes form long, parallel ridges of sand aligned roughly parallel to the prevailing wind direction. They typically form where sand supplies are limited. (b) Longitudinal dunes, 15 m high, in the Gibson Desert, in west-central Australia, are shown in this image. The bright blue areas between the dunes are shallow pools of rainwater, and the darkest patches are areas where Aborigines have set fires to encourage the growth of spring grasses.

(a)

(b)

FIGURE 15.10 (a) Transverse dunes form long ridges of sand that are perpendicular to the prevailing wind direction in areas of little or no vegetation and abundant sand. (b) Aerial view of transverse dunes, Great Sand Dunes National Monument, Colorado.

Parabolic dunes are most common in coastal areas with abundant sand, strong onshore winds, and a partial cover of vegetation (Figure 15.11). Although parabolic dunes have a crescent shape like barchan dunes, their tips point upwind. Parabolic dunes form when the vegetation cover is broken and deflation produces a deflation hollow or blowout. As the wind transports the sand out of the depression, it builds up on the convex downwind dune crest. The central part of the dune is excavated by the wind, while vegetation holds the ends and sides fairly well in place.

LOESS

Windblown silt and clay deposits composed of angular quartz grains, feldspar, micas, and calcite are known as **loess**. The distribution of loess shows that it is derived from three main sources: deserts, Pleistocene glacial outwash deposits, and the floodplains of rivers in semiarid regions. It must be stabilized by moisture and vegetation in order to accumulate. Consequently, loess is not found in deserts, even though they provide much of its material. Because of its unconsolidated nature, loess is easily eroded, and as a result, eroded loess areas are characterized by steep cliffs and rapid lateral and headward stream erosion.

At present, loess deposits cover approximately 10% of Earth's land surface and 30% of the United States. The most extensive and thickest loess deposits occur in northeast China where accumulations greater than 30 m are common. The extensive deserts in central Asia are the source for this loess. Other important loess deposits are on the North European Plain from Belgium eastward to Ukraine, central Asia, and the Pampas of Argentina. In the United States, they occur in the Great Plains, the Midwest, the Mississippi River Valley, and eastern Washington.

Loess-derived soils are some of the world's most fertile. It is therefore not surprising that the world's major grain-producing regions correspond to the distribution of large loess deposits such as the North European Plain, Ukraine, and the Great Plains of North America.

Air-Pressure Belts and Global Wind Patterns

To understand the work of wind and the distribution of deserts, we need to consider the global pattern of air-pressure belts and winds, which are responsible for Earth's atmospheric circulation patterns. Air pressure is the density of air exerted on its surroundings (that is, its weight). When air is heated, it expands and rises, reducing its mass for a given volume and causing a decrease in air pressure. Conversely, when air is cooled, it contracts and air pressure increases. Therefore, those areas of Earth's surface that receive the most solar radiation, such as the equatorial regions, have low air pressure, while the colder areas, such as the polar regions, have high air pressure.

Air flows from high-pressure zones to low-pressure zones. If Earth did not rotate, winds would move in a straight line from one zone to another. Because Earth rotates, however, winds are deflected to the right of their

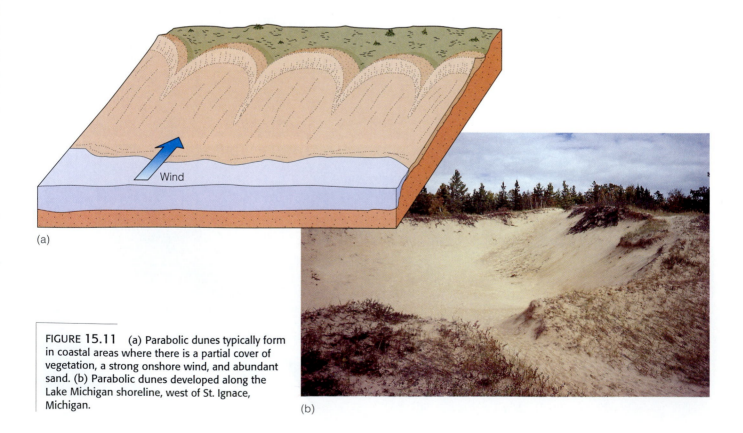

(a)

FIGURE 15.11 (a) Parabolic dunes typically form in coastal areas where there is a partial cover of vegetation, a strong onshore wind, and abundant sand. (b) Parabolic dunes developed along the Lake Michigan shoreline, west of St. Ignace, Michigan.

(b)

direction of motion (clockwise) in the Northern Hemisphere and to the left of their direction of motion (counterclockwise) in the Southern Hemisphere. Such deflection of air between latitudinal zones resulting from Earth's rotation is known as the **Coriolis effect**. Therefore, the combination of latitudinal pressure differences and the Coriolis effect produces a worldwide pattern of east-west–oriented wind belts (Figure 15.12).

Earth's equatorial zone receives the most solar energy, which heats the surface air, causing it to rise. As the air rises, it cools and releases moisture that falls as rain in the equatorial region (Figure 15.12). The rising air is now much drier as it moves northward and southward toward each pole. By the time it reaches 20 to 30 degrees north and south latitudes, the air has become cooler and denser and begins to descend. Compression of the atmosphere warms the descending air mass and produces a warm, dry, high-pressure area, providing the perfect conditions for the formation of the low-latitude deserts of the Northern and Southern Hemispheres.

The Distribution of Deserts

Dry climates occur in the low and middle latitudes, where the potential loss of water by evaporation exceeds the yearly precipitation. Dry climates cover 30% of Earth's land surface and are subdivided into semiarid and arid regions. *Semiarid regions* receive more precipitation than arid regions, yet are moderately dry. Their soils are usually well developed and fertile and support a natural grass cover. *Arid regions,* generally described as **deserts**, are dry; on average they receive less than 25 cm of rain per year, typically have poorly developed soils, and are mostly or completely devoid of vegetation.

The majority of the world's deserts are found in the dry climates of the low and middle latitudes. In North America, most of the southwestern United States and northern Mexico are characterized by this hot, dry climate, while in South America this climate is primarily restricted to the Atacama Desert of coastal Chile and Peru. The Sahara in northern Africa and the Arabian Desert in the Middle East, and the majority of Pakistan and western India, form the largest essentially unbroken desert environment in the Northern Hemisphere. More than 40% of Australia is desert, and most of the rest of it is semiarid.

The remaining dry climates of the world are found in the middle and high latitudes, mostly within continental interiors in the Northern Hemisphere. Many of these areas are dry because of their remoteness from moist maritime air and the presence of mountain ranges that produce a **rainshadow desert** (Figure 15.13). When moist marine air moves inland and meets a mountain range, it is forced upward. As it rises, it cools, forming clouds and producing precipitation that falls on the windward side of the mountains. The air that descends on the leeward side of the mountain range is much warmer and drier, producing a rainshadow desert.

Three widely separated areas are included within the mid-latitude dry-climate zone. The largest of these is the central part of Eurasia extending from just north of the

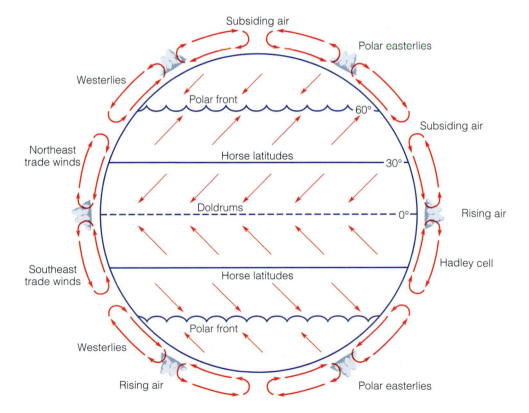

FIGURE 15.12
The general circulation of Earth's atmosphere.

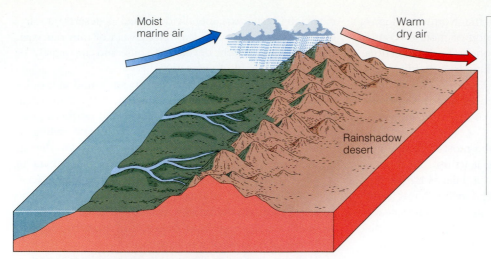

Black Sea eastward to north-central China. The Gobi Desert in China is the largest desert in this region. The Great Basin area of North America is the second largest mid-latitude dry-climate zone and results from the rainshadow produced by the Sierra Nevada. This region adjoins the southwestern deserts of the United States that formed as a result of the low-latitude subtropical high-pressure zone. The smallest of the mid-latitude dry climate areas is the Patagonian region of southern and western Argentina. Its dryness results from the rainshadow effect of the Andes.

The remainder of the world's deserts are found in the cold but dry high latitudes, such as Antarctica.

Characteristics of Deserts

To people who live in humid regions, deserts may seem stark and inhospitable (see Perspective 15.1). Instead of a landscape of rolling hills and gentle slopes with an almost continuous vegetation cover, deserts are dry, have little vegetation, and consist of nearly continuous rock exposures, desert pavement, or sand dunes. And yet despite the great contrast between deserts and more humid areas, the same geologic processes are at work, only operating under different climatic conditions.

TEMPERATURE AND VEGETATION

The heat and dryness of deserts are well known. Many of the deserts of the low latitudes have average summer temperatures ranging between 32° and 38°C. It is not uncommon for some low-elevation inland deserts to record daytime highs of 46° to 50°C for weeks at a time. The highest temperature ever recorded was 58°C in El Azizia, Libya, on September 13, 1922.

During the winter months when the angle of the Sun is lower and there are fewer daylight hours, daytime temperatures average between 10° and 18°C. Winter nighttime lows can be quite cold, with frost and freezing temperatures common in the more poleward deserts.

Deserts display a wide variety of vegetation. While the driest deserts, or those with large areas of shifting sand, are almost devoid of vegetation, most deserts support at least a sparse plant cover. Compared to humid areas, desert vegetation may appear monotonous. A closer examination, however, reveals an amazing diversity of plants that have evolved the ability to live in the near-absence of water.

Desert plants are widely spaced, typically small, and grow slowly. Their stems and leaves are usually hard and waxy to minimize water loss by evaporation and protect the plant from sand erosion. Most plants have a widespread shallow root system to absorb the dew that forms each morning in all but the driest deserts and to help anchor the plant in what little soil there may be. In extreme cases, many plants lie dormant during particularly dry years and spring to life after the first rain shower, with a beautiful profusion of flowers.

WEATHERING AND SOILS

Mechanical weathering is dominant in desert regions. Daily temperature fluctuations and frost wedging are the primary forms of mechanical weathering (see Chapter 6). The breakdown of rocks by roots and from salt crystal growth is of minor importance. Some chemical weathering does occur, but its rate is greatly reduced by aridity and the scarcity of organic acids produced by the sparse vegetation. Most chemical weathering takes place during the winter months when more precipitation occurs, particularly in the mid-latitude deserts.

An interesting feature seen in many deserts is a thin, red, brown, or black shiny coating on the surface of many rocks. This coating, called *rock varnish,* is composed of iron and manganese oxides (Figure 15.14). Because many of the varnished rocks contain little or no iron and manganese oxides, the varnish is thought to result from either windblown iron and manganese dust that settles on the ground or from the precipitated waste of microorganisms.

Desert soils, if developed, are usually thin and patchy because the limited rainfall and the resultant scarcity of

FIGURE 15.14 The shiny black coating on this rock exposed at Castle Valley, Utah, is rock varnish. It is composed of iron and manganese oxides.

vegetation reduce the efficiency of chemical weathering and hence soil formation. Furthermore, the sparseness of the vegetative cover enhances wind and water erosion of what little soil actually forms.

MASS WASTING, STREAMS, AND GROUNDWATER

When traveling through a desert, most people are impressed by such wind-formed features as moving sand, sand dunes, and sand and dust storms. They may also notice the dry washes and dry streambeds. Because of the lack of running water, most people would conclude that wind is the most important erosional geologic agent in deserts. They would be wrong. Running water, even though it occurs infrequently, causes most of the erosion in deserts. The dry conditions and sparse vegetation characteristic of deserts enhance water erosion. If you look closely, you will see the evidence of erosion and transportation by running water nearly everywhere except in areas covered by sand dunes.

Most of a desert's average annual rainfall of 25 cm or less comes in brief, heavy, localized cloudbursts. During these times, considerable erosion takes place because the ground cannot absorb all the rainwater. With so little vegetation to hinder its flow, runoff is rapid, especially on moderately to steeply sloping surfaces, resulting in flash floods and sheetflows. Dry stream channels quickly fill with raging torrents of muddy water and mudflows, which carve out steep-sided gullies and overflow their banks. During these times, a tremendous amount of sediment is rapidly transported and deposited far downstream.

While water is the major erosive agent in deserts today, it was even more important during the Pleistocene Epoch when these regions were more humid. During that time, many of the major topographic features of deserts were forming. Today that topography is being modified by wind and infrequently flowing streams.

Most desert streams are poorly integrated and flow only intermittently. Many of them never reach the sea because the water table is usually far deeper than the channels of most streams, so they cannot draw upon groundwater to replace water lost to evaporation and absorption into the ground. This type of drainage in which a stream's load is deposited within the desert is called *internal drainage* and is common in most arid regions.

While the majority of deserts have internal drainage, some deserts have permanent through-flowing streams such as the Nile and Niger Rivers in Africa, the Rio Grande and Colorado River in the southwestern United States, and the Indus River in Asia. These streams can flow through desert regions because their headwaters are well outside the desert and water is plentiful enough to offset losses resulting from evaporation and infiltration. Demands for greater amounts of water for agriculture and domestic use from the Colorado River, however, are leading to increased salt concentrations in its lower reaches and causing political problems between the United States and Mexico.

WIND

Although running water does most of the erosional work in deserts, wind can also be an effective geologic agent capable of producing a variety of distinctive erosional and depositional features. It is very effective in transporting and depositing unconsolidated sand-, silt-, and dust-sized particles. Contrary to popular belief, most deserts are not sand-covered wastelands, but rather consist of vast areas of rock exposures and desert pavement. Sand-covered regions, or sandy deserts, constitute less than 25% of the world's deserts. The sand in these areas has accumulated primarily by the action of wind.

Desert Landforms

BECAUSE of differences in temperature, precipitation, and wind, as well as the underlying rocks and recent tectonic events, landforms in arid regions vary considerably.

After an infrequent and particularly intense rainstorm, excess water not absorbed by the ground may accumulate in low areas and form *playa lakes*. These lakes are temporary, lasting from a few hours to several months. Most of them are very shallow and have rapidly shifting boundaries as water flows in or leaves by evaporation and seepage into the ground. The water is often very saline.

When a playa lake evaporates, the dry lake bed is called a **playa** or *salt pan* and is characterized by mudcracks and precipitated salt crystals (Figure 15.15). Salts in some playas are thick enough to be mined commercially. For example, borates have been mined in Death Valley, California, for more than 100 years.

Other common features of deserts, particularly in the Basin and Range region of the United States, are alluvial fans and bajadas. **Alluvial fans** form when sediment-laden streams flowing out from generally straight, steep mountain

Petroglyphs

Rock art includes rock paintings (where paints made from natural pigments are applied to a rock surface) and petroglyphs (from the Greek *petro* meaning "rock" and *glyph* meaning "carving or engraving"), which are the abraded, pecked, incised, or scratch marks made by humans on boulders, cliffs, and cave walls. Rock art is found throughout the world and is a valuable archaeological resource that can be used to interpret and increase our knowledge of prehistoric peoples. It provides us with graphical evidence of cultural, social, and religious relationships and practices, as well as changing patterns of trade and communication between different ancient peoples. The oldest known rock art was made by hunters in western Europe and dates back to the Pleistocene Epoch.

In the arid Southwest and Great Basin area of North America where rock art is plentiful, rock paintings and petroglyphs extend back to about 2000 B.C. Here, rock art can be divided into two categories. *Representational art* deals with life-forms such as humans, birds, snakes, and humanlike supernatural beings. Rarely exact replicas, they are more or less stylized versions of the beings depicted. *Abstract art,* on the other hand, bears no resemblance to any real-life images.

Petroglyphs are the most common form of rock art in North America. They are especially abundant in the Southwest and Great Basin area where they occur by the thousands, having been made by Native Americans from many cultures during the past several thousand years. Petroglyphs can be viewed in many of the National Parks and Monuments such as Petrified Forest National Park, Arizona; Dinosaur National Monument, Utah (Figure 1); Canyonlands National Park, Utah; and Petroglyph National Monument, New Mexico—to name a few.

Petroglyphs are made using a variety of techniques. The most common is pecking, in which the surface of the rock is struck by a harder stone, or for more precise control using a stone chisel hit by a hammer stone. This results in a series of dots, which can be made into lines by continued pecking. Differences in pecking techniques characterize different styles of petroglyphs. Another method used is incising or scratching. Here, lines up to several centimeters deep are carved into the rock, using a sharp tool. Sometimes a combination of techniques is used, including painting the figures after they have been carved in the rock.

In arid regions, many rock surfaces display a *patination,* or thin brown or black coating, known as rock varnish, composed of iron and manganese oxides. When this coating is broken by pecking, incising, or scratching, the underlying lighter colored natural-rock surface provides an excellent contrast for the petroglyphs (Figure 2).

One of the best places to view petroglyphs in North America is at Petroglyph National Monument, west of Albuquerque, New Mexico. Established by Congress in 1990, this national monument protects more than 15,000 petroglyphs. While some may be as old as 2000 to 3000 years, the vast majority were created between 1300 and 1650 and reflect the complex culture and religion of the Pueblo peoples.

Figure 1 Various petroglyphs exposed at an outcrop along Club Creek Road, Dinosaur National Monument, Utah.

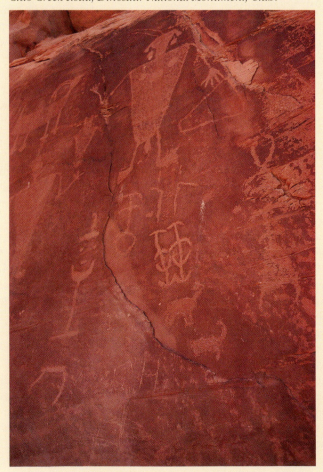

It is important to remember that petroglyphs are a fragile and nonrenewable cultural resource that cannot be replaced if they are damaged or destroyed. A commitment to their preservation is essential so that future generations can study them as well as enjoy their beauty and mystery.

Figure 2 *Humanlike petroglyph exposed at an outcrop along Club Creek Road, Dinosaur National Monument, Utah. Note the contrast between the fresh exposure of the rock where the upper part of the petroglyph's head has been removed, the weathered brown surface of the rest of the petroglyph, and the black rock varnish coating the rock surface.*

FIGURE 15.15 Racetrack Playa, Death Valley, California. The Inyo Mountains can be seen in the background.

FIGURE 15.16 Aerial view of an alluvial fan, Death Valley, California.

fronts deposit their load on the relatively flat desert floor. Once beyond the mountain front where no valley walls contain streams, the sediment spreads out laterally, forming a gently sloping and poorly sorted fan-shaped sedimentary deposit (Figure 15.16). Alluvial fans are similar in origin and shape to deltas (see Chapter 12) but are formed entirely on land. Alluvial fans may coalesce to form a *bajada*. This broad alluvial apron typically has an undulating surface resulting from the overlap of adjacent fans (Figure 15.17).

Large alluvial fans and bajadas are frequently important sources of groundwater for domestic and agricultural use. Their outer portions are typically composed of fine-grained sediments suitable for cultivation, and their gentle slopes allow good drainage of water. Many alluvial fans

FIGURE 15.17 Coalescing alluvial fans forming a bajada at the base of the Black Mountains, Death Valley, California.

and bajadas are also the sites of large towns and cities, such as San Bernardino, California, Salt Lake City, Utah, and Teheran, Iran.

Most mountains in desert regions, including those of the Basin and Range region, rise abruptly from gently sloping surfaces called pediments. **Pediments** are erosional bedrock surfaces of low relief that slope gently away from mountain bases (Figure 15.18). Most pediments are covered by a thin layer of debris, alluvial fans, or bajadas.

Rising conspicuously above the flat plains of many deserts are isolated steep-sided erosional remnants called

inselbergs, a German word meaning "island mountain" (Figure 15.19). Inselbergs have survived for a longer period of time than other mountains because of their greater resistance to weathering.

Other easily recognized erosional remnants common to arid and semiarid regions are mesas and buttes (Figure 15.20). A **mesa** is a broad, flat-topped erosional remnant bounded on all sides by steep slopes. Continued weathering and stream erosion will form isolated pillarlike structures known as **buttes**. Buttes and mesas consist of relatively easily weathered sedimentary rocks capped by

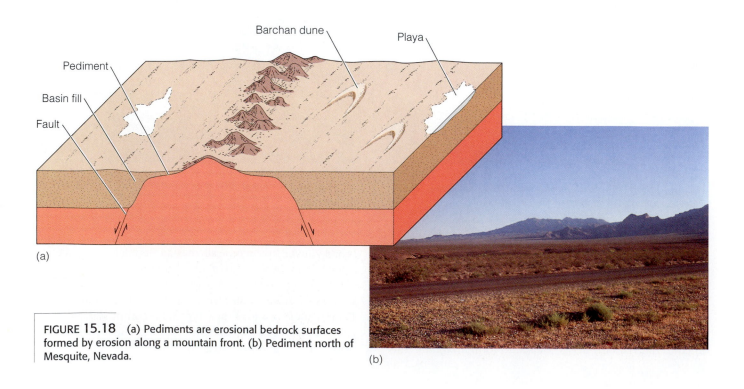

(a)

(b)

FIGURE 15.18 (a) Pediments are erosional bedrock surfaces formed by erosion along a mountain front. (b) Pediment north of Mesquite, Nevada.

FIGURE 15.19 Ayers Rock in central Australia is a good example of an inselberg. This photo shows Ayers Rock at sunset.

nearly horizontal, resistant rocks such as sandstone, limestone, or basalt. They form when the resistant rock layer is breached, allowing rapid erosion of the less resistant underlying sediment. One of the best-known areas of mesas and buttes in the United States is Monument Valley on the Arizona-Utah border.

(a)

(b)

FIGURE 15.20 (a) Mesas southeast of Zuni Pueblo, New Mexico. (b) Butte in Monument Valley, Arizona.

Chapter Summary

1. Wind can transport sediment in suspension or as bed load, which involves saltation and surface creep.

2. Wind erodes material either by abrasion or deflation. Abrasion is a near-surface effect caused by the impact of saltating sand grains. Ventifacts are common wind-abraded features.

3. Deflation is the removal of loose surface material by the wind. Deflation hollows resulting from differential erosion of surface material are common features of many deserts, as is desert pavement, which effectively protects the underlying surface from additional deflation.

4. The two major wind deposits are dunes and loess. Dunes are mounds or ridges of wind-deposited sand, whereas loess is wind-deposited silt and clay.

5. The four major dune types are barchan, longitudinal, transverse, and parabolic. The amount of sand available, the prevailing wind direction, the wind velocity, and the amount of vegetation present determine which type will form.

6. Loess is derived from deserts, Pleistocene glacial outwash deposits, and river floodplains in semiarid regions. Loess covers approximately 10% of Earth's land

surface and weathers to a rich and productive soil.

7. Deserts are very dry (averaging less than 25 cm rain/year), have poorly developed soils, and are mostly or completely devoid of vegetation.

8. The winds of the major east-west–oriented air-pressure belts resulting from rising and cooling air are deflected by the Coriolis effect. These belts help control the world's climate.

9. Dry climates are located in the low and middle latitudes where the potential loss of water by evaporation exceeds the yearly precipitation. Dry climates cover 30% of Earth's surface and are subdivided into semiarid and arid regions.

10. The majority of the world's deserts are in the low-latitude dry-climate zone between 20 and 30 degrees north and south latitudes. Their dry climate results from a high-pressure belt of descending dry air. The remaining deserts are in the middle latitudes where their distribution is related to the rainshadow effect and in the dry polar regions.

11. Deserts are characterized by lack of precipitation and high evaporation rates. Furthermore, rainfall is unpredictable and, when it does occur, tends to be intense and of short duration. As a consequence

of such aridity, desert vegetation and animals are scarce.

12. Mechanical weathering is the dominant form of weathering in deserts. The sparse precipitation and slow rates of chemical weathering result in poorly developed soils.

13. Running water is the dominant agent of erosion in deserts and was even more important during the Pleistocene when wetter climates resulted in humid conditions.

14. Wind is an erosional agent in deserts and is very effective in transporting and depositing unconsolidated fine-grained sediments.

15. Important desert landforms include playas, which are dry lake beds; when temporarily filled with water, they form playa lakes. Alluvial fans are fan-shaped sedimentary deposits that may coalesce to form bajadas. Pediments are erosional bedrock surfaces of low-relief gently sloping away from mountain bases. Inselbergs are isolated steep-sided erosional remnants that rise above the surrounding desert plains. Buttes and mesas are, respectively, pinnacle-like and flat-topped erosional remnants with steep sides.

Important Terms

abrasion
alluvial fan
barchan dune
butte
Coriolis effect
deflation
desert

desert pavement
desertification
dune
inselberg
loess
longitudinal dune
mesa

parabolic dune
pediment
playa
rainshadow desert
transverse dune
ventifact

1. Deserts:
   a. _____ can be found in the low, middle, and high latitudes;
   b. _____ receive more than 25 cm of rain per year;
   c. _____ are mostly or completely devoid of vegetation;
   d. _____ answers (a) and (c);
   e. _____ answers (b) and (c).

2. The Coriolis effect causes wind to be deflected:
   a. _____ to the right in the Northern Hemisphere and the left in the Southern Hemisphere;
   b. _____ to the left in the Northern Hemisphere and the right in the Southern Hemisphere;
   c. _____ only to the left for both hemispheres;
   d. _____ only to the right for both hemispheres;
   e. _____ not at all.

3. Which particle size constitutes most of a wind's suspended load?
   a. _____ sand;
   b. _____ silt;
   c. _____ clay;
   d. _____ answers (a) and (b);
   e. _____ answers (b) and (c).

4. Which of the following is a feature produced by wind erosion?
   a. _____ playa;
   b. _____ loess;
   c. _____ dune;
   d. _____ yardang;
   e. _____ none of these.

5. Which of the following dunes form long ridges perpendicular to the prevailing wind direction in areas where abundant sand is available and little or no vegetation exists?
   a. _____ barchan;
   b. _____ longitudinal;
   c. _____ transverse;
   d. _____ parabolic;
   e. _____ loess.

6. Where are the thickest and most extensive loess deposits in the world?
   a. _____ United States;
   b. _____ Pampas of Argentina;
   c. _____ Belgium;
   d. _____ Ukraine;
   e. _____ northeast China.

7. The major agent of erosion in deserts today is _____; during the Pleistocene Epoch, it was _____.
   a. _____ wind, running water;
   b. _____ wind, wind;
   c. _____ running water, wind;
   d. _____ running water, running water;
   e. _____ wind, glaciers.

8. Poorly sorted, fan-shaped sedimentary deposits with a gentle slope are:
   a. _____ pediments;
   b. _____ inselbergs;
   c. _____ mesas;
   d. _____ buttes;
   e. _____ alluvial fans.

9. What are some of the problems associated with desertification?

10. Describe how the global distribution of air-pressure belts and winds operates.

11. What are the two ways that sediments are transported by wind?

12. Why is desert pavement important in a desert environment?

13. How do sand dunes migrate?

14. Describe the four major dune types and the conditions necessary for their formation.

15. What is loess, and why is it important?

16. Why are most of the world's deserts located in the low latitudes?

17. How are temperature, precipitation, and vegetation interrelated in desert environments?

18. What is the dominant form of weathering in desert regions, and why is it so effective?

Points to Ponder

1. Because much of the recent desertification has been greatly accelerated by human activities, can we do anything to slow the process or restore some type of equilibrium or buffer zone between encroaching deserts and adjacent productive lands?

2. If deserts are very dry regions in which mechanical weathering predominates, why are so many of their distinctive landforms the result of running water and not wind?

World Wide Web Activities

For these web site addresses, along with current updates and exercises, log on to

http://www.wadsworth.com/geo

▶ THE NATIONAL PARK SERVICE—DEATH VALLEY NATIONAL PARK

This site contains general information about Death Valley National Park. What is the general climate of the park? What activities are available? Where is Death Valley National Park located?

▶ DEATH VALLEY NATIONAL PARK

This site—operated by DesertUSA: THE ULTIMATE DESERT RESOURCE—is a compendium of information about Death Valley National Park. Click on *Description of the Park* under the Exploring Death Valley National park heading. What is the physical environment of the area? What types of plants and animals live in this area? What is the geology of the area?

▶ DESERTUSA

This site, devoted to information about deserts, contains a tremendous amount of information about the fauna, flora, geology, physical environment, and other items of interest about deserts. Under the What's New heading, click on *The Desert* site. Read about what deserts are. Click on one of the deserts shown on the map of the United States. Learn about the different deserts in the United States.

CD-ROM Exploration

Explore the following *In-Terra-Active 2.0* CD-ROM module(s) and increase your understanding of key concepts and processes presented in this chapter.

▶ SECTION: SURFACE
MODULE: WINDS AND DESERTS

Two of the driest places on Earth are the Atacama Desert in Chile (bordering the Pacific Ocean) and the Dry Valleys of Antarctica (surrounded by snow and ice). **Question: How can these places be deserts, given the abundant water nearby?**

▶ VIRTUAL REALITY FIELD TRIP: PINNACLES DESERT, AUSTRALIA

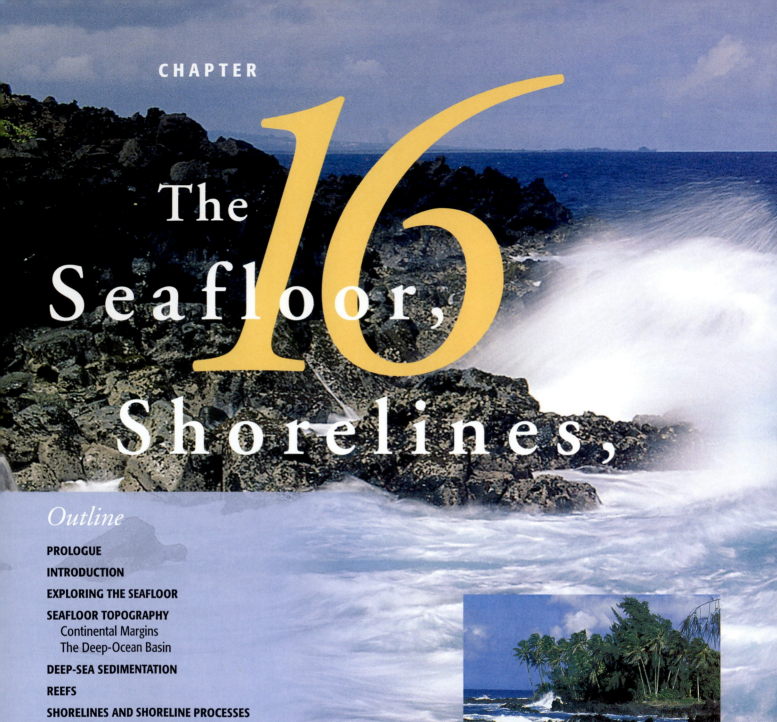

CHAPTER

16

The Seafloor, Shorelines,

Outline

Waves pounding the shoreline of Maui in the Hawaiian Islands.

and

Shoreline

Processes

Most people have heard of the mythical lost continent of Atlantis, but few are aware of the source of the Atlantis legend or the evidence cited for the former existence of this continent. Only two known sources of the Atlantis legend exist, both written in about 350 B.C. by the Greek philosopher Plato. In two of his philosophical dialogues, the Timaeus and the Critias, Plato tells of Atlantis, a large island continent that, according to him, was in the Atlantic Ocean west of the Pillars of Hercules, which we now call the Strait of Gibralter. Plato also wrote that following the conquest of Atlantis by Athens, the continent disappeared:

> There were violent earthquakes and floods and one terrible day and night came when . . . Atlantis . . . disappeared beneath the sea. And for this reason even now the sea there has become unnavigable and unsearchable, blocked as it is by the mud shallows which the island produced as it sank.*

If one assumes that the destruction of Atlantis was a real event, rather than one conjured up by Plato to make a philosophical point, he nevertheless lived long after it was supposed to have occurred. According to Plato, Solon, an Athenian who lived about 200 years before Plato, heard the story from Egyptian priests

*From the *Timaeus*. Quoted in E. W. Ramage, ed., *Atlantis: Fact or Fiction?* (Bloomington: Indiana University Press, 1978), 13.

who claimed the event had taken place 9000 years before their time. Solon told the story to his grandson, Critias, who in turn told it to Plato.

Present-day proponents of the Atlantis legend generally cite two types of evidence to support their claim that Atlantis existed. First, they point to supposed cultural similarities on opposite sides of the Atlantic Ocean basin, such as the similar shapes of the pyramids of Egypt and those of Central and South America. They contend that these similarities are due to cultural diffusion from the highly developed civilization of Atlantis. According to archaeologists, however, few similarities actually exist, and those that do can be explained as the independent development of analogous features by different cultures.

Secondly, supporters of the legend assert that remnants of the sunken continent can be found. No "mud shallows" exist in the Atlantic as Plato claimed, but the Azores, Bermuda, the Bahamas, and the Mid-Atlantic Ridge are alleged to be remnants of Atlantis. If a continent had actually sunk in the Atlantic, it could be easily detected by a gravity survey. Recall that continental crust has a granitic composition and a lower density than oceanic crust. So if a continent were actually present beneath the Atlantic Ocean, a huge negative gravity anomaly would exist, but no such anomaly has been detected. Furthermore, the crust beneath the Atlantic has been drilled in many places, and all the samples recovered indicate that its composition is the same as that of oceanic crust elsewhere.

References to what some people claim as actual remains of buildings and roads constructed by Atlanteans are sometimes seen in the popular literature. For instance, in the Bimini Islands of the Bahamas, "pavementlike" blocks are cited as evidence of an ancient roadway, and cylinders of rocklike material are claimed to be parts of ancient pillars. On close inspection, however, the blocks are nothing more than naturally fractured pieces of limestone. The cylinders appear to be cement manufactured after 1800 for construction that was either dumped in this area or came to rest on the seafloor as a result of shipwreck.

In short, there is no geologic evidence for Atlantis. Nevertheless, some archaeologists think that the legend may be based on a real event. About 1390 B.C., a huge volcanic eruption destroyed the island of Thera in the Mediterranean Sea, which was an important center of early Greek civilization. The eruption was one of the most violent during historic time, and much of the island disappeared when it subsided to form a caldera. Most of the island's inhabitants escaped (Figure 16.1), but the eruption probably contributed to the demise of the Minoan culture on Crete. At least 10 cm of ash fell on parts of Crete, and the coastal areas of the island were probably devastated by tsunami. It is possible that Plato used an account of the destruction of Thera, but fictionalized it for his own purposes, thereby giving rise to Atlantis legend.

FIGURE **16.1** An artist's rendition of the volcanic eruption on Thera that destroyed most of the island in about 1390 B.C. Most of the island's inhabitants escaped the devastation.

Introduction

DURING most of historic time, people knew little of the oceans and, until fairly recently, still believed that the seafloor was flat and featureless. The ancient Greeks determined Earth's size rather accurately, but Western Europeans were not aware of the vastness of the oceans until the fifteenth and sixteenth centuries when various explorers sought new trade routes to the Indies. When Christopher Columbus set sail on August 4, 1492, in an attempt to find a route to the Indies, he greatly underestimated the width of the Atlantic Ocean. Contrary to popular belief, Columbus was not attempting to demonstrate that Earth is spherical—Earth's spherical shape was well accepted by then.

We now know that about 71% of Earth's surface is covered by an interconnected body of salt water we refer to as *oceans* and *seas*. Four very large areas in this body of water are distinct enough to be designated as oceans: the Pacific, the Atlantic, the Indian, and the Arctic (Table 16.1). Seas are simply marginal parts of oceans that occupy an indentation into a continent, for example, the Caribbean Sea and the Sea of Japan.

Research during the last 200 years, and particularly during the last few decades, has shown that the seafloor possesses varied topography including such features as oceanic trenches, submarine canyons and ridges, and broad plateaus, hills, and plains. Furthermore, scientists have come to more fully appreciate the dynamic nature of shorelines and the continuous interplay between their materials and the energy of waves, tides, and nearshore currents.

Exploring the Seafloor

OCEANIC depths were first measured by lowering a weighted line to the seafloor and measuring its length. Now an instrument called an *echo sounder* is used. Depth is calculated by knowing the velocity of sound waves in water and determining how long sound waves from a ship take to reach the seafloor and return to the ship. *Seismic profiling* uses waves that penetrate the layers beneath the seafloor and are reflected back to the source (Figure 16.2). This technique is particularly useful in determining the structure of the oceanic crust beneath seafloor sediments.

Submersibles, both remotely controlled and carrying scientists, are also used to explore the seafloor. In 1985, the *Argo*, towed by a surface vessel and equipped with sonar and television systems, provided the first views of the British ocean liner R.M.S. *Titanic* since it sank in 1912. The U.S. Geological Survey is using a towed device that uses sonar to produce images resembling aerial photographs. Researchers aboard the submersible *Alvin* have observed various parts of the seafloor.

TABLE 16.1				
Numerical Data for the Oceans				
OCEAN*	SURFACE AREA (MILLION KM²)	WATER VOLUME (MILLION KM³)	AVERAGE DEPTH (KM)	MAXIMUM DEPTH (KM)
Pacific	180	700	4.0	11.0
Atlantic	93	335	3.6	9.2
Indian	77	285	3.7	7.5
Arctic	15	17	1.1	5.2
SOURCE: Pinet, P. R. 1992. *Oceanography.*				
*Excludes adjacent seas.				

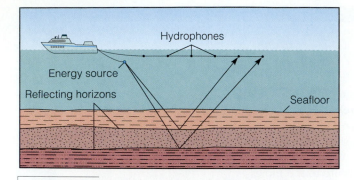

FIGURE 16.2 Diagram showing how seismic profiling is used to detect buried layers at sea. Some of the energy generated at the energy source is reflected from various horizons back to the surface where it is detected by hydrophones.

Deep-sea drilling has also provided a wealth of information about the seafloor. Ships such as the *Glomar Challenger* (now retired) and the JOIDES* *Resolution* are equipped to drill in water more than 6000 m deep and recover long cores of seafloor sediment and oceanic crust. Much of what we know of the composition and structure of the upper oceanic crust was obtained by deep-sea drilling.

*JOIDES is an acronym for Joint Oceanographic Institutions for Deep Earth Sampling.

Seafloor Topography

THE ocean floor is not monotonous and flat as once believed. Indeed, it is as varied as the surface of the continents, but for centuries people were unaware of this fact. In 1959, Maurice Ewing, Bruce Heezen, and Marie Tharp published a spectacular three-dimensional map of the North Atlantic showing vast plains and conical seamounts, as well as the Mid-Atlantic Ridge with its central rift valley. As more of the world's ocean floors were explored, this original map was expanded to reveal numerous other seafloor features (Figure 16.3).

CONTINENTAL MARGINS

The zone separating each continent above sea level from the deep seafloor is the **continental margin**. It consists of a gently sloping *continental shelf,* a more steeply inclined *continental slope,* and, in some cases, a gently inclined *continental rise* (Figure 16.4). At its outer limit, the continental margin merges with the deep seafloor or descends into an oceanic trench.

The width of the continental shelf varies from only a few tens of meters to more than 1000 km. Its outer edge is

FIGURE 16.3 This map of the seafloor resulted from the work of Maurice Ewing, Bruce Heezen, and Marie Tharp.

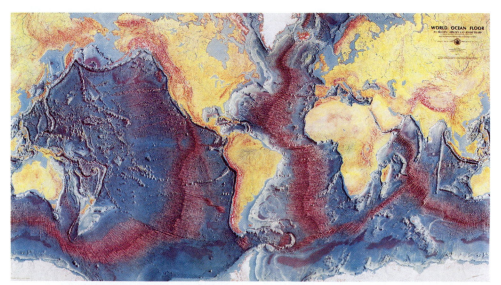

FIGURE 16.4 A generalized profile of the seafloor, showing features of the continental margins. The vertical dimensions of the features in this profile are greatly exaggerated because the vertical and horizontal scales differ.

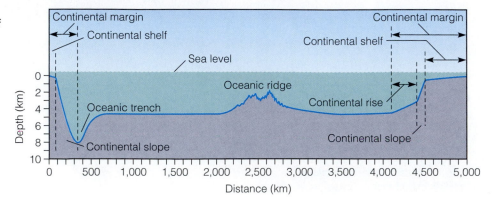

the shelf-slope break, the point at which the inclination of the seafloor increases rather abruptly to several degrees. Rarely is the inclination of the shelf even 1°, whereas the continental slope's inclination is anywhere from several degrees up to 25°. In many places the slope merges with the more gently sloping continental rise, but in some areas it descends directly into an oceanic trench, and a rise is absent (Figure 16.4).

The shelf-slope break, at an average depth of 135 m, serves as an important control on the processes operating on the seafloor. Landward of the break, the shelf is affected by waves and tidal currents, but seaward of the break, the seafloor is unaffected by surface processes. Sediment trans-port onto the slope and rise is therefore controlled by gravity-driven processes, especially turbidity currents, which are sediment-water mixtures. Because they are denser than seawater, turbidity currents flow downslope through *submarine canyons* to the deep seafloor where they typically deposit sequences of graded beds (see Figure 7.12). In fact, where a continental rise is present, it is composed mostly of *submarine fans,* which are simply accumulations of turbidity current deposits (Figure 16.5b).

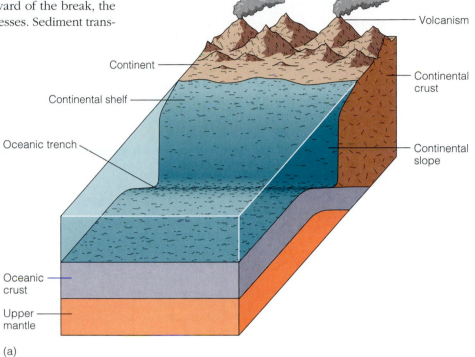

(a)

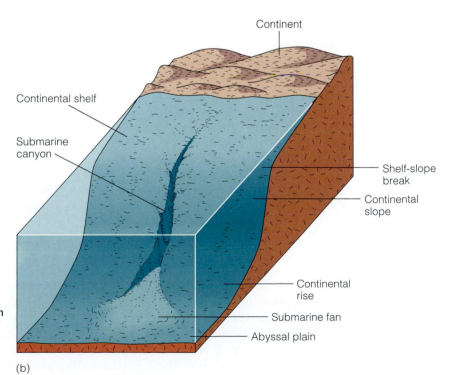

FIGURE **16.5** Diagrammatic views of (a) an active continental margin and (b) a passive continental margin. Submarine fans form by the deposition of sediments carried down submarine canyons by turbidity currents. Much of the continental rise is composed of overlapping submarine fans.

(b)

The *submarine canyons* through which turbidity currents move are deep and steep walled and are best developed on the continental slope. Some of these canyons can be traced across the shelf to streams on land and apparently formed as stream valleys when sea level was lower during the Ice Age. Many have no such association, however, and are thought to have originated through erosion by turbidity currents.

Two types of continental margins are generally recognized, active and passive. An **active continental margin** develops at the leading edge of a continental plate where oceanic lithosphere is subducted (Figure 16.5a). This kind of continental margin is characterized by andesitic volcanism, an inclined seismic zone, and a geologically young mountain range. Furthermore, the continental shelf is quite narrow, and the continental slope descends directly into an oceanic trench, as along the west coast of South America.

A **passive continental margin** occurs at the trailing edge of a continental plate (Figure 16.5b). It possesses a broad shelf and a continental rise that commonly merges with vast, flat abyssal plains (Figure 16.5b). No volcanism and little seismicity occur on this type of continental margin. The eastern margin of North America is a good example of a passive continental margin.

The proximity of the oceanic trench to the continent explains why the shelf is so narrow along an active continental margin. Sediment derived from the continent is simply transported down the slope into the trench. On a passive continental margin, which lacks a trench, the entire continental margin is much wider because land-derived sedimentary deposits build outward into the ocean.

THE DEEP-OCEAN BASIN

Submersibles have carried scientists into the oceanic depths, so some of the seafloor has been observed directly. Nevertheless, most of our knowledge of the deep seafloor comes from deep-sea drilling, echo sounding, seismic profiling, and data gathered by remote devices that descend in excess of 11,000 m.

As noted previously, the deep seafloor is not completely flat and featureless as once believed (Figure 16.3). It possesses a variety of features and has a topography as diverse as that on the continents. Nevertheless, vast flat areas known as **abyssal plains** are present adjacent to the rises of passive continental margins (Figure 16.5b). The oceanic crust beneath these areas is rugged, but it has been buried by land-derived sediments, thus yielding a very flat surface. Abyssal plains are not found along active continental margins where sediments from the land are trapped in an oceanic trench (Figure 16.5a).

Although **oceanic trenches** constitute only a small part of the seafloor, they are very important for it is here that lithospheric plates are consumed by subduction (see Figure 2.16). Oceanic trenches are long, narrow features restricted to active continental margins, so they are common around the margins of the Pacific Ocean basin. The

Peru-Chile Trench, for instance, lies just off the west coast of South America; it is 5900 km long and only about 100 km wide, but more than 8000 m deep. The greatest oceanic depth of more than 11,000 m is in an oceanic trench.

Oceanic ridges are part of a worldwide system of mostly submarine ridges that in some areas have a central rift (Figure 16.6). The Mid-Atlantic Ridge, part of this ridge system, is more than 2000 km wide and rises about 2.5 km above the adjacent seafloor. This oceanic ridge system extends for about 65,000 km, surpassing the length of mountain ranges on land. Oceanic ridges, however, are composed mostly of basalt and have features produced by tensional forces, whereas mountains on land are composed of granitic, metamorphic, and sedimentary rocks that have been deformed by compression. Recall from Chapter 2 that oceanic ridges are the sites of plate divergence and are also referred to as *spreading ridges* (see Figure 2.14). These spreading ridges are sites of shallow-focus earthquakes, high-heat flow, and volcanism. Indeed, a submarine volcano began erupting along the Juan de Fuca Ridge west of Oregon on January 25, 1998.

Oceanic ridges abruptly terminate where they are offset along large fractures in the seafloor. These fractures, better known as *transform faults* (Figure 16.6a), extend for hundreds of kilometers, and many geologists are convinced that some of them continue into continents. Offsets along these fractures yield nearly vertical escarpments more than 3 km high on the seafloor (Figure 16.6a).

The seafloor also has numerous seamounts and guyots. Both are of volcanic origin and rise more than 1 km above the seafloor, but **seamounts** are conical whereas **guyots** are flat topped. Guyots are simply volcanoes that once extended above sea level near a spreading ridge. But as the plate upon which they were situated continued to move away from the ridge and into deeper water, what was an island was eroded by waves as it sank beneath the sea (Figure 16.6b). Other features known as *volcanic hills* are similar to seamounts but average only about 250 m high.

Long, linear ridges and broad plateaus rising above the seafloor are also common in the oceanic basins. They are known as *aseismic ridges* because they lack seismic activity. A few of these features, called *microcontinents,* are thought to be small fragments separated from continents during rifting. Most aseismic ridges form as a linear succession of seamounts and/or guyots extending from an oceanic ridge.

Deep-Sea Sedimentation

DEEP-SEA sediments consist mostly of windblown dust and volcanic ash from the continents and oceanic islands and the shells of microscopic organisms living in the oceans' near-surface waters. Sand and gravel deposits beyond the continental margins or fringes of oceanic islands are rare because the only mechanism that can transport such large

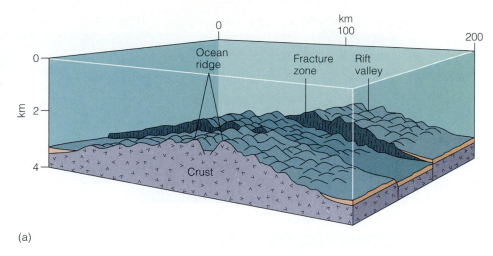

FIGURE **16.6** (a) Diagrammatic view of a fracture or transform fault offsetting a ridge. Earthquakes occur only in the segments between offset ridge crests. (b) Submarine volcanoes may build up above sea level to form seamounts. As the plate that these volcanoes rest on moves away from a spreading ridge, the volcanoes sink beneath sea level and become guyots.

(a)

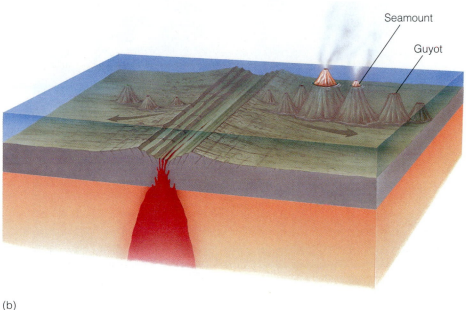

(b)

particles far into the ocean basins is ice rafting, which is effective only adjacent to Greenland and Antarctica.

Most of the sediments on the seafloor are pelagic, meaning that they settled from suspension far from land. Two types of pelagic sediments are generally recognized. **Pelagic clay** covers most of the deeper parts of the ocean basins. It is composed largely of clay-sized particles derived from the continents and oceanic islands. **Ooze** is composed of shells of microscopic marine plants and animals. If it contains mostly calcium carbonate ($CaCO_3$) shells, it is *calcareous ooze;* if it is composed predominantly of silica (SiO_2) shells, it is *siliceous ooze.*

Reefs

MOUNDLIKE, wave-resistant structures composed of the skeletons of various organisms such as corals, clams, sponges, and algae are **reefs**. Reefs grow to a depth of 45 or 50 m and are restricted to shallow tropical seas where

the water is clear and the temperature does not fall below about 20°C.

Three types of reefs are recognized: fringing, barrier, and atoll (Figure 16.7). *Fringing reefs* are solidly attached to the margins of an island or continent. They have a rough, tablelike surface and are up to 1 km wide. *Barrier reefs* are similar except that they are separated from the mainland by a lagoon. The 2000 km long Great Barrier Reef of Australia is a good example of a barrier reef.

An *atoll* is a circular to oval reef surrounding a lagoon (Figure 16.7). They first form as fringing reefs around volcanic islands that are carried below sea level on a moving plate. As the island is carried into progressively deeper water, the reef grows upward so that the living part of the reef remains in shallow water, leaving a circular lagoon surrounded by a more-or-less continuous reef (Figure 16.7). As Figure 16.7 shows, a fringing reef around an island can evolve into a barrier reef and then an atoll as the island subsides below sea level.

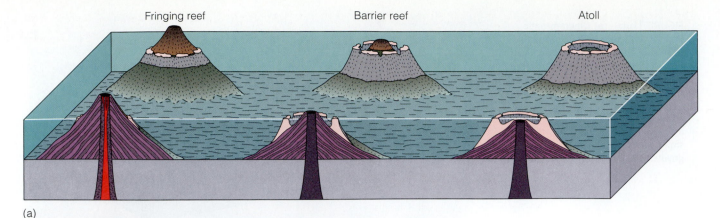

Fringing reef   Barrier reef   Atoll

(a)

(b)

FIGURE 16.7 (a) Three-stage development of an atoll. In the first stage, a fringing reef forms, but as the island sinks, a barrier reef becomes separated from the island by a lagoon. As the island disappears beneath the sea, the barrier reef continues to grow upward, thus forming an atoll. An oceanic island carried into deeper water by plate movement can account for this sequence. (b) View of an atoll in the Pacific Ocean.

Shorelines and Shoreline Processes

SHORELINES are the areas between low tide and the highest level on land affected by storm waves. Here we are concerned mostly with ocean shorelines where waves, tides, and nearshore currents continually modify existing shoreline features. Waves and nearshore currents are also effective geologic agents in large lakes, where the shorelines exhibit many of the same features present along seashores. The most notable differences are that waves and nearshore currents are more energetic on seashores, and even the largest lakes lack appreciable tides.

The continents possess more than 400,000 km of shorelines. They vary from rocky, steep shorelines, such as those in Maine and much of the western United States and Canada, to those with broad sandy beaches as in eastern North America from New Jersey southward. Whatever their type, all shorelines exhibit a continual interplay between the energy levels of shoreline processes and the shoreline materials.

Scientists from several disciplines have contributed to our understanding of shorelines as dynamic systems. Elected officials and city planners of coastal communities must become familiar with shoreline processes so that they can develop policies that serve the public as well as protect the fragile shoreline environment. In short, the study of shorelines is not only interesting but also has many practical applications.

TIDES

Along seacoasts the surface of the ocean rises and falls once or twice daily in response to the gravitational attraction of the Moon and Sun. These regular fluctuations in the sea's surface are **tides**. Most areas experience two high tides and two low tides daily as sea level rises and falls anywhere from a few centimeters to more than 15 m. During rising or *flood tide,* more and more of the nearshore area is flooded until high tide is reached. *Ebb tide* occurs when currents flow seaward during a decrease in the height of the tide.

Both the Moon and the Sun have sufficient gravitational attraction to exert tide-generating forces strong enough to deform the solid body of Earth, but they have a much greater influence on the oceans. The Sun is 27 million times more massive than the Moon, but it is 390 times as far away, and its tide-generating force is only 46% as strong as that of the Moon. Accordingly, the tides are dominated by the Moon, but the Sun plays an important role as well.

If we consider only the Moon acting on a spherical, water-covered Earth, the tide-generating forces produce two bulges on the ocean surface (Figure 16.8). One bulge extends toward the Moon because it is on the side of Earth where the Moon's gravitational attraction is greatest. The other bulge is on the opposite side of Earth, where the Moon's gravitational attraction is least. These two bulges always point toward and away from the Moon (Figure 16.8a), so as Earth rotates and the Moon's position changes, an observer at a particular shoreline location experiences the rhythmic rise and fall of tides twice daily. The heights of two successive high tides may vary depending on the Moon's inclination with respect to the equator.

The Moon revolves around Earth every 28 days, so its position with respect to any latitude changes slightly each day. That is, as the Moon moves in its orbit and Earth rotates on its axis, it takes the Moon 50 minutes longer each day to return to the same position it was in the previous day. Thus, an observer would experience a high tide at 1:00 P.M. on one day, for example, and at 1:50 P.M. on the following day.

Tides are also complicated by the combined effects of the Moon and the Sun. Even though the Sun's tide-generating force is weaker than the Moon's, when the Moon and Sun are aligned every two weeks, their forces are added together and generate *spring tides,* which are about 20% higher than average tides (Figure 16.8b). When the Moon and Sun are at right angles to one another, also at two-week intervals, the Sun's tide-generating force cancels some of that of the Moon, and *neap tides* about 20% lower than average occur (Figure 16.8c).

Tidal ranges are also affected by shoreline configuration. Broad, gently sloping continental shelves as in the Gulf of Mexico have low tidal ranges, whereas steep, irregular shorelines experience a much greater rise and fall of tides. Tidal ranges are greatest in some narrow, funnel-shaped bays and inlets. In the Bay of Fundy in Nova Scotia, a tidal range of 16.5 m has been recorded, and ranges greater than 10 m are known in several other areas.

Tides have an important impact on shorelines because the area of wave attack constantly shifts onshore and offshore as the tides rise and fall. Tidal currents themselves, however, have little modifying effect on shorelines, except in narrow passages where tidal current velocity is great enough to erode and transport sediment. Indeed, if it were not for strong tidal currents, some passageways would be blocked by sediments deposited by nearshore currents.

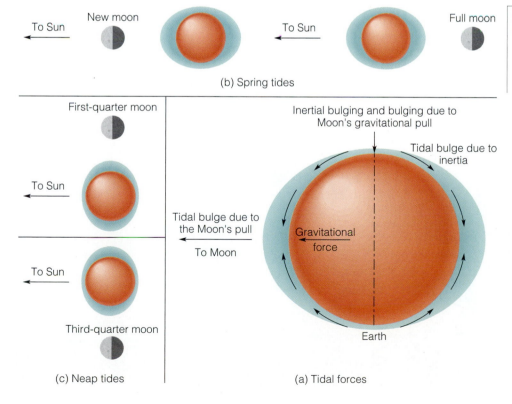

FIGURE **16.8** (a) Tides are caused by the gravitational pull of the Moon and, to a lesser degree, the Sun. Earth-Moon-Sun alignments at the times of the (b) spring and (c) neap tides are shown.

WAVES

Waves, or oscillations of a water surface, can be seen on all bodies of water but are most significant in large lakes and the oceans where they erode, transport, and deposit sediment. The highest part of a wave is its *crest,* whereas the low point between crests is the *trough. Wavelength* is the distance between successive wave crests (or troughs), and *wave height* is the vertical distance from trough to crest (Figure 16.9). As waves move across a water surface, the water "particles" rotate in circular orbits, with little or no net movement in the direction of wave travel (Figure 16.9). They do, however, transfer energy in the direction of wave advance.

The diameters of the orbits followed by water particles in waves diminish rapidly with depth, and at a depth of about one-half wavelength, called **wave base,** they are essentially zero (Figure 16.9). Thus, at depths exceeding wave base, the water and seafloor, or lake floor, are unaffected by surface waves. The significance of wave base will be explored more fully in later sections.

WAVE GENERATION Several processes can generate waves, including displacement of water by landslides, displacement of the seafloor by faulting, and volcanic explosions. But most of the geologic work done on shorelines is accomplished by wind-generated waves. When wind blows over water, some of its energy is transferred to the water, causing the water surface to oscillate. The mechanism that transfers energy from wind to water is related to the frictional drag resulting from one fluid (air) moving over another (water).

As one would expect, the harder and longer the wind blows, the larger the waves generated. Wind velocity and duration, however, are not the only factors controlling the size of waves. High-velocity wind blowing over a small pond will never generate large waves regardless of how long it blows. In fact, waves are present on ponds and most lakes only while the wind is blowing; once the wind stops, the water quickly smooths out. In contrast, the surface of the ocean is always in motion, and waves with heights of 34 m have been recorded during storms.

The reason for the disparity between the wave sizes on ponds and lakes and on the oceans is the **fetch,** which is the distance the wind blows over a continuous water surface. Fetch is limited by the available water surface, so on ponds and lakes it corresponds to their length or width, depending on wind direction.

In the areas where storm waves are generated, waves with different lengths, heights, and periods might merge, making them smaller or larger. Occasionally, two wave crests merge to form *rogue waves* that might be three or four times higher than the average. Indeed, these rogue waves can rise unexpectedly out of an otherwise comparatively calm sea and threaten even the largest ships. In 1942, the R.M.S. *Queen Mary* while carrying 15,000 American soldiers was hit broadside by a rogue wave and nearly capsized.

SHALLOW-WATER WAVES AND BREAKERS Waves moving out from the area of generation form swells and lose only a small amount of energy as they travel across the ocean. In deep-water swells, the water surface oscillates

FIGURE **16.9** Waves and the terminology applied to them. The water in waves moves in circular orbits that decrease in size with depth. At wave base, which equals one-half wavelength, water is not disturbed by surface waves. As deep-water waves move toward shore, the orbital motion of water within them is disrupted when they enter water shallower than wave base. Wavelength decreases while wave height increases, causing the waves to oversteepen and plunge forward as breakers.

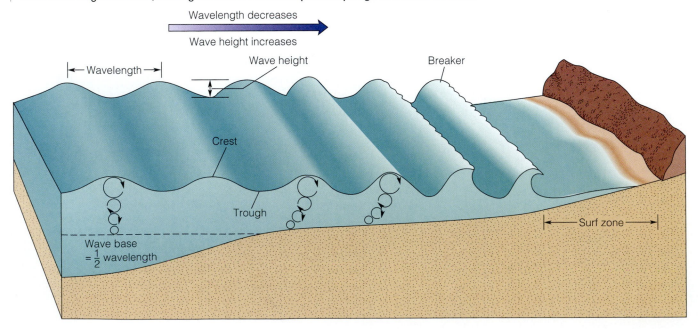

(a)

(b)

FIGURE 16.10 (a) Plunging breaker on the north shore of Oahu, Hawaii. (b) Spilling breaker.

and water particles move in orbital paths, with little net displacement of water taking place in the direction of wave advance (Figure 16.9).

As deep-water waves enter shallow water, they are transformed from broad, undulating swells into sharp-crested waves. This transformation begins at wave base; that is, it begins where wave base intersects the seafloor. At this point, the waves "feel" the bottom, and the orbital motions of water particles within waves are disrupted (Figure 16.9). As they move farther shoreward, the speed of wave advance and wavelength decrease, and wave height increases. In effect, as waves enter shallower water, they become oversteepened; the wave crest advances faster than the wave form, until eventually the crest plunges forward as a **breaker** (Figure 16.9). Breakers are commonly several times higher than deep-water waves, and when they plunge forward, their kinetic energy is expended on the shoreline.

The waves just described are the classic *plunging breakers,* which pound the shorelines of areas with steep offshore slopes, such as on the north shore of Oahu in the Hawaiian Islands (Figure 16.10a). In contrast, shorelines with gentle offshore slopes are characterized by *spilling breakers,* where the waves build up slowly and the wave's crest spills down the front of the wave (Fig. 16.10b). Whether the breakers spill or plunge, the water rushes onto the shore, then returns seaward to become part of the next breaking wave.

The size of breakers releasing their energy on shorelines varies enormously. Waves tend to be larger and more energetic in winter because storms are more common during that season. In addition, waves of various sizes merging not only creates rogue waves but also accounts for variations in the size of waves breaking on the shore. For instance, when waves having different lengths merge, smaller waves result, whereas merging of waves with the same length forms larger waves. As a result, several small waves might break on a beach, followed by a series of larger ones. Surfers commonly take advantage of this phenomenon by

swimming out to sea during a relative calm where they wait for a large set of waves to ride toward shore.

NEARSHORE CURRENTS

It is convenient to identify the *nearshore zone* as the area extending seaward from the shoreline to just beyond the area where waves break. The width of the nearshore zone varies depending on the wave length of the approaching waves, because long waves break at a greater depth, and thus farther offshore, than do short waves. Two types of currents are important in the nearshore zone, longshore currents and rip currents.

WAVE REFRACTION AND LONGSHORE CURRENT Deep-water waves are characterized by long, continuous crests, but rarely are their crests parallel with the shoreline

FIGURE 16.11 Wave refraction. These oblique waves are refracted and more nearly parallel the shoreline as they enter progressively shallower water.

FIGURE **16.12** Suspended sediment, indicated by discolored water, being carried seaward by a rip current.

net effect of this oblique approach is that the waves bend so that they more nearly parallel the shoreline (Figure 16.11). This phenomenon is known as **wave refraction**.

Even though waves are refracted, they still usually strike the shoreline at some angle, causing the water between the breaker zone and the beach to flow parallel to the shoreline. These **longshore currents**, as they are called, are long and narrow and flow in the same general direction as the approaching waves. These currents are particularly important because they transport and deposit large quantities of sediment in the nearshore zone.

RIP CURRENTS Waves carry water into the nearshore zone, so there must be a mechanism for mass transfer of water back out to sea. One way that water moves seaward from the nearshore zone is in **rip currents**, which are narrow surface currents that flow out to sea through the breaker zone (Figure 16.12). Surfers commonly take advantage of rip currents for an easy ride out beyond the breaker zone, but these currents pose a danger to inexperienced swimmers. Rip currents might flow at several kilometers per hour, so if a swimmer is caught in one, it is useless to try to swim directly back to shore. Instead, because rip currents are narrow and usually nearly per-

(Figure 16.11). One part of a wave enters shallow water where it encounters wave base and begins breaking before other parts of the same wave. As a wave begins breaking, its velocity diminishes, but the part of the wave still in deep water races ahead until it too encounters wave base. The

FIGURE **16.13** (a) Small pocket beaches along the California coast. The Golden Gate Bridge and San Francisco are in the background. (b) The Grand Strand of South Carolina, shown here at Myrtle Beach, is 100 km of nearly continuous beach.

(a)

(b)

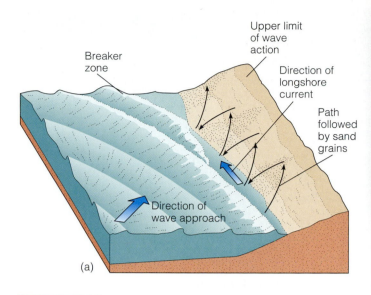

(a)

(b)

FIGURE 16.14 (a) Longshore currents transport sediment along the shoreline between the breaker zone and the upper limit of wave action. Such sediment transport is longshore drift. (b) These groins at Cape May, New Jersey, interrupt the flow of longshore currents so sand is trapped on their upcurrent side. On the downcurrent side of the groins, sand is eroded because of continuing longshore drift.

pendicular to the shore, one can swim parallel to the shoreline for a short distance and then turn shoreward with no difficulty.

Shoreline Deposition

DEPOSITIONAL features of shorelines include beaches, spits, baymouth bars, and barrier islands. The characteristics of these deposits are determined largely by wave energy and longshore currents. Rip currents play only a minor role in the configuration of shorelines.

BEACHES

Beaches are the most familiar of all coastal landforms, attracting millions of visitors each year and providing the economic base for many communities. By definition a **beach** is a deposit of unconsolidated sediment extending landward from low tide to a change in topography such as a line of sand dunes, a sea cliff, or the point where permanent vegetation begins. Depending on shoreline configuration and wave intensity, beaches may be discontinuous, existing only as *pocket beaches* in protected areas such as embayments, or they may be continuous for long distances (Figure 16.13).

Some of the sediment on beaches is derived from weathering and wave erosion of the shoreline, but most of it is transported to the coast by streams and redistributed along the shoreline by longshore currents. **Longshore drift** is the phenomenon involving sand transport along a shoreline by longshore currents (Figure 16.14a). As previously

noted, waves usually strike beaches at some angle, causing the sand grains to move up the beach face at a similar angle; as the sand grains are carried seaward in the backwash, however, they move perpendicular to the long axis of the beach. Thus, individual sand grains move in a zigzag pattern in the direction of longshore currents. This movement is not restricted to the beach; it extends seaward to the outer edge of the breaker zone (Figure 16.14a).

In an attempt to widen a beach or prevent erosion, shoreline residents often build *groins,* structures that project seaward at right angles from the shoreline (Figure 16.14b). Groins interrupt the flow of longshore currents, causing sand to be deposited on their upcurrent side and widening the beach at that location. However, erosion inevitably occurs on the downcurrent side of a groin.

A beach is an area where wave energy is dissipated, so the loose grains composing the beach are constantly affected by wave motion. But the overall configuration of a beach remains unchanged as long as equilibrium conditions persist. A beach's onshore-to-offshore profile can be thought of as a profile of equilibrium; that is, all parts of the beach are adjusted to the prevailing conditions of wave intensity and nearshore currents (Figure 16.15).

Tides and longshore currents affect the configuration of beaches to some degree, but storm waves are by far the most important agent modifying their equilibrium profile. In many areas, beach profiles change with the seasons, so, we recognize *summer beaches* and *winter beaches,* each of which is adjusted to the conditions prevailing at these times. Summer beaches are generally wide and sand covered and possess a smooth offshore profile. During winter,

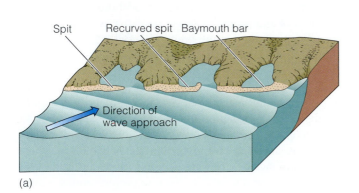

FIGURE 16.15 Seasonal changes in beach profiles.

though, when waves more vigorously attack the beach, a winter beach develops, which is narrower, slopes seaward more gently, and has offshore sand bars paralleling the shoreline (Figure 16.15). Seasonal changes can be so profound that sand moving onshore and offshore yields sand-covered beaches during summer and gravel-covered beaches during winter.

Seasonal changes in beach profiles are related to changing wave intensity. During the winter, energetic storm waves erode the sand from the beach and transport it offshore where it is stored in sand bars. The same sand that was eroded from the beach during the winter returns the next summer when it is driven onshore by the more gentle swells of that season. The volume of sand in the system remains more or less constant; it simply moves offshore or onshore, depending on the energy of waves.

SPITS, BAYMOUTH BARS, AND BARRIER ISLANDS

Other than the beach itself, some of the most common depositional landforms on shorelines are spits and baymouth bars, both of which are variations of the same feature. A **spit** is simply a continuation of a beach forming a point, or "free end," that projects into a body of water, commonly a bay. A **baymouth bar** is a spit that has grown until it completely closes off a bay from the open sea (Figure 16.16).

Both spits and baymouth bars form and grow as a result of longshore currents. Where currents are weak, as in the deeper water at the opening to a bay, longshore current velocity diminishes and sediment is deposited, forming a sand bar. The free ends of many spits are curved by wave refraction or waves approaching from a different direction. Such spits are called *hooks* or *recurved spits* (Figure 16.16a) (Perspective 16.1). A rarer type of spit, a *tombolo,* extends out into the sea and connects an island to the mainland. Tombolos develop on the shoreward sides of islands (Figure 16.16b). Wave refraction around an island causes converging currents that turn seaward and deposit a sand bar connecting the shore with the island.

Barrier islands are long, narrow islands composed of sand and separated from the mainland by a lagoon (Figure 16.17). The origin of barrier islands has been long debated and is still not completely resolved. It is known that they form on gently sloping continental shelves with abundant sand in areas where both tidal fluctuations and wave-energy levels are low. According to one model, barrier is-

FIGURE 16.16 (a) Spits form where longshore currents deposit sand in deeper water, as at the entrance to a bay. A baymouth bar is simply a spit that has grown until it extends across the mouth of a bay. (b) Origin of a tombolo. Wave refraction around an island causes longshore currents to converge and deposit a sand bar that joins the island with the mainland.

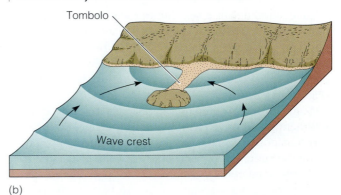

(a)

(b)

lands formed as spits that became detached from the land, while another model proposes that they formed as beach ridges on coasts that were subsequently partly submerged when sea level rose.

As sea level rises, many of the nearly 300 barrier islands along the East and Gulf Coasts of the United States migrate landward during storms. Large waves erode sand from their seaward sides and deposit it on their landward sides, resulting in a gradual landward shift of the entire island complex (Figure 16.18). During the last 120 years, Hatteras Island, North Carolina, migrated 500 m landward so that Cape Hatteras lighthouse, which was built 460 m from the shoreline in 1870, now stands on a promontory in the Atlantic Ocean.

Landward migration of barrier islands would pose few problems if it were not for the numerous communities, resorts, and vacation homes on them. Futile efforts are being made on the northeast end of the Isle of Palms, a barrier island off the coast of South Carolina, to maintain the beach and protect shoreline homes. Following each spring tide, heavy equipment excavates sand from a deposit known as an ebb tide delta and constructs a sand berm to protect the houses from the next spring tide. Two weeks later, the process must be repeated in a never-ending cycle of erosion and artificial replenishment of the beach.

THE NEARSHORE SEDIMENT BUDGET

We can think of the gains and losses of sediment in the nearshore zone in terms of a **nearshore sediment budget** (Figure 16.19). If a nearshore system has a balanced budget, sediment is supplied as fast as it is removed, and the volume of sediment remains more or less constant, although sand may shift offshore and onshore with the changing seasons. A positive budget means gains exceed losses, whereas a negative budget results when losses exceed gains. If a negative budget prevails long enough, a nearshore system is depleted and beaches may disappear.

Erosion of sea cliffs provides some sediment to beaches, but in most areas probably no more than 5 to 10% of the total sediment supply is derived from this source. Most of the sediment on typical beaches is transported to the shoreline by streams and then redistributed along the

shoreline by longshore drift. So, longshore drift also plays a role in the nearshore sediment budget because it continually moves sediment into and away from beach systems.

The primary ways that a nearshore system loses sediment include longshore drift, offshore transport, wind, and deposition in submarine canyons (Figure 16.19). Offshore transport mostly involves fine-grained sediment that is carried seaward where it eventually settles in deeper water.

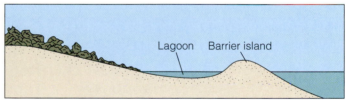

(a)

(b)

FIGURE 16.18 Rising sea level and the landward migration of barrier islands. (a) Barrier island before landward migration in response to rising sea level. (b) Landward movement occurs when storm waves wash sand from the seaward side of the islands and deposit it in the lagoon. (c) Over time, the entire complex migrates landward.

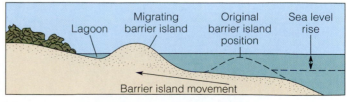

(c)

Cape Cod, Massachusetts

Cape Cod, some of which is designated as a National Seashore, resembles a large human arm extending into the Atlantic Ocean from the coast of Massachusetts (Figure 1). It projects 65 km east of the mainland to the "elbow," and then extends another 65 km northward where it resembles a half-curled hand. Cape Cod and the adjacent Elizabeth Islands, Martha's Vineyard, and Nantucket Island all owe their existence to deposition by late Pleistocene glaciers and the subsequent modification of these glacial deposits by waves. The granite bedrock or foundation upon which the cape and nearby islands were built is at depths of 90 to 150 m.

Between about 23,000 and 16,000 years ago, during the greatest southward advance of the continental glaciers in this area, the ice front was stabilized in the area of present-day Martha's Vineyard and Nantucket Island (Figure 2a). Recall that a stabilized terminus

means the glacier has a balanced budget. Nevertheless, flow continues within the glacier, transporting and depositing sediment as an end moraine. The end moraine that makes up these islands is a terminal moraine because it is the most southerly of all the moraines deposited by this glacier.

As the climate became warmer, the Cape Cod Bay Lobe of this vast glacier began retreating northward. About 14,000 years ago, it once again became stabilized in the area of present-day Cape Cod and the Elizabeth Islands where it deposited a large recessional moraine (Figure 2b). As the ice front continued to retreat northward, meltwater trapped between the ice front and the recessional moraine formed Glacial Lake Cape Cod covering about 1000 km^2 (Figure 2c). Deposits of mud, silt, and sand accumulated in this lake, and these, too, make up part of present-day Cape Cod.

When the continental glaciers withdrew entirely from this region, Cape Cod looked different than it does now. On its east side were several headlands and embayments, but by 6000 years ago sea level had risen enough so that waves began smoothing the shoreline by redistributing the sediment. The headlands were eroded whereas the embayments were filled in, and the shoreline has eroded from 1 to 4 km landward from its former location. Even today many steep wave-cut cliffs are present on the cape, and those facing east are currently eroding at nearly 1 m per year. Most of this erosion takes place during storms, so cliff retreat is episodic rather than continuous.

Some of the sediment eroded from Cape Cod is transported offshore where it settles beyond the reach of waves, but much of it is transported by longshore currents and redistributed along the cape. Hundreds of thousands of cubic meters of sand are transported along the shores of

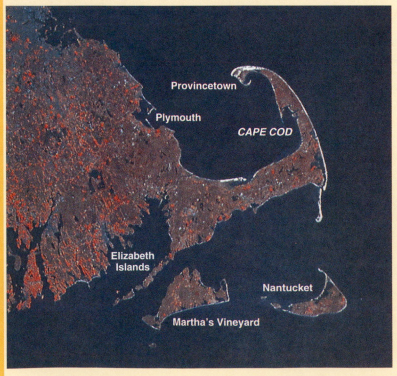

Figure 1 Satellite view of Cape Cod. Notice the long beaches, spits, and baymouth bars in the cape and Martha's Vineyard and Nantucket Island. The dark circular to oval areas on Cape Cod are kettle lakes, or ponds as they are called locally.

the cape and deposited as spits and bay-mouth bars, many of which are still forming. Extending south from the "elbow" is a long spit, and another continues to form at the north where sand is transported around the end of the cape and forms a hood (Figure 1). Wind also transports some of the sand inland where it accumulates as a series of coastal dunes.

Other distinctive features of Cape Cod are its numerous circular to oval ponds. These ponds occupy kettles that formed when large blocks of ice were partly or completely buried by outwash deposits. When these ice masses finally melted, they left depressions measuring up to 0.8 km in diameter that filled with water when the water table rose as a consequence of rising sea level.

Native Americans lived on Cape Cod for thousands of years before the Europeans arrived. Unfortunately, by 1764 these earliest inhabitants had nearly ceased to exist, mostly because of diseases. The first European settlers in this region were the Pilgrims. Despite the persistent myth that they first landed at Plymouth Rock, their first landfall was actually on Cape Cod near what is now Provincetown.

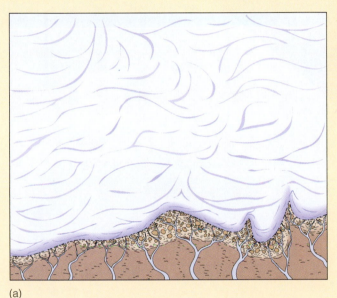

(a)

Figure 2 (a) Position of the Cape Cod Lobe of glacial ice 23,000 to 16,000 years ago when it deposited its terminal moraine that would become Martha's Vineyard and Nantucket Island. (b) Position of the Cape Cod Lobe when the recessional moraine was deposited that now forms Cape Cod. (c) About 5000 years ago, rising sea level covered the lowlands between the moraines, and beaches and sand spits formed.

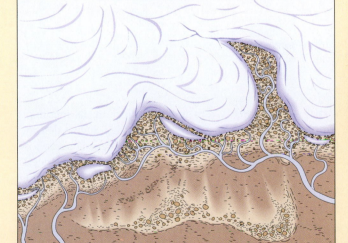

(b)

(c)

Inputs	+	Outputs	=	Balance
Longshore transport into beach River supply Cliff erosion Onshore transport		Longshore transport out of beach Offshore transport Wind transportation into dunes		Accretion Erosion Steady state

(a)

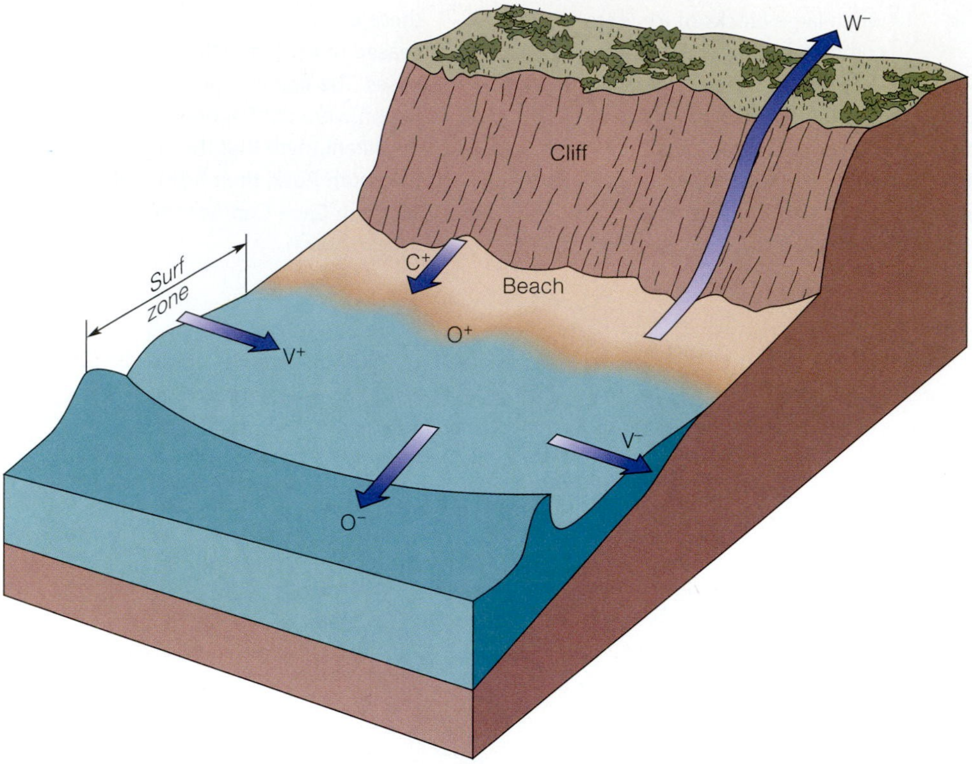

FIGURE 16.19 The nearshore sediment budget. (a) The long-term sediment budget can be assessed by considering inputs versus outputs. If inputs and outputs are equal, a beach is in a steady state or equilibrium. If outputs exceed inputs, the beach has a negative budget and erosion occurs. Accretion takes place when the beach has a positive budget with inputs exceeding outputs. (b) A hypothetical example of a negative nearshore sediment budget. In this example, the beach is losing 5000 m³ a year to erosion.

Inputs
V^+ = longshore transport into beach : + 60,000 m³/yr
C^+ = cliff erosion : + 5,000 m³/yr
O^+ = onshore transport : + 5,000 m³/yr

Outputs
W^- = wind : −1,000 m³/yr
V^- = longshore transport out of beach : − 54,000 m³/yr
O^- = offshore transport (includes transport to submarine canyons) : − 20,000 m³/yr

Balance : − 5,000 m³/yr (net erosion)

(b)

Wind is an important process because it removes sand from beaches and blows it inland where it commonly piles up as sand dunes.

If the heads of submarine canyons are nearshore, huge quantities of sand are funneled into them and deposited in deeper water. La Jolla and Scripps submarine canyons off the coast of southern California funnel off an estimated 2 million m³ of sand each year. In most areas, submarine canyons are too far offshore to interrupt the flow of sand in the nearshore zone.

It should be apparent from the preceding discussion that if a nearshore system is in equilibrium, its incoming supply of sediment exactly offsets its losses. Such a delicate balance tends to continue unless the system is somehow disrupted. One common change that affects this balance is the construction of dams across the streams supplying sand. The sediment contribution from a stream is proportional to its drainage area, but once dams have been built, all sediment from the upper reaches of the drainage systems is trapped in reservoirs and thus cannot reach the shoreline.

Besides building groins to widen beaches, people have

armored shorelines with seawalls and similar structures to protect coastal homes and highways. Unfortunately, these structures have had unanticipated effects on Oahu in the Hawaiian Islands. Erosion of coastal rocks is the primary supply of sand to many beaches on Oahu, but seawalls trap sediment on their landward sides, thus diminishing the amount supplied to beaches. In four areas studied on Oahu, 24 percent of the beaches have either been lost or are much narrower than they were several decades ago.

Shoreline Erosion

ALONG seacoasts where erosion rather than deposition predominates, beaches are lacking or poorly developed, and sea cliffs commonly develop (Figure 16.20). Sea cliffs are erosional features frequently pounded by waves, especially during storms, and the cliff retreats landward as a result of corrosion, hydraulic action, and abrasion. *Corrosion* is an erosional process involving the wearing away of rock by chemical processes, especially the solvent action of seawater. The force of the water itself, called *hydraulic action,* is a particularly effective erosional process. Waves exert tremendous pressure on shorelines by direct impact, but are most effective on sea cliffs composed of unconsolidated sediment or highly fractured rocks. *Abrasion* is an erosional process involving the grinding action of rocks and sand carried by waves.

WAVE-CUT PLATFORMS AND ASSOCIATED LANDFORMS

Sea cliffs erode landward mostly as a result of hydraulic action and abrasion at their bases. As a sea cliff is undercut, the upper part is left unsupported and susceptible to mass wasting processes. Sea cliffs retreat little by little, and as they do, they leave a beveled surface known as a **wave-cut platform** that slopes gently seaward (Figure 16.20). Broad wave-cut platforms exist in many areas, but invariably the water over them is shallow because the abrasive planing

action of waves is only effective to a depth of about 10 m. The sediment eroded from sea cliffs is transported seaward and deposited to form a *wave-built platform,* a seaward extension of the wave-cut platform (Figure 16.20a).

Sea cliffs do not erode and retreat uniformly because some of the materials of which they are composed are more resistant to erosion than others. Those parts of a shoreline consisting of resistant materials might form seaward-projecting *headlands,* whereas less resistant materials erode more rapidly yielding embayments where pocket beaches may be present (Figure 16.13a). Wave refraction around headlands causes them to erode on both sides so that *sea caves* form (Figure 16.21a); if caves on the opposite sides of a headland join, they form a *sea arch* (Figure 16.21b). Continued erosion generally causes the span of an arch to collapse, yielding isolated *sea stacks* on wave-cut platforms (Figure 16.21c). In the long run, shoreline processes tend to straighten an initially irregular shoreline. They do so because wave refraction causes more wave energy to be expended on headlands and less on embayments. As the headlands become eroded, some of the sediment yielded by erosion is deposited in the embayments. The net effect of these processes is to straighten the shoreline.

Types of Coasts

DEPOSITIONAL coasts, such as the U.S. Gulf Coast, are characterized by an abundance of detrital sediment, long and sandy beaches, and the presence of such depositional landforms as deltas and barrier islands. Erosional coasts are generally steep and irregular and typically lack well-developed beaches except in protected areas (Figure 16.13a). They are further characterized by erosional features such as sea cliffs, wave-cut platforms, and sea stacks. Many of the beaches along the west coast of North America fall into this category.

In this section we examine coasts in terms of their changing relationships to sea level. But note that while some coasts, such as those in southern California, are described as

FIGURE 16.20 (a) Wave erosion of a sea cliff produces a gently sloping surface called a wave-cut platform. Deposition at the seaward margin of the wave-cut platform forms a wave-built platform. (b) Sea cliffs and a wave-cut platform.

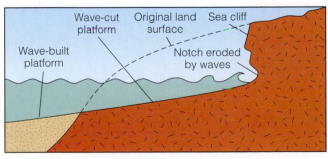

(a)

(b)

FIGURE **16.21** (a) Erosion of a headland and the origin of sea caves, sea arches, and sea stacks. (b) This sea stack in Australia has an arch developed in it. (c) Sea stacks along Australia's south coast.

emergent (uplifted), these same coasts may be erosional as well. In other words, coasts commonly possess features allowing them to be classified in several ways.

If sea level rises with respect to the land or the land subsides, coastal regions are flooded and said to be **submergent** or *drowned*. Much of the East Coast of North America from Maine southward through South Carolina was flooded during the rise in sea level following the Pleistocene Epoch so it is extremely irregular. Recall that during the expansion of glaciers during the Pleistocene, sea level was as much as 130 m lower than at present, and that streams eroded their valleys more deeply as they adjusted to a lower base level. When sea level rose, the lower ends of these valleys were drowned, forming *estuaries* such as Delaware and Chesapeake bays (Figure 16.22). Estuaries are the seaward ends of river valleys where seawater and freshwater mix.

Emergent coasts are found where the land has risen with respect to sea level. Emergence can take place when

water is withdrawn from the oceans, as occurred during the Pleistocene expansion of glaciers. At present, however, coasts are emerging as a result of isostasy or tectonism. In northeastern Canada and the Scandinavian countries, the coasts are irregular because isostatic rebound is elevating formerly glaciated terrain from beneath the sea.

Coasts rising in response to tectonism tend to be rather straight because the seafloor topography being exposed as uplift proceeds is smooth. The west coasts of North and South America are rising as a consequence of plate tectonics. Distinctive features of these coasts are **marine terraces** (Figure 16.23), which are old wave-cut platforms now elevated above sea level. Uplift in these areas appears to be episodic rather than continuous, as indicated by the multiple levels of terraces in some areas. In southern California, several terrace levels are present, each of which probably represents a period of stability followed by uplift. The highest of these terraces is now about 425 m above sea level.

Resources from the Sea

SEAWATER contains many elements in solution, some of which are extracted for various industrial and domestic uses. In many places sodium chloride (table salt) is produced by the evaporation of seawater, and a large proportion of the world's magnesium is produced from seawater. Numerous other elements and compounds can be extracted from seawater, but for many, such as gold, the cost is prohibitive.

In addition to substances in seawater, deposits on the seafloor or within seafloor sediments are becoming increasingly important. Many of these potential resources lie well beyond the margins of the continents, so their ownership is a political and legal problem that has not yet been resolved. Most nations bordering the ocean claim those resources within their adjacent continental margin. The United States, by a presidential proclamation issued on March 10, 1983, claims sovereign rights over an area designated as the **Exclusive Economic Zone (EEZ)**. The EEZ extends seaward 200 nautical miles (371 km) from the coast, giving the United States jurisdiction over an area about 1.7 times larger than its land area. Also included within the EEZ are the areas adjacent to U.S. territories, such as Guam, American Samoa, Wake Island, and Puerto Rico. In short, the United States claims a huge area of the seafloor and any resources on or beneath it.

Numerous resources are found within the EEZ, some of which have been exploited for many years. Sand and gravel for construction are mined from the continental shelf in several areas. About 17% of U.S. oil and natural gas production comes from wells on the continental shelf.

Ancient shelf deposits in the Persian Gulf region contain the world's largest reserves of oil.

A potential resource within the EEZ is methane hydrate, which consists of single methane molecules bound up in networks formed by frozen water. Although methane hydrates have been known since the early part of the last century, they have only recently received much attention.

FIGURE 16.23 This gently sloping surface along the Pacific coast of California is a marine terrace. Notice the old sea stacks rising above this terrace.

Most of these deposits lie within continental margins, but so far it is not known if they can be effectively recovered and used as an energy source. According to one estimate, the amount of carbon in methane hydrates is double that in all coal, oil, and conventional natural gas reserves.

Other resources of interest include deposits that form at submarine hydrothermal vents known as *black smokers* (Figure 16.24). When seawater circulates down through oceanic crust at or near spreading ridges, it is heated to as much as 400°C, reacts with the crust, and is transformed into a metal-bearing solution. As the hot solution rises and discharges onto the seafloor, it cools and forms a deposit containing iron, copper, zinc, and other metals. Hydrothermal vent deposits have been identified within the EEZ at the Gorda Ridge off the coasts of California and Oregon, and similar deposits are present at the Juan de Fuca Ridge within the Canadian EEZ.

Other potential resources include irregular to spherical masses known as *manganese nodules,* which are fairly common in all the ocean basins. These nodules result from chemical reactions in seawater and, in addition to manganese, contain iron oxides, copper, nickel, and cobalt. The United States is interested in this potential resource because it is heavily dependent on imports of manganese and cobalt.

Within the EEZ, manganese nodules are found near Johnston Island in the Pacific Ocean and on the Blake Plateau off the coast of South Carolina and Georgia. In addition, seamounts and seamount chains within the EEZ in the Pacific are known to have metalliferous oxide crusts several centimeters thick from which cobalt and manganese could be mined.

FIGURE 16.24 A hydrothermal vent known as a black smoker on the seafloor. The plume of "black smoke" is simply heated seawater saturated with dissolved minerals. Deposition of these minerals forms a deposit containing iron, copper, zinc, and several other metals.

Chapter Summary

1. Continental margins separate continents above sea level from the deep seafloor. They consist of a continental shelf, a continental slope, and in some places a continental rise.

2. Submarine canyons are best developed on the continental slope, but some extend well up onto shelves and lie offshore from streams. Many submarine canyons were probably eroded by turbidity currents that transport sediments to the rise.

3. Active continental margins are characterized by seismicity, volcanism, a narrow shelf, and a slope that descends into an oceanic trench. Passive continental margins have little seismic activity, no volcanism, and possess broad shelves

and a rise that commonly merges with abyssal plains.

4. Oceanic trenches are long, narrow, deep features where oceanic lithosphere is subducted. Oceanic ridges, also known as spreading ridges, nearly encircle the globe, but they are interrupted and offset by large fractures in the seafloor.

5. Other important features of the seafloor include seamounts, flat-topped seamounts known as guyots, volcanic hills, and aseismic ridges.

6. Deep-sea sediments consist of small particles derived from continents and oceanic islands and the shells of microscopic marine organisms. These sediments are characterized as pelagic clay and ooze.

7. Reefs are wave-resistant structures composed of animal skeletons, particularly those of corals. There are three types of reefs: fringing, barrier, and atoll.

8. Shorelines are continually modified by the energy of nearshore currents and waves, which are oscillations on water surfaces that transmit energy in the direction of wave movement. Surface waves affect the seafloor only to wave base, a depth equal to one-half of the wavelength.

9. Breakers form where waves enter shallow water and the orbital motion of water particles is disrupted. The waves become oversteepened and plunge forward onto the shoreline, thus expending their energy.

10. Longshore currents are generated by waves approaching a shoreline at an angle. These currents are capable of considerable erosion, transport, and deposition.

11. Rip currents are narrow surface currents that carry water from the nearshore zone seaward through the breaker zone.

12. Beaches are the most common shoreline depositional features. They are continually modified by nearshore processes, and their profiles generally exhibit seasonal changes.

13. Spits, baymouth bars, and tombolos all form and grow as a result of longshore current transport and deposition. Barrier islands, which are separated from the mainland by a lagoon, are nearshore sediment deposits of uncertain origin.

14. The volume of sediment in a nearshore system remains rather constant unless the system is somehow disrupted, as when dams are built across the streams that supply sand to the system.

15. Shorelines characterized by erosion have sea cliffs, wave-cut platforms with sea stacks, and discontinuous beaches, whereas depositional shorelines exhibit deltas, long and sandy beaches, and barrier islands.

16. Submergent coasts and emergent coasts are defined on the basis of their relationships to change in sea level. The former have been inundated by rising sea level or subsidence of the coast, and the latter have risen with respect to sea level.

Important Terms

abyssal plain
active continental margin
barrier island
baymouth bar
beach
breaker
continental margin
emergent coast
Exclusive Economic Zone (EEZ)
fetch
guyot

longshore current
longshore drift
marine terrace
nearshore sediment budget
oceanic ridge
oceanic trench
ooze
passive continental margin
pelagic clay
reef
rip current

seamount
shoreline
spit
submergent coast
tide
wave
wave base
wave-cut platform
wave refraction

Review Questions

1. The depth at which the orbital motion in surface waves dies out is known as:
   a. _____ fetch;
   b. _____ wave base;
   c. _____ baymouth bar;
   d. _____ submergent zone;
   e. _____ flood tide.

2. Wave refraction is a phenomenon involving the:
   a. _____ oversteeping and breaking of waves on a beach;
   b. _____ transport of sediment from a shoreline into deeper water;
   c. _____ displacement of water by a submarine landslide;
   d. _____ erosion of canyons on the continental slope and rise;
   e. _____ bending of waves so they more nearly parallel the shoreline.

3. Waves in oceans are larger than those in lakes because of:
   a. _____ deeper water;
   b. _____ larger storms;
   c. _____ greater fetch;
   d. _____ tidal currents;
   e. _____ wave base.

4. Marine terraces are:
   a. _____ coasts where sea level has risen with respect to the land;
   b. _____ beaches on the seaward sides of barrier islands;
   c. _____ the time required for two successive wave crests to pass a given point;
   d. _____ wave-cut platforms now above sea level;
   e. _____ deposits resulting from longshore transport.

5. Waves approaching a shoreline at an angle generate a(n):
   a. _____ longshore current;

   b. _____ marine oscillation;
   c. _____ emergent coast;
   d. _____ continental rise;
   e. _____ submarine fan.

6. That part of a continental margin sloping gently from the shoreline to a marked change in the slope of the seafloor is known as the:
   a. _____ abyssal plane;
   b. _____ continental shelf;
   c. _____ submarine rise;
   d. _____ rift valley;
   e. _____ aseismic ridge.

7. Which one of the following statements is correct?
   a. _____ North America's East Coast is a passive continental margin;
   b. _____ oceanic ridges are composed largely of granite and deformed sedimentary rocks;

c. _____ the deposits of turbidity currents consist of calcareous ooze;

d. _____ most of Earth's volcanism takes place in submarine canyons;

e. _____ the greatest oceanic depths are at continental rises.

8. The type of reef solidly attached to the margin of an island or continent is known as a(n):
   a. _____ atoll;
   b. _____ baymouth bar;
   c. _____ fringing reef;
   d. _____ pelagic clay;
   e. _____ rift.

9. When the Moon and Sun are aligned every two weeks, their combined force generates:
   a. _____ longshore currents;
   b. _____ turbidity currents;
   c. _____ submarine slides;
   d. _____ abyssal flows;
   e. _____ spring tides.

10. The flattest most featureless areas on Earth are:
    a. _____ oceanic ridges;
    b. _____ abyssal plains;
    c. _____ aseismic ridges;
    d. _____ continental margins;
    e. _____ oceanic trenches.

11. Explain the concept of a nearshore sediment budget. Be sure to include a discussion of all inputs and outputs.

12. How does a longshore current originate? What types of deposits are formed by these currents?

13. Describe the continental rise and explain why a rise is present at some continental margins and not at others.

14. Why does an observer at a shoreline experience two high and two low tides each day?

15. Explain how a wave-cut platform originates. What erosional landforms rise above wave-cut platforms?

16. What are rogue waves, and how do they form?

17. What is a rip current, and why do rip currents pose a danger to swimmers?

18. Describe how a fringing reef can evolve into a barrier reef and finally into an atoll.

19. What are the similarities and differences between active and passive continental margins?

20. What is the Exclusive Economic Zone, and what types of resources are found within it?

Points to Ponder

1. Why are long, broad sandy beaches more common in eastern North America than western North America?

2. An initially straight shoreline is composed of granite flanked on both sides by glacial drift. Diagram and explain this shoreline's probable response to erosion.

World Wide Web Activities

▶ **COASTAL PROGRAM DIVISION COASTAL ZONE MANAGEMENT PROGRAM**

This government site is dedicated to the coastal zone management program of the National Ocean Service. What is the mission of the Coastal Zone Management Program?

▶ **U.S. GEOLOGICAL SURVEY: THE NATIONAL MARINE AND COASTAL GEOLOGY PROGRAM**

As stated at the top of this site, The National Marine and Coastal Geology Program is "a plan for geologic research on environmental, hazards, and resources issues affecting the Nation's coastal realms and marine federal lands."

1. Click on the *U.S. Ocean margins; issues for the next decade* site under the Introduction of the Table of Contents. What are some of the issues facing the United States in the next decade concerning its coastal margins? Why should we as citizens be concerned about these issues?

2. Click on the *Mapping the U.S. Exclusive Zone* site under the Profile of the USGS Marine and Coastal Geology Program of the Table of Contents. What is the EEZ? What are the plans for it in the next decade?

3. Click on the *Coastal and Nearshore Erosion* site under the Description of the USGS Marine and Coastal Geology Program of the Table of Contents. What is the scope of the problem concerning coastal and nearshore erosion? What are some of the projects being undertaken in this area?

World Wide **Web** Activities

For these web site addresses, along with current updates and exercises, log on to

http://www.wadsworth.com/geo

▶ WORLD DATA CENTER-A MARINE GEOLOGY & GEOPHYSICS

This site mainly provides databases and database searches on the world's oceans. It also provides links to other related sites concerned with marine geology.

1. Check some of the database sites listed for information and maps about ocean sediments and rocks and sediment thicknesses.

2. Click on the *Ocean Drilling* site and then click on the *Ocean Drilling Program*. What is the history of the Ocean Drilling Program? What are the current legs of the ODP, and where is drilling taking place?

▶ LAMONT—DOHERTY EARTH OBSERVATORY OF COLUMBIA UNIVERSITY

This site contains extensive amounts of information and images of their many and varied ongoing research projects in marine geology and geophysics.

1. Click on the *U.S. Seascapes* site for some spectacular images of the U.S. continental margins. Compare these images with the information presented in this chapter about continental margins.

2. Click on the *Ridge Multibeam Synthesis*. Click on *Seafloor Movie* for movies of seafloor spreading.

▶ THE VENTS IMAGE GALLERY

This site by the National Oceanic and Atmospheric Administration (NOAA) has images of the Juan de Fuca Ridge in the northeast Pacific, taken from the submersible *Alvin*. Click any of the images to enlarge.

▶ LOIHI

Maintained by the Hawaii Center for Volcanology. This site has information on the history of eruptions and development of Loihi volcano, a seamount near the island of Hawaii. Read the history of Loihi and answer the following:

1. How does Loihi volcano differ from other seamounts near Hawaii?

2. How high above the seafloor does Loihi rise, and when did it last erupt?

CD-ROM Exploration

Explore the following *In-Terra-Active 2.0* CD-ROM module(s) and increase your understanding of key concepts and processes presented in this chapter.

▶ SECTION: SURFACE
MODULE: SHORELINES AND COASTS

The photograph of southern Alaska (on your computer screen) documents a range of coastal and glacial landforms. Different parts of this picture preserve clues as to the evolution of the entire landscape. You can see, for example, the beautiful U-shaped valleys in the background, which are clearly of glacial origin. Also notice the distinctive coastline: Whereas most of Alaska's southern coastline is characterized by steep, zigzagging cliffs and rocky shores, this photograph shows a straight coastline with a sandy beach. **Question: What combination of geologic processes can account for this straight and sandy shoreline?**

▶ SECTION: SURFACE
MODULE: THE SEAFLOOR

Mineralogists have it relatively easy when it comes to sampling objects for study—hiking, rock climbing, and excavating to get their specimens. It's not quite the same for geologists studying the seafloor. The difficulty of penetrating the ocean depths and gathering solid information cannot be overestimated. The seafloor is wet and dark, can be more than 30,000 feet deep, and is often covered with sediment that obscures the solid rock beneath. **Question: Given these constraints, what do you think is the most effective way to examine the structure and composition of the seafloor?**

Geologic Time: Concepts

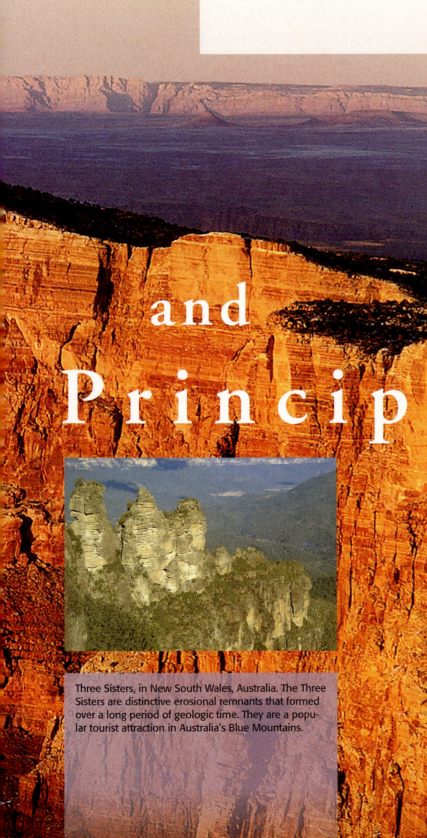

Prologue

and Principles

What is time? We seem obsessed with it and organize our lives around it with the help of clocks, calendars, and appointment books. Yet most of us feel we don't have enough of it—we are always running "behind" or "out of time." According to biologists and psychologists, children less than two years old and animals exist in a "timeless present." Some scientists think that our early ancestors may also have lived in a state of timelessness with little or no perception of a past or future. According to Buddhist, Taoist, and Mayan beliefs, time is circular, and like a circle, all things are destined to return to where they once were. Thus, in these belief systems, there is no beginning or end, but rather a cyclicity to everything.

In some respects, time is defined by the methods used to measure it. Many prehistoric monuments are oriented to detect the summer solstice, and sundials were used to divide the day into measurable units. As civilization advanced, mechanical devices were invented to measure time, the earliest being the water clock, first used by the ancient Egyptians and further developed by the Greeks and Romans. The pendulum

Three Sisters, in New South Wales, Australia. The Three Sisters are distinctive erosional remnants that formed over a long period of geologic time. They are a popular tourist attraction in Australia's Blue Mountains.

clock was invented in the seventeenth century and provided the most accurate timekeeping for the next two and a half centuries.

Today the quartz watch is the most popular timepiece. Powered by a battery, a quartz crystal vibrates approximately 100,000 times per second. An integrated circuit counts these vibrations and converts them into a digital or dial reading on your watch face. An inexpensive quartz watch today is more accurate than the best mechanical watch, and precision-manufactured quartz clocks are accurate to within one second per ten years.

Precise timekeeping is important in our technological world. Ships and aircraft plot their locations by satellite, relying on an extremely accurate time signal. Deep-space probes such as the Voyagers require radio commands timed to billionths of a second, and physicists exploring the motion inside the nucleus of an atom deal in trillionths of a second as easily as we talk about minutes.

To achieve such accuracy, scientists use atomic clocks. First developed in the 1940s, these clocks rely on an atom's oscillating electrons, a rhythm so regular that they are accurate to within a few thousandths of a second per day. An atomic clock accurate to within one second per 3 million years was recently installed at the National Institute of Standards and Technology.

While physicists deal with incredibly short intervals of time, astronomers and geologists are concerned with geologic time measured in millions or billions of years. When astronomers look at a distant galaxy, they are seeing what it looked like billions of years ago. When geologists investigate rocks in the walls of the Grand Canyon, they are deciphering events that occurred over an interval of 2 billion years. Geologists can measure decay rates of such radioactive elements as uranium, thorium, and rubidium to determine how long ago an igneous rock formed. Furthermore, geologists know that Earth's rotational velocity has been slowing down a few thousandths of a second per century as a result of the frictional effects of tides, ocean currents, and varying thicknesses of polar ice. Five hundred million years ago a day was only 20 hours long, and at the current rate of slowing, 200 million years from now a day will be 25 hours long.

Time is a fascinating topic that has been the subject of numerous essays and books. And though we can comprehend concepts like milliseconds and understand how a quartz watch works, deep time, or geologic time, is still very difficult for most people to comprehend.

Introduction

TIME is what sets geology apart from most of the other sciences, and an appreciation of the immensity of geologic time is fundamental to understanding both the physical and biological history of our planet. Most people have difficulty comprehending geologic time because they tend to view time from the perspective of their own existence. Ancient history is what occurred hundreds or even thousands of years ago, but when geologists talk in terms of ancient geologic history, they are referring to events that happened millions or even billions of years ago!

Geologists use two different frames of reference when speaking of geologic time. **Relative dating** involves placing geologic events in a sequential order as determined from their position in the geologic record. Relative dating will not tell us how long ago a particular event occurred, only that one event preceded another. The various principles used to determine relative dating were discovered hundreds of years ago, and since then they have been used to construct the *relative geologic time scale* (Figure 17.1). These principles are still widely used today.

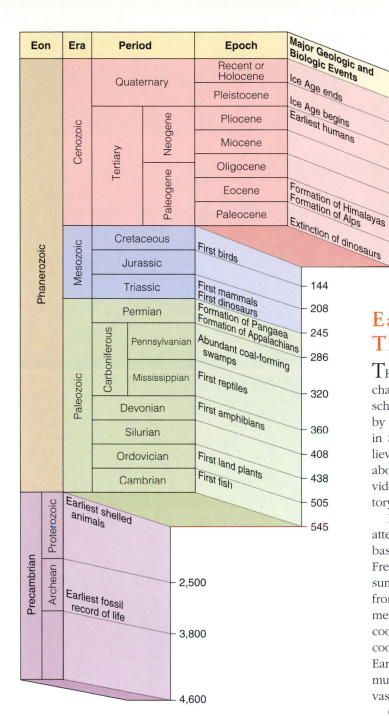

FIGURE **17.1** The geologic time scale. Some of the major biological and geological events are indicated along the right-hand side.

Early Concepts of Geologic Time and the Age of Earth

THE concept of geologic time and its measurement have changed through human history. Many early Christian scholars and clerics tried to establish the date of creation by analyzing historical records and the genealogies found in Scripture. Based on their analyses, they generally believed that Earth and all its features were no more than about 6000 years old. The idea of a very young Earth provided the basis for most Western chronologies of Earth history prior to the eighteenth century.

During the eighteenth and nineteenth centuries, several attempts were made to determine the age of Earth on the basis of scientific evidence rather than revelation. The French zoologist Georges Louis de Buffon (1707–1788) assumed that Earth gradually cooled to its present condition from a molten beginning. To simulate this history, he melted iron balls of various diameters and allowed them to cool to the surrounding temperature. By extrapolating their cooling rate to a ball the size of Earth, he determined that Earth was at least 75,000 years old. While this age was much older than that derived from Scripture, it was still vastly younger than we now know the planet to be.

Other scholars were equally ingenious in attempting to calculate Earth's age. For example, if deposition rates could be determined for various sediments, geologists reasoned that they could calculate how long it would take to deposit any rock layer. They could then extrapolate how old Earth was from the total thickness of sedimentary rock in its crust. Rates of deposition vary, however, even for the same type of rock. Furthermore, it is impossible to estimate how much rock has been removed by erosion or how much a rock sequence has been reduced by compaction. As a result of these variables, estimates ranged from less than 1 million years to more than 1 billion years.

Besides trying to determine Earth's age, the naturalists of the eighteenth and nineteenth centuries were also formulating some of the fundamental geologic principles that

Absolute dating results in specific dates for rock units or events expressed in years before the present. Radiometric dating is the most common method of obtaining absolute-age dates. Such dates are calculated from the natural rates of decay of various radioactive elements present in trace amounts in some rocks. It was not until the discovery of radioactivity near the end of the nineteenth century that absolute ages could be accurately applied to the relative geologic time scale. Today the geologic time scale is really a dual scale: a relative scale based on rock sequences with radiometric dates expressed as years before the present added to it (Figure 17.1).

Early Concepts of Geologic Time and the Age of Earth   363

are used in deciphering Earth history. From the evidence preserved in the geologic record, it was clear to them that Earth is very old and that geologic processes have operated over long periods of time.

James Hutton and the Recognition of Geologic Time

THE Scottish geologist James Hutton (1726–1797) is considered by many to be the father of modern geology. His detailed studies and observations of rock exposures and present-day geological processes were instrumental in establishing the **principle of uniformitarianism** (see Chapter 1), the concept that the same processes have operated over vast amounts of time. Because Hutton relied on known processes to account for Earth history, he concluded that Earth must be very old and wrote that "we find no vestige of a beginning, and no prospect of an end."

Unfortunately, Hutton was not a particularly good writer, so his ideas were not widely disseminated or accepted. In 1830, Charles Lyell published a landmark book, *Principles of Geology,* in which he championed Hutton's concept of uniformitarianism. Instead of relying on catastrophic events to explain various Earth features, Lyell recognized that imperceptible changes brought about by present-day processes could, over long periods of time,

have tremendous cumulative effects. Through his writings, Lyell firmly established uniformitarianism as the guiding philosophy of geology.

Relative-Dating Methods

BEFORE the development of radiometric-dating techniques, geologists had no reliable means of absolute-age dating and therefore depended solely on relative-dating methods. These methods allow events to be placed in sequential order only and do not tell us how long ago an event took place. Though the principles of relative dating may now seem self-evident, their discovery was an important scientific achievement because they provided geologists with a means to interpret geologic history and develop a relative geologic time scale.

Six fundamental geologic principles are used in relative dating: superposition, original horizontality, lateral continuity, cross-cutting relationships, inclusions, and fossil succession.

FUNDAMENTAL PRINCIPLES OF RELATIVE DATING

The seventeenth century was an important time in the development of geology as a science because of the widely circulated writings of the Danish anatomist, Nicolas Steno (1638–1686). Steno observed that when streams flood, they

FIGURE 17.2 The Grand Canyon of Arizona illustrates three of the six fundamental principles of relative dating. The sedimentary rocks of the Grand Canyon were originally deposited horizontally in a variety of marine and continental environments (principle of original horizontality). The oldest rocks are at the bottom of the canyon, and the youngest rocks are at the top, forming the rim (principle of superposition). The exposed rock layers extend laterally for some distance (principle of lateral continuity).

(a)

(b)

FIGURE 17.3   The principle of cross-cutting relationships. (a) A dark-colored dike has been intruded into older light-colored granite, along the north shore of Lake Superior, Ontario, Canada. (b) A small fault displacing tilted beds along Templin Highway, Castaic, California.

spread out across their floodplains and deposit layers of sediment that bury organisms dwelling on the floodplain. Subsequent floods produce new layers of sediments that are deposited or superposed over previous deposits. When lithified, these layers of sediment become sedimentary rock. Thus, in an undisturbed succession of sedimentary rock layers, the oldest layer is at the bottom and the youngest layer is at the top. This **principle of superposition** is the basis for relative-age determinations of strata and their contained fossils (Figure 17.2).

Steno also observed that, because sedimentary particles settle from water under the influence of gravity, sediment is deposited in essentially horizontal layers, illustrating the **principle of original horizontality** (Figure 17.2). Therefore, a sequence of sedimentary rock layers that is steeply inclined from the horizontal must have been tilted after deposition and lithification.

Steno's third principle, the **principle of lateral continuity**, states that a layer of sediment extends laterally in all directions until it thins and pinches out or terminates against the edge of the depositional basin (Figure 17.2).

James Hutton is credited with discovering the **principle of cross-cutting relationships**. Based on his detailed studies and observations of rock exposures in Scotland, Hutton recognized that an igneous intrusion or fault must be younger than the rocks it intrudes or displaces (Figure 17.3).

Another way to determine relative ages is by using the **principle of inclusions**. This principle holds that inclusions, or fragments of one rock contained within a layer of another, are older than the rock layer itself. The batholith shown in Figure 17.4a contains sandstone inclusions, and the sandstone unit shows the effects of baking. Accordingly, we conclude that the sandstone is older than the batholith. In Figure 17.4b, however, the sandstone contains

granite rock fragments, indicating that the batholith was the source rock for the inclusions and is therefore older than the sandstone.

Fossils have been known for centuries, yet their utility in relative dating and geologic mapping was not fully appreciated until the early nineteenth century. William Smith (1769–1839), an English civil engineer involved in surveying and building canals in southern England, independently recognized the principle of superposition by reasoning that the fossils at the bottom of a sequence of strata are older than those at the top of the sequence. This recognition served as the basis for the **principle of fossil succession** or the *principle of faunal and floral succession* as it is sometimes called (Figure 17.5). According to this principle, fossil assemblages succeed one another through time in a regular and predictable order.

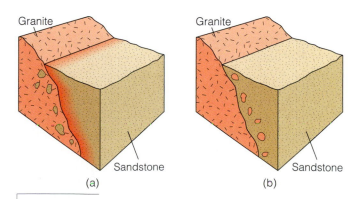
Granite   Granite

Sandstone   Sandstone
(a)   (b)

FIGURE 17.4   (a) The batholith is younger than the sandstone because the sandstone has been baked at its contact with the granite and the granite contains sandstone inclusions. (b) Granite inclusions in the sandstone indicate that the batholith was a source of the sandstone and therefore is older.

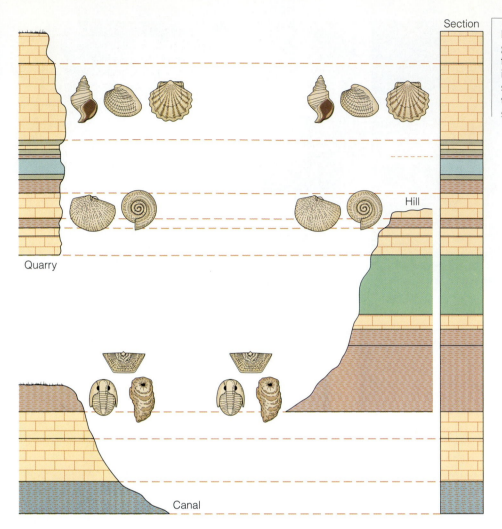

FIGURE **17.5** This generalized diagram shows how William Smith used fossils to identify strata of the same age in different areas (principle of fossil succession). The composite section on the right shows the relative ages of all strata in this area.

Hill

Quarry

Canal

UNCONFORMITIES

Our discussion so far has been concerned with conformable sequences of strata, sequences in which no depositional breaks of any consequence occur. A sharp bedding plane (see Figure 7.11) separating strata may represent a depositional break of minutes, hours, years, or even tens of years, but it is inconsequential when considered in the context of geologic time.

Surfaces of discontinuity that encompass significant amounts of geologic time are **unconformities**, and any interval of geologic time not represented by strata in a particular area is a *hiatus* (Figure 17.6). Thus, an unconformity is a surface of nondeposition or erosion that separates younger strata from older rocks. As such, it represents a break in our record of geologic time. The famous 12-minute gap in the Watergate tapes of Richard Nixon's presidency is somewhat analogous. Just as we have no record of the conversations that were occurring during this period of time, we have no record of the events that occurred during a hiatus.

Three types of unconformities are recognized. A **disconformity** is a surface of erosion or nondeposition between younger and older beds that are parallel with one another (Figure 17.7). Unless a well-defined erosional surface separates the older from the younger parallel beds, the discon-

formity frequently resembles an ordinary bedding plane. Accordingly, many disconformities are difficult to recognize and must be identified on the basis of fossil assemblages.

An **angular unconformity** is an erosional surface on tilted or folded strata over which younger strata have been deposited (Figure 17.8). Both younger and older strata may dip, but if their dip angles are different (generally the older strata dip more steeply), an angular unconformity is present.

The angular unconformity illustrated in Figure 17.8b is probably the most famous in the world. It was here at Siccar Point, Scotland, that James Hutton realized that severe upheavals had tilted the lower rocks and formed mountains that were then worn away and covered by younger, flat-lying rocks. The erosional surface between the older tilted rocks and the younger flat-lying strata meant that a significant gap existed in the rock record. Although Hutton did not use the term *unconformity*, he was the first to understand and explain the significance of such discontinuities in the geologic record.

The third type of unconformity is a **nonconformity**. Here an erosion surface cut into metamorphic or igneous rocks is covered by sedimentary rocks (Figure 17.9). This type of unconformity closely resembles an intrusive ig-

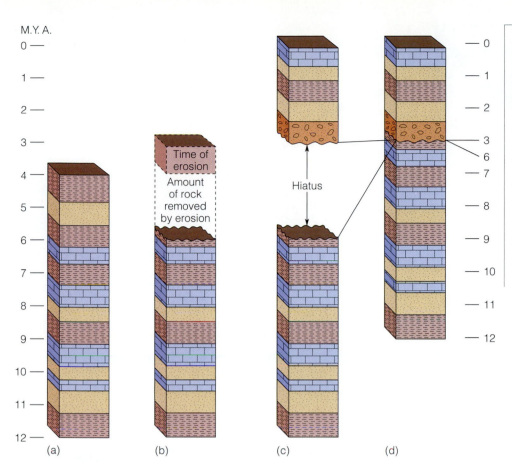

M.Y.A.

(a) (b) (c) (d)

FIGURE **17.6** A simplified diagram showing the development of an unconformity and a hiatus. (a) Deposition began 12 million years ago (M.Y.A.) and continued more-or-less uninterrupted until 4 M.Y.A. (b) A 1-million-year episode of erosion occurred, and during that time rocks representing 2 million years of geologic time were eroded. (c) A hiatus of 3 million years exists between the older strata and the strata that formed during a renewed episode of deposition that began 3 M.Y.A. (d) The actual record. The unconformity is the surface separating the strata and represents a major break in our record of geologic time.

neous contact with sedimentary rocks. The principle of inclusions is helpful in determining whether the relationship between igneous rocks and overlying sedimentary rocks is the result of an intrusion or erosion. In the case of an intrusion, the igneous rocks are younger; but in the case of erosion, the sedimentary rocks are younger. Being able to distinguish between a nonconformity and an intrusive contact is very important because they represent different sequences of events.

Correlation

IF geologists are to decipher Earth history, they must demonstrate the time equivalency of rock units in different areas. This process is known as **correlation**.

If exposures are adequate, units may simply be traced laterally (principle of lateral continuity), even if occasional gaps exist (Figure 17.10). Other criteria used to correlate units are similarity of rock type, position in a sequence, and key beds. *Key beds* are units, such as coal beds or volcanic ash layers, that are sufficiently distinctive to allow identification of the same unit in different areas (Figure 17.10). Besides surface correlation, geologists frequently use well logs, cores, or cuttings to correlate subsurface rock units when exploring for minerals, coal, and petroleum.

Generally, no single location in a region has a geologic record of all the events that occurred during its history; therefore, geologists must correlate from one area to an-

other in order to decipher the complete geologic history of the region. An excellent example is the history of the Colorado Plateau. A record of events occurring over approximately 2 billion years is present in this region. Because of the forces of erosion, the entire record is not preserved at any single location. Within the walls of the Grand Canyon are rocks of the Precambrian and Paleozoic Era, while Paleozoic and Mesozoic Era rocks are found in Zion National Park, and Mesozoic and Cenozoic Era rocks are exposed in Bryce Canyon. By correlating the uppermost rocks at one location with the lowermost equivalent rocks of another area, the history of the entire region can be deciphered.

Although geologists can match up rocks on the basis of similar rock type and superposition, correlation of this type can only be done in a limited area where beds can be traced from one site to another. To correlate rock units over a large area or to correlate age-equivalent units of different composition, fossils and the principle of fossil succession must be used.

Fossils are useful as time indicators because they are the remains of organisms that lived for a certain length of time during the geologic past. Fossils that are easily identified, are geographically widespread, and existed for a rather short geologic time are particularly useful. Such fossils are called **guide fossils** or *index fossils* (Figure 17.11). The trilobite *Isotelus* and the clam *Inoceramus* meet these criteria and are therefore good guide fossils. In contrast, the brachiopod *Lingula* is easily identified

Correlation 367

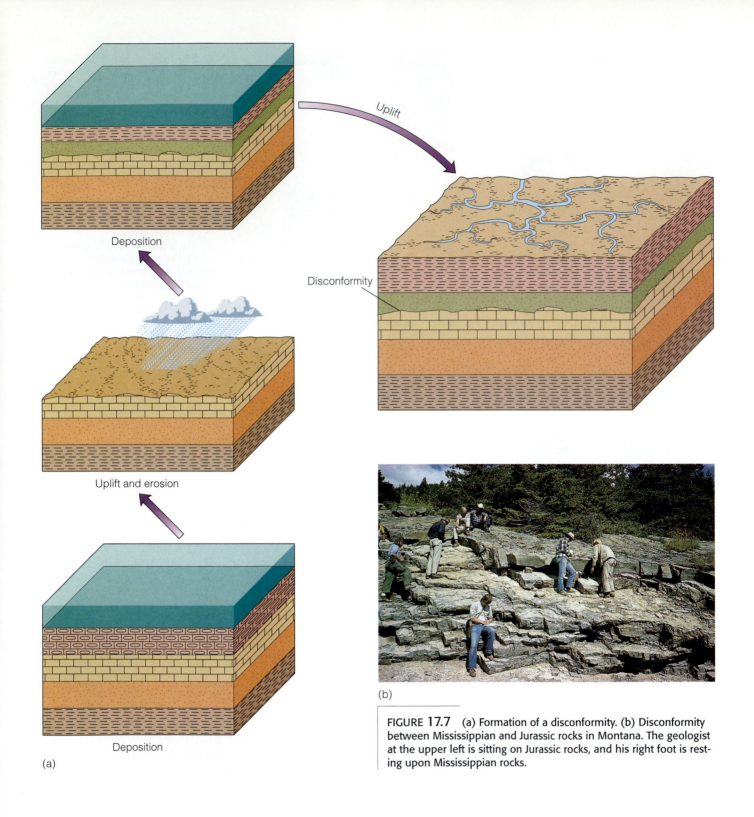

Deposition

Uplift

Disconformity

Uplift and erosion

Deposition

(a)

(b)

FIGURE 17.7 (a) Formation of a disconformity. (b) Disconformity between Mississippian and Jurassic rocks in Montana. The geologist at the upper left is sitting on Jurassic rocks, and his right foot is resting upon Mississippian rocks.

and widespread, but its geologic range of Ordovician to Recent makes it of little use in correlation.

Absolute-Dating Methods

ALTHOUGH most of the isotopes of the 92 naturally occurring elements are stable, some are radioactive and spontaneously decay to other more stable isotopes of el-

ements, releasing energy in the process. The discovery, in 1903 by Pierre and Marie Curie, that radioactive decay produces heat meant that geologists finally had a mechanism for explaining Earth's internal heat that did not rely on residual cooling from a molten origin. Furthermore, geologists had a powerful tool to date geologic events accurately and verify the long time periods postulated by Hutton and Lyell.

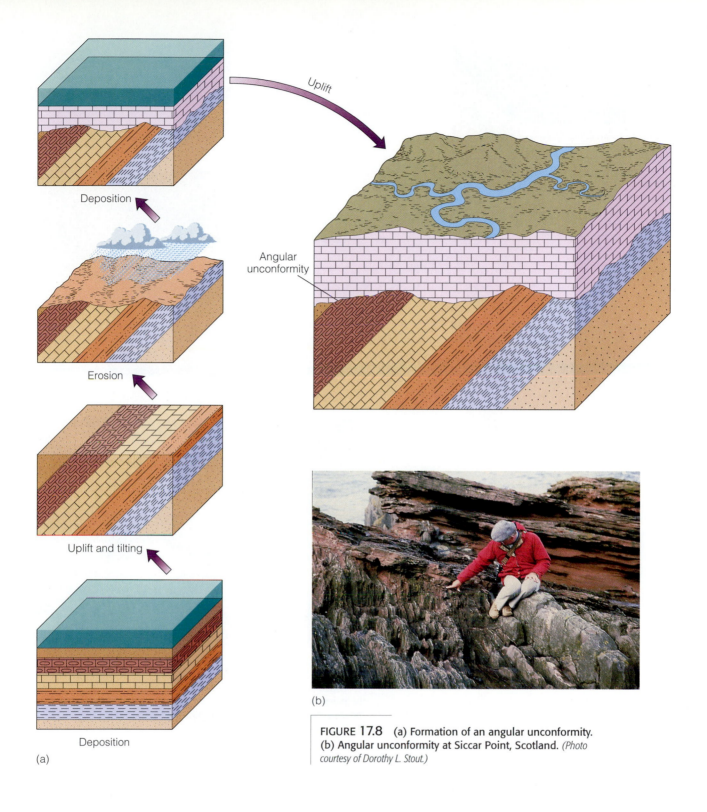

(a)

(b)

FIGURE 17.8 (a) Formation of an angular unconformity. (b) Angular unconformity at Siccar Point, Scotland. (*Photo courtesy of Dorothy L. Stout.*)

Labels in figure: Uplift, Deposition, Erosion, Uplift and tilting, Deposition, Angular unconformity

ATOMS, ELEMENTS, AND ISOTOPES

As we discussed in Chapter 3, all matter is made up of chemical elements, each of which is composed of extremely small particles called *atoms*. The nucleus of an atom is composed of *protons* and *neutrons* with *electrons* encircling it (see Figure 3.3). The number of protons defines an element's *atomic number* and helps determine its properties and characteristics. The combined number of protons and neutrons in an atom is its *atomic mass number*. However, not all atoms of the same element have the same number of neutrons in their nuclei. These variable forms of the same element are called *isotopes* (see Figure 3.4). Most isotopes are stable, but some are unstable and spontaneously decay to a more stable form. It is the decay rate of unstable isotopes that geologists measure to determine the absolute age of rocks.

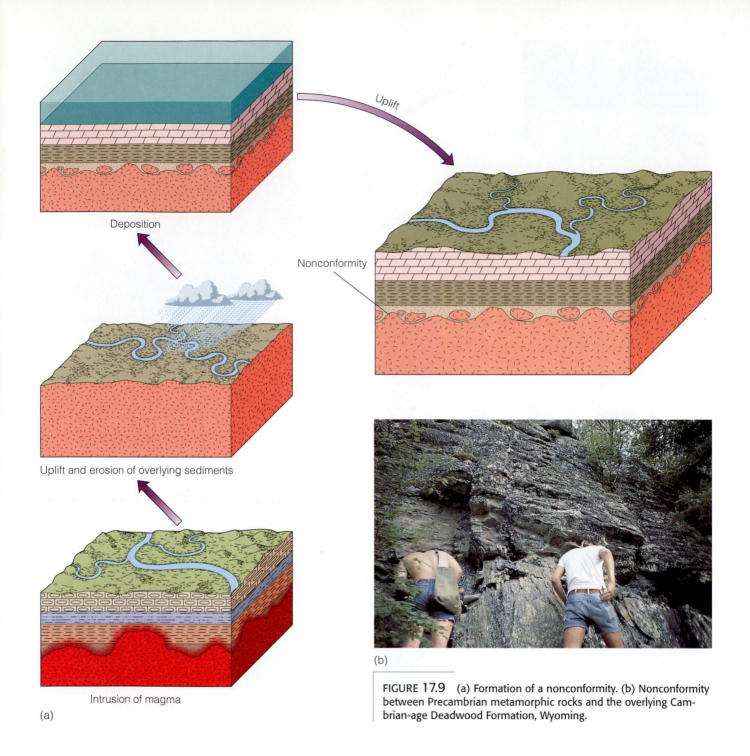

(a)

Deposition

Uplift

Nonconformity

Uplift and erosion of overlying sediments

Intrusion of magma

(b)

FIGURE 17.9 (a) Formation of a nonconformity. (b) Nonconformity between Precambrian metamorphic rocks and the overlying Cambrian-age Deadwood Formation, Wyoming.

RADIOACTIVE DECAY AND HALF-LIVES

Radioactive decay is the process whereby an unstable atomic nucleus is spontaneously transformed into an atomic nucleus of a different element. Three types of radioactive decay are recognized, all of which result in a change of atomic structure (Figure 17.12). In **alpha decay**, two protons and two neutrons are emitted from the nucleus, resulting in a loss of two atomic numbers and four atomic mass numbers. In **beta decay**, a fast-moving electron is emitted from a neutron in the nucleus, changing that neutron to a proton and consequently

increasing the atomic number by one, with no resultant atomic mass number change. **Electron capture** results when a proton captures an electron from an electron shell and thereby converts to a neutron, resulting in a loss of one atomic number but not changing the atomic mass number.

Some elements undergo only one decay step in the conversion from an unstable form to a stable form. For example, rubidium 87 decays to strontium 87 by a single beta emission, and potassium 40 decays to argon 40 by a single electron capture. Other radioactive elements undergo sev-

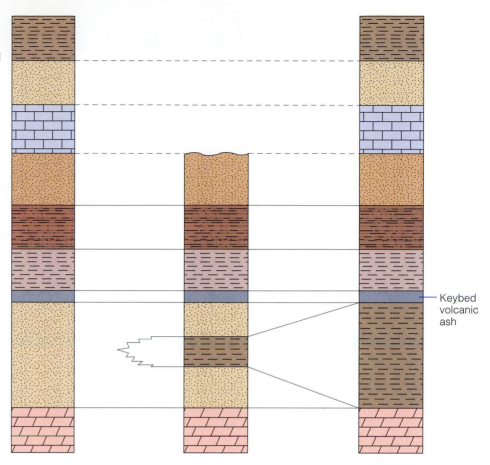

FIGURE 17.10 In areas of adequate exposures, rock units can be traced laterally, even if occasional gaps exist, and correlated on the basis of similarity in rock type and position in a sequence. Rocks can also be correlated by a key bed—in this case, volcanic ash.

Keybed volcanic ash

eral decay steps (see Perspective 17.1). Uranium 235 decays to lead 207 by seven alpha and six beta steps, while uranium 238 decays to lead 206 by eight alpha and six beta steps (Figure 17.13).

When discussing decay rates, it is convenient to refer to them in terms of half-lives. The **half-life** of a radioactive element is the time it takes for one-half of the atoms of the original unstable **parent element** to decay to atoms of a new, more stable **daughter element**. The half-life of a given radioactive element is constant and can be precisely measured. Half-lives of various radioactive elements range from less than a billionth of a second to 49 billion years.

Radioactive decay occurs at a geometric rate rather than a linear rate. Therefore, a graph of the decay rate produces a curve rather than a straight line (Figure 17.14). For example, an element with 1,000,000 parent atoms will have 500,000 parent atoms and 500,000 daughter atoms after one half-life. After two half-lives, it will have 250,000 parent atoms (one-half of the previous parent atoms, which is equivalent to one-fourth of the

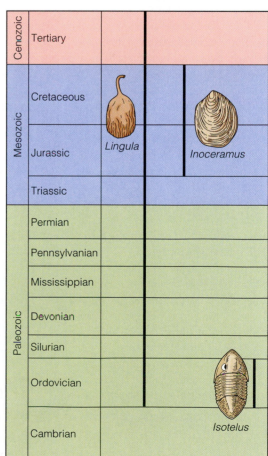

FIGURE 17.11 The geologic ranges of three marine invertebrates. The brachiopod *Lingula* is of little use in correlation because of its long geologic range. The trilobite *Isotelus* and the bivalve *Inoceramus* are good guide fossils because they are geographically widespread, are easily identified, and have short geologic ranges.

Radon: The Silent Killer

Radon is a colorless, odorless, naturally occurring radioactive gas that has a 3.8-day half-life. It is part of the uranium 238–lead 206 radioactive decay series (Figure 17.13) and occurs in any rock or soil that contains uranium 238. Radon concentrations are reported in picocuries per liter (pCi/L) of air (a curie is the standard measure of radiation, and a picocurie is one-trillionth of a curie, or the equivalent of the decay of about two radioactive atoms per minute). Outdoors, radon escapes into the atmosphere where it is diluted and dissipates to harmless levels (0.2 pCi/L is the ambient outdoor level of radon). Radon levels for indoor air range from less than 1 pCi/L to about 3000 pCi/L, but average about 1.5 pCi/L. The Environmental Protection Agency (EPA) considers radon levels exceeding 4 pCi/L to be unhealthy and recommends remedial action be taken to lower them. Continued exposure to elevated levels of radon over an extended period of time is thought by many researchers to increase the risk of lung cancer.

Radon is one of the natural decay products of uranium 238. It rapidly decays by the emission of an alpha particle, producing two short-lived radioactive isotopes—polonium 218 and polonium 214 (Figure 17.13). Both isotopes are solid and can become trapped in your lungs every time you breathe. When polonium decays, it emits alpha and beta particles that can damage lung cells and cause lung cancer.

Your chances of being adversely affected by radon depend on numerous interrelated factors such as geographic location, the geology of the area, the climate, how buildings are constructed, and the amount of time spent in your house. Because radon is a naturally occurring gas, contact with it is unavoidable, but atmospheric concentrations of

it are probably harmless. Only when concentrations of radon build up in poorly ventilated structures does it become a potential health risk.

Concern about the health risks posed by radon first arose during the 1960s when the news media revealed that some homes in the West had been built with uranium mine tailings. Since then, geologists have found that high indoor radon levels can be caused by natural uranium in minerals of the rock and soil on which buildings are constructed. In response to the high cost of energy during the 1970s and 1980s, old buildings were insulated, and new buildings were constructed to be as energy efficient and airtight as possible. Ironically, these energy-saving measures also sealed in radon.

Radon enters buildings through dirt floors, cracks in the floor or walls, joints between floors and walls, floor drains, sumps, and utility pipes as well as any cracks or pores in hollow block walls (Figure 1). Radon can also be released into a building whenever the water is turned on, particularly if the water comes from a private well. Municipal

water is generally safe because it has usually been aerated before it gets to your home.

To find out if your home has a radon problem, you must test for it with commercially available, relatively inexpensive, simple home-testing devices. If radon readings are above the recommended EPA levels of 4 pCi/L, several remedial measures can be taken to reduce your risk. These include sealing all cracks in the foundation, pouring a concrete slab over a dirt floor, increasing the circulation of air throughout the house, especially in the basement and crawl space, providing filters for drains and other utility openings, and limiting time spent in areas with higher concentrations of radon.

It is important to remember that, although the radon hazard covers most of the country, some areas are more likely to have higher natural concentrations of radon than others (Figure 2). Such rocks as uranium-bearing granites, metamorphic rocks of granitic composition, and black shales (high carbon content) are quite likely to cause indoor radon problems. Other rocks such as marine quartz

Figure 1 Some of the common points where radon can enter a house.

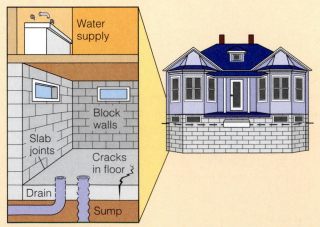

sandstone, noncarbonaceous shales and siltstones, most volcanic rocks, and igneous and metamorphic rocks rich in iron and magnesium typically do not cause radon problems. The permeability of the soil overlying the rock can also affect the indoor levels of radon gas. Some soils are more permeable than others and allow more radon to escape into the overlying structures.

The climate and the type of construction affect not only how much radon enters a structure but also how much escapes. Concentrations of radon are highest during the winter in northern climates because buildings are sealed as tightly as possible. Homes with basements are more likely to have higher radon levels than those built on concrete slabs. While research continues into the sources of indoor radon and ways of controlling it, the most important thing people can do is to test their home, school, or business for radon.

Currently a heated debate is ongoing among scientists concerning the large-scale health hazards resulting from radon exposure and how much money should be spent for its remediation. On the one hand, the EPA and Surgeon General estimate that exposure to high levels of indoor radon cause between 5000 and 20,000 lung cancer deaths each year. Other scientists dispute these figures because of the difficulty of attributing mortality rates for lung cancer directly to radon, particularly when so many other factors, such as smoking, are involved. Central to this debate are two questions: What concentration levels of indoor radon are acceptable, and exactly how serious is the risk from radon at those levels? Unfortunately, the data for making these determinations are simply not available at this time.

Figure 2 *Areas in the United States where granite, phosphate-bearing rocks, carbonaceous shales, and uranium occur. These rocks are all potential sources of radon gas.*

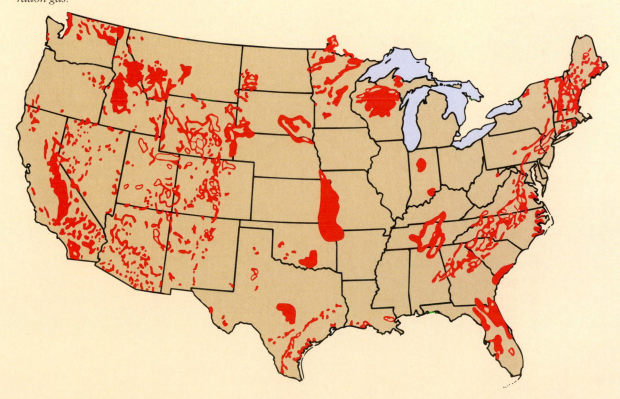

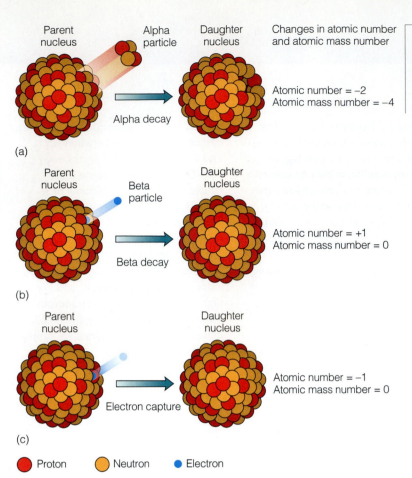

(a)

(b)

(c)

● Proton ● Neutron ● Electron

FIGURE 17.12 Three types of radioactive decay. (a) Alpha decay, in which an unstable parent nucleus emits two protons and two neutrons. (b) Beta decay, in which an electron is emitted from the nucleus. (c) Electron capture, in which a proton captures an electron and is thereby converted to a neutron.

original parent atoms) and 750,000 daughter atoms. After three half-lives, it will have 125,000 parent atoms (one-half of the previous parent atoms or one-eighth of the original parent atoms) and 875,000 daughter atoms, and so on until the number of parent atoms remaining is so few that they cannot be accurately measured by present-day instruments.

By measuring the parent-daughter ratio and knowing the half-life of the parent (determined in the laboratory), geologists can calculate the age of a sample containing the radioactive element. The parent-daughter ratio is usually determined by a *mass spectrometer,* an instrument that measures the proportions of elements of different masses.

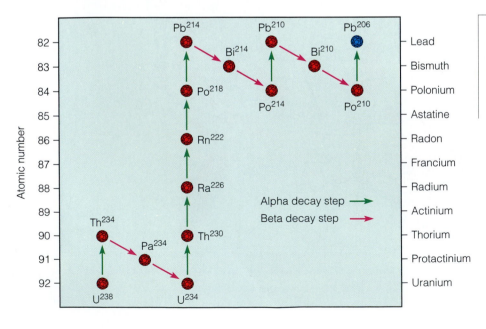

FIGURE 17.13 Radioactive decay series for uranium 238 to lead 206. Radioactive uranium 238 decays to its stable end product, lead 206, by eight alpha and six beta decay steps. A number of different isotopes are produced as intermediate steps in the decay series.

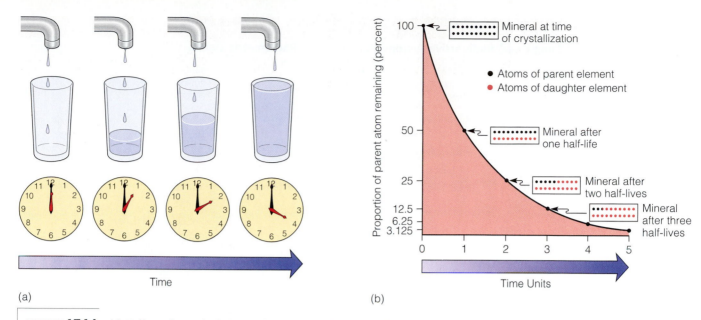

(a) Time

(b) Time Units

100 — Mineral at time of crystallization

● Atoms of parent element
● Atoms of daughter element

50 — Mineral after one half-life

25 — Mineral after two half-lives

12.5
6.25
3.125 — Mineral after three half-lives

Proportion of parent atom remaining (percent)

FIGURE 17.14 (a) Uniform, linear depletion is characteristic of many familiar processes. (b) Geometric radioactive decay curve, in which each time unit represents one half-life, and each half-life is the time it takes for one-half of the parent element to decay to the daughter element.

SOURCES OF UNCERTAINTY The most accurate radiometric dates are obtained from igneous rocks. As a magma cools and begins to crystallize, radioactive parent atoms are separated from previously formed daughter atoms. Because they are the right size, some radioactive parent atoms are incorporated into the crystal structure of certain minerals. The stable daughter atoms, though, are a different size than the radioactive parent atoms and consequently cannot fit into the crystal structure of the same mineral as the parent atoms. Therefore, a mineral crystallizing in a cooling magma will contain radioactive parent atoms but no stable daughter atoms (Figure 17.15). Thus, the time that is being measured is the time of crystallization of the mineral containing the radioactive atoms, and not the time of formation of the radioactive atoms.

To obtain accurate radiometric dates, geologists must be sure that they are dealing with a *closed system,* meaning that neither parent nor daughter atoms have been added or removed from the system since crystallization and that the ratio between them results only from radioactive decay. Otherwise, an inaccurate date will result. If daughter atoms have leaked out of the mineral being analyzed, the calculated age will be too young; if parent atoms have been removed, the calculated age will be too great.

Leakage may occur if the rock is heated, as occurs during metamorphism. If this happens, some of the parent or daughter atoms may be driven from the mineral being analyzed, resulting in an inaccurate age determination. If the daughter product was completely removed, then one would be measuring the time since metamorphism (a useful

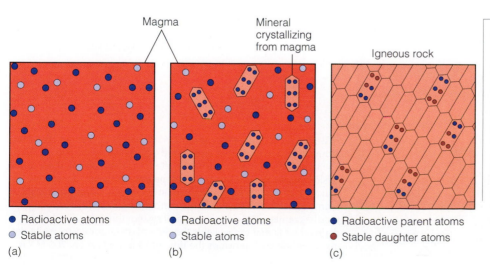

Magma

Mineral crystallizing from magma

Igneous rock

(a) ● Radioactive atoms ○ Stable atoms

(b) ● Radioactive atoms ○ Stable atoms

(c) ● Radioactive parent atoms ● Stable daughter atoms

FIGURE 17.15 (a) A magma contains both radioactive and stable atoms. (b) As the magma cools and begins to crystallize, some radioactive atoms are incorporated into certain minerals because they are the right size and can fit into the crystal structure. Therefore, at the time of crystallization, the mineral will contain 100% radioactive parent atoms and 0% stable daughter atoms. (c) After one half-life, 50% of the radioactive parent atoms will have decayed to stable daughter atoms.

TABLE 17.1

Five of the Principal Long-Lived Radioactive Isotope Pairs Used in Radiometric Dating

ISOTOPES		HALF-LIFE OF PARENT (YEARS)	EFFECTIVE DATING RANGE (YEARS)	MINERALS AND ROCKS THAT CAN BE DATED
Parent	*Daughter*			
Uranium 238	Lead 206	4.5 billion	10 million to 4.6 billion	Zircon
				Uraninite
Uranium 235	Lead 207	704 million		
Thorium 232	Lead 208	14 billion		
Rubidium 87	Strontium 87	48.8 billion	10 million to 4.6 billion	Muscovite
				Biotite
				Potassium feldspar
				Whole metamorphic or igneous rock
Potassium 40	Argon 40	1.3 billion	100,000 to 4.6 billion	Glauconite Hornblende
				Muscovite Whole volcanic rock
				Biotite

measurement itself), and not the time since crystallization of the mineral (Figure 17.16). Because heat affects the parent-daughter ratio, metamorphic rocks are difficult to age-date accurately. Remember that while the parent-daughter ratio may be affected by heat, the decay rate of the parent element remains constant, regardless of any physical or chemical changes.

LONG-LIVED RADIOACTIVE ISOTOPE PAIRS

Table 17.1 shows the five common, long-lived parent-daughter isotope pairs used in radiometric dating. Long-lived pairs have half-lives of millions or billions of years. All of these were present when Earth formed and are still present in measurable quantities. Other shorter-lived radioactive isotope pairs have decayed to the point that only small quantities near the limit of detection remain.

The most commonly used isotope pairs are the uranium-lead and thorium-lead series, which are used principally to date ancient igneous intrusives, lunar samples, and some meteorites. The rubidium-strontium pair is also used for very old samples and has been effective in dating the oldest rocks on Earth as well as meteorites. The potassium-argon method is typically used for dating fine-grained volcanic rocks from which individual crystals cannot be separated; hence, the whole rock is analyzed. Argon is a gas, however, so great care must be taken to ensure that the sample has not been subjected to heat, which would allow argon to escape; such a sample would yield an age that is too young. Other long-lived radioactive isotope pairs exist, but they are rather rare and are used only in special situations.

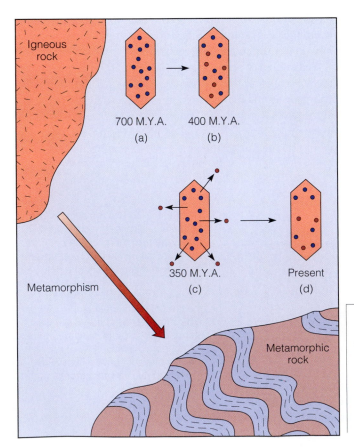

Igneous rock

700 M.Y.A.
(a)

400 M.Y.A.
(b)

350 M.Y.A.
(c)

Present
(d)

Metamorphism

Metamorphic rock

FIGURE 17.16   The effect of metamorphism in driving out daughter atoms from a mineral that crystallized 700 million years ago (M.Y.A.). The mineral is shown immediately after crystallization (a), then at 400 million years (b), when some of the parent atoms had decayed to daughter atoms. (c) Metamorphism at 350 M.Y.A. drives the daughter atoms out of the mineral into the surrounding rock. (d) Assuming the rock has remained a closed chemical system throughout its history, dating the mineral today yields the time of metamorphism, while dating the whole rock provides the time of its crystallization, 700 M.Y.A.

CARBON 14 DATING METHOD

Carbon is an important element in nature and is one of the basic elements found in all forms of life. It has three isotopes; two of these, carbon 12 and 13, are stable, whereas carbon 14 is radioactive (see Figure 3.4). Carbon 14 has a half-life of 5730 years plus or minus 30 years. The **carbon 14 dating technique** is based on the ratio of carbon 14 to carbon 12 and is generally used to date once-living material.

The short half-life of carbon 14 makes this dating technique practical only for specimens younger than about 70,000 years. Consequently, the carbon 14 dating method is especially useful in archaeology and has greatly helped unravel the events of the latter portion of the Pleistocene Epoch.

Carbon 14 is constantly formed in the upper atmosphere when cosmic rays, which are high-energy particles (mostly protons), strike the atoms of upper-atmospheric gases, splitting their nuclei into protons and neutrons. When a neutron strikes the nucleus of a nitrogen atom (atomic number 7, atomic mass number 14), it may be absorbed into the nucleus and a proton emitted. Thus, the atomic number of the atom decreases by one, while the atomic mass number stays the same. Because the atomic number has changed, a new element, carbon 14 (atomic number 6, atomic mass number 14), is formed. The newly formed carbon 14 is rapidly assimilated into the carbon cycle and, along with carbon 12 and 13, is absorbed in a nearly constant ratio by all living organisms (Figure 17.17). When an organism dies, however, carbon 14 is not replenished, and the ratio of carbon 14 to carbon 12 decreases as carbon 14 decays back to nitrogen by a single beta decay step (Figure 17.17).

Currently, the ratio of carbon 14 to carbon 12 is remarkably constant in both the atmosphere and living organisms. There is good evidence, though, that the production of carbon 14, and thus the ratio of carbon 14 to carbon 12, has varied somewhat over the past several thousand years. This was determined by comparing ages established by carbon 14 dating of wood samples against those established by counting annual tree rings in the same samples. As a result, carbon 14 ages have been corrected to reflect such variations in the past.

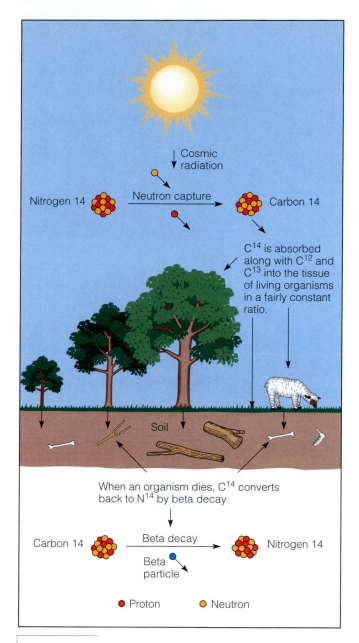

FIGURE 17.17 The carbon cycle showing the formation, dispersal, and decay of carbon 14.

Development of the Geologic Time Scale

THE geologic time scale is a hierarchical scale in which the 4.6-billion-year history of Earth is divided into time units of varying duration (Figure 17.1). It was not developed by any one individual, but rather evolved, primarily during the nineteenth century, through the efforts of many people. By applying relative-dating methods to rock outcrops, geologists in England and western Europe defined the major geologic time units without the benefit of radiometric-dating techniques. Using the principles of superpo-

sition and fossil succession, they could correlate various rock exposures and piece together a composite geologic section. This composite section is in effect a relative time scale because the rocks are arranged in their correct sequential order.

By the beginning of the twentieth century, geologists had developed a relative geologic time scale but did not yet have any absolute dates for the various time-unit boundaries. Following the discovery of radioactivity near the end of the last century, radiometric dates were added to the relative geologic time scale.

Because sedimentary rocks, with rare exceptions, cannot be radiometrically dated, geologists have had to rely on interbedded volcanic rocks and igneous intrusions to apply absolute dates to the boundaries of the various subdivisions of the geologic time scale (Figure 17.18). An ash fall or lava flow provides an excellent marker bed that is a time-equivalent surface, supplying a minimum age for the sedimentary rocks below and a maximum age for the rocks above. Ash falls are particularly useful because they may fall over both marine and nonmarine sedimentary environ-

ments and can provide a connection between these different environments.

Thousands of absolute ages are now known for sedimentary rocks of known relative ages, and these absolute dates have been added to the relative time scale. In this way, geologists have been able to determine the absolute ages of the various geologic periods and to determine their durations (Figure 17.1).

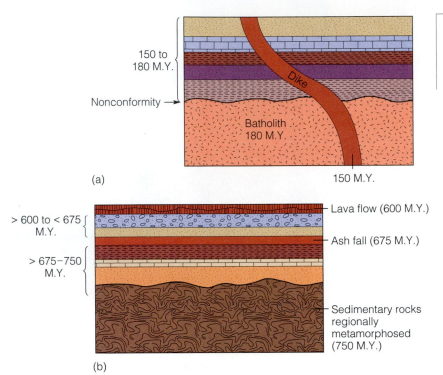

FIGURE **17.18** Absolute ages of sedimentary rocks can be determined by dating associated igneous rocks. In (a) and (b), sedimentary rocks are bracketed by rock bodies for which absolute ages have been determined.

Chapter **Summary**

1. Relative dating involves placing geologic events in a sequential order as determined from their position in the geologic record. Absolute dating results in specific dates for events, expressed in years before the present.

2. During the eighteenth and nineteenth centuries, attempts were made to determine Earth's age based on scientific evidence rather than revelation. While some attempts were quite ingenious, they yielded a variety of ages that now are known to be much too young.

3. James Hutton thought that present-day processes operating over long periods of time could explain all the geologic features of Earth. His observations were instrumental in

establishing the basis for the principle of uniformitarianism.

4. Uniformitarianism, as articulated by Charles Lyell, soon became the guiding principle of geology. It holds that the laws of nature have been constant through time and that the same processes operating today have operated in the past, although not necessarily at the same rates.

5. Besides uniformitarianism, the principles of superposition, original horizontality, lateral continuity, cross-cutting relationships, inclusions, and fossil succession are basic for determining relative geologic ages and for interpreting Earth history.

6. Surfaces of discontinuity encompassing significant amounts of geologic time are common in the geologic record. Such surfaces are unconformities and result from times of nondeposition, erosion, or both.

7. Correlation is the practice of demonstrating equivalency of units in different areas. Time equivalence is most commonly demonstrated by correlating strata containing similar fossils.

8. Radioactivity was discovered during the late nineteenth century, and soon thereafter radiometric-dating techniques allowed geologists to determine absolute ages for geologic events.

9. Absolute-age dates for rock samples are usually obtained by determining how many half-lives of a radioactive parent element have elapsed since the sample originally crystallized. A half-life is the time it takes for one-half of the radioactive parent element to decay to a stable daughter element.

10. The most accurate radiometric dates are obtained from long-lived radioactive isotope pairs in igneous rocks.

11. Carbon 14 dating can be used only for organic matter such as wood, bones, and shells and is effective back to about 70,000 years ago. Unlike the long-lived isotopic pairs, the carbon 14 dating technique determines age by the ratio of radioactive carbon 14 to stable carbon 12.

12. Through the efforts of many geologists applying the principles of relative dating, a relative geologic time scale was established.

13. Most absolute ages of sedimentary rocks and their contained fossils are obtained indirectly by dating associated igneous rocks.

Important Terms

absolute dating
alpha decay
angular unconformity
beta decay
carbon 14 dating technique
correlation
daughter element
disconformity

electron capture
guide fossil
half-life
nonconformity
parent element
principle of cross-cutting relationships
principle of fossil succession
principle of inclusions

principle of lateral continuity
principle of original horizontality
principle of superposition
principle of uniformitarianism
radioactive decay
relative dating
unconformity

Review Questions

1. In which type of unconformity are metamorphic or igneous rocks covered by younger sedimentary rocks?
   a. _____ disconformity;
   b. _____ hiatus;
   c. _____ angular unconformity;
   d. _____ intrusive;
   e. _____ nonconformity.

2. Placing geologic events in sequential order as determined by their position in the geologic record is called:
   a. _____ absolute dating;
   b. _____ uniformitarianism;
   c. _____ relative dating;
   d. _____ correlation;
   e. _____ historical dating.

3. If a rock is heated during metamorphism and the daughter atoms migrate out of a mineral that is subsequently radiometrically dated, an inaccurate date will be obtained. This date will be _____ the actual date.
   a. _____ younger than;
   b. _____ older than;
   c. _____ the same as;
   d. _____ it cannot be determined;
   e. _____ none of these.

4. Which of the following methods can be used to demonstrate age equivalency of rock units?
   a. _____ lateral tracing;
   b. _____ radiometric dating;
   c. _____ guide fossils;
   d. _____ position in a sequence;
   e. _____ all of these.

5. Which fundamental geologic principle states that the oldest layer is on the bottom of a succession of sedimentary rocks and the youngest is on top?
   a. _____ lateral continuity;
   b. _____ fossil succession;
   c. _____ original horizontality;
   d. _____ superposition;
   e. _____ cross-cutting relationships.

6. In which type of radioactive decay is a fast-moving electron emitted from a neutron in the nucleus, resulting in an atomic number increase of one and no change in the atomic mass number?
   a. _____ radiocarbon;
   b. _____ alpha;
   c. _____ fission track;

   d. _____ beta;
   e. _____ electron capture.

7. Which of the following is *not* a long-lived radioactive isotope pair?
   a. _____ uranium-lead;
   b. _____ thorium-lead;
   c. _____ potassium-argon;
   d. _____ carbon-nitrogen;
   e. _____ none of these.

8. What is being measured in radiometric dating?
   a. _____ the time when a radioactive isotope formed;
   b. _____ the time of crystallization of a mineral containing an isotope;
   c. _____ the amount of the parent isotope only;
   d. _____ when the dated mineral became part of a sedimentary rock;
   e. _____ when the stable daughter isotope was formed.

9. If a radioactive element has a half-life of 4 million years, the amount of parent material remaining after 12 million years of decay will be

what fraction of the original amount?

a. _____ 1/32;
b. _____ 1/16;
c. _____ 1/8;
d. _____ 1/4;
e. _____ 1/2.

10. In carbon 14 dating, which ratio is being measured?

a. _____ the parent to daughter isotope;
b. _____ C^{14}/N^{14};
c. _____ C^{12}/C^{13};
d. _____ C^{12}/N^{14};
e. _____ C^{12}/C^{14}.

11. How many half-lives are required to yield a mineral with 625 atoms of U^{238} and 19,375 atoms of Pb^{206}?

a. _____ 4;
b. _____ 5;
c. _____ 6;
d. _____ 8;
e. _____ 10.

12. Which of the following statements about the geologic time scale is *not* correct?

a. _____ the periods are of equal duration;
b. _____ it was originally a relative time scale;
c. _____ it evolved through the efforts of many people;
d. _____ absolute dates were not applied until the twentieth century;
e. _____ none of these.

13. Explain how a geologist would determine the relative ages of a granite batholith and an overlying sandstone formation.

14. Why is the principle of uniformitarianism important to geologists?

15. Are volcanic eruptions, earthquakes, and storm deposits geologic events encompassed by uniformitarianism?

16. Assume a hypothetical radioactive isotope with an atomic number of 150 and an atomic mass number of 300 emits five alpha decay particles and three beta decay particles and undergoes two electron capture steps. What is the atomic number of the resulting stable daughter product? The atomic mass number?

17. Why is it difficult to date sedimentary and metamorphic rocks radiometrically?

18. How does the carbon 14 dating technique differ from the uranium-lead dating method?

Points to Ponder

1. The controversies surrounding the health hazards posed by radon (see Perspective 17.1) and asbestos (see Perspective 8.1) are similar in that massive amounts of money are being spent to remedy what may not be as serious an environmental hazard as policymakers have been led to believe. Using these two highly publicized environmental hazards as examples, discuss how you might objectively assess their potential danger to the public and how you would determine what constitutes an acceptable risk for them.

2. A zircon mineral in an igneous rock contains both uranium 235 and uranium 238. In measuring the parent-daughter ratio for both uranium isotopes in the zircon, it was discovered that uranium 238 had undergone 2 half-lives in decaying to lead 206, while uranium 235 had undergone 10 half-lives during its decay to lead 207. Using Table 17.1, determine the age of the igneous rock based on each long-lived radioactive isotope pair. Are they the same age? If not, why? What explanations can you give for the difference in geologic age between the two samples?

3. Calculate the age of a rock containing the following atoms of radioactive parent element A and stable daughter B. *Radioactive parent element A:* 1,125,000 atoms; *stable daughter element B:* 34,875,000 atoms. The half-life of element A is 6.25 million years. What is the age of the rock containing these parent and daughter elements?

World Wide Web Activities

▶ **UNIVERSITY OF CALIFORNIA MUSEUM OF PALENTOLOGY**

Visit this site for an excellent description of any aspect of geologic time, paleontology, and evolution. Click on the *On-line Exhibits* to go to the *Paleontology without Walls* home page which is an introduction to the UCMP Virtual Exhibits. Click on the *Geologic Time* site. It will take you to the *Geology and Geologic Time* home page, which discusses geologic time. You can then click on one of the eras to learn more about what occurred during that time interval. Click on the *Cenozoic* site to learn more about how geological time scales are constructed.

For these web site addresses, along with current updates and exercises, log on to

http://www.wadsworth.com/geo

▶ UNITED STATES GEOLOGICAL SURVEY: RADON IN EARTH, AIR, AND WATER

Maintained by the U.S. Geological Survey (USGS), this site contains information about the geology of radon, radon potential in the United States, and USGS radon publications, as well as links to other sites with information about radon.

1. Click on *The Geology of Radon* site. This site contains the publication "The Geology of Radon" by James K. Otton, Linda C. S. Gundersen, and R. Randall Schumann, one of a series of general interest publications prepared by the USGS. Click on the various sites listed to learn about what radon is, the geology of radon, and the potential dangers of radon.

2. Click on *Radon Potential of the USA* site. From the map shown, where in the United States is the geologic radon potential greatest? Where is it least? From your study of "The Geology of Radon" in *The Geology of Radon* site, what factors are causing high radon potential in the various areas of the United States?

▶ RADIOCARBON WEB-INFO

Maintained by the radiocarbon labs of Waikato and Oxford Universities, this site contains information about the basis of carbon 14 dating, applications, measurement methods, and other carbon 14 web sites.

1. Click on the *Basis of the Method* site. What is the carbon 14 method? Who developed it? How does it work?

2. Click on the *Measurements methods* site. What are the three principal methods of measuring residual carbon 14 activity?

3. Click on the *Applications* site. What are the various ways carbon 14 can be used to date objects and events? Click on some of the archaeology sites to see how carbon 14 dating is being used in current projects such as the radiocarbon dating of the Dead Sea scrolls.

4. Click on the *Corrections to C14 dates* site. What type of corrections need to be made to carbon 14 dates?

CD-ROM Exploration

Explore the following *In-Terra-Active 2.0* CD-ROM module(s) and increase your understanding of key concepts and processes presented in this chapter.

▶ SECTION: MATERIALS
MODULE: GEOLOGIC TIME: RELATIVE

How old is old? One of geology's greatest legacies is that it answers this question, showing us just how deep-reaching Earth's history really is. In relative time, the immediate goal is not dating an event or object in years, but understanding its place in a geologic sequence. We get our best information from outcrops and other rock records. Study the photo. **Question: Which part of the rock in this picture is oldest? Which is youngest?**

▶ SECTION: MATERIALS
MODULE: GEOLOGIC TIME: ABSOLUTE

In 1903, discovery of radioactivity provided a compelling explanation of where some of Earth's heat comes from. It also gave geologists an extremely useful tool for dating materials, especially minerals. This tool is not, however, foolproof. When you enter the field of radiometric dating, you learn that some specimens yield more reliable dates than others. **Question: What are the preconditions that a mineral must meet for its absolute age dating to be accurate?**

CHAPTER

18

Earth History

Outline

Prologue

Imagine a barren, lifeless, waterless, hot planet with a poisonous atmosphere. Volcanoes erupt nearly continuously, meteorites and comets flash through the atmosphere, and cosmic radiation is intense. The planet's thin, unstable crust is composed entirely of dark-colored igneous rock. Storms form in the turbulent atmosphere; lightning discharges are common, but no rain falls. And because the atmosphere contains no oxygen, nothing burns. Rivers and pools of molten rock emit a continuous reddish glow.

This may sound like a science fiction novel, but it is probably a reasonably accurate description of Earth shortly after it formed (Figure 18.1). We emphasize "probably" because no record exists for the earliest chapter of Earth history, the interval from 4.6 to 3.8 billion years ago, although one area of rocks 3.96 billion years old is now known in Canada. We can only speculate about what Earth was like during this time, based on our knowledge of how planets form and our information about other Earth-like planets.

When Earth formed, it had a tremendous reservoir of primordial heat. This heat was generated by colliding particles as Earth accreted, by gravitational compression, and by the decay of short-lived radioactive elements. Many geologists think that early Earth was so hot that it

Late Proterozoic rocks exposed in Glacier National Park, Montana.

was partly or perhaps almost entirely molten. No one knows what Earth's surface temperature was during its earliest history, but it was almost certainly too hot for liquid water to exist or for any known organism to survive. Volcanism must have been ubiquitous and nearly continuous. Molten rock later solidified to form a thin, discontinuous, dark-colored crust, only to be disrupted by upwelling magmas. Assuming that visitors to early Earth could tolerate the high temperatures, they would also have had to contend with other factors. The atmosphere would be unbreathable by any of today's inhabitants. It probably contained considerable carbon dioxide and water vapor, but little or no oxygen. No ozone layer existed in the upper atmosphere, so our hypothetical visitors, unless protected, would receive a lethal dose of ultraviolet radiation and would be threatened constantly by comet and meteorite impacts. Their view of the Moon would have been spectacular because it was much closer to Earth, but its gravitational attraction would have caused massive Earth tides. And finally, our visitors would experience a much shorter day because Earth rotated on its axis in as little as 10 hours.

Eventually, much of Earth's primordial heat was dissipated into space, and its surface cooled. As it cooled, water vapor began to condense, rain fell, and surface water began to accumulate. The bombardment by comets and meteorites slowed. By 3.8 billion years ago, a few small areas of continental crust existed. The atmosphere still lacked oxygen and an ozone layer, but by as much as 3.5 billion years ago, life appeared.

FIGURE 18.1   Earth as it is thought to have appeared about 4.6 billion years ago.

Introduction

IN this chapter we examine Earth's geologic history, beginning with the origin of continents sometime during the Archean Eon. It was during the Archean and Proterozoic Eons that the earliest continents and ocean basins formed and plate movement began. At the beginning of the Phanerozoic Eon, six major continental landmasses existed, four of which straddled the paleoequator. Plate movements during the Phanerozoic created a changing panorama of continents and ocean basins whose positions affected atmospheric and oceanic circulation patterns and created new environments for habitation by the rapidly evolving biota.

Precambrian Earth History

THE Precambrian encompasses more than 87% of all geologic time! To gain some insight into the duration of the Precambrian, consider the following. If 1 second were to equal 1 year, it would take 127 years of counting to represent the duration of the Precambrian. Nevertheless, our consideration of Precambrian Earth history will be brief. Part of the reason for this disproportionate treatment is the fact that many Precambrian-aged rocks have been complexly deformed and altered by metamorphism, making relative-age determinations difficult, particularly for older Precambrian rocks. In addition, fossils are either rare or absent, so rocks cannot be correlated by age from area to area. In short, deciphering Precambrian Earth history is much more difficult than it is for the more recent Phanerozoic Eon.

Included in the Precambrian are two eons, the *Archean* (3.8 to 2.5 billion years ago) and the *Proterozoic* (2.5 billion to 570 million years ago) (see Figure 17.1). The time preceding the Archean has no formal designation. The beginning of the Archean Eon at 3.8 billion years coincided with the age of the oldest known rocks, but now rocks 3.96 billion years are known in Canada, and rocks in Australia contain detrital minerals dated at 4.2 billion years. Even though rocks of some kind were present at least 4.2 billion years ago, no record is known for the first .4 billion (400 million) years of Earth history. The end of the Archean and beginning of the Proterozoic at 2.5 billion years ago is rather arbitrary but corresponds to a time during which the style of crustal evolution changed.

THE ORIGIN AND EVOLUTION OF CONTINENTS

Rocks 3.8 to 3.96 billion years old thought to represent continental crust are known from several areas, including Minnesota, Greenland, Canada, and South Africa. According to one model for the origin of continents, the earliest crust was thin, unstable, and composed of ultramafic igneous rock. This early crust was disrupted by rising basaltic magmas at ridges and was consumed at subduction zones. Ultramafic crust would have been destroyed because it was dense enough to make recycling by subduction likely. Apparently, only crust of a more granitic composition with its lower density is resistant to destruction by subduction.

A second stage in crustal evolution began when partial melting of earlier-formed basaltic crust resulted in the formation of andesitic island arcs, and partial melting of lower crustal andesites yielded granitic magmas that were emplaced in the crust. As plutons were emplaced in these island arcs, they became more like continental crust. By 3.96 to 3.8 billion years ago, plate movements accompanied by subduction and collisions of island arcs had formed several granitic continental nuclei (Figure 18.2).

SHIELDS AND CRATONS

Each continent processes one or more areas of exposed ancient rocks known as a *shield* (see Figure 8.2). Extending outward from shields are broad *platforms* of ancient rocks buried beneath younger sedimentary rocks and sediments. Shields and platforms collectively make up a continent's **craton**, so a shield is simply the exposed part of a craton. Cratons are composed of a variety of Precambrian rocks and constitute the stable interior parts of continents around which younger continental crust has formed.

In North America, the *Canadian Shield* includes much of Canada; a large part of Greenland; parts of the Lake Superior region in Minnesota, Wisconsin, and Michigan; and parts of the Adirondack Mountains of New York. In general, it is a vast area of subdued topography, numerous lakes, and exposed Precambrian metamorphic, volcanic, plutonic, and sedimentary rocks.

Each continent evolved by accretion along the margins of ancient cratons. To this extent, all continents developed similarly, but the details of each continent's history differ. Several cratons that would eventually become part of the North American craton had formed by 2.5 billion years ago, but they were independent minicontinents that were later assembled into a larger craton. Collisions of these cratons and accretion along their margins eventually gave rise to the present size and shape of North America.

ARCHEAN EARTH HISTORY

Areas underlain by Archean rocks are characterized mostly by rock bodies known as *greenstone belts* and *granite-gneiss complexes* (Figure 18.3). A variety of rocks are found in greenstone belts, but three primary rock units are recognized: The lower and middle units are dominated by volcanic rocks, and the upper unit is sedimentary. The volcanic rocks are typically greenish because of the abundance of the mineral chlorite, which formed during low-grade metamorphism. Most greenstone belts are rather troughlike or synclinal and were intruded by granitic magmas, and many are complexly deformed by folding and thrust faulting.

Some greenstone belts appear to have formed in rifts within continents, but most currently popular models hold that they formed at convergent plate boundaries. In one model they form in back-arc basins that subsequently close, resulting in deformation, metamorphism, and intrusion of magmas. Another model has them forming where island arcs or cratons collide. In either case, they represent accretion at the margin of a craton.

By the latter part of the Archean, several sizable cratons had formed that now constitute the older parts of the Canadian Shield. These cratons were then still independent units or microcontinents, though, that were later amalgamated during the Early Proterozoic to form a larger craton. Unfortunately, the data are so incomplete that the size and configuration of the Archean continent that was to become North America are not known.

Undoubtedly, the present tectonic regime of opening and closing oceans has been a primary agent in Earth evolution since at least the Early Proterozoic. Many geologists are now convinced that some sort of plate tectonics was operating during the Archean as well. However, Earth's radiogenic heat production has diminished through time—thus, during the Archean, when more heat was available,

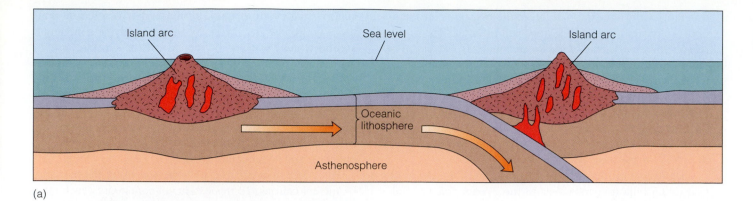

(a)

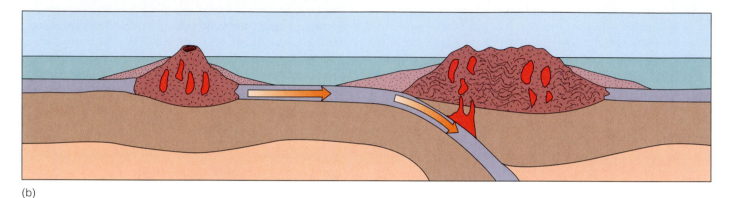

(b)

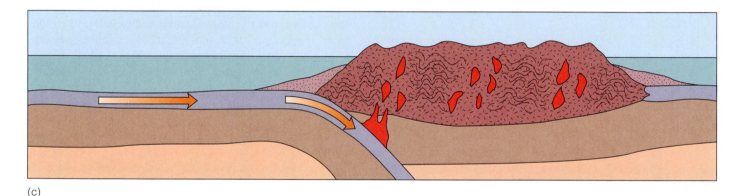

(c)

FIGURE 18.2 Three stages in the origin of granitic continental crust. Andesitic island arcs formed by the partial melting of basaltic oceanic crust are intruded by granitic magmas. As a result of plate movements, island arcs collide and form larger units or cratons. (a) Two island arcs on separate plates move toward one another. (b) The island arcs shown in (a) collide, forming a small craton, and another island arc approaches this craton. (c) The island arc shown in (b) collides with the craton.

seafloor spreading and plate movements occurred faster, and magma was generated more rapidly. In addition, the sedimentary sequences typical of passive continental margins are rare in the Archean but quite common in the Proterozoic. Apparently, continents with adjacent shelves and slopes were either not present or only poorly developed during the Archean. The present-day style of plate tectonics began during the Proterozoic when large, stable cratons were present.

PROTEROZOIC EARTH HISTORY

Although greenstone belts are typical of the Archean, they continued to form during the Proterozoic but at a considerably reduced rate. A notable difference between Archean and Proterozoic rocks is that among the latter are rocks characterized as *quartzite-carbonate-shale assemblages,* which are rare in the Archean (Figure 18.4). These rocks are typical of those that form on passive continental margins, which apparently became widespread during the Proterozoic. The Proterozoic is also characterized by the first known red beds (sandstones and shales with oxidized iron), glacial deposits indicating two times of widespread glaciation, and 92% of Earth's banded iron formations (see Figure 7.18).

The origin of Archean cratons provided the nuclei around which Proterozoic continental crust accreted, thereby forming much larger cratons. One large landmass

Granitic intrusives

Upper sedimentary unit: Sandstones and shales most common

Middle volcanic unit: Mainly basalt

Lower volcanic unit: Mainly peridotite and basalt

Granite-gneiss complex

Greenstone belt succession

(a)

(b)

(c)

FIGURE 18.3 (a) Two adjacent greenstone belts showing their synclinal structure. The two lower units in greenstone belts are mostly volcanic rocks, whereas the upper unit is sedimentary. (b) Pillow lavas in greenstone belts, such as these at Marquette, Michigan, indicate submarine eruptions. (c) Gneiss from a granite-gneiss complex in Canada.

formed by this process, called **Laurentia**, consisted of North America, Greenland, and parts of Scotland and Scandinavia. An important episode of the Proterozoic evolution of Laurentia took place between 2.0 and 1.8 billion years ago. During this time, several Archean cratons were sutured along deformation belts known as *orogens,* giving rise to what is now Greenland, central Canada, and the north-central United States. A good example of this kind of continental growth is provided by the Thelon orogen in Canada, which is the deformed belt where the Slave and Rae cratons collided (Figure 18.5a).

Following the initial Early Proterozoic episode of colliding Archean cratons, considerable accretion took place along Laurentia's southern margin. Between 1.8 and 1.6 billion years ago, growth continued in what is now the southwestern and central United States, as successively younger belts were sutured to Laurentia. Northward-migrating island arcs collided with Laurentia, resulting in deformation, metamorphism, and the emplacement of granitic batholiths in the Central Plains, Yavapai, and

Mazatzal orogens (Figure 18.5b). As a result, a belt of continental crust more than 1000 km wide was accreted to the southern margin of Laurentia.

During part of the Middle Proterozoic, 1.6 to 1.3 billion years ago, Laurentia experienced extensive igneous activity. Magmas were intruded into or erupted onto the surface of already existing crust, so no growth of Laurentia took place during this time. These igneous rocks, consisting of granitic plutons, vast sheets of rhyolite, and ash flows, are now buried beneath Phanerozoic rocks in many areas, but they are exposed in eastern Canada, Greenland, and Scandinavia.

Another episode in the evolution of Laurentia known as the *Grenville orogeny* took place between 1.3 and 1.0 billion years ago (Figure 18.5c). Little agreement exists on the details of the Grenville orogeny, but in any case it represents the final episode of accretion in Laurentia during the Proterozoic. However, contemporaneous with Grenville deformation was an episode of continental rifting that gave rise to the *Midcontinent rift,* a feature with branches

(a)

(b)

FIGURE 18.4 Proterozoic sedimentary rocks of the Great Lakes region. (a) Outcrop of the Sturgeon Quartzite. (b) The Kona Dolomite showing structures known as stromatolites, which are produced by the activities of photosynthesizing bacteria.

extending southwest and southeast from the Lake Superior region. Basalt lava flows and sedimentary rocks of the rift are exposed in Michigan, Wisconsin, and Minnesota but are deeply buried elsewhere.

By about 1 billion years ago, the size and shape of North America were approximately as shown in Figure 18.5c. No further episode of continental accretion took place until the Paleozoic Era.

Precambrian Mineral Resources

ARCHEAN and Proterozoic rocks near Johannesburg, South Africa, have yielded more than 50% of the world's gold since 1886. Gold is also recovered from Precambrian

rocks in Ontario, Canada, and the second largest gold deposit in the United States is in Archean rocks at Lead, South Dakota. Archean rocks in western Australia, Zimbabwe, and Ontario, Canada, contain sulfides of zinc, copper, and nickel. Many of these deposits probably formed near hydrothermal vents on the seafloor just as they do now adjacent to black smokers at or near spreading ridges (see Figure 16.24). About one-fourth of the world's chrome reserves are in Archean rocks, and most of the platinum mined today comes from Proterozoic rocks of South Africa.

Probably the most notable mineral deposits of Proterozoic age are banded iron formations (see Figure 7.18). The richest of these iron ores—those containing up to 70% iron—in the Lake Superior region of Canada and the United States were largely depleted by the time of World War II (1939–1945). Continued mining has been possible because a method was developed to separate the iron ore from unusable rock of lower-grade ores and then shape the iron into pellets containing 65% iron.

Nickel is essential in the production of nickel alloys such as stainless steel and Monel metal (nickel plus copper), which are valued for their strength and resistance to corrosion and heat. The United States must import more than 50% of all nickel used, most of it from the Sudbury mining district of Ontario, Canada, which is also an important area of platinum mining. Economically recoverable oil and gas have been discovered in Proterozoic rocks in China and Siberia, arousing some interest in the midcontinent rift as a potential source of hydrocarbons. And although some rocks within the rift are known to contain petroleum, no producing oil or gas wells are currently operating.

Precambrian pegmatites (see Figure 4.15) are mined for a number of resources. Pegmatites in Maine and South Dakota have yielded magnificent gem-quality specimens of several minerals. These and other pegmatites are also mined for industrial minerals such as quartz, feldspars, micas, and minerals containing such elements as cesium, rubidium, lithium, and beryllium.

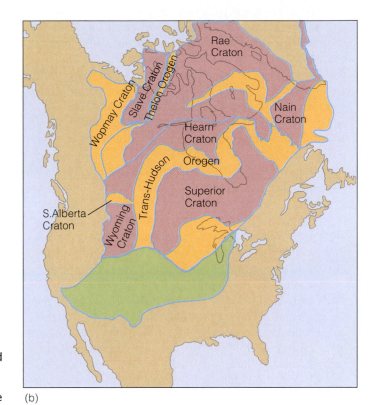

(b)

FIGURE **18.5**   Three stages in the early evolution of the North American craton. (a) By about 1.8 billion years ago, North America consisted of the elements shown here. The various orogens formed when older cratons collided to form a larger craton. (b) and (c) Continental accretion along the southern and eastern margins of North America. By about 1 billion years ago, North America had the size and shape shown diagrammatically in (c).

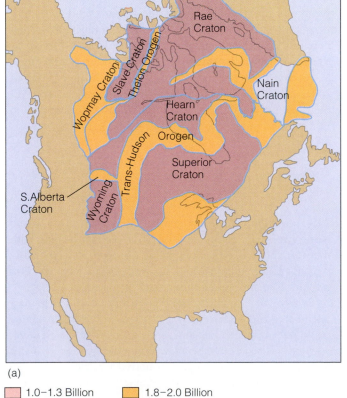

(a)

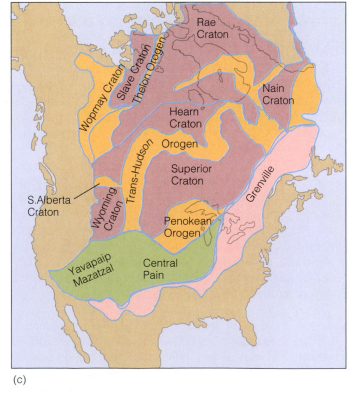

(c)

- 🟪 1.0–1.3 Billion
- 🟩 1.6–1.8 Billion
- 🟧 1.8–2.0 Billion
- 🟫 >2.5 Billion

Paleozoic Paleogeography

By the beginning of the Paleozoic Era, six major continents were present. Besides these large landmasses, geologists have also identified numerous small microcontinents and island arcs associated with various microplates that were present during the Paleozoic. The six major Paleozoic continents are *Baltica* (Russia west of the Ural Mountains and the major part of northern Europe), *China* (a complex area consisting of at least three Paleozoic continents that were not widely separated and are here considered to include China, Indochina, and the Malay Peninsula), *Gondwana* (Africa, Antarctica, Australia, Florida, India, Madagascar, and parts of the Middle East and southern Europe), *Kazakhstania* (a triangular continent centered on Kazakhstan, but considered by some to be an extension of the Paleozoic Siberian continent), *Laurentia* (most of present North America, Greenland, northwestern Ireland, Scotland, and part of eastern Russia), and *Siberia* (Russia east of the Ural Mountains and Asia north of Kazakhstan and south of Mongolia).

In contrast to today's global geography, the Cambrian world consisted of these six continents dispersed around the globe at low tropical latitudes (Figure 18.6a). Water circulated freely among ocean basins, and the polar regions were apparently ice-free. By the Late Cambrian, shallow seas had covered large areas of Laurentia, Baltica, Siberia, Kazakhstania, and China, while major highlands were present in northeastern Gondwana, eastern Siberia, and central Kazakhstania.

During the Ordovician and Silurian periods, plate movement played a major role in the changing global geography. Gondwana moved southward during the Ordovician and began to cross the South Pole as indicated by Upper Ordovician glacial deposits found today in the Sahara (Figure 18.6b). In contrast to the passive continental margin Laurentia exhibited during the Cambrian, an active convergent plate boundary formed along its eastern margin during the Ordovician as indicated by the Late Ordovician *Taconic orogeny* that occurred in New England. During the Silurian, Baltica moved northwestward relative to Laurentia and collided with it to form the larger continent of *Laurasia*. This collision, which closed the northern *Iapetus Ocean,* is marked by the *Caledonian orogeny.* Following this orogeny, the southern part of the Iapetus Ocean still remained open between Laurentia and Gondwana. Siberia and Kazakhstania moved from a southern equatorial position during the Cambrian to north temperate latitudes by the end of the Silurian Period.

During the Devonian, as the southern Iapetus Ocean narrowed between Laurasia and Gondwana, mountain building continued along the eastern margin of Laurasia with the *Acadian orogeny* (Figure 18.7a). The erosion of the resulting highlands provided vast amounts of reddish fluvial sediments that covered large areas of northern Europe and eastern North America. Other Devonian tectonic events, probably related to the collision of Laurentia and Baltica, include the Cordilleran *Antler orogeny* and the change from a passive continental margin to an active convergent plate boundary in the Uralian mobile belt of eastern Baltica. The distribution of reefs, evaporites, and red beds, as well as the existence of similar floras throughout the world, suggests a rather uniform global climate during the Devonian Period.

During the Carboniferous Period, southern Gondwana moved over the South Pole, resulting in extensive continental glaciation. The advance and retreat of these glaciers produced global changes in sea level that affected sedimentation patterns on the cratons. As Gondwana continued moving northward, it first collided with Laurasia during the Early Carboniferous and continued suturing with it throughout the rest of the Carboniferous. The final phase of collision between Gondwana and Laurasia is indicated by the Ouachita Mountains of Oklahoma, which were formed by thrusting during the Late Carboniferous and Early Permian. Elsewhere, Siberia collided with Kazakhstania and moved toward the Uralian margin of Laurasia (Baltica), colliding with it during the Early Permian.

The assemblage of Pangaea was completed during the Permian with an enormous single ocean, *Panthalassa,* surrounding the supercontinent (Figure 18.7b). Waters of this ocean probably circulated more freely than at present, resulting in more equable water temperatures.

The formation of a single landmass had climatic consequences for the terrestrial environment as well. Terrestrial Permian sediments indicate that arid and semiarid conditions were widespread over Pangaea. The mountain ranges produced by the *Hercynian, Alleghenian,* and *Ouachita orogenies* were high enough to create rainshadows that blocked the moist, subtropical, easterly winds—much as the southern Andes Mountains do in western South America today. This produced very dry conditions in North America and Europe, as is evident from the extensive Permian evaporites found in western North America, central Europe, and parts of Russia.

Paleozoic Evolution of North America

It is convenient to divide the Paleozoic history of the North American craton into two parts, the first dealing with the relatively stable continental interior over which shallow seas advanced and retreated, and the second with the mobile belts where mountain building occurred.

In 1963 the American geologist Laurence L. Sloss proposed that the sedimentary-rock record of North America could be subdivided into six cratonic sequences. A *cratonic sequence* is a large-scale rock unit representing a major transgressive-regressive cycle bounded by cratonwide unconformities (Figure 18.8). The transgressive phase, which is usually covered by younger sediments, commonly is well preserved, while the regressive phase of each sequence is

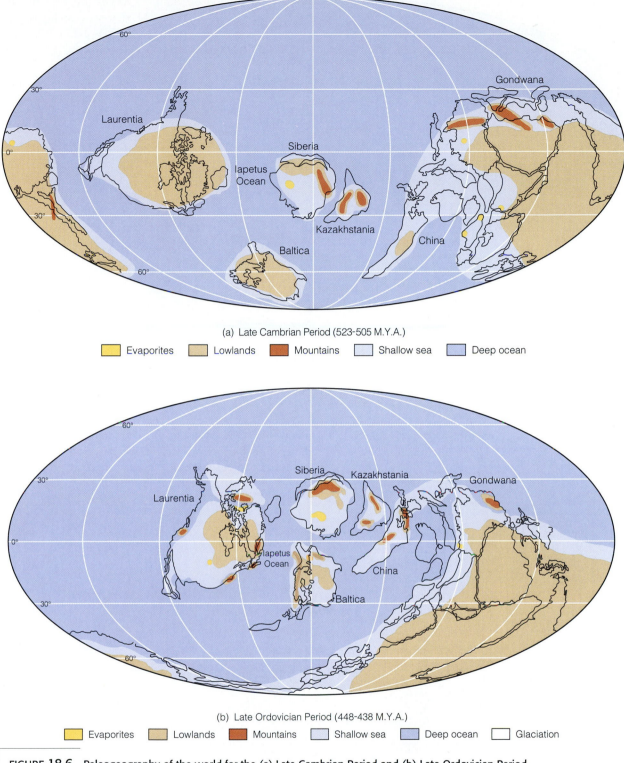

(a) Late Cambrian Period (523-505 M.Y.A.)

| | Evaporites | | Lowlands | | Mountains | | Shallow sea | | Deep ocean |

(b) Late Ordovician Period (448-438 M.Y.A.)

| | Evaporites | | Lowlands | | Mountains | | Shallow sea | | Deep ocean | | Glaciation |

FIGURE 18.6   Paleogeography of the world for the (a) Late Cambrian Period and (b) Late Ordovician Period.

marked by an unconformity. Where rocks of the appropriate age are preserved, each of the six unconformities can be shown to extend across the various sedimentary basins of the North American craton and into the mobile belts along the cratonic margin.

Geologists have also recognized major unconformity-bound sequences in cratonic areas outside North America. Such global transgressive and regressive cycles are caused by changes in sea level and are thought to result from major tectonic and glacial events.

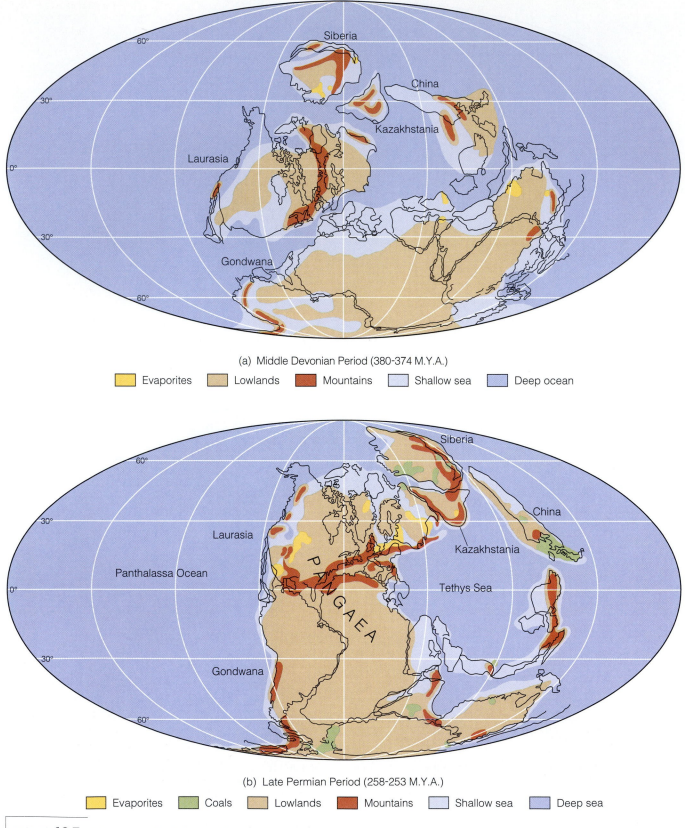

(a) Middle Devonian Period (380-374 M.Y.A.)

Evaporites   Lowlands   Mountains   Shallow sea   Deep ocean

(b) Late Permian Period (258-253 M.Y.A.)

Evaporites   Coals   Lowlands   Mountains   Shallow sea   Deep sea

FIGURE **18.7** Paleogeography of the world for the (a) Middle Devonian Period and (b) Late Permian Period.

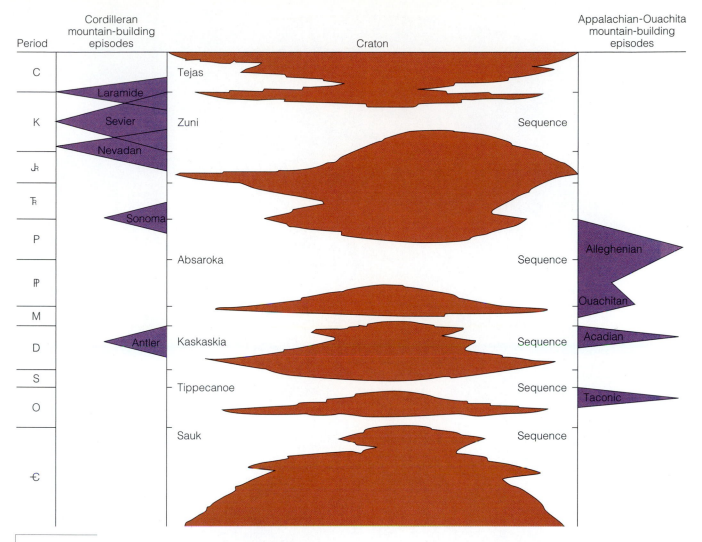

FIGURE 18.8 Cratonic sequences of North America. The white areas represent sequences of rocks that are separated by large-scale unconformities shown as brown areas. The major Cordilleran orogenies are shown on the left side of the figure, and the major Appalachian orogenies are shown on the right side.

THE SAUK SEQUENCE

Rocks of the **Sauk sequence** (Late Proterozoic–Early Ordovician) record the first major transgression onto the North American craton. During the Late Proterozoic and Early Cambrian, deposition of marine sediments was limited to the passive shelf areas of the Appalachian and Cordilleran borders of the craton. The craton itself was above sea level and experiencing weathering and erosion. Because North America was located in a tropical climate at this time (Figure 18.6a) and there is no evidence of any terrestrial vegetation, weathering and erosion of the exposed basement Precambrian rocks must have proceeded rapidly. During the Middle Cambrian, the transgressive phase of the Sauk began with shallow seas encroaching over the craton. By the Late Cambrian, the Sauk Sea had covered most of North America, leaving only a portion of the Canadian Shield and a few large islands above

sea level (Figure 18.9). These islands, collectively named the *Transcontinental Arch,* extended from New Mexico to Minnesota and the Lake Superior region.

THE TIPPECANOE SEQUENCE

As the Sauk Sea regressed from the craton during the Early Ordovician, it revealed a landscape of low relief. The rocks exposed were predominantly limestones and dolostones that were deeply eroded because North America was still located in a tropical environment (Figure 18.6b). The resulting cratonwide unconformity marks the boundary between the Sauk and Tippecanoe sequences.

Like the Sauk sequence, deposition of the **Tippecanoe sequence** (Middle Ordovician–Early Devonian) began with a major transgression onto the craton. This transgressing sea deposited clean quartz sands over most of the craton. The Tippecanoe basal sandstones were followed by widespread carbonate deposition. The limestones were generally the result of deposition by calcium carbonate–secreting marine organisms such as corals and brachiopods. In addition to the limestones, there were also many dolostones.

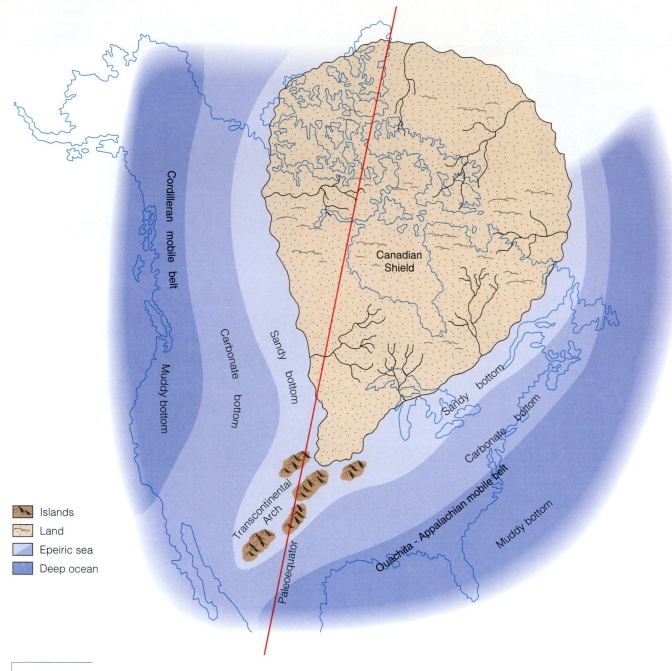

FIGURE 18.9 Paleogeography of North America during the Cambrian Period. Note the position of the Cambrian paleoequator. During this time North America straddled the equator, as indicated in Figure 18.6a.

Legend:
- Islands
- Land
- Epeiric sea
- Deep ocean

Map labels: Cordilleran mobile belt, Carbonate bottom, Sandy bottom, Muddy bottom, Canadian Shield, Sandy bottom, Carbonate bottom, Muddy bottom, Ouachita - Appalachian mobile belt, Transcontinental Arch, Paleoequator

As the Tippecanoe Sea gradually regressed from the craton during the Late Silurian, precipitation of evaporite minerals took place in the Appalachian, Ohio, and Michigan basins (Figure 18.10). In the Michigan Basin alone, approximately 1500 m of sediments were deposited, nearly half of which are halite and anhydrite.

By the Early Devonian, the regressing Tippecanoe Sea had retreated to the craton margin exposing an extensive lowland topography. During this regression, marine deposition was initially restricted to a few interconnected cratonic basins and finally, by the end of the Tippecanoe, to only the margins surrounding the craton.

During the Early Devonian as the Tippecanoe Sea was regressing, the craton experienced mild deformation resulting in the formation of many domes, arches, and basins. These structures were mostly eroded during the time the craton was exposed so that they were eventually covered by deposits from the encroaching Kaskaskia Sea.

THE KASKASKIA SEQUENCE

The boundary between the Tippecanoe sequence and the overlying **Kaskaskia sequence** (Middle Devonian–Middle Mississippian) is marked by a major unconformity. As the

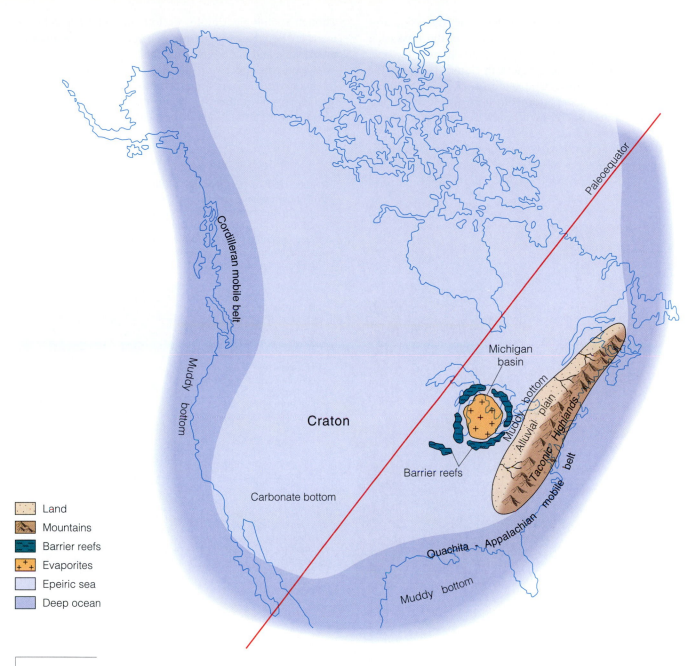

FIGURE 18.10 Paleogeography of North America during the Silurian Period. Note the development of reefs in the Michigan, Ohio, and Indiana-Illinois-Kentucky areas.

Legend:
- Land
- Mountains
- Barrier reefs
- Evaporites
- Epeiric sea
- Deep ocean

Kaskaskia Sea transgressed over the low-relief landscape of the craton, most of the basal beds consisted of clean, well-sorted, quartz sandstones.

Except for widespread Late Devonian and Early Mississippian black shales, the majority of Kaskaskian rocks are carbonates, including reefs, and associated evaporite deposits. In many other parts of the world, such as southern England, Belgium, central Europe, Australia, and Russia, the Middle and early Late Devonian epochs were times of major reef building.

During the Late Mississippian regression of the Kaskaskia Sea from the craton, carbonate deposition was replaced by vast quantities of detrital sediments. The resulting sandstones, particularly in the Illinois Basin, have been studied in great detail because they are excellent petroleum reservoirs. Prior to the end of the Mississippian, the Kaskaskia Sea had retreated to the craton margin, once again exposing the craton to widespread weathering and erosion that resulted in a cratonwide unconformity.

THE ABSAROKA SEQUENCE

The extensive unconformity separating the Kaskaskia and Absaroka sequences essentially divides the rocks into the North American Mississippian and Pennsylvanian systems.

The rocks of the **Absaroka sequence** (Late Mississippian–Early Jurassic) not only differ from those of the Kaskaskia sequence, but they also result from quite different tectonic regimes affecting the North American craton.

One of the characteristic features of Pennsylvanian rocks is their cyclical pattern of alternating marine and nonmarine strata. Such rhythmically repetitive sedimentary sequences are known as **cyclothems** (Figure 18.11). They result from repeated alternations of marine and nonmarine environments, usually in areas of low relief.

Though seemingly simple, cyclothems reflect a delicate interplay between nonmarine deltaic and shallow-marine interdeltaic and shelf environments.

Cyclothems represent transgressive and regressive sequences with an erosional surface separating one cyclothem from another. Thus, idealized cyclothem passes upward from fluvial-deltaic deposits, through coals, to

FIGURE **18.11** (a) Columnar section of a complete cyclothem. (b) Pennsylvanian coal bed, West Virginia. *(Photo courtesy of Wayne E. Moore.)* (c) Reconstruction of the environment of a Pennsylvanian coal-forming swamp. (d) The Okefenokee Swamp, Georgia, is a modern example of a coal-forming environment, similar to those occurring during the Pennsylvanian Period. *(Photo © Patricia Caulfied/Photo Researchers, Inc.)*

Nonmarine environment | Marine environment

Erosion | Terrestrial sedimentation | Coal-forming swamp | Nearshore | Offshore

Progradation

Transgression

(a)

Disconformity
Brackish and
    nonmarine shales
Marine shales
Algal limestones with
    nearshore and brackish
    water invertebrate fossils
Limestones with offshore
    invertebrate fossils
Limestones and shale with
    offshore
    invertebrate fossils

Marine shales with nearshore
    invertebrate fossils

Coal
Underclay
Nonmarine shales and
    sandstones
Nonmarine sandstones
    Disconformity

(b)

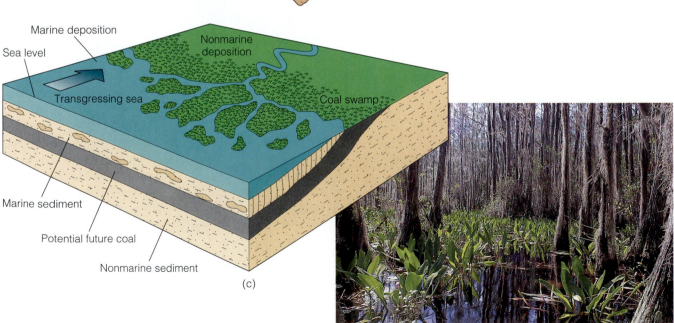

Marine deposition

Sea level

Transgressing sea

Nonmarine deposition

Coal swamp

Marine sediment

Potential future coal

Nonmarine sediment

(c)

(d)

detrital shallow-water marine sediments, and finally to limestones typical of an open marine environment (Figure 18.11a).

Such regularity and cyclicity in sedimentation over a large area requires an explanation. The hypothesis currently favored by most geologists is a rise and fall of sea level related to advances and retreats of Gondwanan continental glaciers. When the Gondwanan ice sheets advanced, sea level dropped, and when they melted, sea level rose. Late Paleozoic cyclothem activity on all of the cratons closely corresponds to Gondwanan glacial-interglacial cycles.

During the Pennsylvanian, the area of greatest deformation occurred in the southwestern part of the North American craton where a series of fault-bounded uplifted blocks formed the *Ancestral Rockies* (Figure 18.12). These mountain ranges had diverse geologic histories and were not all elevated at the same time. Uplift of these mountains, some of which were elevated more than 2 km along near-vertical faults, resulted in the erosion of the overlying Paleozoic sediments and exposure of the Precambrian igneous and metamorphic basement rocks. As the mountains eroded, tremendous quantities of coarse, red-colored sediment were deposited in the surrounding basins.

FIGURE **18.12**   Paleogeography of North America during the Pennsylvanian Period.

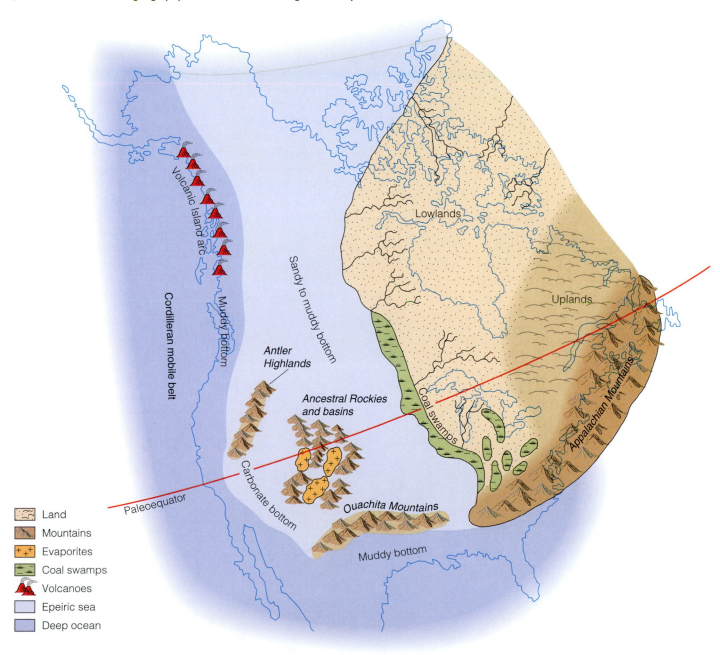

Legend:
- Land
- Mountains
- Evaporites
- Coal swamps
- Volcanoes
- Epeiric sea
- Deep ocean

Currently, these sediments are preserved in many areas, including the Garden of the Gods near Colorado Springs (Figure 18.13).

While the various intracratonic basins were filling with sediment during the Late Pennsylvanian, the Absaroka Sea slowly began retreating from the craton. During the Middle and Late Permian, the Absaroka Sea was restricted to west Texas and southern New Mexico, forming an interrelated complex of lagoonal, reef, and open-shelf environments. Massive reefs grew around the basin margins (Figure 18.14a), while limestones, evaporites, and red beds were deposited in the lagoonal areas behind the reefs.

Spectacular deposits representing the geologic history of this region can be seen today in the Guadalupe Mountains of Texas and New Mexico where the Capitan Limestone forms the caprock of these mountains (Figure 18.14b). These reefs have been extensively studied because of the tremendous oil production that comes from this region.

By the end of the Permian Period, the Absaroka Sea had retreated from the craton, and continental red beds were deposited over most of the southwestern and eastern region.

History of the Paleozoic Mobile Belts

HAVING examined the Paleozoic history of the craton, we now turn to the orogenic activity in the **mobile belts** (elongated areas of mountain-building activity occurring along the margins of continents). The mountain building that occurred during this time had a profound influence on the climate and sedimentary history of the craton. In addition, it was part of the global tectonic regime that sutured the continents together, forming Pangaea by the end of the Paleozoic Era.

APPALACHIAN MOBILE BELT

Throughout Sauk time (Late Proterozoic–Early Ordovician), the Appalachian region was a broad, passive continental margin. Sedimentation was closely balanced by subsidence as thick, shallow marine sands were succeeded by extensive carbonate deposits. During this time, the Iapetus Ocean was widening as a result of movement along a divergent plate boundary (Figure 18.15a).

Beginning with the subduction of the Iapetus plate beneath Laurentia (an oceanic-continental convergent plate boundary), the Appalachian mobile belt was born (Figure 18.15b). The resulting **Taconic orogeny**, named after the present-day Taconic Mountains of eastern New York, central Massachusetts, and Vermont, was the first of several orogenies to affect the Appalachian region.

A large *clastic wedge* (an extensive accumulation of mostly detrital sediments deposited adjacent to an uplifted area) formed in the shallow seas to the west of the Taconic orogeny. These deposits are thickest and coarsest nearest the highland area and become thinner and finer grained away from the source area, eventually grading into carbonates on the craton. The clastic wedge resulting from the erosion of the Taconic Highlands is referred to as the *Queenston Delta*.

The second Paleozoic orogeny to affect Laurentia began during the Late Silurian and concluded at the end of the Devonian Period. The **Acadian orogeny** affected the Appalachian mobile belt from Newfoundland to Pennsylvania as sedimentary rocks were folded and thrust against the craton.

As with the preceding Taconic orogeny, the Acadian orogeny occurred along an oceanic-continental convergent plate boundary. As the northern Iapetus Ocean continued to close during the Devonian, the plate carrying Baltica finally collided with Laurentia, forming a continental-continental convergent plate boundary along the zone of

FIGURE **18.13** Garden of the Gods, view from near Hidden Inn, Colorado Springs, Colorado.

(a)

(b)

FIGURE **18.14** (a) A reconstruction of the Middle Permian Capitan Limestone reef environment. Shown are brachiopods, corals, bryozoans, and large glass sponges. (b) The prominent light-colored Capitan Limestone forms the caprock of the Guadalupe Mountains. The Capitan Limestone is rich in fossil corals and associated reef organisms. *(Photo courtesy of Bill Cornell, University of Texas at El Paso.)*

collision (Figure 18.15c). Weathering and erosion of the Acadian Highlands produced the *Catskill Delta,* a thick clastic wedge named for the Catskill Mountains in northern New York where it is well exposed.

Geologists think that the Taconic and Acadian orogenies were part of the same major orogenic event related to the closing of the Iapetus Ocean (Figure 18.15). This event began with an oceanic-continental convergent plate boundary during the Taconic orogeny and culminated with a continental-continental convergent plate boundary during the Acadian orogeny as Laurentia and Baltica became sutured. Following this, the Hercynian-Alleghenian orogeny began, followed by orogenic activity in the Ouachita mobile belt.

The Hercynian mobile belt of southern Europe and the Appalachian and Ouachita mobile belts of North America mark the zone along which Europe (part of Laurasia) collided with Gondwana. While Gondwana and southern Laurasia collided during the Pennsylvanian and Permian, eastern Laurasia (Europe and southeastern North America) joined together with Gondwana (Africa) as part of the **Hercynian-Alleghenian orogeny** (Figure 18.7b).

CORDILLERAN MOBILE BELT

During the Late Proterozoic and Early Paleozoic, the Cordilleran area was a passive continental margin along which extensive continental shelf sediments were deposited. Beginning in the Middle Paleozoic, an island arc

formed on the western margin of the craton. This eastward-moving island arc collided with the western border of the craton during the Late Devonian and Early Mississippian, resulting in a highland area. This orogenic event, the *Antler orogeny,* was caused by subduction. Erosion of the Antler Highlands produced large quantities of sediment that were deposited to the east in the shallow sea covering the craton and to the west in the deep sea (Figure 18.12).

OUACHITA MOBILE BELT

The Ouachita mobile belt extends for approximately 2100 km from the subsurface of Mississippi to the Marathon region of Texas. Approximately 80% of the former mobile belt is buried beneath a Mesozoic and Cenozoic sedimentary cover. The two major exposed areas in this region are the Ouachita Mountains of Oklahoma and Arkansas and the Marathon Mountains of Texas.

During the Late Proterozoic to Early Mississippian, shallow-water detrital and carbonate sediments were slowly deposited on a broad continental shelf, while in the deeper-water portion of the adjoining mobile belt, bedded cherts and shales were also slowly accumulating. Beginning in the Mississippian Period, the rate of sedimentation increased dramatically as the region changed from a passive continental margin to an active convergent plate boundary. Rapid deposition of sediments continued into the Pennsylvanian with the formation of a clastic wedge that thickened

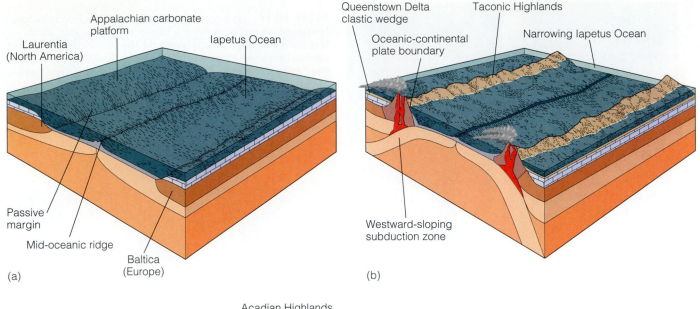

Laurentia
(North America)

Appalachian carbonate platform

Iapetus Ocean

Passive margin

Mid-oceanic ridge

Baltica
(Europe)

(a)

Queenstown Delta clastic wedge

Taconic Highlands

Oceanic-continental plate boundary

Narrowing Iapetus Ocean

Westward-sloping subduction zone

(b)

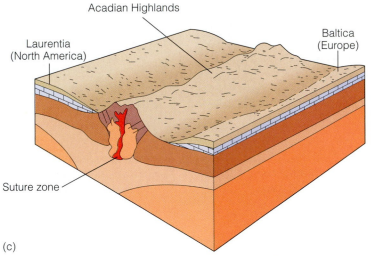

Acadian Highlands

Laurentia
(North America)

Baltica
(Europe)

Suture zone

(c)

to the south. The formation of a clastic wedge marks the beginning of uplift of the area and the formation of a mountain range during the **Ouachita orogeny**. Thrusting of sediments continued throughout the Pennsylvanian and Early Permian as a result of the compressive forces generated along the zone of subduction as Gondwana collided with Laurasia (Figure 18.7b). The collision of Gondwana and Laurasia is marked by the formation of a large mountain range, most of which was eroded during the Mesozoic Era.

Paleozoic Microplates and the Formation of Pangaea

WE have generally presented the geologic history of the mobile belts surrounding the Paleozoic continents in terms of subduction along convergent plate boundaries. It is becoming increasingly clear, however, that accretion along the continental margins is more complicated than the somewhat simple, large-scale plate interactions that we have described. Geologists now recognize that numerous

FIGURE **18.15** Evolution of the Appalachian mobile belt. **(a)** During the Late Proterozoic to the Early Ordovician, the Iapetus Ocean was opening up along a divergent boundary. Both the east coast of Laurentia and the west coast of Baltica were passive continental margins where large carbonate platforms existed. **(b)** Beginning in the Middle Ordovician, the passive margins of Laurentia and Baltica became oceanic-continental plate boundaries, resulting in orogenic activity. **(c)** By the Late Paleozoic, Laurentia and Baltica collided.

terranes or microplates existed during the Paleozoic and were involved in the orogenic events that occurred during that time.

A careful examination of the Paleozoic global paleogeographic maps (Figures 18.6 and 18.7) shows numerous microplates, and their location and role during the formation of Pangaea must be taken into account. Thus, while the basic history of the formation of Pangaea during the Paleozoic remains the same, geologists now realize that microplates also played an important role. Furthermore, the recognition of terranes within mobile belts helps explain some previously anomalous geologic situations.

Paleozoic Mineral Resources

PALEOZOIC-AGED rocks contain a variety of important mineral resources, including energy resources and metallic and nonmetallic mineral deposits. Silica sand, such as from the Upper Cambrian Jordan Sandstone of Minnesota and Wisconsin or the Middle Ordovician St. Peter Sandstone, has a variety of uses, including the manufacture of glass, refractory bricks for blast furnaces, and molds for casting iron, aluminum, and copper alloys.

Thick deposits of Silurian evaporites, mostly rock salt (NaCl) and rock gypsum ($CaSO_4 \cdot H_2O$) altered to rock anhydrite ($CaSO_4$), underlie parts of Michigan, Ohio, New York, and adjacent areas in Ontario, Canada. These rocks are important sources of various salts. In addition, reefs in carbonate rocks associated with these evaporites are reservoirs for oil and gas in Michigan and Ohio.

Metallic mineral resources including tin, copper, gold, and silver are known from Late Paleozoic–aged rocks, especially those that have been deformed during mountain building. The host rocks for deposits of lead and zinc in southeast Missouri are Cambrian dolostones, although some Ordovician rocks contain these metals as well. The Silurian Clinton Formation crops out from Alabama north to New York and has been mined for iron in many places.

Petroleum and natural gas are recovered in commercial quantities from rocks ranging from the Devonian through Permian, while much of the coal in North America and Europe is Pennsylvanian (Late Carboniferous) in age. Large areas in the Appalachian region and the midwestern United States are underlain by vast coal deposits.

The Breakup of Pangaea

JUST as the formation of Pangaea influenced geologic and biologic events during the Paleozoic, the breakup of this supercontinent profoundly affected geologic and biologic events during the Mesozoic. The movement of continents affected global climatic and oceanic regimes as well as the climates of individual continents.

Geologic, paleontologic, and paleomagnetic data indicate that the breakup of Pangaea occurred in four general stages. The first stage involved rifting between Laurasia and Gondwana during the Late Triassic. By the end of the Triassic, the expanding Atlantic Ocean separated North America from Africa (Figure 18.16a). This was followed by the rifting of North America from South America sometime during the Late Triassic and Early Jurassic.

The second stage in Pangaea's breakup involved rifting and movement of the various Gondwana continents during the Late Triassic and Jurassic periods. As early as the Late Triassic, Antarctica and Australia, which remained sutured together, separated from South America and Africa, while India split away from all four Gondwana continents and began moving northward (Figures 18.16a and b).

The third stage of breakup began during the Late Jurassic, when South America and Africa began separating (Figure 18.16b). During this stage, the eastern end of the Tethys Sea began closing as a result of the clockwise rotation of Laurasia and the northward movement of Africa. This narrow Late Jurassic and Cretaceous seaway between Africa and Europe was the forerunner of the present Mediterranean Sea.

By the end of the Cretaceous, Australia and Antarctica had separated, India had nearly reached the equator, South America and Africa were widely separated, and the eastern side of what is now Greenland had begun separating from Europe (Figure 18.16c).

The final stage in Pangaea's breakup occurred during the Cenozoic. During this stage, Australia continued moving northward, and Greenland completely separated from Europe and rifted from North America to form a separate landmass.

The Mesozoic History of North America

THE beginning of the Mesozoic Era was essentially the same in terms of mountain building and sedimentation as the preceding Permian Period in North America. Terrestrial sedimentation continued over much of the craton, while block-faulting and igneous activity began in the Appalachian region as North America and Africa began separating (Figure 18.17). The newly forming Gulf of Mexico was the site of extensive evaporite deposition during the Late Triassic and Jurassic as North America separated from South America (Figure 18.18).

A global rise in sea level during the Cretaceous resulted in worldwide transgressions onto the continents (Figure 18.19). These transgressions were caused by higher heat flow along the oceanic ridges due to increased rifting and the consequent expansion of oceanic crust. By the Middle Cretaceous, sea level probably was as high as at any time since the Ordovician, and approximately one-third of the present land area was covered by shallow seas.

Marine deposition was continuous over much of western North America. A volcanic island arc system that formed off the western edge of the craton during the Permian was sutured to North America sometime during the Permian or Triassic. During the Jurassic, the entire Cordilleran area was involved in a series of major mountain-building episodes that resulted in the formation of the Sierra Nevada, the Rocky Mountains, and other lesser mountain ranges.

EASTERN COASTAL REGION

During the Early and Middle Triassic, coarse detrital sediments derived from the erosion of the recently uplifted Appalachians (Alleghenian orogeny) filled the various intermontane basins and spread over the surrounding areas. As erosion continued during the Mesozoic, this once lofty mountain system was reduced to a low-lying plain. During the Late Triassic, the first stage in the breakup of Pangaea began as North America separated from Africa

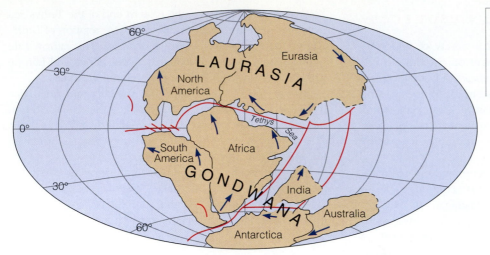

(a) Triassic Period (245–208 M.Y.A)

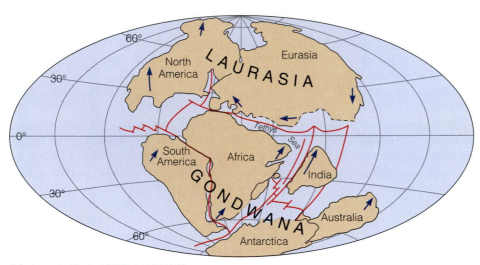

(b) Jurassic Period (208–144 M.Y.A)

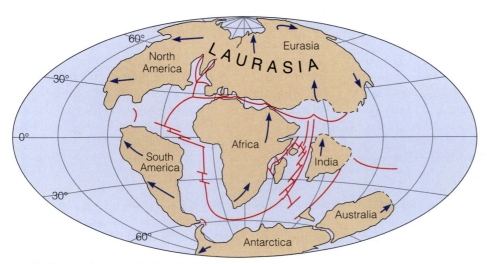

(c) Cretaceous Period (144–66 M.Y.A)

FIGURE 18.16 Paleogeography of the world during the Mesozoic. Blue arrows show the direction of movement for the continents. (a) The Triassic Period, (b) the Jurassic Period, and (c) the Cretaceous Period.

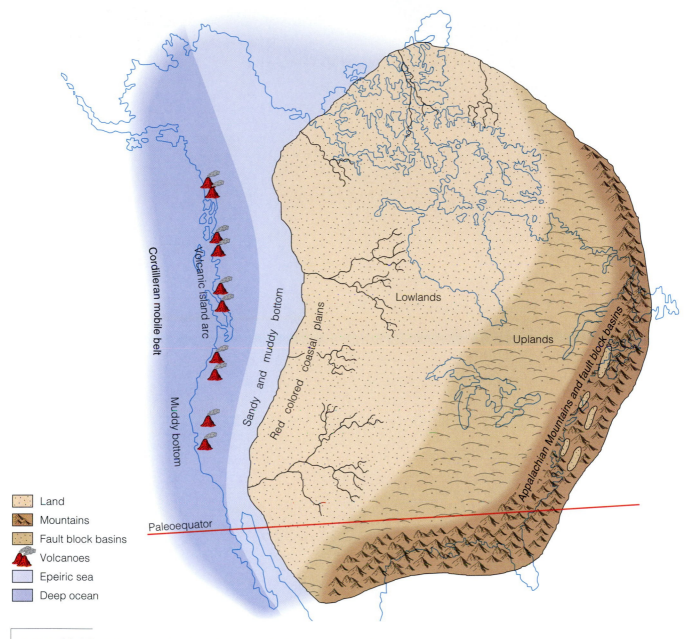

Legend:
- Land
- Mountains
- Fault block basins
- Volcanoes
- Epeiric sea
- Deep ocean

Map labels: Cordilleran mobile belt · Volcanic island arc · Muddy bottom · Sandy and muddy bottom · Red colored coastal plains · Lowlands · Uplands · Appalachian Mountains and fault block basins · Paleoequator

FIGURE 18.17 Paleogeography of North America during the Triassic Period.

and the Atlantic Ocean began to form (Figure 18.16a). Fault-block basins developed in response to upwelling magma beneath Pangaea in a zone stretching from present-day Nova Scotia to North Carolina (Figure 18.17). Erosion of the adjacent fault-block mountains filled these basins with great quantities (up to 6000 m) of poorly sorted, red, nonmarine detrital sediments.

As the Atlantic Ocean grew, rifting ceased along the eastern margin of North America, and this once active plate margin became a passive, trailing continental margin. The fault-block mountains that were produced by this rifting continued to erode during the Jurassic and Early Cretaceous until only a broad, low-lying erosional surface remained. The sediments resulting from erosion contributed to the growing eastern continental shelf.

During the Cretaceous Period, the Appalachian region was reelevated and once again shed sediments onto the continental shelf.

GULF COASTAL REGION

The Gulf Coastal region was above sea level until the Late Triassic. As North America separated from South America during the Late Triassic, the Gulf of Mexico began to form (Figure 18.18). With oceanic waters flowing into this newly formed, shallow, restricted basin, conditions were ideal for evaporite formation. More than 1000 m of evaporites were precipitated at this time, and most geologists think that these Jurassic evaporites are the source for the Tertiary salt domes found today in the Gulf of Mexico and southern Louisiana.

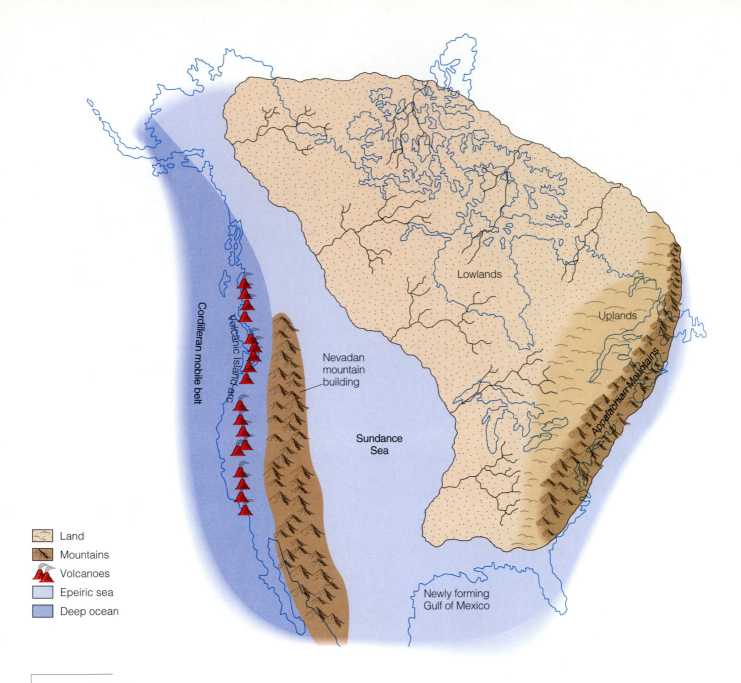

Land

Mountains

Volcanoes

Epeiric sea

Deep ocean

Cordilleran mobile belt

Volcanic island arc

Lowlands

Uplands

Nevadan
mountain
building

Appalachian Mountains

Sundance
Sea

Newly forming
Gulf of Mexico

FIGURE 18.18   Paleogeography of North America during the Jurassic Period.

By the Late Jurassic, circulation in the Gulf of Mexico was less restricted, and evaporite deposition ended. Normal marine conditions returned to the area with alternating transgressing and regressing seas, resulting in the deposition of sandstones, shales, and limestones. These sedimentary rocks were later covered and deeply buried by great thicknesses of Cretaceous and Cenozoic sediments.

During the Cretaceous, the Gulf Coastal region, like the rest of the continental margin, was inundated by northward-transgressing seas as a wide seaway extended from the Arctic Ocean to the Gulf of Mexico (Figure 18.19).

WESTERN REGION

During the Permian, an island arc and ocean basin formed off the western North American craton (Figure 18.17), followed by subduction of an oceanic plate beneath the island arc and the thrusting of oceanic and island arc rocks eastward against the craton margin. This event initiated the *Sonoma orogeny* at or near the Permian-Triassic boundary.

Following the Late Paleozoic–Early Mesozoic destruction of the volcanic island arc during the Sonoma orogeny, the western margin of North America became an oceanic-continental convergent plate boundary. During the Late Triassic, a steeply dipping subduction zone developed

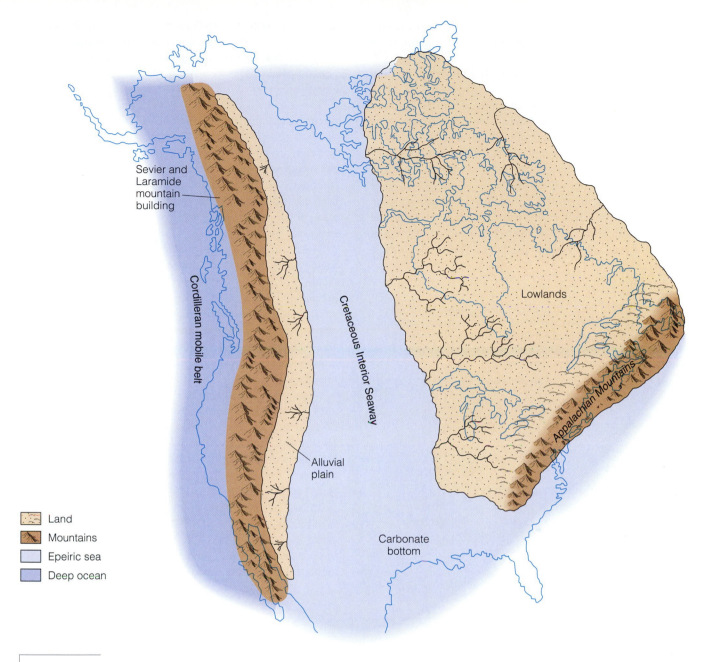

FIGURE 18.19 Paleogeography of North America during the Cretaceous Period.

Land

Mountains

Epeiric sea

Deep ocean

Sevier and Laramide mountain building

Cordilleran mobile belt

Cretaceous Interior Seaway

Alluvial plain

Lowlands

Appalachian Mountains

Carbonate bottom

along the western margin of North America in response to the westward movement of North America over the Pacific plate. This newly created oceanic-continental plate boundary controlled Cordilleran tectonics for the rest of the Mesozoic Era and for most of the Cenozoic Era.

The general term *Cordilleran orogeny* is applied to the mountain-building activity that began during the Jurassic and continued into the Cenozoic (Figure 18.8). The Cordilleran orogeny consisted of a series of individual mountain-building events that occurred in different regions at different times. Most of this Cordilleran orogenic activity is related to the continued westward movement of the North American plate.

The first phase of the Cordilleran orogeny, the **Nevadan** orogeny (Figure 18.8), began during the Late Jurassic and continued into the Cretaceous as large volumes of granitic magma were generated at depth beneath the western edge of North America. These granitic masses ascended as huge batholiths that are now recognized as the Sierra Nevada, southern California, Idaho, and Coast Range batholiths.

The second phase of the Cordilleran orogeny, the **Sevier orogeny**, was mostly a Cretaceous event (Figure 18.8). Subduction of the Pacific plate beneath the North American plate continued during this time, resulting in numerous overlapping, low-angle thrust faults in which blocks of older rocks were thrust eastward on top of younger rocks

(Figure 18.20). This deformation produced generally north-south—trending mountain ranges that stretch from Montana northward into Canada.

During the Late Cretaceous to Early Cenozoic, the final pulse of the Cordilleran orogeny occurred (Figure 18.8). The **Laramide orogeny** developed east of the Sevier orogenic belt in the present-day Rocky Mountain areas of New Mexico, Colorado, and Wyoming. Most of the features of the present-day Rocky Mountains resulted from the Cenozoic phase of the Laramide orogeny, and for that reason, it will be discussed later in this chapter.

Concurrent with the tectonism occurring in the Cordilleran mobile belt, Early Triassic sedimentation on the western continental shelf consisted of shallow-water marine sandstones, shales, and limestones. During the Middle and Late Triassic, the shallow western seas regressed further west, exposing large areas of former seafloor to erosion. Marginal marine and nonmarine Triassic rocks, particularly red beds, contribute to the spectacular and colorful scenery of the region.

These rocks represent a variety of depositional environments, including fluvial, deltaic, floodplain, and desert dunes. The Upper Triassic *Chinle Formation,* for example, is widely exposed over the Colorado Plateau and is probably most famous for its petrified wood in Petrified Forest National Park, Arizona (see Perspective 18.1). This formation, as well as other Triassic formations in the Southwest, also contains the fossilized remains and tracks of amphibians and reptiles.

The Early Jurassic deposits in a large part of the western region consist mostly of clean, cross-bedded sandstones indicative of windblown deposits. The thickest and most prominent of these is the *Navajo Sandstone,* a widespread cross-bedded sandstone that accumulated in a coastal dune environment along the southwestern margin of the craton. The sandstone's most distinctive feature is its large-scale cross-beds, some of which are more than 25 m high (Figure 18.21).

Marine conditions returned to the area during the Middle Jurassic when a wide seaway, the *Sundance Sea,* twice flooded the interior of western North America (Figure 18.18). The resulting deposits were largely derived from tectonic highlands to the west that paralleled the shoreline and were the result of intrusive igneous activity and associated volcanism that began during the Triassic.

During the Late Jurassic, the folding and thrust faulting that began as part of the Nevadan orogeny in Nevada, Utah, and Idaho formed a large mountain chain paralleling the coastline. As the mountain chain grew and shed sediments eastward, the Sundance Sea retreated northward. A

FIGURE **18.20** The Keystone thrust fault, a major fault in the Sevier overthrust belt, is exposed west of Las Vegas, Nevada. The sharp boundary between the light-colored Mesozoic rocks and the overlying dark-colored Paleozoic rocks marks the trace of the Keystone thrust fault. *(Photo © John Shelton.)*

FIGURE **18.21** Large cross-beds of the Jurassic Navajo Sandstone in Zion National Park, Utah.

Petrified Forest National Park

Petrified Forest National Park is located in eastern Arizona about 42 km east of Holbrook. The park consists of two sections: the Painted Desert, which is north of Interstate 40, and the Petrified Forest, which is south of the interstate.

The Painted Desert is a brilliantly colored landscape where colors and hues change constantly throughout the day. The multicolored rocks of the Triassic Chinle Formation have been weathered and eroded to form a badlands topography of numerous gullies, valleys, ridges, mounds, and mesas. The Chinle Formation is composed predominantly of various-colored shale beds that are easily weathered and eroded.

The Petrified Forest was originally set aside as a national monument to protect the large number of petrified logs that lay exposed in what is now the southern part of the park (Figure 1). When the transcontinental railroad constructed a coaling and watering stop in Adamana, Arizona, passengers were encouraged to take excursions to "Chalcedony Park," as the area was then called, to see the petrified forests. In a short time, collectors and souvenir hunters hauled off tons of petrified wood, quartz crystals, and Indian relics. It was not until a huge rock crusher was built to crush the logs for the manufacture of abrasives that the area was declared a national monument and the petrified forests preserved and protected.

During the Triassic Period, the climate of the area was much wetter than today. Numerous fossils of seedless vascular plants such as rushes and ferns as well as gymnosperms such as cycads and conifers are preserved in the Chinle Formation. Most of the logs are conifers, and some were more than 60 m tall and up to 4 m in diameter. Burial of the logs was rapid, and groundwater saturated with silica from the ash of nearby volcanic eruptions quickly preserved the trees.

Deposition continued in the Colorado Plateau region during the Jurassic and Cretaceous, further burying the Chinle Formation. During the Laramide orogeny, the Colorado Plateau area was uplifted and eroded, exposing the Chinle Formation. Because the Chinle is mostly shales, it was easily eroded, leaving the more resistant petrified logs and log fragments exposed on the surface—much as we see them today (Figure 1).

Figure 1 Petrified Forest National Park, Arizona. All logs here are Araucarioxylon, *which is the most abundant tree in the park. The petrified logs have been weathered from the Chinle Formation and are mostly in the position in which they were buried some 200 million years ago.*

large part of the area formerly occupied by the Sundance Sea was then covered by multicolored detrital sediments that comprise the *Morrison Formation,* which contains the world's richest assemblage of Jurassic dinosaur remains (Figure 18.22).

Shortly before the end of the Early Cretaceous, Arctic waters spread southward over the craton. By the beginning of the Late Cretaceous, this incursion joined the northward-transgressing waters from the Gulf area to create an enormous *Cretaceous Interior Seaway* that occupied the area east of the Sevier orogenic belt (Figure 18.19). Extending from the Gulf of Mexico to the Arctic Ocean and more than 1500 km wide at its maximum extent, this seaway effectively divided North America into two large landmasses until just before the end of the Late Cretaceous.

As the Mesozoic Era ended, the Cretaceous Interior Seaway withdrew from the craton. During this regression, marine waters retreated to the north and south, and marginal marine and continental deposition formed widespread coal-bearing deposits on the coastal plain.

Mesozoic Mineral Resources

ALTHOUGH much of the coal in North America is Pennsylvanian or Tertiary in age, important Mesozoic coals occur in the Rocky Mountains states. Mesozoic coals are also known from Alberta and British Columbia, Canada, as well as from Australia, Russia, and China.

Large concentrations of petroleum are found in many areas of the world, but more than 50% of all proven reserves are from Mesozoic-age sediments in the Persian Gulf region. In addition, Mesozoic-age petroleum and natural gas reserves are found in the Gulf Coast region of the United States and Central America. Some of these hydrocarbons are associated with structures formed adjacent to rising salt domes. The salt initially formed in a long, narrow sea when North America separated from Europe and North Africa during the fragmentation of Pangaea.

The richest uranium ores in the United States are widespread in Mesozoic rocks of the Colorado Plateau area of Colorado and adjoining parts of Wyoming, Utah, Arizona, and New Mexico.

South Africa, the world's leading producer of gem-quality diamonds and among the leaders in industrial diamond production, mines these minerals from kimberlite pipes, conical igneous intrusions that form at great depth where pressure and temperature are high. Although kimberlite pipes have formed throughout geologic time, the most intense episode of such activity in South Africa and adjacent countries was during the Cretaceous Period.

The mother lode or source for the placer deposits mined during the 1848–1853 California gold rush is in Jurassic-age intrusive rocks of the Sierra Nevada. During this gold rush, more than $200 million in gold was extracted. Gold placers are also known in Cretaceous-age conglomerates of the Klamath Mountains of California and Oregon.

Porphyry copper was originally named for copper deposits in the western United States, but the term now applies to large, low-grade copper deposits disseminated in a variety of rocks. These prophyry copper deposits are an excellent example of the relationship between convergent plate boundaries and the distribution, concentration, and exploitation of valuable metallic ores (see Chapter 2). The world's largest copper deposits were formed during the Mesozoic and Tertiary along the western margins of North and South America.

Cenozoic Earth History

THE Cenozoic Era, constituting only 1.4% of all geologic time, has a good record of its physical and biological events. Cenozoic rocks are widespread, near or at the surface, and

FIGURE 18.22 The north wall of the visitors' center at Dinosaur National Monument, showing dinosaur bones in bas relief, just as they were deposited 140 million years ago in the Morrison Formation.

have been comparatively little altered by geologic agents such as metamorphism and weathering. Traditionally, the Cenozoic has been divided into two periods, the *Tertiary* (66 to 1.6 million years ago) and the *Quaternary* (1.6 million years to the present). Another scheme for designations of Cenozoic time uses three periods; the Paleogene and Neogene correspond to the Tertiary, and the Quaternary is retained (see Figure 17.1). We will follow the traditional usage of Tertiary and Quaternary.

Many of Earth's features have long histories, but the present distribution of land and sea and the topographic expression of continents and their landforms are all the end products of processes operating during the Cenozoic. The Appalachian Mountain region, for instance, began its evolution during the Precambrian, but its present expression is largely the product of Cenozoic uplift and erosion. In short, Earth's present distinctive aspects developed very recently in the context of geologic time.

CENOZOIC PLATE TECTONICS AND OROGENY

The Late Triassic fragmentation of the supercontinent Pangaea (Figure 18.16a) began an episode of plate motions that continues even now. As a result of these plate motions, Cenozoic orogenic activity was largely concentrated in two major zones or belts, the *Alpine-Himalayan belt* and the *circum-Pacific belt* (Figure 18.23). The Alpine-Himalayan belt includes the mountainous regions of southern Europe and North Africa and extends eastward through the Middle East and India and into southeast Asia, whereas the circum-Pacific belt, as its name implies, nearly encircles the Pacific Ocean basin.

Within the Alpine-Himalayan orogenic belt, the *Alpine orogeny* began during the Mesozoic, but major deformation also occurred from the Eocene to Late Miocene as the African and Arabian plates moved northward against Eurasia. Deformation resulting from plate convergence formed the Pyrenees Mountains between Spain and France, the Alps of mainland Europe, the Apennines of Italy, and the Atlas Mountains of North Africa (Figure 18.23). Active volcanoes in Italy and seismic activity in much of southern Europe and the Middle East indicate that this orogenic belt remains geologically active.

Farther east in the Alpine-Himalayan orogenic belt, the *Himalayan orogen* resulted from the collision of India with Asia (see Figure 10.18). The exact time of this collision is uncertain, but sometime during the Eocene, India's northward drift rate decreased abruptly, indicating the probable time of collision. In any event, as the two continental plates became sutured, an orogeny resulted, accounting for the present-day Himalayas being far inland rather than at a continental margin.

Plate subduction in the circum-Pacific orogenic belt took place throughout the Cenozoic, giving rise to orogenies in the Aleutians, the Philippines, and Japan and along the west coasts of North, Central, and South America. The Andes Mountains in western South America, for example, formed as a result of convergence of the Nazca and South American plates (see Figure 10.17). Spreading at the East Pacific Rise and subduction of the Cocos and Nazca plates beneath Central and South America, respectively, account for continuing orogenic activity in these regions.

FIGURE 18.23 Most of Earth's geologically recent and current mountain building is concentrated in the circum-Pacific and Alpine-Himalayan orogenic belts.

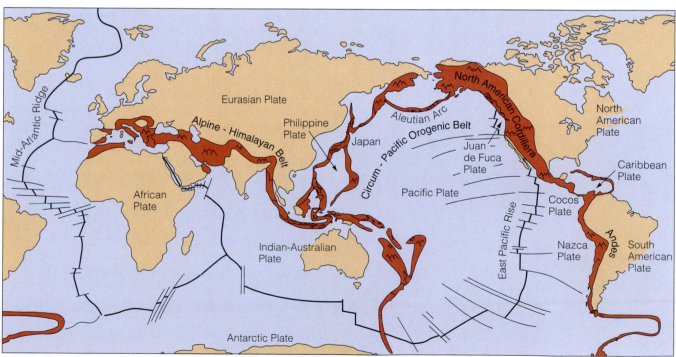

THE NORTH AMERICAN CORDILLERA

The *North American Cordillera* is a complex mountainous region in western North America extending from Alaska into central Mexico (Figure 18.24). It has a long, complex geologic history involving accretion of island arcs along the continental margin, orogeny at an oceanic-continental boundary, vast outpourings of basaltic lavas, and block-faulting. Although the Cordillera has a long history of deformation, the most recent episode of large-scale deformation was the Laramide orogeny, which began during the Late Cretaceous, 85 to 90 million years ago. Like many other orogenies, it occurred along an oceanic-continental boundary. The main

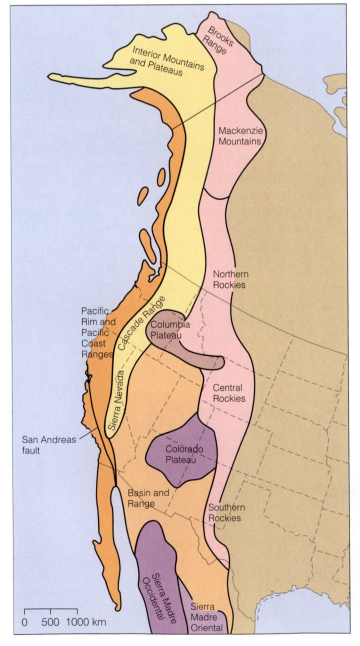

FIGURE 18.24 The North American Cordillera is a complex mountainous region extending from Alaska into central Mexico. It consists of a number of elements as shown here.

Laramide orogeny was centered in the Rocky Mountains of present-day Colorado and Wyoming, but deformation occurred far to the north in Canada and Alaska and as far south as Mexico City. The Lewis overthrust of Montana resulted from Laramide compression, and in the Canadian Rockies, thrust sheets piled one upon another (Figure 18.25).

The Laramide orogeny ceased about 40 million years ago, but since that time the Rocky Mountains have continued to evolve. The mountain ranges formed during the orogeny were eroded, and the valleys between the ranges filled with sediments. Many of the ranges were nearly buried in their own erosional debris, and their present-day elevations are the result of renewed uplift, which is continuing in such areas as the Teton Range of Wyoming.

In other parts of the Cordillera, the Colorado Plateau was uplifted far above sea level, but the rocks were little deformed. In the Basin and Range Province, block-faulting began during the Middle Cenozoic and continues to the present (Figure 18.26). At its western edge, the province is bounded by a large escarpment that forms the east face of the Sierra Nevada. This escarpment resulted from movement on a normal fault that has elevated the Sierra Nevada 3000 m above the basins to the east.

In the Pacific Northwest, an area of about 200,000 km^2, mostly in Washington, is covered by the Cenozoic Columbia River basalts (see Figure 5.18). Issuing from long fissures, these flows overlapped to produce an aggregate thickness of about 1000 m. Widespread volcanism also took place in Oregon, Idaho, California, Arizona, and New Mexico.

The present-day elements of the Pacific coast section of the Cordillera developed as a result of the westward drift of North America, the partial consumption of the oceanic Farallon plate, and the collision of North America with the Pacific-Farallon Ridge. During the Early Cenozoic, the entire Pacific coast was bounded by a subduction zone that stretched from Mexico to Alaska (Figure 18.27). Most of the Farallon plate was consumed at this subduction zone, and now only two small remnants exist—the Juan de Fuca and Cocos plates (Figure 18.27). The continuing subduction of these small plates accounts for seismicity and volcanism in the Cascade Range of the Pacific Northwest and Central America, respectively. Westward drift of the North American plate also resulted in its collision with the Pacific-Farallon Ridge and the origin of the Queen Charlotte and San Andreas transform faults (Figure 18.27).

THE GULF COASTAL PLAIN

The Gulf Coast sedimentation pattern was established during the Jurassic and persisted through the Cenozoic. Much of the Gulf Coastal Plain was dominated by detrital sediment deposition during the Cenozoic. In the Florida section of the coastal plain and the Gulf Coast of Mexico, however, significant carbonate deposition occurred. A carbonate platform was established in Florida during the Cretaceous, and shallow-water carbonate deposition continued through the

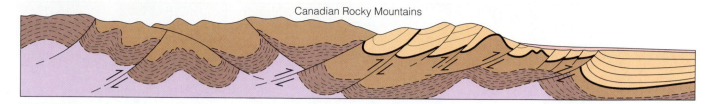

Canadian Rocky Mountains

FIGURE **18.25** Cross section of part of the Canadian Rocky Mountains. The folding and thrust faulting occurred during the Laramide orogeny.

Early Tertiary. Carbonate deposition continues at the present, but now it occurs only in Florida Bay and the Florida Keys.

EASTERN NORTH AMERICA

The eastern seaboard has been a passive continental margin since Late Triassic rifting separated North America from North Africa and Europe. The present distinctive topography of the Appalachian Mountains is the product of Cenozoic uplift and erosion. By the end of the Mesozoic, the Appalachian Mountains had been eroded to a plain. Cenozoic uplift rejuvenated the streams, which responded by renewed downcutting. As the streams eroded downward, they were superposed on resistant strata and cut large canyons across these strata. The distinctive topography of the Valley and Ridge Province is the product of Cenozoic erosion and

FIGURE **18.26** (a) View of the Humboldt Range in Nevada. It is one of many mountain ranges in the Basin and Range Province bounded by normal faults. (b) Cenozoic volcanic rocks of the Snake River Plain exposed at Twin Falls, Idaho. The Snake River Plain is actually a depression that was filled mostly by Pliocene and younger basalt lava flows.

(a)

(b)

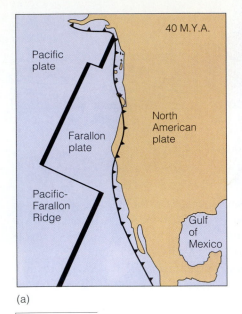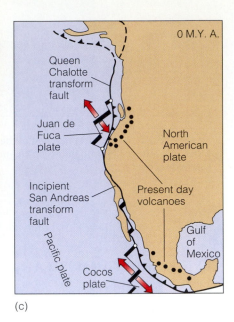

(a) (b) (c)

FIGURE 18.27 (a), (b), and (c) Three stages in the westward drift of North America and its collision with the Pacific-Farallon Ridge. As the North American plate overrode the ridge, its margin became bounded by transform faults rather than a subduction zone.

preexisting geologic structures. It consists of northeast-southwest–trending ridges of resistant upturned strata and intervening valleys eroded into less resistant strata.

Pleistocene Glaciation

WE know today that the last Ice Age began about 1.6 million years ago and consisted of several intervals of glacial expansion separated by warmer interglacial periods. It appears that the present interglacial period began about 10,000 years ago, but geologists do not know whether we are still in an interglacial period or are entering another colder glacial interval.

The climatic effects responsible for Pleistocene glaciation were, as one would expect, worldwide. Nevertheless, the world was not as frigid as it is commonly portrayed in cartoons and movies, nor was the onset of glacial conditions as rapid as many people think. In fact, evidence from various lines of research indicates that the world's climate gradually cooled from the beginning of the Eocene through the Pleistocene, and that during the past 2 million years, at least 20 major warm-cold cycles occurred.

From such glacial features as the distribution of moraines, erratic boulders, and drumlins, Pleistocene glaciers, at their greatest extent, covered about three times as much of Earth's surface as they do now and were up to 3 km thick (Figure 18.28). Geologists have determined that at least four major glacial stages, each followed by an interglacial stage, occurred during the Pleistocene of North America. These four stages, the *Wisconsin, Illinoian, Kansan,* and *Nebraskan,* are named for the states in which the most southerly glacial deposits are well exposed. The

three interglacial stages, the *Sangamon, Yarmouth,* and *Aftonian,* are named for localities of well-exposed interglacial soils and other deposits. In Europe, six or seven major glacial advances and retreats are recognized.

Recent studies indicate that there were an as yet undetermined number of pre-Illinoian glacial events and that the history of glacial advances and retreats in North America is more complex than previously thought. In view of these data, the traditional four-part subdivision of the Pleistocene of North America must be modified.

Cenozoic Mineral Resources

THE United States is the third largest producer of petroleum, accounting for about 17% of the world's total. Much of this production comes from Tertiary reservoirs of the Gulf Coastal Plain and the adjacent continental shelf. Several Tertiary basins in southern California are also important areas of petroleum production. The Green River Formation of Wyoming, Utah, and Colorado has huge reserves of oil shale and evaporites.

Diatomite is a sedimentary rock composed of the microscopic shells of diatoms, which are single-celled marine and freshwater plants that secrete skeletons of silica (SiO_2). This rock, also called *diatomaceous earth,* is used chiefly in gas purification and to filter a number of liquids such as molasses, fruit juices, water, and sewage. The United States is the leader in diatomite production, mostly from Tertiary deposits in California, Oregon, and Washington.

Huge deposits of low-grade coals in the Northern Great Plains are becoming increasingly important resources. These coal deposits are Late Cretaceous to Early Tertiary and are most extensive in the Williston and Powder River basins of Montana, Wyoming, and North and South Dakota.

Gold production from the Pacific Coast, particularly California, comes mostly from Tertiary and Quaternary

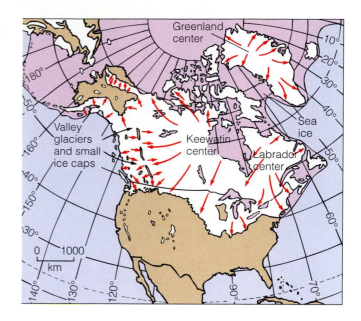

FIGURE 18.28 Centers of ice accumulation and maximum extent of Pleistocene glaciation in North America.

gravels. The gold is found in placer deposits, which formed as concentrations of minerals separated from weathered debris by fluvial processes.

Sand and gravel deposits resulting from glacial activity are a valuable Quaternary resource in many formerly glaciated areas. Most Pleistocene sand and gravel deposits originated as floodplain gravels, outwash sediment, or esker deposits. The bulk of the sand and gravel in the United States and Canada is used in construction and as roadbase and fill for highway and railway construction.

Another Quaternary resource is peat, a vast potential energy resource that has been developed in Canada and Ireland. Peatlands formed from plant assemblages as the result of particular climate conditions.

Chapter Summary

1. A craton is the stable core of a continent. Broad areas where the cratons are exposed are called shields; each continent has at least one shield area.

2. Cratons formed as a result of accretion, a process involving the addition of eroded continental material, igneous rocks, and island arcs to the margin of a craton during orogenies.

3. The North American craton evolved during the Precambrian by collision of smaller cratons along belts of deformation known as orogens and by accretion along its southern and eastern margins. Since then, orogenies have resulted in continental accretion along the craton's margin.

4. Six major continents existed at the beginning of the Paleozoic Era.

5. During the Early Paleozoic (Cambrian-Silurian), Laurentia was moving northward, and Gondwana moved to a south polar location, as indicated by glacial deposits.

6. During the Late Paleozoic, Baltica and Laurentia collided, forming Laurasia. Siberia and Kazakhstania collided and finally were sutured to Laurasia. Gondwana moved over the South Pole and experienced several glacial-interglacial periods, resulting in global changes in sea level and transgres-

sions and regressions along the low-lying craton margins.

7. Laurasia and Gondwana underwent a series of collisions beginning in the Carboniferous. During the Permian, the formation of Pangaea was completed. Surrounding the supercontinent was a global ocean, Panthalassa.

8. The Sauk Sea was the first of several major transgressions onto the North American craton. At its maximum it covered the craton except for the Transcontinental Arch.

9. The Tippecanoe sequence began with deposition of an extensive sandstone over the exposed and eroded Sauk landscape. During Tippecanoe time, extensive carbonate deposition took place.

10. Except for a persistent and widespread black shale during the Late Devonian and Early Mississippian, most of the Kaskaskia sequence was dominated by carbonates and evaporites.

11. Transgressions and regressions over the low-lying craton resulted in cyclothems and the formation of coals during the Pennsylvanian Period. Cratonic mountain building also occurred during the Pennsylvanian, with thick nonmarine detrital rocks and evaporites accumulating in the intervening basins. By the Early Permian, the

Absaroka Sea occupied a narrow zone of the south-central craton. It completely retreated from the craton by the end of the Permian.

12. During Tippecanoe time, an oceanic-continental convergent plate boundary formed, resulting in the Taconic orogeny, the first of several orogenies affecting the Appalachian mobile belt. The newly formed Taconic Highlands shed sediments into the shallow western sea, producing the Queenston Delta, a clastic wedge.

13. The Taconic orogeny was followed by the Acadian orogeny, which began as an oceanic-continental convergent plate boundary until Baltica collided with Laurentia; it then formed a continental-continental convergent plate boundary. The Hercynian-Alleghenian orogeny marks the suturing of Laurasia and Gondwana and the formation of Pangaea.

14. The breakup of Pangaea can be divided into four stages:
   a. The first stage involved the separation of North America from Africa during the Late Triassic, followed by the separation of North America from South America.
   b. The second stage involved the separation of Antarctica, India, and Australia from South

America and Africa during the Jurassic. During this stage, India broke away from the still-united Antarctica and Australia landmass.

   c. During the third stage, South America separated from Africa, while Europe and Africa began converging.

   d. In the last stage, Greenland separated from North America and Europe.

15. The Eastern Coastal region was the initial site of the separation of North America from Africa that began during the Late Triassic.

16. The Gulf Coastal region was the site of major evaporite accumulation during the Jurassic as North America rifted from South America. During the Cretaceous, it was inundated by a transgressing sea, which, at its maximum, connected with a sea transgressing from the north to create the Cretaceous Interior Seaway.

17. Mesozoic rocks of the western region of North America were deposited in a variety of continental and marine environments. In addition, this region was affected by four interrelated orogenies: the Sonoma, Nevadan, Sevier, and Laramide. Each involved igneous intrusions as well as eastward thrust faulting and folding.

18. Cenozoic orogenic activity occurred mostly in two major belts—the Alpine-Himalayan orogenic belt and the circum-Pacific orogenic belt. Each belt is composed of smaller units called orogens.

19. The North American Cordillera is a complex mountainous region extending from Alaska into Mexico. Its Cenozoic evolution included deformation during the Laramide orogeny, extensional tectonics that formed the Basin and Range structures, intrusive and extrusive volcanism, and uplift and erosion.

20. The westward drift of North America resulted in its collision with the Pacific-Farallon Ridge. Subduction ceased, and the continental margin became bounded by major transform faults, except where the Juan de Fuca plate continues to collide with North America.

21. Cenozoic uplift and erosion are responsible for the present topography of the Appalachian Mountains.

22. About 20 warm-cold Pleistocene climatic cycles are recognized, while several intervals of widespread glaciation, separated by interglacial periods, occurred in North America.

Important Terms

Absaroka sequence
Acadian orogeny
craton
cyclothem
Hercynian-Alleghenian orogeny

Kaskaskia sequence
Laramide orogeny
Laurentia
mobile belt
Nevadan orogeny

Ouachita orogeny
Sauk sequence
Sevier orogeny
Taconic orogeny
Tippecanoe sequence

Review Questions

1. Each continent has an exposed shield and platform of ancient rocks. Collectively they make up the stable _____ of a continent.
   a. _____ orogen;
   b. _____ mobile belt;
   c. _____ cyclothem;
   d. _____ craton;
   e. _____ island arc.

2. Greenstone belts and granite-gneiss complexes are the most common rocks of _____ age.
   a. _____ Archean;
   b. _____ Mesozoic;
   c. _____ Cenozoic;
   d. _____ Proterozoic;
   e. _____ Paleozoic.

3. The Proterozoic Eon is the time during which:
   a. _____ most of Earth's banded iron formations were deposited;
   b. _____ greenstone belts formed on passive continental margins;
   c. _____ dinosaurs and other large reptiles were the most common land animals;
   d. _____ the Alpine-Himalayan orogeny took place;
   e. _____ all continents joined to form the supercontinent Pangaea.

4. An elongated area marking the site of former mountain building is a:
   a. _____ craton;
   b. _____ platform;
   c. _____ shield;
   d. _____ shallow sea;
   e. _____ mobile belt.

5. Which one of the following was *not* a Paleozoic continent?
   a. _____ Gondwana;
   b. _____ Baltica;
   c. _____ Kazakhstania;
   d. _____ Eurasia;
   e. _____ Laurentia.

6. The Taconic orogeny resulted from what type of plate movement?
   a. _____ oceanic-oceanic convergent;
   b. _____ oceanic-continental convergent;
   c. _____ continental-continental convergent;
   d. _____ divergent;
   e. _____ transform.

7. Rhythmically repetitive sedimentary sequences are:
   a. _____ tillites;
   b. _____ cyclothems;
   c. _____ orogenies;
   d. _____ reefs;
   e. _____ evaporites.

8. The sedimentary rock that formed during a major transgressive-regressive cycle bounded by unconformities is a(an):
   a. _____ cratonic sequence;
   b. _____ shallow sea;
   c. _____ cyclothem;
   d. _____ biostratigraphic unit;
   e. _____ orogeny.

9. Weathering of which highlands or mountains produced the Queenston Delta?
   a. _____ Acadian;
   b. _____ Laramide;
   c. _____ Nevadan;
   d. _____ Sevier;
   e. _____ Sonoma.

10. The breakup of Pangaea began with initial Triassic rifting between which two continental landmasses?
    a. _____ South America and Africa;
    b. _____ Laurasia and Gondwana;
    c. _____ North America and Eurasia;
    d. _____ Antarctica and India;
    e. _____ India and Australia.

11. The time of greatest post-Paleozoic inundation of the craton occurred during which geologic period?
    a. _____ Triassic;
    b. _____ Jurassic;
    c. _____ Cretaceous;
    d. _____ Cenozoic;
    e. _____ Precambrian.

12. The Cenozoic Era is traditionally divided into two periods, the _____ and the _____.
    a. _____ Mesozoic/Cambrian;
    b. _____ Tertiary/Quaternary;
    c. _____ Cretaceous/Triassic;
    d. _____ Ordovician/Pleistocene;
    e. _____ Jurassic/Pennsylvanian.

13. A complex region of Mesozoic and Cenozoic deformation extending from Alaska into central Mexico is known as the:
    a. _____ Mid-Continent Rift;
    b. _____ Appalachian Mobile Belt;
    c. _____ Gulf Coastal Plain;
    d. _____ North American Cordillera;
    e. _____ Acadian Orogeny.

14. The westward drift of North America and its collision with the Pacific-Farralon Ridge during the Cenozoic Era gave rise to the:
    a. _____ earliest known red beds;
    b. _____ continental glaciers that covered much of North America;
    c. _____ Sauk sequence;
    d. _____ Ouachita orogeny;
    e. _____ San Andreas fault.

15. Describe the structure, composition, and possible origin of a greenstone belt. During which part of geologic time did most greenstone belts form?

16. Give a brief account of how the San Andreas fault came into existence.

17. What are cyclothems? Why are they economically important?

18. Compare the Taconic and Acadian orogenies in terms of the tectonic forces that caused them and the sedimentary deposits that resulted.

19. Give a general global history of the breakup of Pangaea.

20. Briefly discuss the Cenozoic geologic history of the North American Cordillera.

21. Discuss the Mesozoic tectonics of the Cordilleran mobile belt.

22. When was the Proterozoic Eon, and how was it different from the Archean Eon?

23. Explain how granitic continental crust might have originated.

Points to **Ponder**

1. According to estimates made from mapping correlation, the Queenston Delta contains more than 600,000 km³ of rock eroded from the Taconic Highlands. Based on the figure, geologists estimate the Taconic Highlands were at least 4000 m high. It is also estimated that the Catskill Delta contains three times as much sediment as the Queenston Delta. From what you know about the geographic distribution of the Taconic Highlands and the Acadian Highlands, can you estimate how high the Acadian Highland might have been?

2. The breakup of Pangaea influenced the distribution of continental land masses, ocean basins, and oceanic and atmospheric circulation patterns, which in turn affected the distribution of natural resources, landforms, and the evolution of the world's biota. Reconstruct a hypothetical history of the world for a different breakup of Pangaea—one in which the continents separate in a different order or rift apart in a different configuration. How would such a scenario affect the distribution of natural resources? Would the distribution of coal and petroleum reserves be the same? How might evolution be affected? Would human history be different?

3. During the Paleozoic Era, eastern North America experienced considerable deformation, whereas during the Mesozoic and Cenozoic eras, deformation took place mostly in the western part of the continent. What accounts for this changing pattern of deformation?

For these web site addresses, along with current updates and exercises, log on to

http://www.wadsworth.com/geo

▶ PALEOMAGNETISM AT INDIANA STATE UNIVERSITY

This site, maintained by Joe Meert, contains reconstructions of the positions of continents during the Precambrian.

▶ STRATIGRAPHY OF THE BLACK HILLS

This site contains information about the geology of the Black Hills, South Dakota. You can view a nice relief image of the Black Hills, as well as a geologic map of South Dakota. In addition, a stratigraphic column of the Black Hills is provided, with links to the various formations of the Black Hills. Under the Cambrian to Jurassic stratigraphic column, click on any of the Paleozoic formation to learn more about that particular formation.

▶ UNION COLLEGE GEOLOGY DEPARTMENT

This site, maintained by the Union College Geology Department, is located in Schenectady, New York. This particular site provides information about the sediments that were deposited during the Middle and Late Ordovician Period in the present site of the Mohawk River Valley of New York. It also contains sites showing the paleogeography of the world and New York State during the Ordovician Period.

1. Click on the *New York* icon to see what the area looked like during the Late Ordovician. How does it compare to the paleogeographic map of North America during this time period?

2. Click on the *Paleocurrent indicators* icon. What was the general direction of the currents during this time? How was this determined?

▶ MORRISON RESEARCH INITIATIVE

The Morrison Research Initiative is a multidisciplinary study of the Upper Jurassic Morrison Formation that is funded by the National Park Service and the U.S. Geological Survey. What are the project objectives? Why is the Morrison Formation an important formation?

For these web site addresses, along with current updates and exercises, log on to

http://www.wadsworth.com/geo

▶ ILLINOIS STATE MUSEUM

This site, maintained by the Illinois State Museum, contains information about their programs, collections, exhibits, and calendar of events.

1. Click on the *Exhibits* icon. Click on the *Ice Ages* site under the Online Exhibits category. At this site you will learn about the ice ages. What are ice ages? When did they occur? Why do ice ages occur?

2. Click on the *Exhibits* icon. Click on *The Midwestern U.S. 16,000 Years Ago* site under the Online Exhibits category. This site shows what the Midwest looked like 16,000 years ago, Click on the *wander through the exhibit* site to find out more about the environment and plants and animals of the midwestern United States at this time. You can also click on the *lists of the topics* covered in this exhibit to learn more about the Ice Age and its biota.

▶ THE CENOZOIC ERA

This site has a summary of Cenozoic climatic change and the ice ages. It also has a review of Cenozoic mammals and birds with many restorations.

1. What kinds of evidence indicate that Earth's climate changed during the Cenozoic Era?

2. What are the names of the four ice ages recognized in North America, and when did each begin and end?

3. What was the maximum drop in sea level during the Pleistocene Epoch?

▶ MILWAUKEE PUBLIC MUSEUM

Milwaukee Public Museum's *Continents, Oceans, and Life in Motion—A New View of the Third Planet* is an excellent overview of Earth and life history. Take *A Walk Through Geologic Time* and *The Cenozoic Era* and see maps showing plate positions, dioramas of fossil plants and animals of western Nebraska, the Ice Age in Wisconsin, and information about and images of Ice Age animals. This site also has links to other sites of geologic interest.

CD-ROM Exploration

Explore the following *In-Terra-Active 2.0* CD-ROM module(s) and increase your understanding of key concepts and processes presented in this chapter.

▶ SECTION: INTERIOR
MODULE: PLANETARY GEOLOGY

The inner terrestrial planets (Mercury, Venus, Earth, Mars) have many surface features in common. It appears that they share some of the same geologic processes. This image of the surface of Venus has been compiled from radar images from NASA. The actual surface is completely obscured by thick clouds. **Question: What can we know about the geology of Venus from this sort of picture?**

CHAPTER

19

Life History

Outline

Diorama of the environment and biota of the Phyllopod bed of the Burgess Shale, British Columbia, Canada. In the background is the vertical wall of a submarine escarpment with algae growing on it. The large cylindrical ribbed organisms on the muddy bottom in the foreground are sponges.

Prologue

On August 30 and 31, 1909, near the end of the summer field season, Charles D. Walcott, geologist and head of the Smithsonian Institution, was searching for fossils along a trail on Burgess Ridge between Mount Field and Mount Wapta, near Field, British Columbia, Canada. On the west slope of the ridge, he discovered the first soft-bodied fossils from the Burgess Shale, a discovery of immense importance in deciphering the early history of life. During the following week, Walcott and his collecting party split open numerous blocks of shale, many of which yielded the impressions of a number of soft-bodied organisms beautifully preserved on bedding planes (Figure 19.1). Returning the next summer, Walcott quarried the site and shipped back thousands of fossil specimens to the United States National Museum of Natural History where he later cataloged and studied them.

Walcott had discovered more than just another collection of well-preserved Cambrian fossils; his find provided a rare glimpse into a world previously almost unknown—that of the soft-bodied animals that lived some 530 million years ago. The beautifully preserved fossils from the Burgess Shale present a much more complete picture of a Middle Cambrian community than deposits containing only fossils of the hard parts

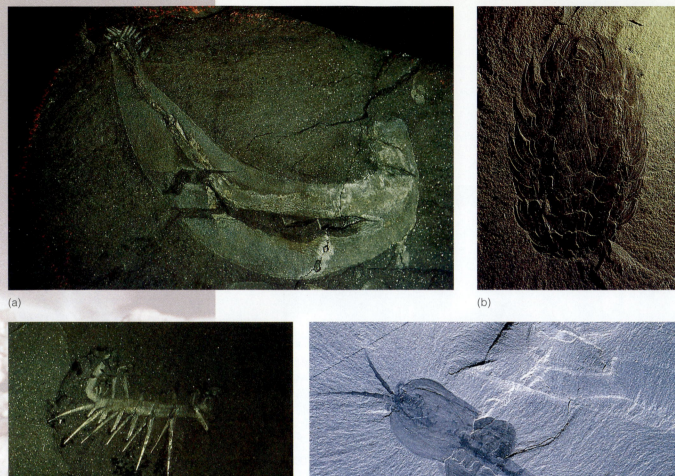

FIGURE 19.1 Some of the fossil animals preserved in the Burgess Shale. (a) *Ottoia*, a carnivorous worm. (b) *Wiwaxia*, a scaly armored sluglike creature whose affinities remain controversial. (c) *Hallucigenia*, a velvet worm. (d) *Waptia*, an anthropod.

of organisms. Specifically, the Burgess Shale contains species of trilobites, sponges, brachiopods, mollusks, and echinoderms, all of which have hard parts and are characteristic of Cambrian faunas throughout the world. But in addition to the diverse skeletonized fauna, a large and varied fossil assemblage of soft-bodied animals is also present. In all, more than 100 genera of animals, at least 60 of which were soft-bodied and preserved as impressions, have been recovered from the Burgess Shale.

The animals whose exquisitely preserved fossil remains are found in the Burgess Shale lived in and on mud banks that formed the top of a steep escarpment. Periodically, this area would slide down the escarpment carrying the mud and animals to the base where they were deposited in a deep-water anaerobic environment devoid of life. Here, bacterial degradation did not destroy the buried animals, and they were compressed by the weight of the overlying sediments, eventually resulting in their preservation as carbonaceous remains.

Introduction

RECALL from Chapter 7 that **fossils** are the remains or traces of prehistoric organisms and that the only evidence for prehistoric life consists of various types of body fossils and trace fossils (see Figure 7.15). The quality of the record of prehistoric life, or simply the *fossil record,* varies considerably depending on the types of organisms living at a particular time and their environment. For example, clams and corals have very good fossil records because both have easily preserved skeletons and both live in areas of active sediment accumulation. Marine worms also live where sedimentation takes place, but they lack skeletons and thus have a very poor fossil record; trace fossils of worms are common, though. In short, the fossil record has a bias toward marine organisms with preservable skeletons. Nevertheless, fossils of a number of other organisms are much more common than most people realize. Despite its shortcomings, the fossil record does provide us with an overview of the history of life.

Precambrian Life

PRIOR to the mid-1950s, we had very little knowledge of fossils older than Paleozoic. Scientists had long assumed that the fossils so abundant in Cambrian sedimentary rocks must have had a long earlier history, but no record of these earlier organisms was known. Some enigmatic Precambrian fossils had been reported, but they were mostly dismissed as some type of inorganic features. In fact, the Precambrian was once referred to as the *Azoic,* meaning devoid of life.

In the early 1900s, Charles Walcott, who discovered the Burgess Shale fauna (see the Prologue), described layered, moundlike structures from the Precambrian of Ontario, Canada, that are now called **stromatolites**. Walcott proposed that they represented reefs constructed by algae, but paleontologists did not demonstrate that stromatolites are the products of organic activity until 1954. Studies of present-day stromatolites in such areas as Shark Bay, Australia, show that they originate when sediment grains are trapped on sticky mats of photosynthesizing cyanobacteria, commonly called blue-green algae.

The oldest known stromatolites are in 3.3- to 3.5-billion-year-old rocks in Australia. Indirect evidence for even more ancient life comes from 3.8-billion-year-old rocks in Greenland. These rocks contain small carbon spheres that may be of biologic origin, but the evidence is currently not conclusive.

The earliest life-forms were all varieties of single-celled bacteria that lacked a cell nucleus and apparently reproduced asexually, much as bacteria do today (Figure 19.2). During this time in Earth history (3.8 to about 1.4 billion years ago),

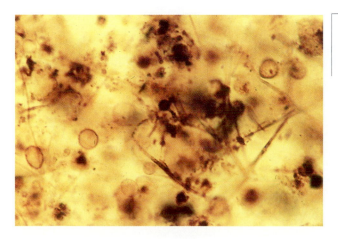

FIGURE **19.2** Photomicrographs of spheroidal and filamentous microfossils from stromatolitic chert of the Gunflint Iron Formation, Ontario.

evolution was a comparatively slow process, and organic diversity was limited. About 1.4 billion years ago, more complex cells with a distinct nucleus and capable of sexual reproduction appeared in the fossil record. These organisms were still single celled, but once they evolved, organic diversity increased markedly. Multicelled algae are found in rocks at least 1 billion years old, and during the Late Precambrian, about 700 million years ago, the first multicelled animals evolved (Figure 19.3).

These multicelled animals are collectively referred to as the *Ediacaran fauna,* and their fossils are found on all continents except Antarctica. The type of animals comprising the Ediacaran fauna is currently the subject of debate. Some investigators are of the opinion that these oldest known animals include jellyfish and sea pens, segmented worms, and primitive arthropods. One worm-like fossil (Figure 19.3b) has even been cited as a possible ancestor of the trilobites that were so common during the Early Paleozoic. Other researchers disagree and think that these animals represent an early evolutionary development quite distinct from the ancestry of any present-day animals.

Some investigators think that some of the Ediacaran animals were not animals at all but rather single-celled organisms that had the capability of inflating their bodies with water. Regardless of the affinities of these Ediacaran organisms, all agree that complex, multicelled soft-bodied animals were not only present but also distributed nearly worldwide by the Late Proterozoic. One good example is the fossil known as *Kimberella* from Russia that was very likely some kind of mollusk, perhaps a sluglike creature. In addition, numerous tracks and trails in Late Proterozoic rocks provide compelling evidence for the presence of complex animals such as worms.

Their fossil record is poor because they lacked durable skeletons. Near the end of the Proterozoic, several animals possessing skeletons did evolve. Evidence for their existence comes from minute scraps of shell-like material and spicules, presumably from sponges. Nevertheless, animals with durable skeletons of chitin (a complex organic substance), silica (SiO_2), and calcium carbonate ($CaCO_3$) were not abundant until the beginning of the Paleozoic.

Paleozoic Life

At the beginning of the Paleozoic Era, animals with skeletons appeared rather abruptly in the fossil record. In fact, their appearance is sometimes described as an explosive development of new types of animals. Nevertheless, their rapid appliance in the fossil record was rapid only in the context of geologic time, having taken place over millions of years during the Cambrian Period. The earliest of these animals with skeletons were *invertebrates*—that is, animals lacking a segmented vertebral column. Rather than focusing on the history of each invertebrate group (Table 19.1), we survey the evolution of the Paleozoic marine invertebrate communities, concentrating on their major features and the changes that occurred.

(a)

FIGURE 19.3 The Ediacaran fauna of Australia. Impressions of multicelled animals: (a) *Tribrachidium heraldicum,* a possible primitive echinoderm; (b) *Spriggina floundersi,* a possible ancestor of trilobites; and (c) *Parvancorina minchami.* (d) Reconstruction of the Ediacaran environment. *(Photos courtesy of Neville Pledge, South Australian Museum.)*

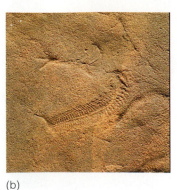
(b)

(c)

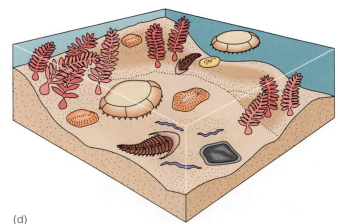
(d)

TABLE 19.1

The Major Invertebrate Groups and Their Geologic Ranges

Phylum Protozoa	Cambrian-Recent	**Phylum Mollusca**	Cambrian-Recent
Class Sarcodina	Cambrian-Recent	Class Monoplacophora	Cambrian-Recent
Order Foraminifera	Cambrian-Recent	Class Gastropoda	Cambrian-Recent
Order Radiolaria	Cambrian-Recent	Class Bivalvia	Cambrian-Recent
Phylum Porifera	Cambrian-Recent	Class Cephalopoda	Cambrian-Recent
Class Demospongea	Cambrian-Recent	**Phylum Annelida**	Precambrian-Recent
Order Stromatoporoida	Cambrian-Oligocene	**Phylum Arthropoda**	Cambrian-Recent
Phylum Archaeocyatha	Cambrian	Class Trilobita	Cambrian-Permian
Phylum Cnidaria	Cambrian-Recent	Class Crustacea	Cambrian-Recent
Class Anthozoa	Ordovician-Recent	Class Insecta	Silurian-Recent
Order Tabulata	Ordovician-Permian	**Phylum Echinodermata**	Cambrian-Recent
Order Rugosa	Ordovician-Permian	Class Blastoidea	Ordovician-Permian
Order Scleractinia	Triassic-Recent	Class Crinoidea	Cambrian-Recent
Phylum Bryozoa	Ordovician-Recent	Class Echinoidea	Ordovician-Recent
Phylum Brachiopoda	Cambrian-Recent	Class Asteroidea	Ordovician-Recent
Class Inarticulata	Cambrian-Recent	**Phylum Hemichordata**	Cambrian-Recent
Class Articulata	Cambrian-Recent	Class Graptolithina	Cambrian-Mississippian

CAMBRIAN MARINE INVERTEBRATE COMMUNITY

Although almost all the major invertebrate groups evolved during the Cambrian Period, many were represented by only a few species. While trace fossils are common and echinoderms diverse, trilobites, brachiopods, and archaeocyathids (bottom-dwelling organisms that constructed reeflike structures and lived only during the Cambrian) comprised the majority of Cambrian skeletonized life (Figure 19.4). It is important to remember, however, that the fossil record is biased toward organisms with durable skeletons and that we generally know little about the soft-bodied organisms of that time (see the Prologue). At the end of the Cambrian Period, trilobites suf-

fered mass extinctions, and even though they persisted until the end of the Paleozoic Era, their numbers were considerably diminished.

ORDOVICIAN MARINE INVERTEBRATE COMMUNITY

A major transgression that began during the Middle Ordovician (Tippecanoe sequence) resulted in a widespread inundation of the craton. This vast shallow sea opened numerous new marine habitats that were soon filled by a variety of organisms such that the Ordovician is characterized by a dramatic increase in the diversity of the total shelly fauna (Table 19.1). The end of the Ordovician, however,

FIGURE 19.4 Reconstruction of a Cambrian marine community. Floating jellyfish, swimming arthropods, sponges, and scavenging trilobites are shown.

was a time of mass extinctions in the marine realm. More than 100 families of marine invertebrates did not survive into the Silurian, and many geologists think that these extinctions were the result of the extensive glaciation that occurred in Gondwana at the end of the Ordovician Period (see Chapter 18).

SILURIAN AND DEVONIAN MARINE INVERTEBRATE COMMUNITY

The mass extinction at the end of the Ordovician was followed by rediversification and recovery of many of the decimated groups. Brachiopods, bryozoans, gastropods, bivalves, corals, crinoids, and graptolites were just some of the groups that rediversified beginning during the Silurian. In fact, the Silurian and Devonian were times of major reef building in which organic reef builders diversified in new ways, building massive reefs larger than any produced during the Cambrian or Ordovician (Figure 19.5). Another mass extinction occurred near the end of the Devonian and resulted in a worldwide near-total collapse of the massive reef communities.

CARBONIFEROUS AND PERMIAN MARINE INVERTEBRATE COMMUNITY

The Carboniferous invertebrate marine community responded to the Late Devonian extinctions in much the same way the Silurian invertebrate marine community responded to the Late Ordovician extinctions—that is, by renewed diversification. Large organic reefs like those existing earlier in the Paleozoic virtually disappeared, however, and were replaced by small patch reefs that flourished during the Late Paleozoic.

The Permian invertebrate marine faunas resembled those of the Carboniferous. They were not distributed as widely, though, due to the restricted size of the shallow seas on the cratons and the reduced shelf space along the continental margins (Figure 19.6).

THE PERMIAN MARINE INVERTEBRATE EXTINCTION EVENT

The greatest recorded mass extinction event to affect the marine invertebrate community occurred at the end of the Permian Period (Figure 19.7). Before the Permian ended, roughly one-half of all marine invertebrate families and per-

FIGURE 19.5 Reconstruction of a Middle Devonian reef from the Great Lakes area. Shown are corals, cephalopods, trilobites, and brachiopods. (From the Field Museum, Chicago # Geo80821c.)

FIGURE 19.6 A Permian patch-reef community from the Glass Mountains of west Texas. Shown are algae, brachiopods, cephalopods, and corals.

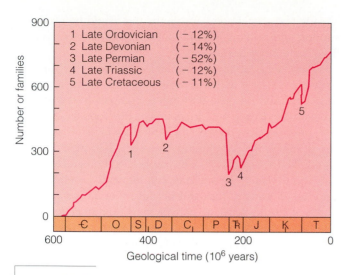

1	Late Ordovician	(−12%)
2	Late Devonian	(−14%)
3	Late Permian	(−52%)
4	Late Triassic	(−12%)
5	Late Cretaceous	(−11%)

FIGURE 19.7 Phanerozoic diversity for marine invertebrate and vertebrate families. Note the three episodes of Paleozoic mass extinctions, with the greatest occurring at the end of the Permian Period.

haps 90% of all marine invertebrate species became extinct. Among these extinct invertebrates were two groups of primitive corals, two groups of bryozoans, many brachiopods, and all remaining trilobites; several other groups either became extinct or were greatly reduced in number. On land, many amphibian and reptile families also died out.

What caused such a crisis for the marine invertebrates? Many hypotheses have been proposed, but no completely satisfactory answer has been found. One hypothesis relates the extinctions to a reduction of living area related to widespread regression of the seas and suturing of the continents when Pangaea formed. Continental convergence resulted in regression of the shallow seas from the cratons and reduction of the shallow-water shelf area surrounding each continent.

The Permian mass extinctions were probably caused by a combination of many interrelated geologic and biologic factors. In any case, the surviving marine invertebrate faunas of the Early Triassic were of very low diversity and were widely distributed around the world.

VERTEBRATE EVOLUTION

Besides the numerous invertebrate groups, **vertebrates** (animals with a segmented vertebral column) also evolved and diversified during the Paleozoic. Remains of the most primitive vertebrates, the fishes, are found in Late Cambrian-aged marine rocks in Wyoming. These fish, referred to as *ostracoderms,* were jawless and had poorly developed fins and an external covering of bony armor (Figure 19.8). Except for their distant relatives, the lamprey and slime hag, ostracoderms are now all extinct.

During the Silurian, two more groups of fish evolved. One of these groups, the *placoderms* (heavily armored jawed fish), included one of the largest marine predators that ever existed (Figure 19.8). The other group, the *acanthodians,* were enigmatic fish characterized by large spines, scales covering much of the body, jaws, teeth, and reduced bony armor (Figure 19.8). Many scientists think the acanthodians included the probable ancestors of the present-day bony and cartilaginous fish groups (Figure 19.8).

Among the bony fishes, one group known as lobe-fins was particularly important because they included the ancestor of the amphibians, the first land-dwelling vertebrate animals (Figure 19.9). In fact, the structural similarity between the group of lobe-finned fishes that gave rise to the amphibians and the earliest amphibians is striking and is one of the better documented transitions from one major group to another.

Although amphibians were living on land by the Devonian, they were not the first land-dwelling organisms.

FIGURE 19.8 Recreation of a Devonian sea floor showing (a) an ostracoderm *(Hemicyclaspis),* (b) a placoderm *(Bothriolepis),* (c) an acanthodian *(Parexus),* and (d) a ray-finned fish *(Cheirolepis).*

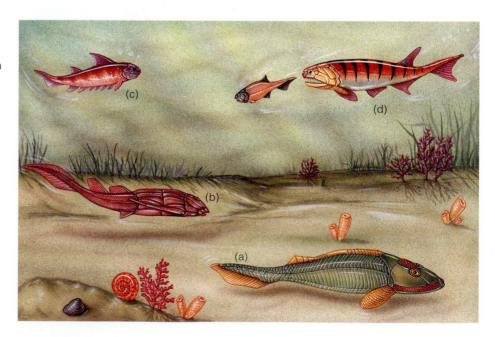

FIGURE **19.9** A Late Devonian landscape in the eastern part of Greenland. Shown is *Ichthyostega,* an amphibian that grew to a length of about 1 m. The flora of the time was diverse, consisting of a variety of small and large seedless vascular plants.

Plants made the transition to land during the Ordovician, and several invertebrates including insects, millipedes, spiders, and snails invaded the land before amphibians. All these organisms encountered several problems during the transition from water to land. The most critical obstacles for animals were drying out, reproduction, the effects of gravity, and the extraction of oxygen from the atmosphere by lungs rather than from water by gills. These problems were partly solved by some of the lobe-finned fishes; they already had a backbone and limbs that could be used for support and walking on land and lungs to extract oxygen from the atmosphere.

Amphibians' ability to colonize the land was limited both because they had to return to water to lay their gelatinous eggs and because they never completely solved the problem of drying out. In contrast, the reptiles evolved skin or scales to preserve their internal moisture and an egg in which the developing embryo is surrounded by a liquid-filled sac and provided with both a food and a waste sac. The evolution of such an egg allowed vertebrates to colonize all parts of the land because they no longer had to return to the water as part of their reproductive cycle. The oldest known reptiles evolved during the Mississippian Period and were small, agile animals.

One of the descendant groups of these early reptiles was the *pelycosaurs,* which became the dominant reptiles by the Permian (Figure 19.10). The pelycosaurs were the first land-dwelling vertebrates to become diverse and widespread. Moreover, they were the ancestors of the *therapsids,* the advanced mammal-like reptiles that gave rise to mammals during the Triassic Period.

PLANT EVOLUTION

Plants encountered many of the same problems animals faced when they made the transition to land: drying out, the effects of gravity, and reproduction. Plants adapted by evolving a variety of structural features that allowed them to invade the land during the Ordovician and later periods. Most experts agree that the ancestors of land plants were green algae that first evolved in a marine environment, then moved into freshwater, and finally onto land.

The commonest and most widespread land plants are vascular plants, which have a tissue system of specialized cells for the movement of water and nutrients. Nonvascular

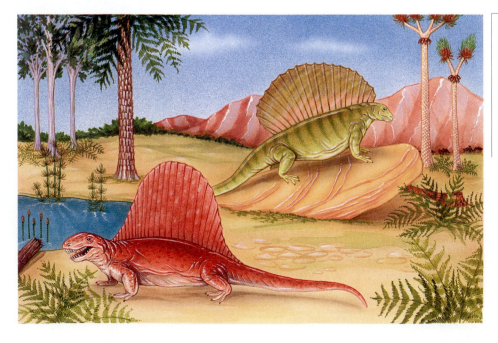

FIGURE **19.10** Most pelycosaurs have a characteristic sail on their back. One hypothesis explains the sail as a type of thermoregulatory device. Other hypotheses are that it was a type of sexual display or a device to make the reptile look more intimidating. Shown here are (foreground) the carnivore *Dimetrodon* and the herbivore *Edaphosaurus.*

FIGURE 19.11   The earliest known fertile land plant was *Cooksonia,* seen in this fossil from the Upper Silurian of South Wales. *Cooksonia* consisted of upright, branched stems terminating in sporangia (spore-producing structures). It also had a resistant cuticle and produced spores typical of a vascular plant. These plants probably lived in moist environments such as mud flats. This specimen is 1.49 cm long. *(Photo courtesy of Dianne Edwards, University College, England.)*

plants, such as mosses and fungi, lack these specialized cells and are typically small and usually live in low, moist areas. Nonvascular plants were probably the first to make the transition to land, but their fossil record is poor.

The earliest known, well-preserved fossils of vascular land plants are small, leafless, Y-shaped stems from the Middle Silurian of Wales and Ireland (Figure 19.11). They are known as **seedless vascular plants** because they did not produce seeds. Although these plants lived on land, they never completely solved the problem of drying out and were thus restricted to moist areas. Even their living descendants, such as ferns, are usually found in moist areas. During the Pennsylvanian Period, the seedless vascular plants became very abundant and diverse because of the widespread coal-forming swamps that were ideally suited to their lifestyle (Figure 19.12).

Along with the evolution of diverse seedless vascular plants, another significant floral event occurred by the Late Devonian. The evolution of the seed at this time liberated vascular plants from their dependence on moist conditions and allowed them to spread over all parts of the land. The first to do so were the flowerless seed plants, or **gymnosperms**, which include the living cycads, conifers, and ginkgoes. While the seedless vascular plants dominated the flora of the Pennsylvanian coal-forming swamps, the gymnosperms made up an important element of the Late Paleozoic flora, particularly in the non-swampy areas.

FIGURE 19.12   Reconstruction of a Pennsylvanian coal swamp with its characteristic vegetation. The amphibian is *Eogyrinus.*

Mesozoic Life

THE Mesozoic Era is designated as the "Age of Reptiles," alluding to the fact that various reptiles were the most common land-dwelling vertebrate animals. Among the reptiles were dinosaurs, flying reptiles, and marine reptiles, thus accounting for the fact that many people find the Mesozoic the most interesting time in the history of life. Although dinosaurs and their relatives evoke considerable interest, many other groups of organisms were not only present but also quite common. Many invertebrate animals were common in the seas as well as on land, and vertebrates other than reptiles proliferated. For instance, mammals evolved from mammal-like reptiles during the Triassic, and birds evolved probably from small carnivorous dinosaurs during the Jurassic. Another wave of extinctions took place at the end of the Mesozoic.

MARINE INVERTEBRATES

Following the wave of extinctions at the end of the Paleozoic, the Mesozoic was a time when marine invertebrates repopulated the seas. Among the mollusks, the clams (Figure 19.13), oysters, and snails became increasingly diverse and abundant, and the cephalopods were among the most important Mesozoic invertebrate groups. On the other hand, the brachiopods never completely recovered from their near extinction and have remained a minor invertebrate group ever since. In areas of warm, clear, shallow marine waters, corals again proliferated, but these corals were of a new and more familiar type.

Single-celled animals known as foraminifera (Table 19.1) were also important, and they diversified tremendously during the Jurassic and Cretaceous periods. Floating

or planktonic forms in particular became extremely common, but many of them became extinct at the end of the Mesozoic, and only a few types survived into the Cenozoic.

REPTILES

Reptile diversification began during Late Mississippian time. From this basic stock of *stem reptiles* all other reptiles evolved. Birds and mammals, too, have their ancestors among the reptiles, so they are a part of this major evolutionary diversification.

ARCHOSAURS AND THE ORIGIN OF DINOSAURS
A group of reptiles known as **archosaurs** (*archo* meaning "ruling," and *sauros* meaning "lizard") includes crocodiles, pterosaurs (flying reptiles), dinosaurs, and birds. The inclusion of such diverse animals in a single group implies they share a common ancestor, and indeed they possess several characteristics that unite them. For instance, teeth set in individual sockets are found in all of these animals except present-day birds, but even the earliest birds possessed this feature.

Dinosaurs are characterized by many shared characteristics, but two orders, the **Saurischia** and **Ornithischia** are recognized based on their pelvic structure (Figure 19.14). Saurischian dinosaurs had a lizardlike pelvis and are therefore referred to as lizard-hipped dinosaurs. Ornithischians had a birdlike pelvis; hence they are called bird-hipped dinosaurs. The traditional interpretation is that each order evolved independently during the Late Triassic, but it is now agreed that they had a single common ancestor much like the genus *Lagosuchus,* an archosaur from Middle Triassic rocks of Argentina. This animal was a small (less than 1 m long), long-legged, carnivorous animal that walked and ran on its hind limbs, so it was **bipedal** as opposed to a **quadrupedal** animal that moves on all four limbs.

DINOSAURS
A common but erroneous perception of **dinosaurs** is that they were poorly adapted animals. True, they became extinct, but to consider this a failure is to ignore the fact that for more than 140 million years they were the dominant land vertebrates. During their existence, they diversified into numerous types, adapted to a wide variety of environments, and some may have been warm-blooded and much more active than previously thought. Eventually, the dinosaurs did die out, an event that then enabled mammals to become the dominant land vertebrates.

Two groups of saurischian dinosaurs are recognized: theropods and sauropods (Figure 19.14). Theropods were carnivorous bipeds (Figure 19.15) that ranged in size from tiny *Compsognathus,* which was the size of a chicken, to the largest land-dwelling carnivores known. Until recently, the largest of these was 3- to 5-ton *Tyrannosaurus,* but in 1995 similar theropod dinosaurs that may have weighed 7 to 8 tons were discovered in Argentina and Africa.

Included among the sauropods were the giant, quadrupedal herbivores such as *Apatosaurus, Diplodocus,* and

FIGURE 19.13 Bivalves, represented here by two Cretaceous forms, were particularly diverse and abundant during the Mesozoic. *(Photo courtesy of Sue Monroe.)*

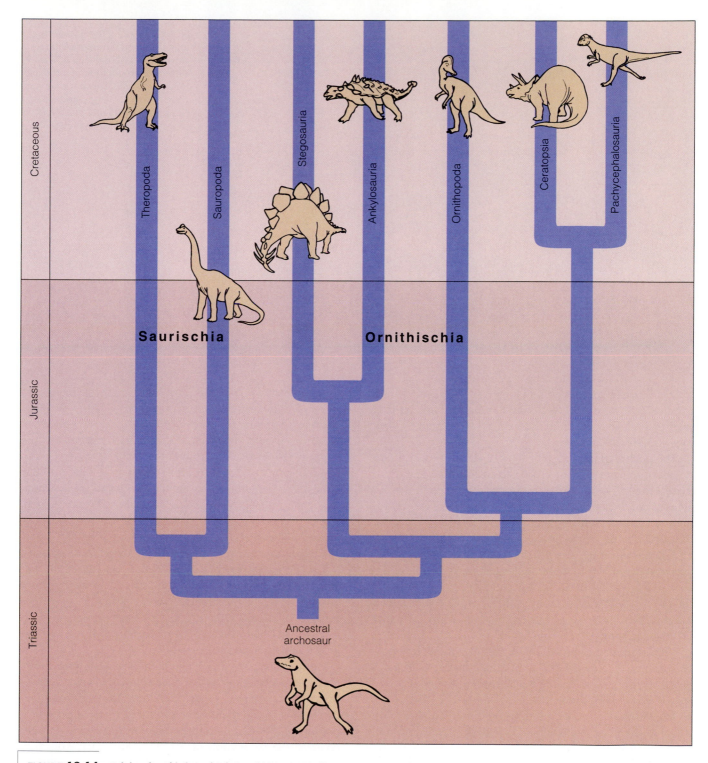

FIGURE 19.14 Origin of and inferred relationships among dinosaurs.

Brachiosaurus (Figure 19.14), the largest known land animals of any kind; some sauropods weighed several tens of tons. Evidence from fossil trackways indicates that sauropods moved in herds. They depended on their size and herding behavior rather than speed as their primary protection from predators.

The great diversity of ornithischian dinosaurs is manifested by the fact that five distinct groups are recognized: ornithopods, pachycephalosaurs, ankylosaurs, stegosaurs, and ceratopsians (Figure 19.14). Ornithopods include duck-billed dinosaurs, which had flattened, bill-like mouths (Figure 19.14). These dinosaurs were particularly varied and abundant during the Cretaceous, and some species were characterized by head crests that may have functioned as resonating chambers to amplify bellowing. Some duck-billed dinosaurs practiced colonial nesting and

FIGURE 19.15 The 3- to 4-m-long theropod dinosaur *Deinonychus,* shown here dining on the carcass of an ornithopod, probably hunted in packs. *Deinonychus* was characterized by large, sickle-like claws on its hind feet that were likely used to slash its prey.

may have cared for their young long after they hatched. All ornithopods were herbivores and primarily bipedal, but their well-developed forelimbs also allowed them to walk in a quadrupedal fashion.

The pachycephalosaurs constitute a most peculiar group of ornithischian dinosaurs. The most distinctive feature of these bipedal herbivores is the dome-shaped skull that resulted from thickening of the bones (Figure 19.14). According to one hypothesis, these domed skulls were used in intraspecific butting contests for dominance and mates.

Ankylosaurs were heavily armored, quadrupedal herbivores, and some were quite large (Figure 19.14). Bony armor protected the back, flanks, and top of the head, and the tail ended in a large, bony clublike growth. No doubt a blow delivered by the powerful tail could seriously injure an attacking predator.

The stegosaurs, represented by the familiar genus *Stegosaurus* (Figure 19.14), were quadrupedal herbivores with bony spikes on the tail, which were undoubtedly used for defense, and body plates on the back. The exact arrangement of these plates is debated, but many paleontologists think they functioned as a device to absorb and dissipate heat.

A rather good fossil record indicates that large, Late Cretaceous ceratopsians evolved from small, Early Cretaceous ancestors (Figure 19.14). The later ceratopsians were characterized by huge heads, a large bony frill over the top of the neck, and a large horn or horns on the skull. Fossil trackways indicate that these large, quadrupedal herbivores moved in herds.

WARM-BLOODED DINOSAURS All living reptiles are *ectotherms*—that is, cold-blooded animals whose body temperature varies in response to the outside temperature.

Endotherms, warm-blooded animals such as birds and mammals, are capable of maintaining a rather constant body temperature regardless of the outside temperature. Some investigators think that dinosaurs, or at least some dinosaurs, were endotherms.

Proponents of dinosaur endothermy note that dinosaur bones are penetrated by numerous passageways that, when the animals were living, contained blood vessels. Bones of endotherms typically have this structure, but considerably fewer of these passageways are found in bones of ectotherms. Living crocodiles and turtles have this so-called endothermic bone structure, yet they are ectotherms. And in some small mammals the bone structure is more typical of ectotherms, yet we know that they are capable of maintaining a constant body temperature. It may be that bone structure is more related to body size and growth patterns than to endothermy.

Because endotherms have high metabolic rates, they must eat more than ectotherms of comparable size. Consequently, endothermic predators require large prey populations. They would therefore constitute a much smaller proportion of the total animal population than their prey. In contrast, the proportion of ectothermic predators to their prey population is much greater. Where data are sufficient to allow an estimate, dinosaur predators appear to have made up 3 to 5% of the total population. These figures are comparable to present-day mammalian populations. However, a number of uncertainties about the composition of fossil communities make this argument for endothermy unconvincing to many paleontologists.

Living endotherms have a large brain in relation to body size. A relatively large brain is not necessary for endothermy, but endothermy does seem to be a prerequisite for having a large brain because a complex nervous system requires a rather constant body temperature. Some dinosaurs, particularly the small carnivores, did have a large brain in relation to their body, but many did not. That the small carnivorous ones had a large brain seems to be a good argument for endothermy, but there is an even more compelling argument. The relationship of birds to small carnivorous dinosaurs implies that these dinosaurs were endothermic or at least trending in that direction.

The large sauropods were probably not endothermic but may nevertheless have been able to maintain their body temperatures within narrow limits as endotherms do. A large animal heats up and cools down slowly because it has a small surface area compared to its volume. With proportionately less surface area to allow heat loss, sauropods probably retained body heat more efficiently than smaller dinosaurs.

Obviously, considerable disagreement exists on dinosaur endothermy. In general, a fairly good case can be made for endothermic, small, carnivorous dinosaurs, but for the others the question is still open.

FLYING REPTILES Pterosaurs, the first vertebrate animals to fly, evolved during the Triassic and were abundant until their extinction at the end of the Mesozoic (Figure 19.16). Pterosaur flight adaptations include a wing membrane supported by an elongate fourth finger, light hollow bones, and development of those parts of the brain associated with muscular coordination and sight.

Size varied considerably. Some early species ranged from sparrow to robin size, while one Cretaceous pterosaur from Texas had a wingspan of at least 12 m. The fact that at least one pterosaur species had a coat of hair or hairlike feathers and was a flier strongly suggests that it, and perhaps all pterosaurs, were endotherms.

Studies of fossils and experiments with scale models indicate that larger pterosaurs such as *Pteranodon* (Figure 19.16b) took advantage of thermal updrafts to stay airborne, mostly by soaring but occasionally flapping their wings for maneuvering. In contrast, small pterosaurs remained aloft by vigorously flapping their wings just as present-day small birds do.

MARINE REPTILES Ichthyosaurs are probably the most familiar of the Mesozoic marine reptiles (Figure 19.17a) (see Perspective 19.1). Most of these animals were about 3 m long and were completely aquatic. One species, however, measured 12 m long. Aquatic adaptations included a streamlined, somewhat fishlike body, a powerful tail for propulsion, and flipperlike forelimbs for maneuvering. The numerous sharp teeth indicate that ichthyosaurs were fish eaters. Some fossils with young ichthyosaurs within the body cavity support the interpretation that female ichthyosaurs retained the eggs in their bodies and gave birth to live young.

A second group of Mesozoic marine reptiles, the **plesiosaurs**, consisted of two varieties: short necked and long necked (Figure 19.17b). Most plesiosaurs were between 3.6 and 6 m long, but one species from Antarctica measures 15 m. Long-necked plesiosaurs were heavy-bodied animals with mouthfuls of sharp teeth and limbs specialized into oarlike paddles. They probably rowed themselves through the water and may have used their long necks in snakelike fashion to capture fish. Plesiosaurs probably came ashore to lay their eggs.

The mosasaurs were a group of Late Cretaceous marine lizards related to the present-day Komodo dragon or monitor lizard. Mosasaurs ranged from small species only 2.5 m long to giants measuring more than 9 m. Mosasaur limbs resembled paddles and were probably used mostly for maneuvering, while their long tail provided propulsion. All were predators.

FIGURE **19.16** (a) Long-tailed pterosaur from the Jurassic of Europe. *Pteranodon* (b) was a short-tailed Cretaceous pterosaur with a wingspan of more than 6 m.

(a)

(b)

Mary Anning's Contributions to Paleontology

The early history of paleontology is dominated by Western European males, a situation that no longer prevails. Now men and women from all continents are making notable contributions. But perhaps the most notable early exception is Mary Anning (1799–1847), who began a remarkable career as a fossil collector when she was only 11 years old.

Mary Anning was born in Lyme Regis on England's southern coast. When only 15 months old she survived a lightning strike that, according to one report, killed three girls, and according to another, killed a nurse tending her. In 1810 Mary's father, a cabinetmaker who also sold fossils part time, died leaving the family nearly destitute. Mary Anning (Figure 1) expanded the fossil business and became a professional fossil collector known to the paleontologists of her

time, some of whom visited her shop to buy fossils or gather information. She collected fossils from the Dorset coast near Lyme Regis and is reported to have been the inspiration for the tongue twister, "She sells sea shells sitting on the sea shore."

Soon after her father's death, she made her first important discovery, a nearly complete skeleton of a Jurassic ichthyosaur, which was described in 1814 by Sir Everard Home. The sale of this fossil specimen provided considerable financial relief for her family. In 1821 she made a second major discovery and excavated the remains of a plesiosaur. And in 1828 she found the first pterosaur in England, which was sent to the eminent geologist William Buckland at Oxford University.

By 1830 Mary Anning's fortunes began declining as collectors and museums

had less discretionary money to spend on fossils. Indeed, she may once again have become destitute were it not for her geologist friend Henry Thomas de la Beche, also a resident of Lyme Regis. De la Beche drew a fanciful scene called *Duria antiquior*, meaning "An earlier Dorset," in which he brought to life the fossils Mary Anning had collected (Figure 2). The scene was made into a lithograph that was printed and sold widely, the proceeds of which went directly to Mary Anning.

Mary Anning died of cancer in 1847, and although only 48 years old, she had a fossil-collecting career that spanned 36 years. Her contributions to paleontology are now widely recognized but, unfortunately, soon after her death she was mostly forgotten. Apparently, people who purchased her fossils were credited with finding them. "It didn't occur to

Figure 1 Mary Anning (1799–1847), who lived in Lyme Regis on England's southern coast, began collecting and selling fossils when she was only 11 years old.

them to credit a woman from the lower classes with such astonishing work. So an uneducated little girl, with a quick mind and an accurate eye, played a key role in setting the course of the 19th century geologic revolution. Then—we simply forgot about her."*

*John Lienhard, University of Houston.

Figure 2 *This fanciful scene, titled* Duria antiquior *by Henry De la Beche, features restorations of some of the fossils collected by Mary Anning. Note especially the ichthyosaurs, plesiosaurs, and pterosaurs.*

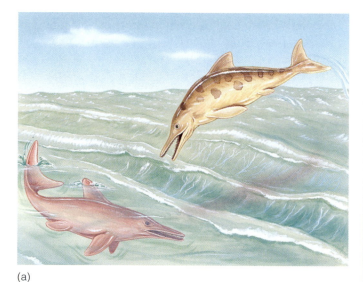

(a)

(b)

FIGURE 19.17 Mesozoic marine reptiles. (a) Ichthyosaurs and (b) a long-necked plesiosaur.

BIRDS

Fossils from the Jurassic Solnhofen Limestone of Germany show the first features we associate with birds. Several fossil specimens showing feather impressions have been discovered, but in almost every other known physical feature, these fossils are more similar to small carnivorous dinosaurs. This birdlike creature known as *Archaeopteryx* re-

tained dinosaur-like teeth, tail, hind-limb structure, and brain size but also possessed feathers and a wishbone, characteristics typical of birds (Figure 19.18).

Until recently, *Archaeopteryx* was the only known pre-Cretaceous bird, but the discovery of fossils of two

FIGURE 19.18 *Archaeopteryx,* a Jurassic-age animal from Germany, has feathers and a wishbone and is therefore classified as a bird. In almost all other anatomical features, though, it more closely resembles theropod dinosaurs. Notice the teeth, claws on the wings, and long tail, none of which are features of present-day birds.

crow-sized individuals called *Protoavis* has perhaps changed that situation. These fossils are from Triassic rocks, so they predate *Archaeopteryx,* and some investigators think they were more birdlike than *Archaeopteryx.* *Protoavis* had hollow bones and the breastbone structure of birds, but because no impressions of feathers were found on these specimens, some investigators think they may simply have been small carnivorous dinosaurs. If *Protoavis* proves to be a bird rather than a reptile, it would imply that birds evolved earlier than originally thought.

MAMMALS

In a previous section, we briefly mentioned therapsids, or advanced mammal-like reptiles. One particular group of therapsids was the most mammal-like of all and during the Late Triassic gave rise to mammals. This transition from mammal-like reptile to mammals is especially well documented by fossils and is so gradational that classification of some fossils as either reptile or mammal is difficult. In fact, the first mammals retained several reptilian characteristics, but had mammalian features as well. A good example is their lower jaw. In typical mammals the lower jaw is a single bone, whereas in reptiles it consists of several bones. The first mammals retained more than one bone in the lower jaw but had teeth and a jaw-skull joint characteristic of mammals. In short, some mammalian features evolved more rapidly than others, thereby accounting for animals possessing characteristics of both reptiles and mammals. Even though mammals appeared at the same time as dinosaurs, their diversity remained low, and all of them were small animals during the rest of the Mesozoic Era (Figure 19.19).

FIGURE 19.19 The oldest known placental mammals were members of the order Insectivora such as those in this scene from the Late Cretaceous. These animals probably fed on insects, worms, and grubs.

PLANTS

Triassic and Jurassic land-plant communities, like those of the Late Paleozoic, were composed of seedless vascular plants and various flowerless seed plants, or what are known as *gymnosperms*. Among the gymnosperms the conifers continued to diversify, and cycads that superficially resemble palm trees made their appearance. Both seedless vascular plants and gymnosperms are well represented in the present-day flora, but neither group is as abundant as it once was.

The long dominance of seedless plants and gymnosperms ended during the Early Cretaceous when many were replaced by **angiosperms,** or flowering plants. Angiosperms probably evolved from specialized gymnosperms. Indeed, recent studies have identified both fossil and living gymnosperms of this ancestral group that shows close relationships to angiosperms. In any case, since they first evolved, angiosperms have adapted to nearly every terrestrial habitat from mountains to deserts. Some have even adapted to shallow coastal waters. Their reproduction involving flowers to attract animal pollinators and the evolution of enclosed seeds largely accounts for their success. Another measure of their success is that they now account for about 96% of all vascular plant species.

CRETACEOUS MASS EXTINCTION

The mass extinction at the close of the Mesozoic was second in magnitude only to the extinctions at the end of the Paleozoic. Casualties of the Mesozoic extinction include dinosaurs, flying reptiles, marine reptiles, and several kinds of marine invertebrates. Among the latter were the ammonites, which had been so abundant through the Mesozoic, a type of reef-building clam, and some floating, marine organisms.

Numerous ideas have been proposed to explain Mesozoic extinctions, but most have been dismissed as improbable or untestable. A new proposal was made several years ago based on a discovery at the Cretaceous-Tertiary boundary in Italy—a clay layer 2.5 cm thick, with an abnormally high concentration of the platinum group element iridium. Since this discovery, high iridium concentrations have been identified at many other Cretaceous-Tertiary boundary sites (Figure 19.20). The significance of this discovery lies in the fact that iridium is rare in crustal rocks but occurs in much higher concentrations in some meteorites. Several investigators proposed a meteorite impact to explain this iridium anomaly and further postulated that the impact of a large meteorite, perhaps 10 km in diameter, set in motion a chain of events that led to extinctions. Some Cretaceous-Tertiary boundary sites also contain soot and shock-metamorphosed quartz grains, both of which are cited as further evidence of an impact.

The meteorite-impact scenario goes something like this: Upon impact, about 60 times the mass of the meteorite was blasted from Earth's crust high into the atmosphere, and the heat generated at impact started raging fires that added more particulate matter to the atmosphere. Sunlight was blocked for several months, causing a temporary cessation of photosynthesis; food chains collapsed and extinctions followed. In addition, with sunlight greatly diminished, Earth's surface temperatures were drastically reduced and could have added to the biologic stress.

Some investigators now claim that they have found the probable impact site centered on the town of Chicxulub on the Yucatán Peninsula of Mexico. The structure is about 180 km in diameter and lies beneath layers of sedimentary rock, so it has been detected only in drill holes and by geophysical work. Although an impact origin for this structure is gaining acceptance, some geologists think it is some kind of volcanic feature.

Even if a meteorite did hit Earth, did it lead to extinctions? Some paleontologists think that dinosaurs, some marine invertebrates, and many plants were already on the decline and headed for extinction before the end of the Cretaceous. A meteorite impact, if one actually occurred, may have simply hastened the process.

FIGURE 19.20 Close-up of the iridium-rich clay layer at the cretaceous-tertiary boundary in the Raton Basin, Colorado. *(Photo courtesy of D. J. Nichols, U.S. Geological Survey.)*

Cenozoic Life

THE world's flora and fauna continued to evolve during the Cenozoic Era as more familiar types of plants and animals appeared. The flowering plants continued to diversify, but some gymnosperms, especially conifers, remained abundant, and seedless vascular plants still occupied many habitats. The Cenozoic marine environment was populated by those plants and animals that survived the Mesozoic extinction event. The planktonic foraminifera comprised a major component of the marine invertebrate community, but corals, clams, and snails also proliferated. Following the extinction of the ammonites at the end of the Mesozoic, only their relatives—the nautiloids, squids, and octopuses—were present.

MAMMALS

For more than 100 million years, mammals had coexisted with dinosaurs, yet their fossil record indicates that even during the Cretaceous Period only a limited number of varieties existed. The mass extinction eliminated the dinosaurs and their relatives, thereby creating numerous adaptive opportunities that were quickly exploited by mammals. In fact, the Cenozoic Era is called the "Age of Mammals."

Among the mammals, only *monotremes* lay eggs, whereas the marsupials and placentals give birth to live young. **Marsupials** are born in a very immature, almost embryonic condition and then undergo further development in the mother's pouch. In the **placentals**, a placenta nourishes the embryo, permitting the young to develop much more fully before birth. The fossil record of monotremes is so poor that their relationship to other mammals is not clear. The marsupials and placentals, on the other hand, diverged from a common ancestor during the Cretaceous and since then have had separate evolutionary histories.

The phenomenal success of placental mammals is related in part to their reproductive method. A measure of this success is that more than 90% of all mammals, fossil and living, are placental. The only long-term success for marsupials has been in the Australian region where most present-day species live. Marsupials were also common in South America during much of the Cenozoic before a land connection existed between that continent and North America. Once a land connection formed a few million years ago, many placentals migrated south, and all species of indigenous marsupials except opossums became extinct.

Although a major diversification of placental mammals began early in the Cenozoic, these mammals are considered archaic because they were primitive and included several varieties that were holdovers from the Mesozoic (Figure 19.21). Most of these Early Cenozoic mammals had not yet become clearly differentiated from their ancestors, and the differences between herbivores and carnivores were slight. If we could somehow go back and visit this time, we would probably not recognize many of the animals. Some would be at least vaguely familiar, but the ancestors of horses, camels, whales, and elephants would bear little resemblance to their living descendants. By the Middle Cenozoic, all the major groups of living placental mammals had evolved, and many of these would be easily recognized.

Numerous groups of mammals evolved during the

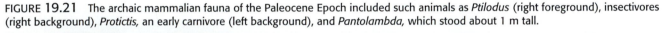

FIGURE 19.21 The archaic mammalian fauna of the Paleocene Epoch included such animals as *Ptilodus* (right foreground), insectivores (right background), *Protictis,* an early carnivore (left background), and *Pantolambda,* which stood about 1 m tall.

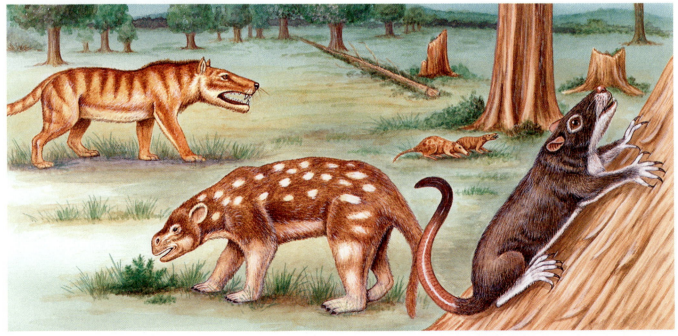

Cenozoic, but some such as camels and horses and their relatives have excellent fossil records. Camels evolved from small, four-toed ancestors and were particularly abundant in North America where most of their evolutionary history is recorded. They became extinct in North America during the Pleistocene, but not before some species migrated to South America and Asia where they still survive. Horses and their living relatives, the rhinoceroses and tapirs, also evolved from small Early Cenozoic ancestors (Figure 19.22). Horses and rhinoceroses were also common in North America, and like camels, they died out here but survived in the Old World.

One of the most remarkable aspects of the Cenozoic history of mammals is that so many very large species existed during the Pleistocene Epoch. In North America there were mastodons and mammoths, giant bison, huge ground sloths, giant camels, and beavers nearly 2 m tall at the shoulder. Kangaroos 3 m tall, wombats the size of rhinoceroses, leopard-sized marsupial lions, and large platypuses existed in Australia. In Europe and parts of Asia, cave bears, elephants, and the giant deer commonly called the Irish elk (Figure 19.23) were abundant. In addition to mammals, gi-

ant birds up to 3.5 m tall and weighing 585 kg existed in New Zealand, Madagascar, and Australia, and giant vultures with a wingspan of 3.6 m are known from California.

Many smaller mammals were also present, many of which still exist. The major evolutionary trend in mammals, however, was toward large body size. Perhaps this was an adaptation to the cooler temperatures of the Pleistocene. Large animals have proportionately less surface area compared to their volume and therefore retain heat more effectively than do small animals.

About 10,000 years ago, almost all of the large terrestrial mammals of North America, South America, and Australia became extinct. Extinctions also occurred on the other continents, but they were of considerably lesser impact. This extinction was modest by comparison to earlier ones, but it was unusual in that it affected, with few exceptions, only large terrestrial mammals. The debate over the cause of this extinction continues between those who think that the large mammals could not adapt to the rapid climatic changes at the end of the Ice Age and those who think that these mammals were killed off by human hunters, a hypothesis known as *prehistoric overkill*.

FIGURE 19.22 Evolution of horses. (a) Summary chart showing the recognized genera of horses and their evolutionary relationships. Note that during the Oligocene, two separate lines emerged, one leading to three-toed browsing horses and the other to one-toed grazers, which includes all present-day horses. (b) Simplified diagram showing some of the evolutionary trends from *Hyracotherium* to the present-day horse, *Equus*. Important trends shown here include an increase in size, loss of toes, and development of long, complex chewing teeth.

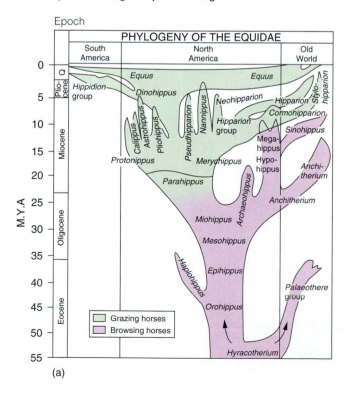

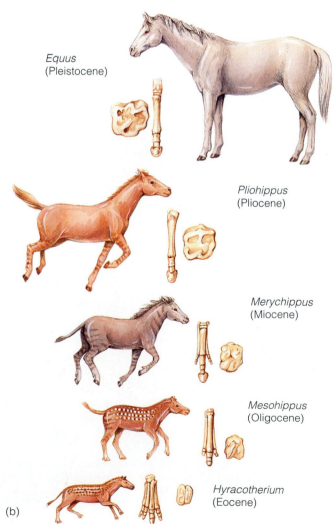

(a)

(b)

FIGURE **19.23** Restoration of the giant deer *Megalocereos giganteus,* commonly called the Irish elk. It lived in Europe and Asia during the Pleistocene. Large males had an antler spread of about 3.35 m. (From the Field Museum, Chicago, Neg. # CK30T.)

PRIMATE EVOLUTION

Several evolutionary trends in the order **primates** help define the order and are related to its *arboreal,* or tree-dwelling, ancestry. These include changes in the skeleton and mode of locomotion; an increase in brain size; a shift toward smaller, fewer, and less specialized teeth; and the evolution of stereoscopic vision and a grasping hand with opposable thumb.

Primitive primates may have evolved by the Late Cretaceous, but they were undoubtedly present by the Early Paleocene. By the Eocene, large primates had appeared; by

FIGURE **19.24** Recreation of a Pliocene landscape showing *Australopithecus afarensis,* a species of early hominids, gathering and eating various fruits and seeds.

the Oligocene, primitive New and Old World monkeys had evolved in South America and Africa, respectively.

The **hominids** (family Hominidae), the primate family that includes present-day humans and their extinct ancestors, have a fossil record extending back 4.4 million years. Several features distinguish them from other primates: Hominids walk on two legs rather than four; they show a trend toward a large, complex brain; and they manufacture and use sophisticated tools.

Ardipithecus ramidus is the oldest known hominid. Recently discovered by Tim White and his Ethiopian colleagues at Aramis, Ethiopia, this nearly complete skeleton has been dated at 4.4 million years. As its bones are studied in more detail, its place in the hominid lineage will become clearer.

Australopithecine is a collective term for all members of the genus *Australopithecus,* which currently includes five species (Figure 19.24). Many paleontologists accept the evolutionary scheme in which humans and the later australopithicines split from an earlier and common ancestor of *Australopithecus,* less than 3 million years ago.

The earliest member of our own genus *Homo* is *Homo habilis,* who existed from about 2.4 to 1.4 million years ago. Evolving from *H. habilis* was *H. Erectus,* a widely distributed species that migrated from Africa during the Pleistocene Epoch. The archaeological record indicates that *H. erectus* made tools, probably used fire, and lived in caves.

Currently, a heated debate surrounds the transition from *H. erectus* to our own species, *Homo sapiens.* Paleoanthropologists are split into two camps. On the one side are those who support the "Out of Africa" view, in which our ancestors, *H. erectus,* migrated from Africa between 1 and 2 million years ago and spread throughout Eurasia. Opposing this viewpoint are the "multiregionalists" who think that *H. erectus* migrated from Africa between 1 and 2 million years ago and established separate populations throughout Eurasia. At this time, not enough evidence points to which theory is correct, and the emergence of *H. sapiens* from ancestral *H. erectus* is still not resolved.

The most famous of all fossil humans are the Neanderthals who inhabited Europe and the Near East from about 230,000 to 30,000 years ago. Based on specimens from more than 100 sites, we now know that Neanderthals were not much different from us, only more robust.

Chapter Summary

1. During much of the Precambrian, all known life-forms were single-celled varieties of bacteria. Multicelled algae are known from rocks at least 1 billion years old, and multicelled animals first appeared about 700 million years ago.

2. Marine invertebrate animals with durable skeletons appeared in abundance during the Cambrian and Ordovician Periods and diversified throughout the rest of the Paleozoic.

3. The Cambrian invertebrate community was dominated by three major groups—trilobites, brachiopods, and archaeocyathids. The trilobites suffered mass extinctions at the end of the Cambrian.

4. The Ordovician marine invertebrate community marked the beginning of dominance by the shelly fauna and the start of large-scale reef building. The end of the Ordovician Period was a time of major extinctions for many invertebrate groups.

5. The Silurian and Devonian Periods were times of diverse faunas dominated by reef-building animals, while the Carboniferous and Permian Periods saw a great decline in invertebrate diversity and mass extinctions at the end of the Permian.

6. Fish are the earliest known vertebrates, with their first fossil occurrence in Upper Cambrian rocks. They have had a long and varied history, including jawless and jawed armored forms (ostracoderms and placoderms), cartilaginous forms, and bony forms. A group of lobe-finned fish gave rise to the amphibians.

7. The evolution of an egg that could be laid on land was the critical factor in the reptiles' ability to colonize all parts of the land. Pelycosaurs and therapsids were the dominant reptile groups during the Permian.

8. Plants had to overcome the same basic problems as the animals, namely, drying out, reproduction, and gravity—in making the transition from water to land.

9. The ancestor of terrestrial vascular plants was probably some type of green alga. The earliest seedless vascular plants were small, leafless stalks, with spore-producing structures on their tips. From this simple beginning, plants evolved many of the major structural features characteristic of today's plants.

10. Among the marine invertebrates, survivors of the Permian extinction diversified and gave rise to increasingly complex Mesozoic marine invertebrate communities.

11. Triassic and Jurassic land-plant communities were composed of seedless plants and gymnosperms. Angiosperms appeared during the Early Cretaceous, diversified rapidly, and soon became the dominant land plants.

12. Dinosaurs evolved during the Late Triassic, but were most abundant and diverse during the Jurassic and Cretaceous. Based on pelvic structure, two distinct orders of dinosaurs are recognized—saurischians (lizard-hipped) and ornithischians (bird-hipped).

13. Pterosaurs were the first flying vertebrate animals. At least one pterosaur species had hair or feathers, so it was very likely endothermic.

14. The fish-eating, porpoiselike ichthyosaurs were thoroughly adapted to an aquatic life. Female ichthyosaurs probably retained eggs within their bodies and gave birth to live young. Plesiosaurs were heavy-bodied marine reptiles that probably came ashore to lay eggs.

15. Birds probably evolved from small carnivorous dinosaurs during the Jurassic, while the earliest mammals evolved during the Late Triassic. Several types of Mesozoic mammals existed, but all were small, and their diversity was low.

16. Mesozoic mass extinctions account for the disappearance of dinosaurs, several other groups of reptiles, and a number of marine invertebrates at the end of the Cretaceous. One hypothesis holds that the extinctions were caused by the impact of a large meteorite or comet with Earth.

17. Marine invertebrate groups that survived the Mesozoic extinctions continued to expand and diversify during the Cenozoic.

18. The Early Cenozoic mammalian fauna was composed of Mesozoic holdovers and a number of new groups. This was a time of diversification among mammals. Placental mammals owe much of their success to their method of reproduction.

19. One of the most remarkable aspects of the Cenozoic history of mammals is that so many very large species existed during the Pleistocene Epoch. Perhaps this was an adaptation to the cooler temperatures that existed during this time.

20. The primates probably evolved during the Late Cretaceous. Hominids, the group to which humans and their ancestors belong, exhibit several traits that characterize and differentiate them from other mammal groups. These include walking on two legs rather than four, an increase in brain size, and the evolution of a grasping hand with opposable thumb.

21. The earliest hominids evolved 4.4 million years ago, and members of our genus *Homo* migrated from Africa between land 2 million years ago. Neanderthals inhabited Europe and the Near East between 230,000 and 30,000 years ago.

Important Terms

angiosperm
archsaur
bipedal
dinosaur
fossil
gymnosperm
hominid

ichthyosaur
marsupial
Ornithischia
placental
plesiosaur
primate
pterosaur

quadrupedal
Saurischia
seedless vascular plant
stromatolite
vertebrate

Review Questions

1. The greatest recorded mass extinction to affect the marine invertebrate community occurred at the end of which period?
   a. _____ Cambrian;
   b. _____ Ordovician;
   c. _____ Silurian;
   d. _____ Devonian;
   e. _____ Permian.

2. Jawed armored fish are:
   a. _____ ostracoderms;
   b. _____ placoderms;
   c. _____ cartilaginous;
   d. _____ bony;
   e. _____ lobe-finned.

3. Amphibians evolved from which fish group?
   a. _____ ostracoderms;
   b. _____ placoderms;

   c. _____ cartilaginous;
   d. _____ mosasaurs;
   e. _____ lobe-finned.

4. Which algal group was the probable ancestor of vascular plants?
   a. _____ red;
   b. _____ blue-green;
   c. _____ green;
   d. _____ brown;
   e. _____ yellow.

5. The most significant evolutionary change that allowed reptiles to colonize all parts of the land was:
   a. _____ endothermy;
   b. _____ origin of limbs capable of supporting the animals on land;
   c. _____ evolution of an egg that contained a food and waste sac and surrounded the embryo in a fluid-filled sac;
   d. _____ evolution of a watertight skin;
   e. _____ evolution of tear ducts.

6. Which of the following evolutionary trends characterize primates?
   a. _____ grasping hand with opposable thumb;
   b. _____ stereoscopic vision;
   c. _____ increase in brain size;
   d. _____ change in overall skeletal structure;
   e. _____ all of these.

7. Which was the first hominid to migrate out of Africa?
   a. _____ *Ardipithecus ramidus;*
   b. _____ *Australopithecus afarensis;*
   c. _____ *Homo habilis;*
   d. _____ *Homo erectus;*
   e. _____ *Homo sapiens.*

8. The large body size of Pleistocene mammals may have been an adaptation to:
   a. _____ increased predation;
   b. _____ more seasonal climates;
   c. _____ cooler temperatures;
   d. _____ higher elevations;
   e. _____ longer summers.

9. Which of the following is a hypothesis for Pleistocene extinctions?
   a. _____ meteorite impact;
   b. _____ prehistoric overkill;
   c. _____ reduced area of continental shelves;
   d. _____ freezing;
   e. _____ extensive volcanism.

10. Which one of the following occurred during the Jurassic Period?
    a. _____ Acadian orogeny;
    b. _____ origin of birds;
    c. _____ evolution of giant mammals;
    d. _____ collision of Gondwana with Laurasia;
    e. _____ extinction of dinosaurs.

11. The largest dinosaurs were among the:
    a. _____ placoderms;
    b. _____ plesiosaurs;
    c. _____ monotremes;
    d. _____ pelycosaurs;
    e. _____ sauropods.

12. Structures common in some Precambrian rocks that were produced by the activities of photosynthesizing cyanobacteria or blue-green algae are:
    a. _____ stromatolites;
    b. _____ ichthyosaurs;
    c. _____ gymnosperms;
    d. _____ quadrupeds;
    e. _____ theropods.

13. The Ediacaran fauna contains:
    a. _____ especially well-preserved dinosaurs;
    b. _____ numerous fish skeletons;
    c. _____ the first-known fossil animals;
    d. _____ amphibians that were ancestors of reptiles;
    e. _____ seedless vascular plants and flowering plants.

14. Which one of the following statements is correct?
    a. _____ during the Cenozoic gymnosperms largely replaced flowering plants;
    b. _____ all carnivorous dinosaurs were stegosaurs;
    c. _____ flying reptiles were the ancestors of birds;
    d. _____ cephalopods were particularly abundant during the Mesozoic;
    e. _____ the most abundant present-day mammals are marsupials.

15. Describe the problems that had to be overcome before organisms could inhabit the land.

16. Why were the reptiles so much more successful at extending their habitat than the amphibians?

17. What are the names of the different groups of dinosaurs, and how are they differentiated from one another?

18. What is the evidence for endothermy in dinosaurs and pterosaurs?

19. Briefly summarize the evidence for the proposal that a meteorite impact caused Mesozoic mass extinctions.

20. How do placental mammals differ from marsupials and monotremes?

21. Briefly summarize the Cenozoic evolutionary history of mammals.

Points to Ponder

1. Discuss the implications for humans if a meteorite the size that struck Earth 66 million years ago should again hit the planet.

2. Based on the history of life as preserved in the geologic record, can you make any predictions about the future direction of life? What factors do you think will affect future evolutionary events?

 ## CD-ROM Exploration

Explore the following *In-Terra-Active 2.0* CD-ROM module(s) and increase your understanding of key concepts and processes presented in this chapter.

▶ **VIRTUAL REALITY FIELD TRIP: METEOR CRATER, ARIZONA**

World Wide **Web** Activities

For these web site addresses, along with current updates and exercises, log on to

http://www.wadsworth.com/geo

► KEVIN'S TRILOBITE HOME PAGE

This site contains information, photos, and line drawings of trilobites. Besides the Table of Contents, it has links to other paleontology sites, a trilobite literature section, trilobite collectors and specialists, and trilobite classification—in other words, just about anything you would want to know about trilobites. It is maintained by Kevin Brett at the University of Alberta, Alberta, Canada.

1. Click on the *Trilobite Papers VII* site. What are the current areas of research concerning trilobites?

2. In scrolling through the images of trilobites from the various Paleozoic periods, can you discern any general evolutionary trends from these images?

► ILLINOIS STATE MUSEUM MAZON CREEK FOSSILS

This site contains some of the more interesting and dramatic types of fossils recovered from the Francis Creek Shale in Illinois.

1. Click on the *Where are Mazon Creek Fossils found* site. Where can Mazon Creek fossils be found? How did they form? What are the different types of fossils recovered from the Francis Creek Shale? Click on the names of some of these fossils to see what they look like.

2. Click on *The Importance of Mazon Creek Fossils* site. Why are these fossils important?

► UNIVERSITY OF CALIFORNIA MUSEUM OF PALEONTOLOGY

This is an excellent site to visit for any aspect of geologic time, paleontology, and evolution.

1. Click on the *On-line Exhibits* to go to the Paleontology without Walls home page, which is an introduction to the UCMP Virtual Exhibits. Click on the *Phylogeny* icon. This will take you to the Phylogeny of Life home page. Click on the *Metazoa, All Animals* icon. This will take you to the Introduction to the Metazoa home page. From here you can click on one of four sites to learn more about the life history and ecology of any animal group, its morphology or systematics, and its fossil record. Pick a fossil group and find out as much as you can about that group. For example, learn about trilobites and compare the information at this site with that presented in Kevin's Trilobite Home Page.

2. Click on the *Phylogeny* icon. This will take you to the Phylogeny of Life home page. Click on the *Vertebrates* icon. This will take you to the Introduction to the Vertebrates home page. Click on the *Systematics* icon. At this site, click on the *Tetrapoda* icon. At this site, click on the *Dinosaurs* site. This page gives a lot of information about dinosaurs, including other web sites worth visiting.

   While on the Vertebrates home page, click the following icons: *Fossil Record, Diapsids, Birds,* and *Fossil Record.* See the restoration of the oldest-known bird. Scroll down and see *Protoavis.* Why is there some doubt that *Protoavis* is actually a bird?

3. Click on the *Phylogeny* icon. This takes you to the Phylogeny of Life home page. Click on the *Three Domains of Life* icon. At this site, click on the *Introduction to the Bacteria* icon. Next click on the *Bacteria: Fossil Record* icon. Why is cyanobacteria an easy microfossil to recognize? Check and see how stromatolites form and click on the *Shark Bay* icon to see examples of present-day stromatolites.

4. Click on *The Biosphere, All Life* icon on The Phylogeny of Life home page. This will take you to the Three Domains of Life home page. Click on the *Eucaryota* icon. This takes you to the Introduction to the Eukaryota home page. Click on the *Plants* site. This will take you to the Introduction to the Plantae home page. Click on the *Fossil Record* icon. At this site, you can then learn more about the fossil history of plants. Check out the Paleozoic history of the plants.

For these web site addresses, along with current updates and exercises, log on to

http://www.wadsworth.com/geo

▶ ROYAL TYRRELL MUSEUM WEB SITE

This site contains a tremendous amount of information about dinosaurs and the museum in particular. It is supported by the Royal Tyrrell Museum Cooperating Society, "a non-profit society funded to support the scientific, educational, recreational and public operations of the Royal Tyrrell Museum of Palaeontology." the home page contains icons for Directions to the Museum, Admissions, Virtual Tour, Kong Long, Explorers Programmes, and Educational Programming.

1. Click on the *Virtual Tour* icon. It takes you to a map of the museum as well as an index of the various museum exhibits and locations. Click on *The Origin of Dinosaurs* site. What is the origin of the dinosaurs? Where is the oldest dinosaur found?

2. Click on the *Ichthyosaurs* site. What are ichthyosaurs? Where have they been found in Canada?

3. Check out the various sites and the What's New section to find out about life in the Mesozoic.

4. Click on the *Age of Mammals* icon. This site has a summary of Cenozoic life, with emphasis on mammals. A variety of restorations of extinct mammals are shown along with explanations of how they lived and what descendants they have, if any. Check out the extinct mammals *Uintatherium* and *Dinictis*.

▶ ILLINOIS STATE GEOLOGICAL SURVEY—DINOSAURS AND VERTEBRATE PALEONTOLOGY LINKS

This is one of the best starting points to access dinosaur information on the Internet. Maintained by Russell J. Jacobson—also known as Dino Russ, keeper of the Lair—this site lists hundreds of links to all manner of dinosaur information. The site is broken down into categories such as Dinosaur Art, Dinosaur Digs, Dinosaur Eggs, Dinosaur Information, and Dinosaur News and Commentaries, to name a few. Each category has highlighted links for you to click on and which will take you to that particular site. Each site listed has a short description of what will be found at that site.

▶ THE MAMMOTH SAGA

This site—maintained by the Department of Information Technology, Swedish Museum of Natural History, and authored by Dr. Ulf Carlber—is a virtual exhibition of mammoths and other animals and plants of the ice ages. It is based on an exhibition held at the Swedish Museum of Natural History, Stockholm from May 11 to September 18, 1994.

1. Click on the *The Ice Ages* site from the Table of Contents. When did the last Ice Age occur? How much of the Earth's surface was covered by continental ice masses?

2. Click on the *Sabretooth Cats* site from the Table of Contents. Are sabretooth cats related to tigers? What are the characteristics of sabretooth cats?

3. Click on *The Mammoth* site from the Table of Contents. What are mammoths? How many species of mammoths are there? What was their geographic range?

4. Click on the other sites from the Table of Contents to learn more about the Ice Age.

▶ ORIGINS OF HUMANKIND

This site is a comprehensive Internet resource concerning human origins. Because this field of research is so broad and changing so rapidly, rather than have specific sites to visit, click on any of the sites listed on the home page to learn more about controversial theories and to find links to other sites concerned about human evolution.

Epilogue

A theme of this book is that Earth is a complex, dynamic planet that has changed continuously since its origin some 4.6 billion years ago. These changes and the present-day features we observe are the result of interactions between the various interrelated internal and external Earth systems and cycles.

The rock cycle (see Figure 1.12), with its recycling of Earth materials to form the three major rock groups, illustrates the interrelationships between Earth's internal and external processes. The hydrologic cycle (see Figure 12.3) is the continuous recycling of water from the oceans, to the atmosphere, to the land, and eventually back to the oceans. Changes within this cycle can have profound effects on Earth's topography as well as its biota. For example, a rise in global temperature will cause the ice caps to melt, contributing to rising sea level, which will greatly affect coastal areas where many of the world's large population centers are located.

We have also seen how intimately intertwined Earth's various surface systems are. To those unfamiliar with geology, building a dam across a stream to control floodwaters downstream might seem like a good idea. Because of the interconnections between Earth's various systems, the construction of that dam can have profound effects on the landscape upstream from the dam as well as on beaches along the shoreline where the stream finally drains. The reservoir that accumulates behind the dam results in a new temporary base level, causing the stream to readjust itself and its gradient to the new conditions. Sand that was formerly carried by the stream and deposited in the ocean where it was transported by longshore currents along the shoreline to form a beach is now trapped behind the dam. As a result, the beach is reduced or even eliminated.

On a larger scale, the movement of plates has had a profound effect on the formation of landscapes, the distribution of mineral resources, and atmospheric and oceanic circulation patterns, as well as the evolution and diversification of life.

The launching of Sputnik I, the world's first artificial satellite in 1957, ushered in a new global consciousness in terms of how we view Earth and our place in the global ecosystem. Satellites have provided us with the ability to view not only the beauty of our planet, but also the fragility of Earth's biosphere, and the role humans play in shaping and modifying the environment. The pollution of the atmosphere, oceans, and many of our lakes and streams, as well as the denudation of huge areas of tropical forests, plus the scars from strip mining, and the depletion of the ozone layer are all visible in the satellite images beamed back from space and attest to the impact humans have had on the ecosystem.

Accordingly, we must understand that changes we make in the global ecosystem can have wide-ranging effects that we might not be aware of. For this reason, an understanding of geology, and science in general, is of paramount importance so that disruption to the ecosystem is minimal. On the other hand, we must also remember that humans are part of the ecosystem and like all other life-forms, our presence alone affects the ecosystem. We must therefore act in a responsible manner, based on sound scientific knowledge, so future generations will inherit a habitable environment.

This is the reasoning behind the concept of sustainable development discussed in Chapter 1. By redefining 'wealth' to include such natural capital as clean air and water, as well as productive land, appropriate measures can be taken to ensure future generations have sufficient natural resources to maintain and improve their standard of living.

If we are to have a world in which poverty is not widespread, then we must develop policies that ensure continuing economic development along with management of our natural resources. Meeting the needs of a growing global population will result in increased demand for food, water, and natural resources, particularly nonrenewable mineral and energy resources. As the demand for these resources increases, geologists will play an increasingly important role in locating them, as well as ensuring protection of the environment for the benefit of future generations.

Probably the greatest challenge to the environment is population growth. With 5.7 billion people in 1995, it is projected that the world's population will grow by an additional 1.7 billion people during the next two decades, bringing Earth's human population to 7.4 billion. Though this may not seem to be a geologic problem, we must remember that these people must be fed, housed, and clothed, and all with a minimal impact on the environment. Some of this population growth will be in areas that are already at risk from such geologic hazards as earthquakes, volcanic eruptions, and mass wasting. Safe and adequate water supplies must be found and kept from being polluted. More oil, gas, coal, and alternative energy resources must be discovered and utilized to provide the energy to fuel the economies of nations with ever increasing populations. New mineral resources must be found. In addition, ways to reduce usage and reuse materials must be found so as to decrease dependency on new sources of these materials.

When such environmental issues as acid rain, the greenhouse effect and global warming, and the depletion of the ozone layer are discussed and debated, it is important to remember that they are not isolated topics, but are part of a larger system that involves the entire Earth. Furthermore, it is important to remember that Earth goes through cycles of much longer duration than the human perspective of time. Although they may have disastrous effects on the human species, global warming and cooling are part of a larger cycle that has resulted in numerous glacial advances and retreats during the past 1.6 million years. In fact, geologists can make important contributions to the debate on global warming because of their geologic perspective. Long-term trends can be studied by analyzing deep-sea sediments, ice cores,

changes in sea level during the geologic past, and the distribution of plants and animals through time.

Most scientists would argue that the greatest environmental problem facing the world today is overpopulation. As the global human population increases, nations are finding it increasingly difficult to maintain adequate food supplies and fresh water. Crop yields can only be increased so much, and as the population continues to grow, productive farmlands are being taken over by villages, towns, and cities, which only heightens the pressure on the remaining land. To expand food production on the land that remains, farmers must increasingly rely on the use of fertilizers and pesticides, which leads to pollution of water supplies by runoff and depletion of natural minerals in the soil.

While more than two-thirds of Earth's surface is water, only about one percent of that water is available for human use. Just as with other natural resources, useable water is unevenly distributed throughout the world. What this means, is that in much of the world today, there isn't enough water, and where there are sufficient fresh water supplies, they frequently are being wasted and polluted.

Water tables are falling throughout the world as underground aquifers are being drawn down and are not recharged (see Chapter 13). Lakes are shrinking and becoming saline as water is diverted for irrigation and other uses (see Perspective 1.1). Pollution from untreated sewage, pesticide and fertilizer runoff, and industrial waste are all contributing to the pollution of the world's water supply. Because of their specialized training, geologists will play an important role in terms of finding, managing, and ensuring the world has sufficient useable and nonpolluted water resources.

In addition to food and water, human populations require a great many other resources, such as energy and building materials. Because many natural resources, such as oil, gas, coal, and minerals, essential to society today and in the future are finite, we must find economical ways to recy-

cle these resources and use a larger percentage of renewable resources.

The problem of overpopulation and its effects on the global ecosystem are varied. For many of the poor and nonindustrialized countries, the problem is too many people and not enough food. For the more developed and industrialized countries, it is too many people rapidly depleting both the nonrenewable and renewable natural resource base. And in the most industrialized countries, it is people producing more pollutants than the environment can safely recycle on a human time scale. In all cases, it is environmental imbalance created by a human population exceeding Earth's carrying capacity.

One result of an environmental imbalance is global warming caused by the greenhouse effect. Carbon dioxide is produced as a by-product of respiration and the burning of organic material. As such, it is a component of the global ecosystem and is constantly being recycled as part of the carbon cycle. The concern in recent years over the increase in atmospheric carbon dioxide has to do with its role in the greenhouse effect. The recycling of carbon dioxide between the crust and atmosphere is an important climatic regulator because carbon dioxide, as well as other gases such as methane, nitrous oxide, chlorofluorocarbons, and water vapor, allow sunlight to pass through them but trap the heat reflected back from Earth's surface. Heat is thus retained, causing the temperature of Earth's surface and, more importantly, the atmosphere to increase, producing the greenhouse effect.

Until the Industrial Revolution began during the mid-eighteenth century, humans' contribution to the global temperature pattern was negligible. With industrialization and its accompanying burning of tremendous amount of fossil fuels, carbon dioxide levels in the atmosphere have been steadily increasing since about 1880. In fact, atmospheric levels of carbon dioxide are currently almost 30% higher than a century ago, while nitrous oxide levels have climbed 15% and methane has increased 100% since the Industrial Revolution. At their current yearly rate of increase,

greenhouse gas concentrations will probably triple from their present concentrations within the next one hundred years, unless something is done to reduce their production. The first steps in reducing greenhouse-causing emissions were taken in 1997 in Kyoto, Japan, where a treaty was signed binding nations to reduce their emissions to earlier levels by the next century.

Research also indicates that deforestation of large areas, particularly in the tropics, is another cause of increased levels of carbon dioxide. This is because plants use carbon dioxide in photosynthesis and thus remove it from the atmosphere. With a decrease in the global vegetation cover, less carbon dioxide is removed from the atmosphere.

Because of the increase in human-produced greenhouse gases during the last two hundred years, many scientists are concerned that a global warming trend has already begun and will result in severe global climatic shifts. Most computer models based on the current rate of increase in greenhouse gases show Earth warming as a whole by as much as 5°C during the next one hundred years. Such a temperature change will be uneven, however, with the greatest warming occurring in the higher latitudes. As a consequence of this warming, rainfall patterns will shift dramatically. This will have a major effect on the largest grain-producing areas of the world, such as the American Midwest. Drier and hotter conditions will intensify the severity and frequency of droughts, leading to more crop failures and higher food prices. With such shifts in climate, Earth may experience an increase in desertification (see the Prologue to Chapter 15).

We cannot leave the subject of global warming without pointing out that many scientists are not convinced that the global warming trend is the direct result of increased human activity related to industrialization. They point out that while there has been an increase in greenhouse gases, there is still uncertainty about their rate of generation and rate of removal and about whether the 0.5°C rise in global temperature during the past century is the result of normal climatic variations through time or the result of human activity. Furthermore, they point out that even if there is a general global warming during the next 100 years, it is not certain that the dire predictions made by proponents of global warming will come true. Earth, as we know, is a remarkably complex system, with many feedback mechanisms and interconnections throughout its various subsystems. It is very difficult to predict all of the consequences that global warming would have for atmospheric and oceanic circulation patterns.

In conclusion, the most important lesson to be learned from the study of geology is that Earth is an extremely complex planet with interactions between its various systems. If the human species is to survive, we must understand how the various Earth systems work and interact with each other, and more importantly, how our actions affect the delicate balance between these systems.

The study of geology is more than learning numerous facts about Earth. Geology is an integral part of our lives. As individuals and societies, the standard of living that we enjoy is directly dependent on the consumption of natural resources and interaction with the environment. An appreciation of geology and its relationship to the environment is critical if we, as a species, are to continue to exist on this planet.

 ## CD-ROM Exploration

Explore the following *In-Terra-Active 2.0* CD-ROM module(s) and increase your understanding of key concepts and processes presented in this chapter.

▶ **SECTION: INTO THE FUTURE**
MODULE: INTO THE FUTURE

In this module, you will have a chance to explore three environmental problems that threaten our quality of life: global warming, depletion of energy resources, and controlling waste management. You will be able to track these topics through recent history and into the future. Finally, you will be able to apply what you have learned to control the destiny of the planet. As time progresses, you will make decisions on how we dispose of waste, utilize our resources, and affect our atmosphere. After you have made your decisions, you will receive feedback on the consequences of your choices.

Appendix A

ENGLISH-METRIC CONVERSION CHART

	ENGLISH UNIT	CONVERSION FACTOR	METRIC UNIT	CONVERSION FACTOR	ENGLISH UNIT
Length	Inches (in.)	2.54	Centimeters (cm)	0.39	Inches (in.)
	Feet (ft)	0.305	Meters (m)	3.28	Feet (ft)
	Miles (mi)	1.61	Kilometers (km)	0.62	Miles (mi)
Area	Square inches (in.^{2})	6.45	Square centimeters (cm^2)	0.16	Square inches (in.^{2})
	Square feet (ft^2)	0.093	Square meters (m^2)	10.8	Square feet (ft^2)
	Square miles (mi^2)	2.59	Square kilometers (km^2)	0.39	Square miles (mi^2)
Volume	Cubic inches (in.^{3})	16.4	Cubic centimeters (cm^3)	0.061	Cubic inches (in.^{3})
	Cubic feet (ft^3)	0.028	Cubic meters (m^3)	35.3	Cubic feet (ft^3)
	Cubic miles (mi^3)	4.17	Cubic kilometers (km^3)	0.24	Cubic miles (mi^3)
Weight	Ounces (oz)	28.3	Grams (g)	0.035	Ounces (oz)
	Pounds (lb)	0.45	Kilograms (kg)	2.20	Pounds (lb)
	Short tons (st)	0.91	Metric tons (t)	1.10	Short tons (st)
Temperature	Degrees Fahrenheit (°F)	$-32° \times 0.56$	Degrees Celsius (Centigrade)(°C)	$\times 1.80 + 32°$	Degrees Fahrenheit (°F)

EXAMPLES:
10 inches = 25.4 centimeters; 10 centimeters = 3.9 inches
100 square feet = 9.3 square meters; 100 square meters = 1080 square feet
50°F = 10.08°C; 50°C = 122°F

PERIODIC TABLE OF THE ELEMENTS

Legend:

47 — Atomic Number
Ag — Symbol of Element
silver — Name of Element
107.9 — Atomic Mass Number (rounded to four significant figures)

Representative Elements | Transition Elements | Inner-transition Elements | Noble Gases

Period	(1)* I A	(2) II A	(3) III B	(4) IV B	(5) V B	(6) VI B	(7) VII B	(8) VIII B	(9) VIII B
1	1 **H** hydrogen 1.008								
2	3 **Li** lithium 6.941	4 **Be** beryllium 9.012							
3	11 **Na** sodium 22.99	12 **Mg** magnesium 24.31							
4	19 **K** potassium 39.10	20 **Ca** calcium 40.08	21 **Sc** scandium 44.96	22 **Ti** titanium 47.90	23 **V** vanadium 50.94	24 **Cr** chromium 52.00	25 **Mn** manganese 54.94	26 **Fe** iron 55.85	27 **Co** cobalt 58.93
5	37 **Rb** rubidium 85.47	38 **Sr** strontium 87.62	39 **Y** yttrium 88.91	40 **Zr** zirconium 91.22	41 **Nb** niobium 92.91	42 **Mo** molybdenum 95.94	43 **Tc**^{x} technetium 98.91	44 **Ru** ruthenium 101.1	45 **Rh** rhodium 102.9
6	55 **Cs** cesium 132.9	56 **Ba** barium 137.3	57 **La** lanthanum 138.9	72 **Hf** hafnium 178.5	73 **Ta** tantalum 180.9	74 **W** tungsten 183.9	75 **Re** rhenium 186.2	76 **Os** osmium 190.2	77 **Ir** iridium 192.2
7	87 **Fr**^{x} francium (223)	88 **Ra**^{x} radium 226.0	89 **Ac**^{x} actinium (227)	104 **Unq**^{x} (261)	105 **Unp**^{x} (262)	106 **Unh**^{x} (263)	107 **Uns**^{x} (262)	108 **Uno**^{x} (265)	109 **Une**^{x} (266)

Lanthanides

58 **Ce** cerium 140.1	59 **Pr** praseodymium 140.9	60 **Nd** neodymium 144.2	61 **Pm**^{x} promethium (147)	62 **Sm** samarium 150.4

Actinides

90 **Th**^{x} thorium 232.0	91 **Pa**^{x} protactinium 231.0	92 **U**^{x} uranium 238.0	93 **Np**^{x} neptunium 237.0	94 **Pu**^{x} plutonium (244)

x: All isotopes are radioactive.

( ) Indicates mass number of isotope with longest known half-life.

* Number in ( ) heading each column represents the group designation recommended by the American Chemical Society Committee on Nomenclature.

			(13) III A	(14) IV A	(15) V A	(16) VI A	(17) VII A	(18) Noble Gases
								2 **He** helium 4.003
			5 **B** boron 10.81	6 **C** carbon 12.01	7 **N** nitrogen 14.01	8 **O** oxygen 16.00	9 **F** fluorine 19.00	10 **Ne** neon 20.18
(10)	(11) I B	(12) II B	13 **Al** aluminum 26.98	14 **Si** silicon 28.09	15 **P** phosphorus 30.97	16 **S** sulfur 32.06	17 **Cl** chlorine 35.45	18 **Ar** argon 39.95
28 **Ni** nickel 58.71	29 **Cu** copper 63.55	30 **Zn** zinc 65.37	31 **Ga** gallium 69.72	32 **Ge** germanium 72.59	33 **As** arsenic 74.92	34 **Se** selenium 78.96	35 **Br** bromine 79.90	36 **Kr** krypton 83.80
46 **Pd** palladium 106.4	47 **Ag** silver 107.9	48 **Cd** cadmium 112.4	49 **In** indium 114.8	50 **Sn** tin 118.7	51 **Sb** antimony 121.8	52 **Te** tellurium 127.6	53 **I** iodine 126.9	54 **Xe** xenon 131.3
78 **Pt** platinum 195.1	79 **Au** gold 197.0	80 **Hg** mercury 200.6	81 **Tl** thallium 204.4	82 **Pb** lead 207.2	83 **Bi** bismuth 209.0	84 **Po**^{x} polonium (210)	85 **At**^{x} astatine (210)	86 **Rn**^{x} radon (222)

63 **Eu** europium 152.0	64 **Gd** gadolinium 157.3	65 **Tb** terbium 158.9	66 **Dy** dysprosium 162.5	67 **Ho** holmium 164.9	68 **Er** erbium 167.3	69 **Tm** thulium 168.9	70 **Yb** ytterbium 173.0	71 **Lu** lutetium 175.0

95 **Am**^{x} americium (243)	96 **Cm**^{x} curium (247)	97 **Bk**^{x} berkelium (247)	98 **Cf**^{x} californium (251)	99 **Es**^{x} einsteinium (254)	100 **Fm**^{x} fermium (257)	101 **Md**^{x} mendelevium (258)	102 **No**^{x} nobelium (255)	103 **Lr**^{x} lawrencium (256)

MINERAL IDENTIFICATION TABLES

Metallic Luster					
MINERAL	CHEMICAL COMPOSITION	COLOR	HARDNESS SPECIFIC GRAVITY	OTHER FEATURES	COMMENTS
Chalcopyrite	$CuFeS_2$	Brassy yellow	$3\frac{1}{2}$–4 4.1–4.3	Usually massive; greenish black streak; iridescent tarnish	The most common copper mineral and an important source of copper. Mostly in hydrothermal rocks.
Galena	PbS	Lead gray	$2\frac{1}{2}$ 7.6	Cubic crystals; 3 cleavages at right angles	The ore of lead. Mostly in hydrothermal rocks.
Graphite	C	Black	1–2 2.09–2.33	Greasy feel; writes on paper; 1 direction of cleavage	Used for pencil "leads" and as a dry lubricant. Mostly in metamorphic rocks.
Hematite	Fe_2O_3	Red brown	6 4.8–5.3	Usually granular or massive; reddish brown streak	Most important ore of iron. An accessory mineral in many rocks.
Magnetite	Fe_3O_4	Black	$5\frac{1}{2}$–$6\frac{1}{2}$ 5.2	Strong magnetism	An important ore of iron. An accessory mineral in many rocks.
Pyrite	FeS_2	Brassy yellow	$6\frac{1}{2}$ 5.0	Cubic and octahedral crystals	The most common sulfide mineral. Found in some igneous and hydrothermal rocks and in sedimentary rocks associated with coal.
Nonmetallic Luster					
MINERAL	CHEMICAL COMPOSITION	COLOR	HARDNESS SPECIFIC GRAVITY	OTHER FEATURES	COMMENTS
Anhydrite	$CaSO_4$	White, gray	$3\frac{1}{2}$ 2.9–3.0	Crystals with 2 cleavages; usually in granular masses	Found in limestones, evaporite deposits, and the cap rocks of salt domes. Used as a soil conditioner.
Apatite	$Ca_5(PO_4)_3F$	Blue, green, brown, yellow, white	5 3.1–3.2	6-sided crystals; in massive or granular masses	An accessory mineral in many rocks. The main constituent of bone and dentine. A source of phosphorous for fertilizer.
Augite	$Ca(Mg,Fe,Al)(Al,Si)_2O_6$	Black, dark green	6 3.25–3.55	Short 8-sided crystals; 2 cleavages; cleavages nearly at right angles	The most common pyroxene mineral. Found mostly in mafic igneous rocks.
Barite	$BaSO_4$	Colorless, white, gray	3 4.5	Tabular crystals; high specific gravity for a nonmetallic mineral	Commonly found with ores of a variety of metals and in limestones and hot spring deposits. A source of barium.

Nonmetallic Luster					
MINERAL	CHEMICAL COMPOSITION	COLOR	HARDNESS SPECIFIC GRAVITY	OTHER FEATURES	COMMENTS
Biotite (Mica)	$K(Mg,Fe)_3AlSi_3O_{10}(OH)_2$	Black, brown	2½ 2.9–3.4	1 cleavage direction; cleaves into thin sheets	Occurs in both felsic and mafic igneous rocks, in metamorphic rocks, and in clay-rich sedimentary rocks.
Calcite	$CaCO_3$	Colorless, white	3 2.71	3 cleavages at oblique angles; cleaves into rhombs; reacts with dilute hydrochloric acid	The most common carbonate mineral. Main component of limestone and marble. Also common in hydrothermal rocks.
Cassiterite	SnO_2	Brown to black	6½ 7.0	High specific gravity for a nonmetallic mineral	The main ore of tin. Most is concentrated in alluvial deposits because of its high specific gravity.
Chlorite	$(Mg,Fe)_3(Si,Al)_4O_{10}$ $(Mg,Fe)_3(OH)_6$	Green	2 2.6–3.4	1 cleavage; occurs in scaly masses	Common in low-grade metamorphic rocks such as slate.
Corundum	Al_2O_3	Gray, blue, pink, brown	9 4.0	6-sided crystals and great hardness are distinctive	An accessory mineral in some igneous and metamorphic rocks. Used as a gemstone and for abrasives.
Dolomite	$CaMg(CO_3)_2$	White, yellow, gray, pink	3½–4 2.85	Cleavage as in calcite; reacts with dilute hydrochloric acid when powdered	The main constituent of dolostone. Also found associated with calcite in some limestones and marble.
Fluorite	CaF_2	Colorless, purple, green, brown	4 3.18	4 cleavage directions; cubic and octahedral crystals	Occurs mostly in hydrothermal rocks and in some limestones and dolostones. Used in the manufacture of steel and the preparation of hydrofluoric acid.
Garnet	$Fe_3Al_2(SiO_4)_3$	Dark red	7–7½ 4.32	12-sided crystals common; uneven fracture	Found mostly in gneiss and schist. Used as a semi-precious gemstone and for abrasives.
Gypsum	$CaSO_4 \cdot 2H_2O$	Colorless, white	2 2.32	Elongate crystals; fibrous and earthy masses	The most common sulfate mineral. Found mostly in evaporite deposits. Used to manufacture plaster of Paris and cements.

Nonmetallic Luster					
MINERAL	CHEMICAL COMPOSITION	COLOR	HARDNESS SPECIFIC GRAVITY	OTHER FEATURES	COMMENTS
Halite	$NaCl$	Colorless, white	3–4 2.2	3 cleavages at right angles; cleaves into cubes; cubic crystals; salty taste	Occurs in evaporite deposits. Used as a source of chlorine and in the manufacture of hydro-chloric acid, many sodium compounds, and food seasoning.
Hornblende	$NaCa_2(Mg,Fe,Al)_5$ $(Si,Al)_8O_{22}(OH)_2$	Green, black	6 3.0–3.4	Elongate, 6-sided crystals; 2 cleavages intersecting at 56° and 124°	A common rock-forming amphibole mineral in igneous and metamorphic rocks.
Illite	$(Ca,Na,K)(Al,Fe^{+3},Fe^{+2},$ $Mg)_2(Si,Al)_4O_{10}(OH)_2$	White, light gray, buff	1–2 2.6–2.9	Earthy masses; particles too small to observe properties	A clay mineral common in soils and clay-rich sedimentary rocks.
Kaolinite	$Al_2Si_4O_{10}(OH)_8$	White	2 2.6	Massive; earthy odor	A common clay mineral formed by chemical weathering of aluminum-rich silicates. The main ingredient of kaolin clay used for the manufacture of ceramics.
Muscovite (Mica)	$KAl_2Si_3O_{10}(OH)_2$	Colorless	2–2½ 2.7–2.9	1 direction of cleavage; cleaves into thin sheets	Common in felsic igneous rocks, metamorphic rocks, and some sedimentary rocks. Used as an insulator in electrical appliances.
Olivine	$(Fe,Mg)_2SiO_4$	Olive green	6½ 3.3–3.6	Small mineral grains in granular masses; conchoidal fracture	Common in mafic igneous rocks.
Plagioclase feldspars	Varies from $CaAl_2Si_2O_8$ to $NaAlSi_3O_8$	White, gray, brown	6 2.56	2 cleavages at right angles	Common in igneous rocks and a variety of meta-morphic rocks. Also in some arkoses.
Potassium feldspar — Microcline	$KAlSi_3O_8$	White, pink, green	6 2.56	2 cleavages at right angles	Common in felsic igneous rocks, some metamorphic rocks, and arkoses. Used in the manufacture of porcelain.
Potassium feldspar — Orthoclase	$KAlSi_3O_8$	White, pink			
Quartz	SiO_2	Colorless, white, gray, pink, green	7 2.67	6-sided crystals; no cleavage; conchoidal fracture	A common rock-forming mineral in all rock groups and hydrothermal rocks. Also occurs in varieties known as chert, flint, agate, and chalcedony.

		Nonmetallic Luster			
MINERAL	CHEMICAL COMPOSITION	COLOR	HARDNESS SPECIFIC GRAVITY	OTHER FEATURES	COMMENTS
Siderite	$FeCO_3$	Yellow, brown	4 3.8–4.0	3 cleavages at oblique angles; cleaves into rhombs	Found mostly in concretions and sedimentary rocks associated with coal.
Smectite	$(Al,Mg)_8(Si_4O_{10})_3(OH)_{10} \cdot 12H_2O$	Gray, buff, white	$1–1\frac{1}{2}$ $2\frac{1}{2}$	Earthy masses; particles too small to observe properties	A clay mineral with the unique property of swelling and contracting as it absorbs and releases water.
Sphalerite	ZnS	Yellow, brown, black	$3\frac{1}{2}–4$ 4.0–4.1	6 cleavages; cleaves into dodecahedra	The most important ore of zinc. Commonly found with galena in hydrothermal rocks.
Talc	$Mg_3Si_4O_{10}(OH)_2$	White, green	1 2.82	1 cleavage direction; usually in compact masses	Formed by the alteration of magnesium silicates. Mostly in metamorphic rocks. Used in ceramics, cosmetics, and as a filler in paints.
Topaz	$Al_2SiO_4(OH,F)$	Colorless, white, yellow, blue	8 3.5–3.6	High specific gravity; 1 cleavage direction	Found in pegmatites, granites, and hydrothermal rocks. An important gemstone.
Zircon	Zr_2SiO_4	Brown, gray	$7\frac{1}{2}$ 3.9–4.7	4-sided, elongate crystals	Most common as an accessory in granitic rocks. An ore of zirconium and used as a gemstone.

Appendix D

TOPOGRAPHIC MAPS

NEARLY everyone has used a map of one kind or another and is probably aware that a map is a scaled down version of the area depicted. For a map to be of any use, however, one must understand what is shown on a map and how to read it. A particularly useful type of map for geologists, and people in many other professions, is a *topographic map,* which shows the three-dimensional configuration of Earth's surface on a two-dimensional sheet of paper.

Maps showing relief—differences in elevation in adjacent areas—are actually models of Earth's surface. Such maps are available for some areas, but they are expensive, difficult to carry, and impossible to record data on. Thus, paper sheets showing relief by using lines of equal elevation known as contours are most commonly used. Topographic maps depict (1) relief, which includes hills, mountains, valleys, canyons, and plains; (2) bodies of water such as rivers, lakes, and swamps; (3) natural features such as forests, grasslands, and glaciers; and (4) various cultural features including communities, highways, railroads, land boundaries, canals, and power transmission lines.

Topographic maps known as quadrangles are published by the U.S. Geological Survey (USGS). The area depicted on a topographic map can be identified by referring to the map's name in the upper right and lower right corners, which is usually derived from some prominent geographic feature (Lincoln Creek Quadrangle, Idaho) or community (Mt. Pleasant Quadrangle, Michigan). In addition, most maps have a state outline map along the bottom margin, and shown within the outline is a small black rectangle indicating the part of the state represented by the map.

CONTOURS

Contour lines, or simply contours, are lines of equal elevation used to show topography. Think of contours as the lines formed where imaginary horizontal planes intersect Earth's surface at specific elevations. On maps, contours are brown, and every fifth contour, called an *index contour,* is darker than adjacent ones and labeled with its elevation (Figure D1). Elevations on most USGS topographic maps are in feet, although a few use meters; in either case, the specified elevation is above or below mean sea level. Because contours are defined as lines of equal elevation, they cannot divide or cross one another, although they will converge and appear to join in areas with vertical or overhanging cliffs. Notice in Figure D1 that where contours cross a stream they form a V that points upstream toward higher elevations.

The vertical distance between contours is the *contour interval.* If an area has considerable relief, a large contour interval is used, perhaps 80 or 100 feet, whereas a small interval such as 5, 10, or 20 feet is used in areas with little relief. The values recorded on index contours are always multiples of the map's contour interval, shown at the bottom of the map. For instance, if a map has a contour interval of 10 feet, index contour values such as 3600, 3650, and 3700 feet might be shown (Figure D1). In addition to contours, specific elevations are shown at some places on maps and may be indicated by a small ×, next to which is a number. A specific elevation might also be shown adjacent to the designation BM (benchmark), a place where the elevation and location are precisely known.

Contour spacing depends on slope, so in areas with steep slopes, contours are closely spaced because there is a considerable increase in elevation in a short distance. In contrast, if slopes are gentle, contours are widely spaced (Figure D1). Furthermore, if contour spacing is uniform, the slope angle remains constant, but if spacing changes, slope angle changes. However, one must be careful in comparing slopes on maps with different contour intervals or different scales.

Topographic features such as hills, valleys, plains, and so on can easily be shown by contours. For instance, a hill is shown by a concentric pattern of contours with the highest elevation in the central part of the pattern. All contours must close on themselves, but they may do so beyond the confines of a particular map. A concentric contour pattern also might show a closed depression, but in this case special hachured contours, which have short bars perpendicular to the contour pointing toward the central part of the depression, are used (Figure D1).

MAP SCALES

Any map is a scaled down version of the area shown, so to be of any use a map must have a scale. Highway maps, for example, commonly have a scale such as "1 inch equals 10 miles," by which one can readily determine distances. Two types of scales are used on topographic maps. The first and most easily understood is a graphic scale, which is simply a bar subdivided into appropriate units of length (Figure D1). This scale appears at the bottom center of the map and may show miles, feet, kilometers, or meters. Indeed, graphic scales on USGS topographic maps generally show both English and metric distance units.

A ratio or fractional scale, which represents the degree of reduction of the area depicted, appears above the graphic scale. On a map with a ratio scale of 1:24,000, for

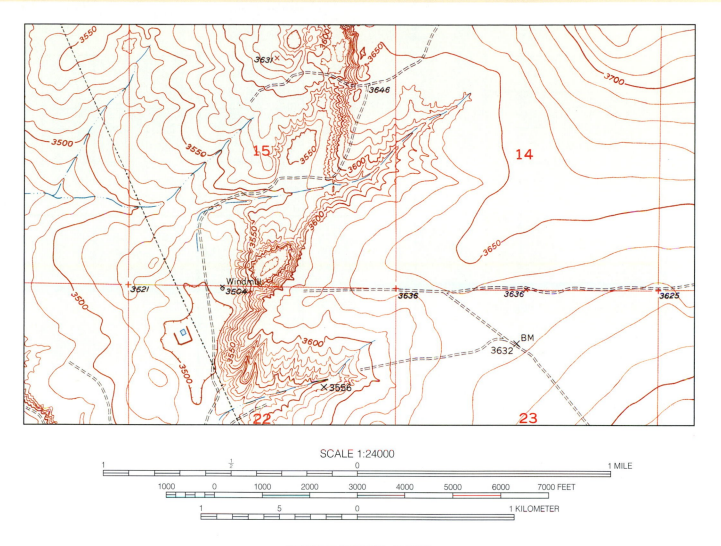

SCALE 1:24000

CONTOUR INTERVAL 10 FEET
DATUM IS MEAN SEA LEVEL

FIGURE **D1** Part of the Bottomless Lakes Quadrangle, New Mexico, which has a contour interval of 10 feet; every fifth contour is darker and labeled with its elevation. Notice that contours are widely spaced where slopes are gentle and more closely spaced where they are steeper, as in the central part of the map. Hills are shown by contours that close on themselves, whereas depressions are indicated by contours with hachure marks pointing toward the center of the depression. The dashed blue lines on the map represent intermittent streams; notice that where contours cross a stream's channel they form a V that points upstream.

instance, the area shown on the map is 1/24,000th the size of the actual land area (Figure D1). Another way to express this relationship is to say that any unit of length on the map equals 24,000 of the same units on the ground. Thus, 1 inch on the map equals 24,000 inches on the ground, which is more meaningful if one converts inches to feet, making 1 inch equal to 2000 feet. A few maps have scales of 1:63,360, which converts to 1 inch equals 5280 feet, or 1 inch equals 1 mile.

USGS topographic maps are published in a variety of scales such as 1:50,000, 1:62,500, 1:125,000, and 1:250,000. One should also realize that large-scale maps cover less area than small-scale maps, and the former show much more detail than the latter. For example, a large scale map (1:24,000) shows more surface features in greater detail than does a small scale map (1:125,000) for the same area.

MAP LOCATIONS

Location on topographic maps can be determined in two ways. First, the borders of maps correspond to lines of latitude and longitude. Latitude is measured north and south of the equator in degrees, minutes, and seconds, whereas the same units are used to designate longitude east and

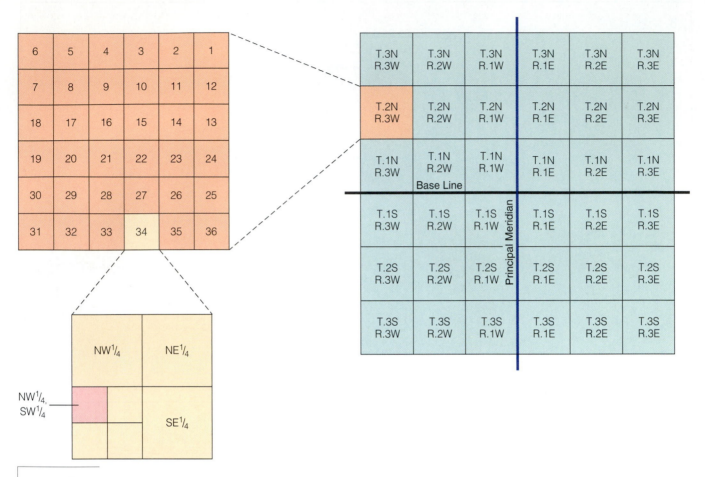

FIGURE D2 The General Land Office Grid System. Each 36-square-mile township is designated by township and range numbers. Townships are subdivided into sections, which can be further subdivided into quarter sections and quarter-quarter sections.

west of the prime meridian, which passes through Greenwich, England. Maps depicting all areas within the United States are noted in north latitude and west longitude. Latitude and longitude are noted in degrees and minutes at the corners of maps, but usually only minutes and seconds are shown along the margins. Many USGS topographic maps cover 7½ or 15 minutes of latitude and longitude and are thus referred to as 7½ and 15 minute quadrangles.

Beginning in 1812, the General Land Office (now known as the Bureau of Land Management) developed a standardized method for accurately defining the location of property in the United States. This method, known as the General Land Office Grid System has been used for all states except those along the eastern seaboard (except Florida), parts of Ohio, Tennessee, Kentucky, West Virginia, and Texas.

As new land acquired by the United States was surveyed, the surveyors laid out north–south lines they called *principal meridians* and east–west lines known as *base lines*. These intersecting lines form a set of coordinates for the location of specific pieces of property. The basic unit in the General Land Office Grid System is the *township*, an area measuring 6 miles on a side, and thus covering 36 square miles (Figure D2). Townships are numbered north and south of base lines and are designated as T.1N., T.1S., and so on. Rows of townships known as *ranges* are numbered east and west of principal meridians; R.2W and R.4E, for example. Note in Figure D2 that each township has a unique designation of township and range numbers.

Townships are subdivided into 36 1-square-mile (640-acre) *sections* numbered from 1 to 36. Because of surveying errors and the adjustments necessary to make a grid system conform to Earth's curved surface, not all sections are exactly 1 mile square. Nevertheless, each section can be further subdivided into half sections and quarter sections designated NE¼, NW¼, SE¼, and SW¼, and each quarter section can be further divided into quarter–quarter sections. To show the complete designation for an area, the smallest unit is noted first (quarter–quarter section) followed by quarter section, section number, township, and range. For example, the area indicated in Figure D2 is the NW¼, SW¼, Sec. 34, T.2N., R.3W.

Because only a few principal meridicans and base lines were established, they do not appear on most topographic maps. Nevertheless, township and range numbers are printed along the margins of 7½- and 15-minute quadrangles, and a grid consisting of red land boundaries depicts sections. In addition, each section number is shown in red within the map. However, small scale maps show only township and range.

WHERE TO OBTAIN TOPOGRAPHIC MAPS

Many people find topographic maps useful. Land-use planners, personnel in various local, state, and federal agencies, as well as engineers and real estate developers might use these maps for a variety of reasons. In addition, hikers, backpackers, and others interested in exploring undeveloped areas commonly use topographic maps because trails are shown by black dashed lines. Furthermore, map users can readily determine their location by interpreting the topographic features depicted by contours, and they can anticipate the type of terrain they will encounter during offroad excursions.

Topographic maps for local areas are available at some sporting goods stores, at National Park Visitor Centers, and from some state geologic surveys. Free index maps showing the names and locations of all quadrangles for each state are available from the USGS to anyone uncertain of which specific map is needed. Any published topographic map can be purchased from two main sources. For maps of areas east of the Mississippi River write to:

Branch of Distribution
U.S. Geological Survey
1200 S. Eads Street
Arlington, Virginia 22202

Maps for areas west of the Mississippi River can be obtained from:

Branch of Distribution
U.S. Geological Survey
Box 25286 Federal Center
Denver, Colorado 80225

Answers

MULTIPLE-CHOICE REVIEW QUESTIONS

CHAPTER 1
1. b; 2. a; 3. d; 4. a; 5. d; 6. c; 7. a; 8. c; 9. c; 10. e; 11. b.

CHAPTER 2
1. d; 2. a; 3. e; 4. d; 5. b; 6. c; 7. b; 8. b; 9. c; 10. a; 11. b; 12. c; 13. d.

CHAPTER 3
1. c; 2. d; 3. b; 4. a; 5. e; 6. c; 7. e; 8. d; 9. c; 10. b.

CHAPTER 4
1. b; 2. d; 3. a; 4. a; 5. b; 6. c; 7. e; 8. a; 9. e; 10. b.

CHAPTER 5
1. e; 2. a; 3. b; 4. d; 5. c; 6. a; 7. e; 8. c; 9. c; 10. d.

CHAPTER 6
1. a; 2. c; 3. e; 4. a; 5. c; 6. b; 7. a; 8. c; 9. a; 10. e.

CHAPTER 7
1. b; 2. a; 3. e; 4. b; 5. a; 6. e; 7. c; 8. d; 9. b; 10. e.

CHAPTER 8
1. b; 2. e; 3. a; 4. c; 5. a; 6. c; 7. c; 8. d; 9. d; 10. d; 11. a; 12. d; 13. b.

CHAPTER 9
1. e; 2. b; 3. a; 4. b; 5. d; 6. a; 7. c; 8. b; 9. c; 10. d.

CHAPTER 10
1. b; 2. d; 3. a; 4. e; 5. c; 6. c; 7. a; 8. b; 9. b; 10. e.

CHAPTER 11
1. e; 2. e; 3. b; 4. a; 5. e; 6. a; 7. e; 8. d; 9. c; 10. e.

CHAPTER 12
1. c; 2. c; 3. a; 4. e; 5. d; 6. c; 7. b; 8. e; 9. d; 10. b.

CHAPTER 13
1. e; 2. d; 3. e; 4. d; 5. e; 6. c; 7. e; 8. e; 9. e; 10. a; 11. c.

CHAPTER 14
1. d; 2. a; 3. b; 4. e; 5. a; 6. c; 7. b; 8. d; 9. a; 10. c.

CHAPTER 15
1. d; 2. a; 3. e; 4. d; 5. c; 6. e; 7. d; 8. e.

CHAPTER 16
1. b; 2. e; 3. c; 4. d; 5. a; 6. b; 7. a; 8. c; 9. e; 10. b.

CHAPTER 17
1. e; 2. c; 3. a; 4. e; 5. d; 6. d; 7. d; 8. b; 9. c; 10. e; 11. b; 12. a.

CHAPTER 18
1. d; 2. a; 3. a; 4. e; 5. d; 6. c; 7. b; 8. a; 9. a; 10. a; 11. c; 12. b; 13. d; 14. e.

CHAPTER 19
1. e; 2. b; 3. e; 4. c; 5. c; 6. e; 7. d; 8. c; 9. b; 10. b; 11. e; 12. a; 13. c; 14. d.

Glossary

A

aa A lava flow with a surface of rough, angular blocks and fragments.

abrasion The wearing and scraping of exposed rock surfaces by the impact of solid particles.

Absaroka sequence A widespread sequence of Pennsylvanian and Permian sedimentary rocks bounded above and below by unconformities; deposited during a transgressive-regressive cycle of the Absaroka Sea.

absolute dating The process of assigning ages in years before the present to geologic events. Various radioactive decay–dating techniques yield absolute ages. See also *relative dating*.

abyssal plain Vast, flat area on the seafloor adjacent to the continental rises of passive continental margins.

Acadian orogeny A Devonian orogeny in the northern Appalachian mobile belt resulting from a collision of Baltica with Laurentia.

active continental margin A continental margin characterized by volcanism and seismicity at the leading edge of a continental plate where oceanic lithosphere is subducted. See also *passive continental margin*.

aftershock An earthquake following a main shock resulting from adjustments along a fault. Aftershocks are common after a large earthquake, but most are smaller than the main shock.

alluvial fan A cone-shaped alluvial deposit formed where a stream flows from mountains onto an adjacent lowland.

alluvium A collective term for all detrital material transported and deposited by streams.

alpha decay A type of radioactive decay involving the emission of a particle consisting of two protons and two neutrons from the nucleus of an atom; decreases the atomic number by two and the atomic mass number by four.

angiosperm Any of the vascular plants having flowers and seeds; the flowering plants.

angular unconformity An unconformity below which older strata dip at a different angle (usually steeper) than the overlying strata. See also *disconformity* and *nonconformity*.

anticline An up-arched fold in which the oldest exposed rocks coincide with the fold axis, and all strata dip away from the axis.

aphanitic An igneous rock texture in which individual mineral grains are too small to be seen without magnification; results from rapid cooling and generally indicates an extrusive origin.

aquiclude Any material that prevents the movement of groundwater.

aquifer A permeable layer that allows the movement of groundwater.

archosaur One of a group of animals including dinosaurs, flying reptiles (pterosaurs), and birds.

arête A narrow, serrated ridge separating two glacial valleys or adjacent cirques.

artesian system A confined groundwater system in which high hydrostatic pressure builds, causing water in the system to rise above the level of the aquifer.

ash Pyroclastic material measuring less than 2 mm.

assimilation A process in which magma changes composition by reacting with country rock.

asthenosphere The part of the upper mantle over which lithospheric plates move.

atom The smallest unit of matter that retains the characteristics of an element.

atomic mass number The total number of protons and neutrons in the nucleus of an atom.

atomic number The number of protons in the nucleus of an atom.

aureole A zone surrounding a pluton in which contact metamorphism has taken place.

B

barchan dune A crescent-shaped sand dune with the tips of the crescent pointing downwind.

barrier island A long, narrow island composed of sand oriented parallel to the shoreline but separated from the mainland by a lagoon.

basal slip A type of glacial movement in which a glacier slides over its underlying surface.

basalt plateau A plateau built up by numerous lava flows from fissure eruptions.

base level The lowest limit to which a stream can erode.

basin The circular equivalent of a syncline. All strata in a basin dip toward a central point, and the youngest exposed rocks are in the center of the fold.

batholith A discordant, irregularly shaped pluton composed chiefly of granitic rocks. Has a surface area of at least 100 km^2.

baymouth bar A spit that has grown until it cuts off a bay from the open ocean or a lake.

beach A deposit of unconsolidated sediment extending landward from low tide to a change in topography or where permanent vegetation begins.

bed (bedding) Bed refers to an individual layer of rock, especially sedimentary rock, whereas bedding refers to the layered arrangement of rocks. See also *strata (stratification)*.

bedding plane The surface between one layer of sediment or sedimentary rock and another.

bed load The part of a stream's sediment load that is transported along its bed; consists of sand and gravel.

beta decay A type of radioactive decay during which a fast-moving electron is emitted from a neutron, thereby converting it to a proton; results in an increase of one atomic number, but does not change atomic mass number.

biochemical sedimentary rock A sedimentary rock resulting from the chemical processes of organisms.

bipedal Walking on two legs as a means of locomotion.

bonding The process whereby atoms are joined to other atoms.

Bowen's reaction series A mechanism accounting for the derivation of intermediate and felsic magmas from a mafic magma. It has a discontinuous branch of ferromagnesian silicates that change from one to another over specific temperature ranges and a continuous branch of plagioclase feldspars whose composition changes as the temperature decreases.

braided stream A stream possessing an intricate network of dividing and rejoining channels. Braiding occurs when sand and gravel bars are deposited within channels.

breaker A wave that oversteepens as it enters shallow water until its crest plunges forward.

butte An isolated, steep-sided, pinnacle-like erosional feature found in arid and semiarid regions.

C

caldera A large, steep-sided circular to oval volcanic depression usually formed by summit collapse resulting from partial draining of an underlying magma chamber.

carbon 14 dating technique An absolute age–dating method relying on determining the ratio of C^{14} to C^{12} in a sample; useful back to about 70,000 years ago; can be applied only to organic substances.

carbonate mineral A mineral containing the negatively charged carbonate ion $(CO_3)^{-2}$, e.g., calcite $(CaCO_3)$.

carbonate rock A sedimentary rock containing mostly carbonate minerals (e.g., limestone and dolostone).

cave A naturally formed subsurface opening that is generally connected to the surface and is large enough for a person to enter.

cementation The precipitation of minerals as binding material between and around sediment grains, thus converting sediment to sedimentary rock.

chemical sedimentary rock Rock formed of minerals derived from materials taken into solution during chemical weathering.

chemical weathering The decomposition of rock materials by chemical alteration of parent material.

cinder cone A small steep-sided volcano composed of pyroclastic materials that accumulated around a vent.

circum-Pacific belt A zone of seismic and volcanic activity that nearly encircles the Pacific Ocean basin.

cirque A steep-walled, bowl-shaped depression formed by erosion at the upper end of a glacial trough.

cleavage The breaking or splitting of mineral crystals along planes of weakness.

columnar joint A type of joint in some igneous rocks in which six-sided columns form as a result of cooling.

compaction A method of lithification whereby the pressure exerted by the weight of overlying sediment reduces the amount of pore space and thus the volume of a deposit.

complex movement A combination of different types of mass movements in which one type is not dominant; most complex movements involve sliding and flowing.

composite volcano A volcano composed of pyroclastic layers, lava flows typically of intermediate composition, and mudflows; also called stratovolcano.

compound A substance resulting from the bonding of two or more different elements, e.g., water (H_2O) and quartz (SiO_2).

compression Stress resulting when rocks are squeezed by external forces directed toward one another.

concordant Refers to plutons whose boundaries are parallel to the layering in the country rock. See also *discordant*.

cone of depression The cone-shaped depression in the water table around a well; results from pumping water from an aquifer faster than it can be replenished.

contact metamorphism Metamorphism of country rock adjacent to a pluton.

continental-continental plate boundary A convergent plate boundary along which two continental lithospheric plates collide, e.g., the collision of India with Asia. See also *convergent plate boundary, divergent plate boundary, oceanic-continental plate boundary,* and *oceanic-oceanic plate boundary*.

continental drift The theory that the continents were once joined into a single landmass that broke apart with the various fragments moving with respect to one another.

continental glacier A glacier covering at least 50,000 km^2 and unconfined by topography. Also called an ice sheet.

continental margin The area separating the part of a continent above sea level from the deep seafloor. Consists of a continental shelf and slope and, in some places, a rise.

convergent plate boundary The boundary between two plates that are moving toward one another See also *continental-continental plate boundary, divergent plate boundary, oceanic-continental plate boundary,* and *oceanic-oceanic plate boundary*.

core Earth's innermost part below the mantle at about 2900 km; divided into an outer liquid core and an inner solid core.

Coriolis effect The deflection of winds to the right of their direction of motion (clockwise) in the Northern Hemisphere and to the left (counterclockwise) in the Southern Hemisphere due to Earth's rotation.

correlation The demonstration of time equivalency of rock units in different areas.

covalent bond A bond formed by the sharing of electrons between atoms.

crater A circular or oval depression at the summit of a volcano resulting from the extrusion of gases, pyroclastic materials, and lava.

craton The relatively stable part of a continent; consists of a shield and a buried extension of a shield known as a platform; the ancient nucleus of a continent.

creep A type of mass wasting involving slow downslope movement of soil or rock.

cross-bedding Layers in sedimentary rocks that were deposited at an angle to the surface upon which they were accumulating.

crust The outermost part of Earth overlying the mantle.

crystal settling The physical separation and concentration of minerals in the lower part of a magma chamber by crystallization and gravitational settling.

crystalline solid A solid in which the constituent atoms are arranged in a regular, three-dimensional framework.

Curie point The temperature at which iron-bearing minerals in a cooling magma attain their magnetism.

cyclothem A vertical sequence of cyclically repeated sedimentary rocks resulting from alternating periods of marine and nonmarine deposition; commonly contain a coal bed.

D

daughter element An element formed by the radioactive decay of another element, e.g., argon 40 is the daughter element of potassium 40. See also *parent element*.

debris flow A mass wasting process; much like a mudflow but more viscous and containing larger particles.

deflation The removal of loose surface sediment by the wind.

deformation A general term referring to any change in shape or volume, or both, of rocks in response to stress. Deformation involves folding and fracturing.

delta An alluvial deposit formed where a stream discharges into a lake or the sea.

depositional environment Any area where sediment is deposited such as on a stream's floodplain or on a beach.

desert Any area that receives less than 25 cm of rain per year and has a high evaporation rate.

desert pavement A surface mosaic of close-fitting gravel particles formed by the removal of sand-sized and smaller particles by the wind.

desertification The expansion of deserts into formerly productive lands.

detrital sedimentary rock Sedimentary rock consisting of the solid particles (detritus) of preexisting rocks, e.g., sandstone and conglomerate.

differential pressure Pressure that is not applied equally to all sides of a rock body.

differential weathering Weathering of rock at different rates, producing an uneven surface.

dike A tabular or sheetlike discordant pluton.

dinosaur Any of the Mesozoic reptiles belonging to the groups designated as ornithischians and saurischians.

dip A measure of the maximum angular deviation of an inclined plane from horizontal.

dip-slip fault A fault on which all movement is parallel with the dip of the fault plane. See also *normal fault* and *reverse fault*.

discharge The volume of water in a stream moving past a particular point in a given period of time.

disconformity An unconformity above and below which the strata are parallel. See also *angular unconformity* and *nonconformity*.

discontinuity A boundary across which seismic wave velocity abruptly changes, e.g., the mantle-core boundary.

discordant Refers to plutons with boundaries cutting across the layering in the country rock. See also *concordant*.

dissolved load The part of a stream's load consisting of ions in solution.

divergent plate boundary The boundary between two plates that are moving apart. See also *continental-continental plate boundary, convergent plate boundary, oceanic-continental plate boundary,* and *oceanic-oceanic plate boundary*.

divide A topographically high area that separates adjacent drainage basins.

dome A circular equivalent of an anticline. All strata dip outward from a central point, and the oldest exposed rocks are in the center of the dome.

drainage basin The surface area drained by a stream and its tributaries.

drainage pattern The regional arrangement of stream channels in a drainage system.

dripstone Various cave deposits resulting from the deposition of calcite.

drumlin An elongate hill of till formed by the movement of a continental glacier or floods.

dune A mound or ridge of wind-deposited sand.

dynamic metamorphism Metamorphism occurring in fault zones where rocks are subjected to high differential pressure.

E

earthflow A type of mass wasting process; involves downslope flowage of water-saturated soil.

earthquake Vibrations caused by the sudden release of energy, usually as a result of displacement of rocks along faults.

elastic rebound theory A theory that explains the sudden release of energy when rocks are deformed by movement on a fault.

elastic deformation A type of deformation in which the material returns to its original shape when stress is relaxed.

electron A negatively charged particle of very little mass that orbits the nucleus of an atom.

electron capture A type of radioactive decay in which a proton captures an electron and is thereby converted to a neutron; results in a loss of one atomic number but no change in atomic mass number.

electron shell Electrons orbit rapidly around the nuclei of atoms at specific distances known as electron shells.

element A substance composed of all the same atoms; it cannot be changed into another element by ordinary chemical means.

emergent coast A coast where the land has risen with respect to sea level.

end moraine A pile of rubble deposited at the terminus of a glacier. See also *medial moraine, lateral moraine, recessional moraine,* and *terminal moraine.*

epicenter The point on Earth's surface directly above the focus of an earthquake.

erosion The removal of weathered materials from their source area.

esker A long, sinuous ridge of stratified drift formed by deposition by running water in a tunnel beneath stagnant ice.

evaporite A sedimentary rock formed by inorganic chemical precipitation of minerals from solution, e.g., rock salt and rock gypsum.

Exclusive Economic Zone An area extending 371 km seaward from the coast of the United States and its territories in which the United States claims all sovereign rights.

exfoliation The process whereby slabs of rock bounded by sheet joints slip or slide off the host rock.

exfoliation dome A large rounded dome of rock resulting from the process of exfoliation.

F

fault A fracture along which movement has occurred parallel to the fracture surface.

felsic magma A type of magma containing more than 65% silica and considerable sodium, potassium, and aluminum, but little calcium, iron, and magnesium. See also *intermediate magma* and *mafic magma.*

ferromagnesian silicate A silicate mineral containing iron and magnesium or both. See also *nonferromagnesian silicate.*

fetch The distance the wind blows over a continuous water surface.

fiord A glacial valley below sea level.

firn Granular snow formed by the partial melting and refreezing of snow.

fissure eruption An eruption in which lava or pyroclastic materials are emitted from a long, narrow fissure or group of fissures.

floodplain A low-lying, relatively flat area adjacent to a stream that is partly or completely covered by water when the stream overflows its banks.

fluid activity An agent of metamorphism in which water and carbon dioxide promote metamorphism by increasing the rate of chemical reactions.

focus The place within Earth where an earthquake originates and energy is released.

foliated texture A texture of metamorphic rocks in which platy and elongate minerals are arranged in a parallel fashion.

footwall block The block of rock that lies beneath a fault plane.

fossil Remains or traces of prehistoric organisms preserved in rocks of the crust.

fracture A break in rock resulting from intense applied pressure.

frost action The mechanical weathering process that disaggregates rocks by repeated freezing and thawing of water in cracks and crevices.

frost heaving The process whereby a mass of sediment or soil undergoes freezing, expansion, and actual lifting, followed by thawing, contraction, and lowering of the mass.

frost wedging The opening and widening of cracks by the repeated freezing and thawing of water.

G

geologic time scale A chart with the designation for the earliest interval of geologic time at the bottom, followed upward by designations for more recent intervals of time.

geology The science concerned with the study of Earth; includes studies of Earth materials (minerals and rocks), surface and internal processes, and Earth history.

geothermal energy Energy that comes from the steam and hot water trapped within the crust.

geothermal gradient The temperature increase with depth; it averages about 25°C/km near the surface.

geyser A hot spring that intermittently ejects hot water and steam.

glacial budget The balance between accumulation and wastage in a glacier.

glacial drift A collective term for all sediment deposited by glacial activity, including till deposited directly by glacial ice and outwash deposited by streams derived from melting ice.

glacial ice Ice that has formed from firn.

glacier A mass of ice on land that moves by plastic flow and basal slip.

Glossopterisflora A Late Paleozoic association of plants found only on the Southern Hemisphere continents and India.

Gondwana One of six large Paleozoic continents; composed mostly of present-day South America, Africa, Antarctica, Australia, and India.

graded bedding A type of sedimentary bedding in which an individual bed shows a decrease in grain size from bottom to top.

graded stream A stream possessing an equilibrium profile in which a delicate balance exists between gradient, discharge, flow velocity, channel characteristics, and sediment load so that neither significant erosion nor deposition occurs within the channel.

gradient The slope over which a stream flows; usually expressed in m/km or ft/mi.

gravity anomaly Departure from the expected force of gravity. Anomalies may be positive or negative, indicating a mass excess and mass deficiency, respectively.

ground moraine A deposit formed of sediment liberated from melting ice as a glacier's terminus retreats.

groundwater The underground water stored in the pore spaces in rocks, sediment, or soil.

guide fossil Any fossil that can be used to determine the relative geologic ages of rocks and to correlate rocks of the same age in different areas.

guyot A flat-topped seamount of volcanic origin rising more than 1 km above the seafloor.

gymnosperm The flowerless, seed-bearing land plants.

H

half-life The time required for one-half of the original number of atoms of a radioactive element to decay to a stable daughter product, e.g., the half-life of potassium 40 is 1.3 billion years.

hanging valley A tributary glacial valley whose floor is at a higher level than that of the main glacial valley.

hanging wall block The block of rock that lies above a fault plane.

heat An agent of metamorphism.

Hercynian-Alleghanian orogeny Pennsylvanian to Permian orogeny in the Hercynian mobile belt of southern Europe and the Appalachian mobile belt from New York to Alabama.

hominid Abbreviated form of Hominidae, the family to which humans belong. Bipedal primates such as *Australopithecus* and *Homo* are hominids.

horn A steep-walled, pyramidal peak formed by the headward erosion of cirques.

hot spot A localized zone of melting below the lithosphere.

hot spring A spring in which the water temperature is warmer than the temperature of the human body (37°C).

humus The material in soils derived by bacterial decay of organic matter.

hydraulic action The power of moving water.

hydrologic cycle The continuous recycling of water from the oceans, through the atmosphere, to the continents, and back to the oceans.

hydrolysis The chemical reaction between the hydrogen (H^+) ions and hydroxyl (OH^-) ions of water and a mineral's ions.

hypothesis A provisional explanation for observations; subject to continual testing and modification. If well supported by evidence, hypotheses are then generally called theories.

I

ichthyosaur Any of the porpoiselike, Mesozoic marine reptiles.

igneous rock Any rock formed by cooling and crystallization of magma or lava, or by the accumulation and consolidation of pyroclastic materials.

incised meander A deep, meandering canyon cut into bedrock by a stream.

index mineral A mineral that forms within a specific temperature and pressure range during metamorphism.

infiltration capacity The maximum rate at which sediment or soil can absorb water.

inselberg An isolated steep-sided erosional remnant rising above a desert plain.

intensity The subjective measure of the kind of damage done by an earthquake as well as people's reaction to it.

intermediate magma A magma having a silica content of 53 to 65% and an overall composition intermediate between felsic and mafic magmas. See also *felsic magma* and *mafic magma*.

intrusive igneous rock See *plutonic rock*.

ion An electrically charged atom produced by adding or removing electrons from its outermost electron shell.

ionic bond A bond resulting from the attraction of positively and negatively charged ions.

isostatic rebound The phenomenon in which unloading of the crust causes it to rise upward until equilibrium is again attained. See also *principle of isostasy*.

J

joint A fracture along which no movement has occurred parallel with the fracture surface.

K

kame Conical hill of stratified drift originally deposited in a depression on a glacier's surface.

karst topography A topography with numerous caves, springs, sinkholes, solution valleys, and disappearing streams developed by groundwater erosion.

Kaskaskia sequence A widespread sequence of Devonian and Mississippian sedimentary rocks bounded above and below by unconformities; deposited during a transgressive-regressive cycle of the Kaskaskia Sea.

L

laccolith A concordant pluton with a mushroomlike geometry.

lahar A mudflow consisting of volcanic materials such as ash.

Laramide orogeny A Late Cretaceous to Early Cenozoic episode of deformation in the area of the present-day Rocky Mountains.

lateral moraine The sediment deposited as a long ridge of till along the margin of a valley glacier.

laterite A red soil rich in iron or aluminum or both that forms in the tropics by intense chemical weathering.

Laurasia A Late Paleozoic Northern hemisphere continent consisting of present-day continents of North America, Greenland, Europe, and Asia.

Laurentia The name given to a Proterozoic continent composed mostly of North America and Greenland and parts of Scotland and Scandinavia.

lava Magma at the surface; the molten rock material that flows from a volcano or a fissure.

lava dome A bulbous, steep-sided mass of very viscous magma forced upward through a volcanic conduit.

lava flow A stream of magma issuing from a volcano or fissure.

leaching The dissolution or removal of soluble minerals from a soil or rock by percolating water.

lithification The process of converting sediment into sedimentary rock.

lithosphere Earth's outer, rigid part consisting of the upper mantle, oceanic crust, and continental crust.

lithostatic pressure Pressure exerted on rock by the weight of overlying rock; it is applied equally in all directions.

loess Windblown silt and clay deposits derived from deserts, Pleistocene glacial outwash, and floodplains of streams in semiarid regions.

longitudinal dune A long ridge of sand aligned generally parallel to the direction of the prevailing wind.

longshore current A current between the breaker zone and the beach, flowing parallel to the shoreline and produced by wave refraction.

longshore drift The movement of sediment along a shoreline by longshore currents.

low-velocity zone The zone within the mantle between 100 and 250 km deep where the velocity of both P- and S-waves decreases markedly; it corresponds closely to the asthenosphere.

M

mafic magma A silica-poor magma containing 45 to 52% silica and proportionately more calcium, iron, and magnesium than intermediate and felsic magmas. See also *felsic magma* and *intermediate magma.*

magma Molten rock material generated within Earth.

magma chamber A cavity containing a reservoir of magma within the upper mantle or lower crust.

magma mixing The process of mixing magmas of different composition, thereby producing a modified version of the parent magmas.

magnetic anomaly Any change, such as a change in average strength, of the magnetic field.

magnetic reversal The phenomenon in which the north and south magnetic poles are completely reversed.

magnitude The total amount of energy released by an earthquake at its source. See also *Richter Magnitude Scale.*

mantle The thick layer between Earth's crust and core.

mantle plume A stationary column of magma originating deep within the mantle that slowly rises to the surface to form volcanoes or flood basalts.

marine terrace A wave-cut platform now elevated above sea level.

marsupial Any of the pouched mammals such as opossums, kangaroos, and wombats. At present, marsupials are most common in Australia.

mass wasting The downslope movement of rock, sediment, or soil under the influence of gravity.

meandering stream A stream possessing a single channel with broadly looping curves.

mechanical weathering The disaggregation of rock materials by physical forces yielding smaller pieces that retain the chemical composition of the parent material.

medial moraine A moraine formed where lateral moraines of two valley glaciers merge. See also *end moraine, lateral moraine, recessional moraine,* and *terminal moraine.*

Mediterranean belt A zone of seismic and volcanic activity extending westerly from Indonesia through the Himalayas, across Iran and Turkey, and through the Mediterranean region of Europe.

mesa A broad, flat-topped erosional remnant bounded on all sides by steep slopes and capped by resistant rock.

metamorphic rock Any rock that has been altered by heat, pressure, or chemical fluids or a combination of these agents of metamorphism.

metamorphic zone The region between lines of equal metamorphic intensity.

microplate A small lithospheric plate that is clearly of different origin than rocks of the surrounding area.

Milankovitch theory A theory that explains cyclic variations in climate and the onset of glacial episodes as a consequence of irregularities in Earth's rotation and orbit.

mineral A naturally occurring, inorganic, crystalline solid having characteristic physical properties and a narrowly defined chemical composition.

mobile belt An elongated area of deformation as indicated by folds and faults; generally adjacent to a craton.

Modified Mercalli Intensity Scale A scale having values ranging from I to XII that is used to characterize earthquake intensity based on damage.

Mohorovičić discontinuity (Moho) The boundary between Earth's crust and mantle.

monocline A simple bend or flexure in otherwise horizontal or uniformly dipping rock layers.

mud crack A sedimentary structure found in clay-rich sediment that has dried out. When drying occurs, the sediment shrinks and intersecting fractures form.

mudflow A mass wasting process; a flow consisting of mostly clay- and silt-sized particles and more than 30% water.

N

native element A mineral composed of a single element, e.g., gold and silver.

natural levee A ridge of sandy alluvium deposited along the margins of a stream channel during floods.

nearshore sediment budget The balance between additions and losses of sediment in the nearshore zone.

neutron An electrically neutral particle found in the nucleus of an atom.

Nevadan orogeny Late Jurassic to Cretaceous deformation that strongly affected the western part of North America.

nonconformity An unconformity in which stratified rocks above an erosion surface overlie igneous or metamorphic rocks. See also *angular unconformity* and *disconformity*.

nonferromagnesian silicate A silicate mineral that does not contain iron or magnesium. See also *ferromagnesian silicate*.

nonfoliated texture A metamorphic texture in which there is no discernible preferred orientation of minerals.

normal fault A dip-slip fault on which the hanging wall block has moved down relative to the footwall block. See also *dip-slip fault* and *reverse fault*.

nucleus The central part of an atom consisting of one or more protons and neutrons.

nuée ardente A mobile dense cloud of hot pyroclastic materials and gases ejected from a volcano.

O

oblique-slip fault A fault having both dip-slip and strike-slip movement.

oceanic-continental plate boundary A type of convergent plate boundary along which oceanic lithosphere and continental lithosphere collide; characterized by subduction of the oceanic plate beneath the continental plate and by volcanism and seismicity. See also *continental-continental plate boundary, convergent plate boundary, divergent plate boundary,* and *oceanic-oceanic plate boundary.*

oceanic-oceanic plate boundary A type of convergent plate boundary along which two oceanic lithospheric plates collide and one is subjected beneath the other; characterized by seismicity, volcanism, and the origin of a volcanic island arc. See also *continental-continental plate boundary, convergent plate boundary, divergent plate boundary,* and *oceanic-continental plate boundary.*

oceanic ridge A submarine mountain system found in all of the oceans; it is composed of volcanic rock (mostly basalt) and displays features produced by tension.

oceanic trench A long, narrow depression in the seafloor where subduction occurs.

ooze Deep-sea pelagic sediment composed mostly of shells of marine animals and plants.

Ornithischia An order of dinosaurs that includes ornithopods, stegosaurs, ankylosaurs, pachycephalosaurs, and ceratopsians. Dinosaurs characterized by a birdlike pelvis.

orogeny The process of forming mountains, especially by folding and thrust faulting; an episode of mountain building.

Ouachita orogeny An orogeny involving deformation of the Ouachita mobile belt during the Pennsylvanian Period.

outwash plain The sediment deposited by the meltwater discharging from the terminus of a continental glacier.

oxbow lake A cutoff meander filled with water.

oxidation The reaction of oxygen with other atoms to form oxides or, if water is present, hydroxides.

P

pahoehoe A type of lava flow with a smooth ropy surface.

paleomagnetism The study of remanent magnetism in rocks so that the intensity and direction of Earth's past magnetic field can be determined.

Pangaea The name proposed by Alfred Wegener for a supercontinent consisting of all landmasses that existed at the end of the Paleozoic Era.

parabolic dune A crescent-shaped dune in which the tips of the crescent point upwind.

parent element An unstable element that changes by radioactive decay into a stable daughter element. See also *daughter element*.

parent material The material that is mechanically and chemically weathered to yield sediment and soil.

passive continental margin The trailing edge of a continental plate consisting of a broad continental shelf and a continental slope and rise, commonly with an abyssal plain adjacent to the rise. Passive continental margins lack volcanism and intense seismic activity. See also *active continental margin*.

pedalfer A soil formed in humid regions with an organic-rich A horizon and aluminum-rich clays and iron oxides in horizon B.

pediment An erosion surface of low relief gently sloping away from a mountain base.

pedocal A soil of arid and semiarid regions with a thin A horizon and a calcium carbonate-rich B horizon.

pelagic clay Generally brown or reddish deep-sea sediment composed of clay-sized particles derived from the continents and oceanic islands.

permeability A material's capacity for transmitting fluids.

phaneritic A coarse-grained texture in igneous rocks in which the minerals are easily visible without magnification; results from slow cooling and generally indicates an intrusive origin.

pillow lava Bulbous masses of basalt resembling pillows formed when lava is rapidly chilled underwater.

placental Any of the mammals that have a placenta to nourish the embryo; most living and fossils mammals are placentals.

plastic flow The flow that occurs in response to pressure and causes permanent deformation.

plastic deformation The result of stress in which a material cannot recover its original shape and retains the configuration produced by the stress such as folding of rocks.

plate An individual piece of lithosphere that moves over the asthenosphere.

plate tectonic theory The theory that large segments or plates of lithosphere move relative to one another.

playa A dry lake bed found in deserts.

plesiosaur A type of Mesozoic marine reptile.

plunging fold A fold with an inclined axis.

pluton An intrusive igneous body that forms when magma cools and crystallizes within the crust, e.g., batholith and sill.

plutonic (intrusive igneous) rock Igneous rock that crystallizes from magma intruded into or formed in place within the crust.

point bar The sedimentary body deposited on the gently sloping side of a meander loop.

porosity The percentage of a material's total volume that is pore space.

porphyritic An igneous texture with mineral grains of markedly different sizes.

pressure release A mechanical weathering process in which rocks formed far below the surface expand upon being exposed at the surface due to release of pressure.

primate Any of the mammals belonging to the order Primates; characteristics include large brain, stereoscopic vision, and grasping hand.

principle of cross-cutting relationships A principle used to determine the relative ages of events; holds that an igneous intrusion or fault must be younger than the rocks that it intrudes or cuts.

principle of fossil succession A principle holding that fossils, and especially assemblages of fossils, succeed one another through time in a regular and determinable order.

principle of inclusions A principle that holds that inclusions, or fragments, in a rock unit are older than the rock unit itself, e.g., granite fragments in a sandstone are older than the sandstone.

principle of isostasy The theoretical concept of Earth's crust "floating" on a denser underlying layer.

principle of lateral continuity A principle that holds that sediment layers extend outward in all directions until they terminate.

principle of original horizontality A principle that holds that sediment layers are deposited horizontally or very nearly so.

principle of superposition A principle that holds that younger layers of strata are deposited on top of older strata.

principle of uniformitarianism A principle that holds that we can interpret past events by understanding present-day processes; based on the assumption that natural laws have not changed through time.

proton A positively charged particle found in the nucleus of an atom.

pterosaur Any of the Mesozoic flying reptiles.

P-wave A compressional, or push-pull, wave; the fastest seismic wave and one that can travel through solids, liquids, and gases; also known as a primary wave.

P-wave shadow zone The area between 103° and 143° from an earthquake focus where little P-wave energy is recorded.

pyroclastic materials Fragmental material such as ash explosively ejected from a volcano.

pryoclastic (fragmental) texture A fragmental texture found in igneous rocks composed of pyroclastic materials.

pyroclastic sheet deposit Vast, sheet-like deposits of felsic pyroclastic materials erupted from fissures.

Q

quadrupedal Walking on four legs as a means of locomotion.

quick clay A clay that spontaneously liquefies and flows like water when disturbed.

R

radioactive decay The spontaneous change of an atom to an atom of a different element.

rainshadow desert A desert found on the leeward side of a mountain range; forms because moist marine air moving inland yields precipitation on the windward side of the mountain range and the air descending on the leeward side is much warmer and drier.

rapid mass movement A type of mass movement involving a visible movement of material.

recessional moraine A type of end moraine formed when a glacier's terminus retreats, then stabilizes and deposits till. See also *end moraine, lateral moraine, medial moraine,* and *terminal moraine.*

reef A moundlike, wave-resistant structure composed of the skeletons of organisms.

reflection The return to the source of some of a seismic wave's energy when it encounters a boundary separating materials of different density or elasticity.

refraction The change in direction and velocity of a seismic wave when it travels from one material into another of different density or elasticity.

regional metamorphism Metamorphism that occurs over a large area resulting from high temperature and pressure, and the action of chemical fluids within the crust.

regolith The layer of unconsolidated rock and mineral fragments and soil that covers much of Earth's surface.

relative dating The process of determining the age of an event relative to other events; involves placing geologic events in their correct chronological order, but involves no consideration of when the events occurred in terms of number of years ago. See also *absolute dating.*

reserve The part of the resource base that can be extracted economically.

resource A concentration of naturally occurring solid, liquid, or gaseous material in or on Earth's crust in such form and amount that economic extraction of a commodity from the concentration is currently or potentially feasible.

reverse fault A dip-slip fault in which the hanging wall block moves upward relative to the footwall block. See also *dip-slip fault* and *normal fault*.

Richter Magnitude Scale An open-ended scale that measures the amount of energy released during an earthquake.

rill erosion Erosion by running water that scours small channels in the ground.

rip current A narrow surface current that flows out to sea through the breaker zone.

ripple mark Wavelike (undulating) structure produced in granular sediment such as sand by unidirectional wind and water currents, or by oscillating wave currents.

rock An aggregate of one or more minerals, as in limestone and granite, or a consolidated aggregate of particles of other rocks, as in sandstone and conglomerate; although exceptions to this definition, coal and natural glass are considered rocks.

rock cycle A sequence of processes through which Earth materials may pass as they are transformed from one rock type to another.

rockfall A common type of extremely rapid mass wasting in which rocks fall through the air.

rock-forming mineral A common mineral that comprises a significant portion of a rock.

rock slide A type of rapid mass wasting in which rocks move downslope along a more or less planar surface.

rounding The process by which the sharp corners and edges of sedimentary particles are abraded during transport.

runoff The surface flow of streams.

S

salt crystal growth A mechanical weathering process in which rocks are disaggregated by the growth of salt crystals in crevices and pores.

Sauk sequence A widespread sequence of sedimentary rocks bounded above and below by unconformities; deposited during a latest Proterozoic to Early Ordovician transgressive-regressive cycle of the Sauk Sea.

Saurischia An order of dinosaurs that includes sauropods and theropods. Dinosaurs characterized by a lizardlike pelvis.

scientific method A logical, orderly approach that involves gathering data, formulating and testing hypotheses, and proposing theories.

seafloor spreading The theory that the seafloor moves away from spreading ridges and is eventually consumed at subduction zones.

seamount A structure of volcanic origin rising more than 1 km above the seafloor.

sediment Loose aggregate of solids derived from preexisting rocks, or solids precipitated from solution by inorganic chemical processes or extracted from solution by organisms.

sedimentary rock Any rock composed of sediment, e.g., sandstone and limestone.

sedimentary structure Any structure in sedimentary rock such as cross-bedding, mud cracks, and ripple marks that formed at the time of deposition or shortly thereafter.

seedless vascular plant A type of land plant with vascular tissues for transport of fluids and nutrients throughout the plant; reproduces by spores rather than seeds, e.g., ferns and horsetail rushes.

seismology The study of earthquakes.

Sevier orogeny Cretaceous deformation that affected the continental shelf and slope areas of the Cordilleran mobile belt.

shear The result of forces acting parallel to one another but in opposite directions; results in deformation by displacement of adjacent layers along closely spaced planes.

shear strength The resisting forces helping to maintain slope stability.

sheet erosion Erosion that is more or less evenly distributed over the surface and removes thin layers of soil.

sheet joint A large fracture more or less parallel to a rock surface resulting from pressure released by expansion of the rock.

shield volcano A large, dome-shaped volcano with a low rounded profile built up mostly of overlapping basalt lava flows (e.g., Mauna Loa and Kilauea on the island of Hawaii).

shoreline The line of intersection between the sea or a lake and the land.

silica A compound of silicon and oxygen atoms.

silica tetrahedron The basic building block of all silicate minerals. It consists of one silicon atom and four oxygen atoms.

silicate A mineral containing silica (e.g., quartz [SiO_2] and orthoclase [$KAlSi_3O_8$]).

sill A tabular or sheetlike concordant pluton.

sinkhole A depression in the ground that forms in karst regions by the solution of the underlying carbonate rocks or by the collapse of a cave roof.

slide A type of mass movement involving movement of material along one or more surfaces of failure.

slow mass movement Mass movement that advances at an imperceptible rate and is usually only detectable by its effects.

slump A type of mass wasting that occurs along a curved surface of failure and results in the backward rotation of the slump block.

soil Regolith consisting of weathered material, water, air, and humus that can support plants.

soil degradation Any processes leading to a loss of soil productivity; may involve erosion, chemical pollution, and compaction.

soil horizon A distinct soil layer that differs from other soil layers in texture, structure, composition, and color.

solifluction A type of mass wasting involving the slow downslope movement of water-saturated surface materials.

solution A reaction in which the ions of a substance become dissociated in a liquid, and the solid substance dissolves.

sorting A term referring to the degree to which all particles in sediment or sedimentary rock are about the same size.

spheroidal weathering A type of chemical weathering in which corners and sharp edges of angular rocks weather more rapidly than flat surfaces, thus yielding spherical shapes.

spit A continuation of a beach forming a point of land that projects into a body of water, commonly a bay.

spring A place where groundwater flows or seeps out of the ground. Springs occur where the water table intersects the ground surface.

stock An irregularly-shaped discordant pluton with a surface area less than 100 km².

stoping A process in which rising magma detaches and engulfs pieces of country rock.

strata (stratification) Strata (singular *stratum*) refers to the layers in sedimentary rocks, whereas stratification refers to the layered aspect of sedimentary rocks. See also *bed (bedding)*.

stratified drift Glacial drift displaying both sorting and stratification.

stream Runoff confined to channels regardless of size.

stream terrace An erosional remnant of a floodplain that formed when a stream was flowing at a higher level.

strike The direction of a line formed by the intersection of a horizontal plane with an inclined plane, such as a rock layer.

strike-slip fault A fault involving horizontal movement so that blocks on opposite sides of a fault plane slide sideways past one another.

stromatolite A structure in sedimentary rocks, especially limestones, produced by entrapment of sediment grains on sticky mats of photosynthesizing bacteria.

subduction zone A long, narrow zone at a convergent plate boundary where an oceanic plate descends beneath another plate, e.g., the subduction of the Nazca plate beneath the South American plate.

submergent coast A coast along which sea level rises with respect to the land or the land subsides.

surface wave An earthquake wave that travels along or just below Earth's surface. Produces a rolling or swaying motion rather than the sharp jolts caused by body waves.

suspended load The smallest particles carried by a stream, such as silt and clay, which are kept suspended by fluid turbulence.

s-wave A shear wave that moves material perpendicular to the direction of travel, thereby producing shear stresses in the material it moves through; also known as a secondary wave.

s-wave shadow zone Those areas more than 103° from an earthquake focus where no S-waves are recorded.

syncline A down-arched fold in which the youngest exposed rocks coincide with the fold axis, and all strata dip inward toward the axis.

system A combination of related parts that interact in an organized fashion. Earth systems include the atmosphere, hydrosphere, biosphere, and solid Earth.

T

Taconic orogeny An Ordovician orogeny that resulted in deformation of the Appalachian mobile belt.

talus An accumulation of angular pieces of mechanically weathered rock at the base of a slope.

tension A type of stress in which forces act in opposite directions but along the same line and tend to stretch an object.

terminal moraine A type of end moraine; the outermost moraine marking the greatest extent of a glacier. See also *end moraine, lateral moraine, medial moraine,* and *recessional moraine.*

theory An explanation for some natural phenomenon that has a large body of supporting evidence; to be considered scientific, a theory must be testable, e.g., plate tectonic theory.

thermal convection cell A type of circulation of material in the asthenosphere during which hot material rises, moves laterally, cools and sinks, and is reheated and repeats the cycle.

thermal expansion and contraction A type of mechanical weathering in which the volume of rock changes in response to heating and cooling.

tide The regular fluctuation in the sea's surface in response to the gravitational attraction of the Moon and Sun.

till All sediment deposited directly by glacial ice.

Tippecanoe sequence A widespread sequence of sedimentary rocks bounded above and below by unconformities; deposited during an Ordovician to Early Devonian transgressive-regressive cycle of the Tippecanoe Sea.

transform fault A type of fault that changes one type of motion between plates into another type of motion.

transform plate boundary Plate boundary along which plates slide past one another, and crust is neither produced nor destroyed; on land recognized as a strike-slip fault.

transport The mechanism by which weathered material is moved from one place to another, commonly by running water, wind, or glaciers.

transverse dune A long ridge of sand perpendicular to the prevailing wind direction.

tsunami A destructive sea wave that is usually produced by an earthquake but can also be caused by submarine landslides or volcanic eruptions.

U

unconformity   An erosion surface separating younger strata from older rocks. See also *angular unconformity, disconformity,* and *nonconformity.*

U-shaped glacial trough   A valley with very steep or vertical walls and a broad, rather flat floor. Formed by the movement of a glacier through a stream valley.

V

valley glacier   A glacier confined to a mountain valley or to an interconnected system of mountain valleys.

valley train   A long, narrow deposit of stratified drift confined within a glacial valley.

velocity   A measure of the downstream distance water travels per unit of time.

ventifact   A stone whose surface has been polished, pitted, grooved, or faceted by wind abrasion.

vertebrate   Any animal having a segmented vertebral column; includes fish, amphibians, reptiles, mammals, and birds.

vesicle   A small hole or cavity formed by gas trapped in cooling lava.

viscosity   A fluid's resistance to flow.

volcanic island arc   A curved chain of volcanic islands parallel to a deep-sea trench where oceanic lithosphere is subducted causing volcanism and the origin of volcanic islands.

volcanic neck   An erosional remnant of the material that solidified in a volcanic pipe.

volcanic pipe   The conduit connecting the crater of a volcano with an underlying magma chamber.

volcanic (extrusive igneous) rock   Igneous rock formed when magma is extruded onto the surface where it cools and crystallizes, or when pyroclastic materials become consolidated.

volcanism   The process whereby magma and its associated gases rise through the crust and are extruded onto the surface or into the atmosphere.

volcano   A conical mountain formed around a vent as a result of the eruption of lava and pyroclastic materials.

W

water table   The surface separating the zone of aeration from the underlying zone of saturation.

water well   A well made by digging or drilling into the zone of saturation.

wave   An undulation on a water surface.

wave base   A depth of about one-half wavelength, where the diameter of the orbits of water particles in waves is essentially zero; the depth below which water is not affected by surface waves.

wave-cut platform   A beveled surface that slopes gently in a seaward direction; formed by erosion and landward retreat of a sea cliff.

wave refraction   The bending of waves so that they more nearly parallel the shoreline.

weathering   The physical breakdown and chemical alteration of rocks and minerals at or near Earth's surface.

Z

zone of accumulation   In soil terminology, another name for horizon B where soluble minerals leached from horizon A accumulate as irregular masses. In glacial terminology, the part of a glacier where additions exceed losses and the glacier's surface is perennially covered by snow.

zone of aeration   The zone above the water table that contains both water and air within the pore spaces of the rock, sediment, or soil.

zone of saturation   The zone below the water table in which all pore spaces are filled with groundwater.

zone of wastage   The part of a glacier where losses from melting, sublimation, and calving of icebergs exceed the rate of accumulation.

Additional Readings

CHAPTER 1

Dietrich, R. V. 1989. Rock music. *Earth Science* 42, no. 2: 24–25.

———. 1991. How can I get others interested in rocks? *Rocks & Minerals* 67, no. 2: 124–128.

Dietrich, R. V., and B. J. Skinner. 1990. *Gems, granites, and gravels*. New York: Cambridge University Press.

Ellis, W. S. 1990. A Soviet sea lies dying. *National Geographic* 177, no. 2: 73–93.

Francis, P., and S. Self. 1983. The eruption of Krakatau. *Scientific American* 249, no. 5: 172–187.

Holloway, M. 1993. Sustaining the Amazon. *Scientific American* 269, no. 1: 90–100.

Micklin, P. P. 1993. The shrinking Aral Sea. *Geotimes* 38, no. 4: 14–18.

Mirsky, A. 1989. Geology in our everyday lives. *Journal of Geological Education* 37, no. 1: 9–12.

Monroe, J. S., and R. Wicander. 1997. *The changing Earth: Exploring geology and evolution*. 2d ed. Belmont, Calif.: West/Wadsworth.

Officer, C., and J. Page. 1993. *Tales of the Earth*. New York: Oxford University Press.

Pestrong, R. 1994. Geoscience and the arts. *Journal of Geological Education* 42, no. 3: 249–257.

World Commission on Environment and Development. 1987. *Our common future*. New York: Oxford University Press.

CHAPTER 2

Boling, R. 1996. How to move a continent. *Earth* 5, no. 2: 14.

Bonatti, E. 1987. The rifting of continents. *Scientific American* 256, no. 3: 96–103.

Brimhall, G. 1991. The genesis of ores. *Scientific American* 264, no. 5: 84–91.

Condie, K. 1989. *Plate tectonics and crustal evolution*. 3d ed. New York: Pergamon.

Cromie, W. J. 1989. The roots of midplate volcanism. *Mosaic* 20, no. 4: 19–25.

Dalziel, I. W. D. 1995. Earth before Pangea. *Scientific American* 272, no. 1: 58–63.

Kearey, P., and F. J. Vine. 1996. *Global tectonics*. 2d ed. Palo Alto, Calif.: Blackwell Scientific.

Klein, G. D., ed. 1994. Pangaea: Paleoclimate, tectonics, and sedimentation during accretion, zenith, and breakup of a supercontinent. *Geological Society of America Special Paper 288*. Boulder, Colo.: Geological Society of America.

Luhmann, J. G., J. B. Pollack, and L. Colin. 1994. The Pioneer mission to Venus. *Scientific American* 270, no. 4: 90–97.

Murphy, J. B., and R. D. Nance. 1992. Mountain belts and the supercontinent cycle. *Scientific American* 266, no. 4: 84–91.

Nance, R. D., T. R. Worsley, and J. B. Moody. 1988. The supercontinent cycle. *Scientific American* 259, no. 1: 72–79.

Parks, N. 1994. Exploring Loihi: The next Hawaiian Island. *Earth* 5, no. 5: 56–63.

CHAPTER 3

Berry, L. G., B. Mason, and R. V. Dietrich. 1983. *Mineralogy*. 2d ed. San Francisco: Freeman.

Blackburn, W. H., and W. H. Dennen. 1988. *Principles of mineralogy*. Dubuque, Iowa: Brown.

Cepeda, J. C. 1994. *Introduction to minerals and rocks*. New York: Macmillan.

Dietrich, R. V., and B. J. Skinner. 1979. *Rocks and rock minerals*. New York: Wiley.

———. 1990. *Gems, granites, and gravels: Knowing and using rocks and minerals*. New York: Cambridge University Press.

Hochleitner, R. 1994. *Minerals: Identifying, classifying, and collecting them*. Hauppauge, N.Y.: Barron's Educational Series.

Klein, C., and C. S. Hurlbut, Jr. 1985. *Manual of mineralogy* (after James D. Dana). 20th ed. New York: Wiley.

Pough, F. H. 1987. *A field guide to rocks and minerals*. 4th ed. Boston: Houghton Mifflin.

Schumann, W. 1992. *Handbook of rocks, minerals, and gemstones*. New York: Houghton Mifflin.

———. 1995. *Gemstones of the world*. New York: Sterling.

———. 1997. *Minerals of the world*. New York: Sterling.

Sofianides, A. S., and G. E. Harlow. 1990. *Gems and crystals*. New York: Simon & Schuster.

CHAPTER 4

Baker, D. S. 1983. *Igneous rocks*. Englewood Cliffs, N.J.: Prentice-Hall.

Best, M. G. 1982. *Igneous and metamorphic petrology*. San Francisco: Freeman.

Dietrich, R. V., and R. Wicander. 1983. *Minerals, rocks, and fossils*. New York: Wiley.

Hall, A. 1987. *Igneous petrology*. Essex, Eng.: Longman Scientific and Technical.

Hess, P. C. 1989. *Origins of igneous rocks*. Cambridge, Mass.: Harvard University Press.

MacKenzie, W. S., C. H. Donaldson, and C. Guilford. 1982. *Atlas of igneous rocks and their textures*. New York: Halsted Press.

McBirney, A. R. 1984. *Igneous petrology*. San Francisco: Freeman, Cooper.

Middlemost, E. A. K. 1985. *Magma and magmatic rocks*. London: Longman Group.

Nichols, J. W., and J. K. Russell. 1990. *Modern methods of igneous petrology: Understanding magmatic processes*. Washington, D.C.: Mineralogical Society of America.

Raymond, L. A. 1995. *Petrology: The study of igneous, sedimentary, and metamorphic rocks*. Dubuque, Iowa: Brown.

CHAPTER 5

Aylesworth, T. G., and V. Aylesworth. 1983. *The Mount St. Helens disaster: What we've learned*. New York: Franklin Watts.

Brantley, S. R., and W. E. Scott. 1993. The danger of collapsing lava domes: Lessons for Mount Hood, Oregon. *Earthquakes and Volcanoes* 24, no. 6: 244–268.

Chester, D. K. 1993. *Volcanoes and society*. London: Arnold.

Claque, D. A., and C. Heliker. 1992. The ten-year eruption of Kilauea volcano. *Earthquakes and Volcanoes* 23, no. 6: 244–254.

Coffin, M. F., and O. Eldholm. 1993. Large igneous provinces. *Scientific American* 269, no. 4: 42–49.

Decker, R. W., and B. B. Decker. 1997. *Volcanoes*. 3d ed. New York: Freeman.

Fisher, R. V., G. Heiken, and J. B. Hulen. 1997. *Volcanoes*. Princeton, N.J.: Princeton University Press.

Francis, P. 1993. *Volcanoes: A planetary perspective*. New York: Oxford University Press.

Gore, R. 1998. "Cascadie: Living on fire." *National Geographic,* 193, no. 5:6-37.

Harris, S. L. 1976. *Fire and ice: The Cascade volcanoes*. Seattle, Wash.: The Mountaineers.

Lipman, P. W., and D. R. Mullineaux, eds. 1981. The 1980 eruptions of Mount St. Helens, Washington. *United States Geological Survey Professional Paper 1250*.

Pendick, D. 1994. Under the volcano. *Earth* 3, no. 3: 34–39.

———. 1995. Return to Mount St. Helens. *Earth* 4, no. 2: 24–33.

Rampino, M. R., S. Self, and R. B. Strothers. 1988. Volcanic winters. *Annual Review of Earth and Planetary Sciences* 16: 73–99.

Ritche, D. 1994. *The encyclopedia of earthquakes and volcanoes*. New York: Facts on File.

Scarth, A. 1994. *Volcanoes: An introduction*. College Station, Tex.: Texas A&M University Press.

Simkin, T., and L. Siebert. 1994. *Volcanoes of the world*. Tucson, Ariz.: Geosciences Press.

Tilling, R. I. 1987. *Eruptions of Mount St. Helens: Past, present, and future*. U.S. Geological Survey.

Tilling, R. I., C. Heliker, and T. L. Wright. 1987. *Eruptions of Hawaiian volcanoes: Past, present, and future*. U.S. Geological Survey.

Wenkam, R. 1987. *The edge of fire: Volcano and earthquake country in western North America and Hawaii*. San Francisco: Sierra Club Books.

Wolfe, G. W. 1992. The 1991 eruptions of Mount Pinatubo, Philippines. *Earthquakes and Volcanoes* 23, no. 1: 5–37.

Wright, T. L., and T. C. Pierson. 1992. Living with volcanoes: The U.S. Geological Survey's volcano hazards program. *U.S. Geological Survey Circular 1073*.

CHAPTER 6

Bear, F. E. 1986. *Earth: The stuff of life*. 2d rev. ed. Norman: University of Oklahoma Press.

Birkeland, P. W. 1984. *Soils and geomorphology*. New York: Oxford University Press.

Buol, S. W., F. D. Hole, and R. J. McCracken. 1980. *Soil genesis and classification*. Ames: Iowa State University Press.

Carroll, D. 1970. *Rock weathering*. New York: Plenum Press.

Coughlin, R. C. 1984. *State and local regulations for reducing agricultural erosion*. American Planning Association, Planning Advisory Service Report No. 386.

Courtney, F. M., and S. T. Trudgill. 1984. *The soil: An introduction to soil study*. 2d ed. London: Arnold.

Gibbons, B. 1984. Do we treat our soil like dirt? *National Geographic* 166, no. 3: 350–389.

Parfit, M. 1989. The dust bowl. *Smithsonian* 20, no. 3: 44–54, 56–57.

Robinson, D. A., and R. B. G. Williams, eds. 1994. *Rock weathering through geologic time: Principles and applications*. New York: Wiley.

Wild, A. 1993. *Soils and the environment: An introduction*. New York: Cambridge University Press.

CHAPTER 7

Blatt, H., G. Middleton, and R. Murray. 1980. *Origin of sedimentary rocks*. New York: Freeman.

Boggs, S., Jr. 1995. *Principles of sedimentology and stratigraphy* 2d ed. Columbus, Ohio: Merrill.

Collinson, J. D., and D. B. Thompson. 1989. *Sedimentary structures*. 2d ed London: Allen & Unwin.

Eldridge, N. 1997. *Fossils*. Princeton, N.J.: Princeton University Press.

Fritz, W. J., and J. N. Moore. 1988. *Basics of physical stratigraphy and sedimentology*. New York: Wiley.

Julien, P. Y. 1995. *Erosion and sedimentation*. New York: Cambridge University Press.

LaPorte, L. F. 1979. *Ancient environments*. 2d ed. Englewood Cliffs, N.J.: Prentice-Hall.

Moody, R. 1986. *Fossils*. New York: Macmillan Publishing Co.

Selley, R. C. 1982. *An introduction to sedimentology*. 2d ed. New York: Academic Press.

Simpson, G. G. 1983. *Fossils and the history of life*. New York: Scientific American Books.

CHAPTER 8

Abelson, F. H. 1990. The asbestos fiasco. *Science* 247, no. 4946: 1017.

Best, M. G. 1982. *Igneous and metamorphic petrology*. San Francisco: Freeman.

Bowes, D. R., ed. 1989. *The encyclopedia of igneous and metamorphic petrology*. New York: Van Nostrand Reinhold.

Cepeda, J. C. 1994. *Introduction to rocks and minerals*. New York: Macmillan.

Gillen, C. 1982. *Metamorphic geology*. London: Allen & Unwin.

Gunter, M. E. 1994. Asbestos as a metaphor for teaching risk perception. *Journal of Geological Education* 42, no. 1: 17–24.

Hyndman, D. W. 1985. *Petrology of igneous and metamorphic rocks*. 2d ed. New York: McGraw-Hill.

Kokkou, A. 1993. *The Getty kouros colloquium*. Athens: Kapon Editions.

Margolis, S. V. 1989. Authenticating ancient marble sculpture. *Scientific American* 260, no. 6: 104–111.

CHAPTER 9

Anderson, D. L., and A. M. Dziewonski. 1984. Seismic tomography. *Scientific American* 251, no. 4: 60–68.

Bolt, B. A. 1996. *Earthquakes*. New York: Freeman.

Bonatti, E. 1994. The Earth's mantle below the ocean. *Scientific American* 270, no. 3: 44–51.

Canby, T. Y. 1990. California earthquake—prelude to the big one? *National Geographic* 177, no. 5: 76–105.

Dawson, J. 1993. CAT scanning the Earth. *Earth 2*, no. 3: 36–41.

Fischman, J. 1992. Falling into the gap: A new theory shakes up earthquake predictions. *Discover* October 1992: 56–63.

Fowler, C. M. R. 1990. *The solid Earth*. New York: Cambridge University Press.

Frohlich, C. 1989. Deep earthquakes. *Scientific American* 260, no. 1: 48–55.

Hanks, T. C. 1985. *National earthquake hazard reduction program: Scientific status*. U.S. Geological Survey Bulletin 1659.

Jeanloz, R. 1990. The nature of the Earth's core. *Annual Review of Earth and Planetary Sciences,* 18: 357–86.

Jeanloz, R. and T. Lay. 1993. The core-mantle boundary. *Scientific American* 268, no. 5: 48–55.

Johnston, A. C., and L. R. Kanter. 1990. Earthquakes in stable continental crust. *Scientific American* 262, no. 3: 68–75.

Monastersky, R. 1996. Core concerns. *Science News* 150, no. 16: 250–251.

———. 1998. The mush zone: A slurpy layer lurks deep inside the planet. *Science News* 153: 109–111.

Murck, B. W., B. J. Skinner, and S. C. Porter. 1997. *Dangerous Earth: An introduction to geologic hazards*. New York: Wiley.

Wesson, R. L., and R. E. Wallace. 1985. Predicting the next great earthquake in California. *Scientific American* 252, no. 2: 35–43.

Wyession, M. E. 1995. The inner workings of the Earth. *American Scientist* 83, no. 2: 134–146.

———. 1996. Journey to the center of the Earth. *Earth* 5, no. 6: 46–49.

CHAPTER 10

Davis, G. H., and S. J. Reynolds. 1996. *Structural geology of rocks and regions*. 2d ed. New York: Wiley.

Hatcher, R. D., Jr. 1990. *Structural geology: Principles, concepts, and problems*. Columbus, Ohio: Merrill.

Howell, D. G. 1985. Terranes. *Scientific American* 253, no. 5: 116–125.

———. 1989. *Tectonics of suspect terranes: Mountain building and continental growth*. London: Chapman and Hall.

Jones, D. L., A. Cox, P. Coney, and M. Beck. 1982. The growth of western North America. *Scientific American* 247, no. 5: 70–84.

Kearney, P., and F. J. Vine. 1996. *Global tectonics*. 2d ed. London: Blackwell Science.

Keller, E. A., and N. Pinter. 1996. *Active tectonics: Earthquakes, uplift, and landscapes*. Upper Saddle River, N.J.: Prentice Hall.

Lisle, R. J. 1988. *Geological structures and maps: A practical guide*. New York: Pergamon.

Molnar, P. 1986. The geologic history and structure of the Himalaya. *American Scientist* 74, no. 2: 144–154.

———. 1986. The structure of mountain ranges. *Scientific American* 255, no. 1: 70–79.

Moores, E. M., and R. J. Twiss. 1995. *Tectonics*. New York: Freeman.

CHAPTER 11

Crozier, M. J. 1989. *Landslides: Causes, consequences, and environment*. Dover, N.H.: Croom Helm.

Fleming, R. W., and F. A. Taylor. 1980. *Estimating the cost of landslide damage in the United States*. U.S. Geological Survey Circular 832.

Highland, L. M., and W. M. Brown III. 1996. Landslides—natural hazard sleepers. *Geotimes* 41, no. 1: 16–19.

McPhee, J. 1989. *The control of nature.* New York: Farrar, Straus & Giroux.

Nuther, E. B., R. T. Proctor, and P. H. Moser. 1993. *The citizen's guide to the geologic hazards.* Arvada, Colo.: American Institute of Professional Geologists.

Parks, N. 1993. The fragile volcano. *Earth* 6, no. 4: 42–49.

Plant, N., and G. B. Griggs. 1990. Coastal landslides and the Loma Prieta earthquake. *Earth Science* 43: 12–18.

Small, R. J., and M. J. Clark. 1982. *Slopes and weathering.* New York: Cambridge University Press.

Zaruba, Q., and V. Mencl. 1982. *Landslides and their control.* 2d ed. Amsterdam: Elsevier.

CHAPTER 12

Baker, V. R. 1982. *The channels of Mars.* Austin: University of Texas Press.

Beven, K., and P. Carling, eds. 1989. *Floods.* New York: Wiley.

Frater, A., ed. 1984. *Great rivers of the world.* Boston: Little, Brown.

Knighton, D. 1984. *Fluvial forms and processes.* London: Arnold.

Leopold, L. B. 1994. *A view of the river.* Cambridge, Mass.: Harvard University Press.

Marison, A. 1994. The great flood of '93. *National Geographic* 185, no. 1: 42–81.

Mayer, L., and D. Nash, eds. 1987. *Catastrophic flooding.* Boston: Allen & Unwin.

McPhee, J. 1989. *The control of nature.* New York: Farrar, Straus & Giroux.

Patrick, R. 1995. *Rivers of the United States.* New York: Wiley.

Petts, G., and I. Foster. 1985. *Rivers and landscape.* London: Arnold.

Rachocki, A. 1981. *Alluvial fans.* New York: Wiley.

CHAPTER 13

Bryan, T. S. 1992. Valley of geysers. *Earth* 1, no. 4: 20–29.

Courbon, P., C. Chabert, P. Bosted, and K. Lindslay. 1989. *Atlas of the great caves of the world.* St. Louis: Cave Books.

Dietrich, R. V. 1993. How are caves formed? *Rocks and Minerals* 68, no. 4: 264–268.

Dolan, R., and H. G. Goodell. 1986. Sinking cities. *American Scientist* 74, no. 1: 38–47.

Fetter, C. W. 1988. *Applied hydrogeology.* 2d ed. Columbus, Ohio: Merrill.

Grossman, D., and S. Schulman. 1994. Verdict at Yucca Mountain. *Earth* 3, no. 2: 54–63.

Jennings, J. N. 1983. Karst landforms. *American Scientist* 71, no. 6: 578–586.

———. 1985. *Karst geomorphology.* 2d ed. Oxford, Eng.: Basil Blackwell.

Monastersky, R. 1988. The 10,000-year test. *Science News* 133: 139–141.

Price, M. 1985. *Introducing groundwater.* London: Allen & Unwin.

Rinehart, J. S. 1980. *Geysers and geothermal energy.* New York: Springer-Verlag.

Sloan, B., ed. 1977. *Caverns, caves, and caving.* New Brunswick, N.J.: Rutgers University Press.

Whipple, C. G. 1996. Can nuclear waste be stored safely at Yucca Mountain? *Scientific American* 274, no. 6: 72–79.

CHAPTER 14

Anderson, B. G., and H. W. Borns, Jr. 1994. *The ice age world.* Oslo-Copenhagen-Stockholm: Scandinavian University Press.

Broecker, W. S., and G. H. Denton. 1990. What drives glacial cycles? *Scientific American* 262, no. 1: 49–56.

Carozzi, A. V. 1984. Glaciology and the ice age. *Journal of Geological Education* 32: 158–170.

Covey, C. 1984. The Earth's orbit and the ice ages. *Scientific American* 250, no. 2: 58–66.

Drewry, D. J. 1986. *Glacial geologic processes.* London: Arnold.

Grove, J. M. 1988. *The Little Ice Age.* London: Methuen.

John, B. S. 1977. *The ice age. Past and present.* London: Collins.

———. 1979. *The winters of the world.* London: David & Charles.

Kurten, B. 1988. *Before the Indians.* New York: Columbia University Press.

Oeland, G. 1997. Iceland's trial by fire. *National Geographic* 191, no. 5: 58–70.

Schneider, S. H. 1990. *Global warming: Are we entering the greenhouse century?* San Francisco: Sierra Club Books.

Sharp, R. P. 1988. *Living ice: Understanding glaciers and glaciation.* New York: Cambridge University Press.

Williams, R. S., Jr. 1983. *Glaciers: Clues to future climate?* United States Geological Survey.

CHAPTER 15

Agnew, C., and A. Warren. 1990. Sand trap. *The Sciences* March/April: 14–19.

Chasan, D. J. 1997. Will the dunes march once again? *Smithsonian* 28, no. 9: 70–79.

Conway, T. 1993. *Painted dreams: Native American rock art.* Minocqua, Wis.: Northword Press.

Cooke, R. A., A. Warren, and A. Goudie. 1993. *Desert geomorphology.* London: UCL Press.

Dorn, R. I. 1991. Rock varnish. *American Scientist* 79, no. 6: 542–553.

Ellis, W. S. 1987. Africa's Sahel: The stricken land. *National Geographic* 172, no. 2: 140–179.

Greeley, R., and J. Iversen. 1985. *Wind as a geologic process*. Cambridge, Mass.: Cambridge University Press.

MacKinnon, D. J., and P. S. Chavez, Jr. 1993. Observing dust storms from space. *Earth* 2, no. 3: 60–65.

Sletto, B. 1997. Desert in disguise. *Earth* 6, no. 1: 42–49.

Somerville, D. 1994. Into the Red Center. *Earth* 3, no. 1: 32–41.

Thomas, D. S. G., ed. 1989. *Arid zone geomorphology*. New York: Halsted Press.

Waters, T. 1993. Dunes. *Earth* 2, no. 1: 44–51.

Whitney, M. A. 1985. Yardangs. *Journal of Geological Education* 33, no. 2: 93–96.

CHAPTER 16

Bird, E. C. F. 1984. *Coasts: An introduction to coastal geomorphology*. New York: Blackwell.

Bird, E. C. F., and M. L. Schwartz. 1985. *The world's coastline*. New York: Van Nostrand Reinhold.

Flanagan, R. 1993. Beaches on the brink. *Earth* 2, no. 6: 24–33.

Fox, W. T. 1983. *At the sea's edge*. Englewood Cliffs, N.J.: Prentice-Hall.

Garrett, C., and L. R. M. Maas. 1993. Tides and their effects. *Oceanus* 36, no. 1: 27–37.

Hecht, J. 1988. America in peril from the sea. *New Scientist* 118: 54–59.

Monastersky, R. 1996. The mother lode of natural gas. *Science News* 150, no. 19: 298–299.

Pethick, J. 1984. *An introduction to coastal geomorphology*. London: Arnold.

Pinet, P. 1996. *Invitation to oceanography*. St. Paul: West.

Snead, R. 1982. *Coastal landforms and surface features*. Stroudsburg, Pa.: Hutchinson Ross.

Sunamura, T. 1992. *Geomorphology of rocky coasts*. New York: Wiley.

Viles, H., and T. Spencer. 1995. *Coastal problems*. London: Arnold.

Walden, D. 1990. Raising Galveston. *American Heritage of Invention & Technology* 5: 8–18.

Williams, S. J., K. Dodd, and K. K. Gohn. 1990. Coasts in crisis. *U.S. Geological Survey Circular 1075*.

Yulsman, T. 1996. The seafloor laid bare. *Earth* 5, no. 3: 42–51.

CHAPTER 17

Berry, W. B. N. 1987. *Growth of a prehistoric time scale*. 2d ed. Palo Alto, Calif.: Blackwell Scientific.

Boslough, J. 1990. The enigma of time. *National Geographic* 177, no. 3: 109–132.

Geyh, M. A., and H. Schleicher. 1990. *Absolute age determination*. New York: Springer-Verlag.

Gould, S. J. 1987. *Time's arrow, time's cycle*. Cambridge, Mass.: Harvard University Press.

Gunderson; L. C. S. 1992. Hidden hazards of radon. *Earth* 1, no. 6: 55–61.

Harland, W. B., R. L. Armstrong, A. V. Cox, L. E. Craig, A. G. Smith, and D. G. Smith. 1990. *A geologic time scale 1989*. New York: Cambridge University Press.

Itano, W. M., and N. F. Ramsey. 1993. Accurate measurement of time. *Scientific American* 269, no. 1: 56–57.

Ramsey, N. F. 1988. Precise measurement of time. *American Scientist* 76, no. 1: 42–49.

CHAPTER 18

Bally, A. W., and A. R. Palmer, eds. 1989. *The geology of North America: An overview*. The Geology of North America, Vol. A. Boulder, Colo.: Geological Society of America.

Catacosinos, P. A., and P. A. Daniels, Jr., eds. 1991. *Early sedimentary evolution of the Michigan Basin*. Geological Society of America Special Paper 256. Boulder, Colo.: Geological Society of America.

Dallmeyer, R. D., ed. 1989. *Terranes in the Circum-Atlantic Paleozoic orogens*. Geological Society of America Special Paper 230. Boulder, Colo.: Geological Society of America.

Dawson, A. G. 1992. *Ice Age Earth: Late Quaternary geology and climate*. New York: Routledge.

Hatcher, R. D., Jr., W. A. Thomas, and G. W. Viele, eds. 1989. *The Appalachian-Ouachita orogen in the United States*. The Geology of North America, Vol. F-2. Boulder, Colo.: Geological Society of America.

McKerrow, W. S., and C. R. Scotese, eds. 1990. *Palaeozoic palaeogeography and biogeography*. The Geological Society Memoir No. 12. London: Geological Society of London.

Morrison, R. B., ed. 1991. *Quaternary nonglacial geology: Conterminous U.S.* The geology of North America, Vol. K-2. Boulder, Colo.: Geological Society of America.

Moullade, M., and A. E. M. Nairn, eds. 1991. *The Phanerozoic geology of the world I: The Palaeozoic*. New York: Elsevier Science.

Nations, J. D., and J. G. Eaton, eds. 1991. *Stratigraphy, depositional environments, and sedimentary tectonics of the western margin, Cretaceous Western Interior Seaway*. Geological Society of America Special Paper 260. Boulder, Colo.: Geological Society of America.

Nisbet, E. G. 1987. *The young Earth: An introduction to Archean geology*. Boston: Allen & Unwin.

Osborne, R., and D. Tarling, eds. 1996. *The historical atlas of the Earth*. New York: Henry Holt.

Salvador, A., ed. 1991. *The Gulf of Mexico Basin*. The Geology of North America, Vol. J. Boulder, Colo.: Geological Society of America.

Sloss, L. L., ed. 1988. *Sedimentary cover–North American Craton: U.S.* The Geology of North America, Vol. D-2. Boulder, Colo.: Geological Society of America.

Smith, A. G. 1994. *Atlas of Mesozoic and Cenozoic coastlines*. New York: Cambridge University Press.

Woodrow, D. L., and W. D. Sevan. 1985. *The Catskill Delta*. Geological Society of America Special Paper 201. Boulder, Colo.: Geological Society of America.

Ziegler, P. A. 1989. *Evolution of Laurussia: A study in Late Palaeozoic plate tectonics*. Boston: Kluwer.

CHAPTER 19

Archibald, J. D. 1996. *Dinosaur extinction and the end of an era: What the fossils say*. New York: Columbia University Press.

Bakker, R. T. 1993. Bakker's field guide to Jurassic dinosaurs. *Earth* 2, no. 5: 33–43.

———. 1993. Jurassic sea monsters. *Discover* 14, no. 9: 78–85.

Briggs, D. E. G., D. H. Erwin, and F. J. Collier. 1994. *The fossils of the Burgess Shale*. Washington, D.C.: Smithsonian Institution Press.

Carroll, R. L. 1992. The primary radiation of terrestrial vertebrates. *Annual Review of Earth and Planetary Sciences* 20: 45–84.

Colbert, E. H., and M. Morales. 1991. *Evolution of the vertebrates*. 4th ed. New York: Wiley.

Cowen, R. 1995. *History of life*. 2d ed. Palo Alto, Calif.: Blackwell Scientific.

Currie, P. J. 1996. The great dinosaur egg hunt. *National Geographic* 189, no. 5: 96–111.

Currie, P. J., and K. Padian, eds. 1997. *Encyclopedia of dinosaurs*. New York: Academic Press.

Dingus, L., and T. Rowe. 1998. *The mistaken extinction: Dinosaur evolution and the origin of birds*. New York: Freeman.

Dodson, P. 1997. *The horned dinosaurs*. Princeton, N.J.: Princeton University Press.

Droser, M. L., R. Fortey, and Xing Li. 1996. The Ordovician radiation. *American Scientist* 84, no. 2: 122–131.

Erwin, D. E. 1996. The mother of mass extinctions. *Scientific American* 275, no. 1: 72–78.

Farlow, J. O., and M. K. Brett-Surman, eds. 1997. *The complete dinosaur*. Bloomington: Indiana University Press.

Fastovsky, D. E., and D. B. Weishampel. 1996. *The evolution and extinction of the dinosaurs*. New York: Cambridge University Press.

Feduccia, A. 1996. *The origin and evolution of birds*. New Haven, Conn.: Yale University Press.

Flanagan, R. 1996. Out of Africa. *Earth* 5, no. 1: 26–35.

Fortey, R. 1998. *Life: A natural history of the first four billion years on Earth*. New York: Knopf.

Gordon, M. S., and E. C. Olson. 1995. *Invasions of the land*. New York: Columbia University Press.

Gore, R. 1993. Dinosaurs. *National Geographic* 183, no. 1: 2–53.

———. 1993. The Cambrian Period explosion of life. *National Geographic* 184, no. 4: 120–136.

———. 1996. Neanderthals. *National Geographic* 189, no. 1: 2–35.

———. 1997. The dawn of humans: The first steps. *National Geographic* 191, no. 2: 72–99.

Gray, J., and W. Shear. 1992. Early life on land. *American Scientist* 80, no. 5: 444–456.

Horner J. R., and E. Dobb. 1997. *Dinosaur lives*. New York: HarperCollins.

Johanson, D. C. 1996. Face-to-face with Lucy's family. *National Geographic* 189, no. 3: 96–117.

Larick, R., and R. L. Ciochon. 1996. The African emergence and early Asian dispersals of the genus *Homo*. *American Scientist* 84, no. 6: 538–551.

Leakey, M. 1995. The dawn of humans. *National Geographic* 188, no. 3: 38–51.

Monastersky, R. 1998. The rise of life on Earth. *National Geographic* 193, no. 3: 54–81.

Morris, S. C. 1998. *The crucible of creation: The Burgess Shale and the rise of animals*. NY: Oxford University Press.

Padian, K., and L. M. Chiappe. 1998. The origin of birds and their flight. *Scientific American* 278, no. 2: 38–47.

Schopf, J. W., ed. 1992. *Major events in the history of life*. Boston: James & Bartlett.

Sereno, P. C. 1996. Africa's dinosaur castaways. *National Geographic* 189, no. 6: 106–119.

Shipman, P. 1998. *Taking wing: Archaeopteryx and the evolution of flight*. New York: Simon & Schuster.

Shreeve, J. 1997. Uncovering Patagonia's lost world. *National Geographic* 192, no. 6: 120–137.

Stewart, W. N., and G. W. Rothwell. 1993. *Paleobotany and the evolution of plants*. 2d ed. New York: Cambridge University Press.

Thomson, K. W. 1991. Where did tetrapods come from? *American Scientist* 79, no. 6: 488–490.

Tudge, C. 1996. The future of *Homo sapiens*. *Earth* 5, no. 1: 36–40.

Webster, D. 1996. Dinosaurs of the Gobi. *National Geographic* 190, no. 1: 70–89.

Wellenhofer, P. 1991. *The illustrated encyclopedia of pterosaurs*. New York: Crescent Books.

Credits

3.13	Photo courtesy of Sue Monroe
3.14b	Photo courtesy of Sue Monroe
3.15a, b	Photos courtesy of Sue Monroe
3.16	Precision Graphics
3.17	Precision Graphics
3.18	Photo courtesy of Sue Monroe
3.19a, b	Photos by Stew Monroe

CHAPTER 4

Two-page opener

© 1997 Photo Disc, Inc.

Inset opener Photo by Stew Monroe

4.1	Photo of painting by Herbert Collins, courtesy Devil's Tower National Monument
4.2	Precision Graphics
4.3	Precision Graphics
4.4	Carlyn Iverson
4.5	Precision Graphics
4.6	Precision Graphics
4.7	Photo courtesy of David J. Matty
4.8	Precision Graphics
4.9	Precision Graphics
4.10a, b, c	Photos courtesy of Sue Monroe
4.11a, b	Photos courtesy of Sue Monroe
4.12	Precision Graphics. Modified from R. V. Dietrich, *Geology and Michigan: Fortynine Questions and Answers.* 1979.
4.13a, b	Photos courtesy of Sue Monroe
4.14a, b	Photos courtesy of Sue Monroe
4.15	Photo courtesy of Sue Monroe
4.16	Precision Graphics
4.17a, b, c	Photo courtesy of Sue Monroe
4.18	Precision Graphics
4.19	Photo courtesy of Richard L. Chambers
4.20	Precision Graphics
4.21	Precision Graphics

CHAPTER 5

Two-page opener

© Krafft/Explorer/Photo Researchers

5.1	Precision Graphics
5.2	© 1995 Stephen Cottrell
5.3	U.S. Geological Survey
5.4	Photo by Stew Monroe
5.5a	© G. Brad Lewis/Omjalla Images
5.5b	J. B. Stokes, U.S. Geological Survey
5.6a, b	Photos by Stew Monroe
5.7	Reproduced by permission of Marie Tharp, 1 Washington Avenue, South Nyack, NY 10960
5.8	University of Colorado
5.9	Photo courtesy of Sue Monroe
5.10a, b, c, d	Precision Graphics. From Howell Williams, *Crater Lake: The Story of It's Origin* (Berkeley, Calif.: University of California Press): Illustrations from p. 84. Copyright © 1941 Regents of the University of California, © renewed 1969 Howell Williams.
5.11a	Precision Graphics
5.11b	Photo by Stew Monroe
5.12a	Precision Graphics
5.12b	Solarfilma/GeoScience Features
5.13a	Precision Graphics
5.13b	Tom & Pat Leeson/Photo Researchers, Inc.

Perspective 5.1 Figure 2

© 1990 Richard During/Tony Stone Images

5.14a	Reuters/Bettmann Archive
5.14b	U.S. Department of Interior, U.S. Geological Survey, David A. Johnston Cascades Volcano Observatory, Vancouver, WA.
5.15	Precision Graphics
5.16	UPI/Corbis-Bettmann
5.17	Precision Graphics. From R. I. Tilling, U.S. Geological Survey
5.18	Ward's Natural Science Establishment, Inc.
5.20	Precision Graphics. Modified from R. I. Tilling, C. Heliker, and T. L. Wright, *Eruptions of Hawaiian Volcanoes: Past, Present, and Future.* 1987. U.S. Geological Survey.
5.21	Carlyn Iverson

CHAPTER 6

Two-page opener

Jack Dykinga/Tony Stone Images

Inset opener Photo courtesy of Frank Hannah

6.01a, b	Photos by Stew Monroe
6.02a, b	Photos courtesy of Sue Monroe
6.03	Photo by Stew Monroe
6.04	Precision Graphics. From A. Cox and R. R. Doell, "Review of Paleomagnetism." *GSA Bulletin* 71 (1960): 758, Figure 33.
6.05	Photo by Stew Monroe
6.06	Photo by Stew Monroe
6.07	© Mark Gibson/Visuals Unlimited
6.08	Photo courtesy of W. D. Lowry
6.10	Precision Graphics
6.11	Bill Beatty/Visuals Unlimited
6.12	Precision Graphics
6.13	Precision Graphics
6.13a, b, c	Precision Graphics
6.13d	Photo courtesy of Sue Monroe
6.14a	Photo courtesy of John S. Shelton
6.14b	John D. Cunningham/Visuals Unlimited
6.15	Precision Graphics
6.16	Precision Graphics
6.17	Photos courtesy of Sue Monroe
6.18	Kansas State Historical Society
6.19	Photo by Stew Monroe
6.20	© Frank Lembrecht/Visuals Unlimited
6.21	Science VU/Visuals Unlimited

CHAPTER 7

Two-page opener

© 1996 Photo Disc, Inc.

Inset opener Photo by Stew Monroe

7.1	Photo courtesy of Sue Monroe
7.2	R. L. Elderkin, U.S. Geological Survey
7.3	Precision Graphics
7.4a, b	Photos courtesy of R. V. Dietrich
7.5	Carlyn Iverson
7.6	Precision Graphics
7.7	Photo courtesy of Sue Monroe
7.8a, b	Photos courtesy of Sue Monroe
7.9a, b, c, d	Photos courtesy of Sue Monroe
7.10	Photo courtesy of Sue Monroe
7.11	Photo by Stew Monroe

10.16	Precision Graphics
10.17	Precision Graphics
10.18	Precision Graphics. From Peter Molnar, "The Geologic History and Structure of the Himalayas," *American Scientist* 74: 148–149, Figure 4. Reprinted by permission of *American Scientist*, journal of Sigma Xi, The Scientific Research Society.
10.19	Precision Graphics
10.20	Precision Graphics
10.21	Precision Graphics
10.22	Precision Graphics
Perspective 10.1 Figure 1	Photo by Stew Monroe
Perspective 10.1 Figure 2	Photo by Sue Monroe

CHAPTER 11

Two-page opener

Rod Rolle/Gamma Liaison, Inc.

Inset opener	Photo by Stew Monroe
11.1a	Precision Graphics
11.1b	George Plafker, U.S. Geological Survey
11.2	Precision Graphics
11.3a, b	Precision Graphics
11.4	Precision Graphics
11.5	Boris Yaro, *Los Angeles Times*
11.6	Precision Graphics
11.7	Precision Graphics
11.8	Photo courtesy of R. V. Dietrich
11.9	Precision Graphics
11.10	Photo courtesy of John S. Shelton
11.10 inset	Precision Graphics
11.11	Precision Graphics
11.12	Precision Graphics
11.12a	Photo courtesy of Eleanora I. Robbins, U.S. Geological Survey
11.13	Photo by Stew Monroe
11.14a	Precision Graphics
11.14b	Photo by Reed Wicander
11.15a	Precision Graphics
11.15b	Photograph from "Alaska Earthquake Collection," no. 43ct, U.S. Geological Survey
11.16a	Precision Graphics. From O. J. Ferrians, Jr., R. Kachadoorian, and G. W. Greene, *U.S. Geological Survey Professional Paper 678.* 1969.
11.16b	B. Bradley and the University of Colorado's Geology Department. National Geophysical Data Center, NOAA, Boulder, CO.
11.17a	Precision Graphics
11.17b	B. Bradley and the University of Colorado's Geology Department. National Geophysical Data Center, NOAA, Boulder, CO.
11.18a	Precision Graphics
11.18b	Photo by Reed Wicander
11.19	Precision Graphics
11.20a	Precision Graphics
11.20b	Dell R. Foutz/Visuals Unlimited
Perspective 11.1 Figure 1, left	Precision Graphics
Perspective 11.1 Figure 1, right	T. Spencer/Colorific

CHAPTER 12

Two-page opener

© 1997 Photo Disc, Inc.

Inset opener	Photo courtesy of R. V. Dietrich
12.1a	Photo courtesy of Michael Lawton
12.1b	Photo courtesy of Sue Monroe
12.2	Precision Graphics
12.3	Carlyn Iverson
12.4	Precision Graphics
12.5	Precision Graphics
12.6	Precision Graphics
12.7a, b	Photo courtesy of R. V. Dietrich
12.8	Photo by Stew Monroe
12.9	Photo courtesy of John S. Shelton
12.10a	Precision Graphics
12.10b	Photo by Stew Monroe
12.11	Precision Graphics
12.12	Precision Graphics
12.13a, c	Precision Graphics
12.13b	Photo by Stew Monroe
12.14	Photo courtesy of John S. Shelton
12.15a	Photo by Stew Monroe
12.15b	Precision Graphics
12.16	Precision Graphics
12.17a, b	Precision Graphics
12.17c	Photo courtesy of the Kentucky Department of Parks
12.18	Precision Graphics
12.19	Precision Graphics
12.20	Precision Graphics
12.21	Precision Graphics
12.22	Precision Graphics
12.23a	Photo courtesy of John S. Shelton
12.23b	Photo courtesy of Sue Monroe
Perspective 12.1 Figure 1	Precision Graphics
Perspective 12.1 Figure 2	Precision Graphics
Perspective 12.1 Figure 3	Cameron Davidson/Comstock, Inc.

CHAPTER 13

Two-page opener

© Stock Boston/David Bartruff/PNI

Inset opener	British Tourist Authority
13.1	Photo courtesy of Ed Cooper
13.2	Precision Graphics
13.3	Precision Graphics
13.4	Precision Graphics
13.5	Precision Graphics
13.6	Precision Graphics
13.7	Precision Graphics
13.8	Frank Kujawa, University of Central Florida, GeoPhoto Publishing Company
13.9	Precision Graphics
13.10a	Photo by Reed Wicander
13.10b	Photo courtesy of John S. Shelton
13.11	Precision Graphics
13.12	Daniel W. Gotshall/Visuals Unlimited
13.13	Precision Graphics. From J. B. Weeks et al., *U.S. Geological Survey Professional Paper 1400-A.* 1988.

16.14a	Precision Graphics
16.14b	Photo courtesy of John S. Shelton
16.15	Precision Graphics
16.16	Precision Graphics
16.17	NASA
16.18	Precision Graphics
16.19	Precision Graphics
16.20a	Precision Graphics
16.20b	Photo courtesy of John S. Shelton
16.21a	Precision Graphics
16.21b	Photo courtesy of Nick Harvey
16.21c	Suzanne and Nick Geary/Tony Stone Images
16.22	GEOPIC, Earth Satellite Corporation
16.23	Photo courtesy of Sue Monroe
16.24	Peter Ryan/Scripps/Science Photo Library/Photo Researchers, Inc.

Perspective 16.1 Figure 1

Department of the Interior/USGS. EROS Data Center

Perspective 16.1 Figure 2

Precision Graphics

CHAPTER 17

Two-page opener

© 1997 Photo Disc, Inc.

Inset opener Photo by Reed Wicander

17.1	Precision Graphics. From A. R. Palmer, "The Decade of North American Geology, 1983 Geologic Time Scale," *Geology* (Boulder, Colo.: Geological Society of America, 1983): 504. Reprinted by permission of the Geological Society of America.
17.2	Photo by Reed Wicander
17.3a	Photo by Stew Monroe
17.3b	Photo by Reed Wicander
17.4	Precision Graphics
17.5	Precision Graphics. From *Giants of Geology* by C. L. Fenton and M. A. Fenton. Used by permission of Doubleday, a division of Bantam Doubleday Dell Publishing Group, Inc.
17.6	Precision Graphics
17.7a	Precision Graphics
17.7b	Photo by Stew Monroe
17.8a	Precision Graphics
17.08b	Photo courtesy of Dorothy L. Stout
17.09a	Precision Graphics
17.09b	Photo by Stew Monroe
17.10	Precision Graphics. From *History of the Earth: An Introduction to Historical Geology, 2/E,* by Bernhard Kummer. © 1970 by W. H. Freeman and Company; reprinted by permission.
17.11	Precision Graphics
17.12	Precision Graphics
17.13	Precision Graphics. Data from S. M. Richardson and H. Y. McSween, Jr., *Geochemistry—Pathways and Processes* (Englewood Cliffs, NJ: Prentice-Hall, 1989).

Perspective 17.1 Figure 1

Precision Graphics

Perspective 17.1 Figure 2

Precision Graphics. Data from Environmental Protection Agency

17.14	Precision Graphics
17.15	Precision Graphics
17.16	Precision Graphics
17.17	Precision Graphics
17.18	Precision Graphics. From L. W. Mintz, *Historical Geology: The Science of a Dynamic Earth, 3/E.* (Westerville, Ohio: Charles E. Merrill Publishing Company, 1981): 27, Figure 2.18.

CHAPTER 18

Two-page opener

© Jim Steinberg/Photo Researchers

18.1	Herb Orth, Life Magazine. © Time Warner Inc.: painting by Chesley Bonestell, © The Estate of Chesley Bonestell.
18.2	Precision Graphics
18.3a	Precision Graphics.
18.3c	Photo courtesy of R. V. Dietrich
18.5	Precision Graphics. Reproduced with permission from P. F. Hoffman, "United Plates of America, The Birth of a Craton: Early Proterozoic Assembly and Growth of Laurentia," *Annual Review of Earth and Planetary Sciences*, V. 16, p. 544 © 1988 by Annual Reviews, Inc.
18.6	Carto-Graphics. Topography reprinted with permission from Bambach, Scotese, and Zeigler, "Before Pangaea: The Geographies of the Paleozoic World," *American Scientist*, v. 68, no. 1, 1980.
18.7	Carto-Graphics. Topography reprinted with permission from Bambach, Scotese, and Zeigler, "Before Pangaea: The Geographies of the Paleozoic World," *American Scientist*, v. 68, no. 1, 1980.
18.8	Precision Graphics. Reprinted with permission from L.L. Sioss, "Sequences in the Cratonic Interior of North America," *GSA Bulletin*, v. 74, 1963, p. 110 (Fig. 6), Geological Society of America.
18.9	Carto-Graphics
18.10	Carto-Graphics
18.11a, c	Precision Graphics
18.11b	Photo courtesy of Wayne E. Moore
18.11d	Photo © Patricia Caulfield/Photo Researchers
18.12	Carto-Graphics
18.13	Photo © Tom Bean, 1993/Tom & Susan Bean, Inc.
18.14a	Rubin's Studio of Photography, the Petroleum Museum, Midland, Texas
18.14b	Photo courtesy of Bill Cornell, The University of Texas at El Paso
18.15	Precision Graphics
18.16	Carto-Graphics. Reprinted with permission from Dietz and Holden, "Reconstruction of Pangaea: Breakup and Dispersion of Continents, Permian to Present," *Journal of Geophysical Research*, 1970, v. 75, no. 26, pp. 4939–4956, © 1970 by the American Geophysical Union.
18.17	Carto-Graphics
18.18	Carto-Graphics
18.19	Carto-Graphics
18.20	Photo courtesy John S. Shelton
18.21	Photo by Reed Wicander
18.23	Precision Graphics. Reprinted by permission of the Geological Society and A. M. Spencer, ed., *Mesozoic-Cenozoic Orogenic Belts* (Bath: Geological Society Publishing House, 1974).

18.24	Precision Graphics
18.25	Precision Graphics. Figure 175 (page 386) from *Structural Geology of North America, 2/E* by A. J. Eardley. Copyright © 1962 by A. J. Eardley. Copyright renewed. Reprinted by permission of HarperCollins, Publishers, Inc.
18.26a	Photo by Stew Monroe
18.27	Precision Graphics. Reprinted with permission from W. R. Dickinson, "Cenozoic Plate Tectonic Setting of the Cordilleran Region in the U.S.," in *Cenozoic Paleogeography of the Western U.S.*, Pacific Coast, Symposium 3, 1979, pp. 10–11 and pp. 2, 4.
18.28	Precision Graphics

Perspective 18.1 Figure 1

© Stephen J. Kraseman/Photo Researchers

CHAPTER 19

Two-page opener

Jonathan Blair/Woodfin Camp

Inset opener	Smithsonian Institution. Transparency No 86-13471A
19.1a, b, c	Douglas H. Erwin/National Museum of Natural History
19.1d	© Chip Clark, 1995/National Museum of Natural History
19.2	Photo by Reed Wicander
19.3a, b, c	Photos courtesy of Nevill Pledge, South Australian Museum
19.3d	Precision Graphics
19.4	Carnegie Museum of Natural History
19.5	From the Field Museum, Chicago #Geo80821c
19.6	© American Museum of Natural History, Trans. #K10269

19.7	Precision Graphics. Reprinted with permission from Raup and Sekoski, "Mass Extinctions in the Marine Fossil Record," *Science*, v. 215, 1982, p. 1502, © 1982 by the American Association for the Advancement of Science.
19.8	Carlyn Iverson
19.9	Carlyn Iverson
19.10	Carlyn Iverson
19.11	Photo courtesy of Dianne Edwards, University College, England
19.12	Carlyn Iverson
19.13	Photo courtesy of Sue Monroe
19.14	Precision Graphics
19.15	© Francois Gohier/Photo Researchers, Inc.
19.16	Carlyn Iverson
19.17	Carlyn Iverson

Perspective 19.1 Figure 1

Copyright, The Natural History Museum, London

Perspective 19.1 Figure 2

Watercolor by Henry De la Beche, © National Museum of Wales, Cardiff

19.18	© Tom McHugh/Photo Researchers
19.19	Carlyn Iverson
19.20	Photo courtesy of D. J. Nichols, U.S. Geological Survey
19.21	Carlyn Iverson
19.22a	Publication Services. Reprinted with permission from B. J. MacFadden. "Patterns of Phylogeny and Rates of Evolution in Fossil Horses," *Paleobiology*, v. 11, no. 3, 1985, p. 247 Fig. 1, © The Paleontological Society.
19.22b	Carlyn Iverson
19.23	From the Field Museum, Chicago, Neg. # CK30T
19.24	Darwen and Valley Hennings

Index

Bipedal animals, 430
Bird-hipped dinosaurs, 430
Birds, Mesozoic, 435–36
Bird's-foot delta, 256
Birthstones, 48
Bituminous coal, 143, 149
Black Hills (South Dakota), 56, 206–7, 274
Black Mountains (California), 328
Black smokers, 356
Block-faulting, 210–11
Blocks, pyroclastic, 95
Block slide, 231
Bloodstone, 48
Blowouts, 319
Blue asbestos, 160
Body fossils, 146
Body waves, 181
Bombs, pyroclastic, 95
Bonding, 51–52
    covalent, 53
    ionic, 52–53
    metallic, 53
    van der Waals or residual, 53–54
Bootlegger Cove Clay, Alaska, 234, 235
Bornite, 170
Bothriolepis, 427
Bottomless Lakes Quadrangle (New Mexico), 458
Bottomset beds, 253
Bowen's reaction series, 72–73
Brachiosaurus, 431
Braided streams, 250
Brain, of endotherms, 432–33
Breakers (waves), 344, 345
Breaker zone, 346, 347
Breccia
    sedimentary, 140–41
    tuff-breccia, 82
    volcanic, 79
Bridalveil Falls (California), 302
Brule Formation, 114
Bryce Canyon National Park (Utah), 113–15
Buildings, earthquake-resistant, 188–89
Bulldozing, 299
Burgess Ridge (Canada), 419–20
Burgess Shale fauna, 419–20, 423
Buttes, 328–29

C

Calcareous ooze, 341
Calcareous tufa, 286
Calcite
    characteristics of, 57, 59, 60, 62, 275, 454
    dissolution of, 120
Calcium, in Earth's crust, 57
Calcium carbonate, 139, 424
Caldera, 97
Caledonian orogeny, 390
Caliche, 127
Calico Mountains (California), 204
California
    Black Mountains, 328
    Bridalveil Falls, 302

Calico Mountains, 204
Castaic, 365
Death Valley. *See* Death Valley (California)
Devil's Postpile National Monument, 95
    earthquakes in, 175–77
El Capitan, 73, 81
    gold rush, 56
J. Paul Getty Museum, 158–59
Lamar, 128
Lassen Volcanic National Park, 92, 94
Loma Prieta earthquake, 186, 187
Long Beach, 280, 281
Mount Lassen, 102
Mount Whitney, 201, 303
Northridge earthquake, 186
Pebble Beach, 116
Peninsular Ranges Batholith, 163
Salton Sea, 321
San Francisco, 185, 286, 346
San Joaquin Valley, 280
Uniform Building Code, 188
Calving, 295
Cambrian Period
    continents, 390, 391
    fossils, 419
    marine invertebrates, 425
    paleogeography, 394
Camel Rock (New Mexico), 116
Camels, 439
Canada
    Arctomys Cave, 277
    Athabaska Tar Sands, 150
    Burgess Ridge, 419–20
    gold rush, 56
    Gunflint Iron Formation, 423
    mineral and energy resources, 64
    Queen Charlotte Islands, 56
    Yukon Territory, 56
Canadian Rocky Mountains, 411
Canadian Shield, 385
Canyons. *See also* Grand Canyon
    Bryce Canyon National Park (Utah), 113–15
    submarine, 339, 340
Capacity, 250
Cape Cod (Massachusetts), 350–51
Cape Hatteras (North Carolina), 349
Capillary fringe, 272
Capitan Limestone reef, Permian Period, 399
Cap rock, 149
Carbonate, 57
Carbonate minerals, 59
Carbonate rocks, 142
Carbon 14 dating technique, 377
Carbon dioxide, 448
    carbonic acid formation and, 120
    from volcanic eruptions, 93–94
Carbonic acid, 275
    chemical weathering and, 120
    groundwater erosion and, 275
Carboniferous marine invertebrates, 426
Carnotite, 150
Cascade Range, volcanic eruptions in, 96–97

Cassiterite, 170, 454
Castaic (California), 365
Catskill Delta, 399
Caverns, 277
Caves, 276–77, 278
    Mammoth Cave, Kentucky, 269–70, 273
    sea, 353
Cementation, 139, 140
Cenozoic Era, 408–12
    mammals, 438–40
    mineral resources, 412–13
    primates, 440–41
Ceratopsians, 432
Chadron Formation, 114
Chalcedony Park, 407
Chalcopyrite, 170, 453
Chalk, 142
Channel flow, 247
Channel roughness, 247–48
Charleston earthquake (South Carolina), 182
Chattanooga shale, 151
Cheirolepis, 427
Chemical sediment, 137
Chemical sedimentary rocks, 141–43
Chemical weathering, 116–23
    controlling factors, 121, 124–25
    definition of, 119
    hydrolysis and, 121
    oxidation and, 120–21
    solution and, 119–20
Chert, 142, 143
Chesapeake Bay, 355
China
    earthquake prediction program, 191
    Tangshan earthquake, 191
China Paleozoic Era, 390
Chinle Formation, 406
Chitin, 424
Chlorite, 454
Chrysotile asbestos, 160
Cinder cones, 99
Circum-Pacific orogenic belt, 409
    seismic activity in, 180, 181
    volcano distribution in, 106, 107
Cirques, 302–3
Clastic texture, 140
Clastic wedge, 398
Clasts, 140
Clay, 138
    deposits, 130
    minerals, 59
    pelagic, 341
    porosity of, 271
    quick, 234
Clay minerals, 125
Claystone, 141
Cleavage, of minerals, 54, 61, 165
Climate
    chemical weathering and, 121, 124
    mass wasting and, 226–27
    short-term events, glaciation and, 310
    soil formation and, 126–27
Closed system, 375

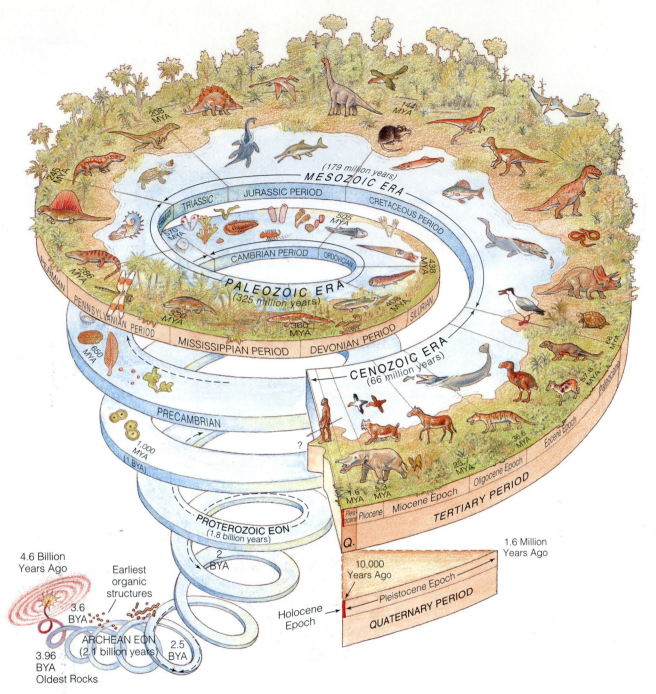

Geologic Time Depicted in a Spiral History of the Earth